AF789282

Condition Monitoring and Diagnostic Engineering Management

Condition Monitoring and Diagnostic Engineering Management

Proceedings of COMADEM 90:
The Second International Congress
on Condition Monitoring and
Diagnostic Engineering Management

Brunel University 16–18 July 1990

Co-sponsoring Organizations

British Institute of NDT
Institution of Production Engineers
Institute of Measurement and Control
Institution of Electrical Engineers
Institute of Acoustics
British Society for Strain Measurement
Safety and Reliability Society
Society of Environmental Engineers
Institution of Plant Engineers
Royal Institution of Naval Architects
British Computer Society
Institution of Nuclear Engineers
Institute of Petroleum
Institute of Energy
ISA International – England Section
Institution of Corrosion Science and Technology
Institution of Diagnostic Engineers

Proceedings Editors

Raj B.K.N. Rao, Joe Au and Brian Griffiths

CHAPMAN AND HALL

LONDON • NEW YORK • TOKYO • MELBOURNE • MADRAS

UK Chapman and Hall, 11 New Fetter Lane, London EC4P 4EE

USA Van Nostrand Reinhold, 115 5th Avenue, New York NY10003

JAPAN Chapman and Hall Japan, Thomson Publishing Japan,Hirakawacho Nemoto Building, 7F, 1-7-11 Hirakawa-cho, Chiyoda-ku, Tokyo 102

AUSTRALIA Chapman and Hall Australia, Thomas Nelson Australia,480 La Trobe Street, PO Box 4725, Melbourne 3000

INDIA Chapman and Hall India, R. Seshadri, 32 Second Main Road, CIT East, Madras 600 035

First edition 1990

© 1990 Chapman and Hall
Softcover reprint of the hardcover lst edition 1990

T.J. Press (Padstow) Ltd, Padstow, Cornwall

ISBN-13: 978-94-010-6678-5 e-ISBN-13: 978-94-009-0431-6
DOI: 10.1007/978-94-009-0431-6

All rights reserved. No part of this publication may be reproduced or transmitted, in any form or by any means, electronic, mechanical, photocopying, recording or otherwise, or stored in any retrieval system of any nature, without the written permission of the copyright holder and the publisher, application for which shall be made to the publisher.

British Library Cataloguing in Publication Data
COMADEN 90

Condition monitoring and diagnostic engineering management.
1. Buildings. Engineering services. Monitoring
1. Title II. Rao, Raj B. K. N. III. Au, Joe IV. Griffiths, Brian
696

ISBN-13: 978-94-010-6678-5

Library of Congress Cataloguing-in-Publication Data
Available

Contents

Introduction xi
Raj B.K.N. Rao

Keynote

Condition monitoring today and tomorrow – an airline perspective 1
A.C.D. Cumming

Condition monitoring today and tomorrow – a manufacturing perspective 8
B.E. Jones

Machine/Plant/Process Health Monitoring

Dynamic performance of a hydrostatic thrust bearing with an ERFS and grooved lands 14
D. Ashman, E.W. Parker and A. Cowley

The use of stress wave sensors for the diagnosis of bearing faults 20
M.N.M. Badi, D.E. Johnson and G.J. Trmal

Monitoring of check valves in reciprocating pumps 26
T. Berther and P. Davies

Signature analysis of a gear box of turbine generator set 32
M.A. Rao, D.A. Rao and K.R.D. Roy

Condition monitoring of wear using acoustic emission techniques 38
R.J. Boness, S.L. McBride and M. Sobczyk

The condition monitoring of machine tools 44
A. Davies and J.H. Williams

Condition monitoring systems for machining applications 49
M. Raghunandan and R. Krishnamurthy

Tool wear monitoring by the analysis of surface roughness 56
W.P. Dong and W.Y. Wang

Characterization of the acoustic emission behaviour during quasi static punching tests 62
P. Souquet, C. Bouhelier, Dr E. Grosset, A. Maillard, L. Li, Dr J. Au and A. Mardapittas

The determination of cutting tool condition using vibration signals 68
T.N. Moore and Z.F. Reif

Monitoring tool wear during the turning process 74
S. Taibi, J.E.T. Penny, J.D. Maiden and M. Bennouna

Monitoring of a production grinding process 80
G. Trmal, C.B. Zhu and P.S. Midha

The technique of mesh obscuration for debris monitoring – the COULTER LCM II 86
M.P. Cowan and R.A. Wenman

Finding what matters – health monitoring by FCM 92
T.M. Hunt

Grease – a suitable case for condition monitoring 98
M.H. Jones

Engine monitors for efficient take-off performance monitors 108
R. Khatwa

Condition/performance monitoring of internal combustion engine by continuous measurements of crankshaft twist angle 114
J.T. Lenaghan and A.I. Khalil

Rotating machinery noise and vibration study 120
G.H. Lim

Vibration analysis as tool for computerised machine monitoring 126
G. Luelf and Dr R. Vogel

Using a PC based system for machinery transient analysis 132
C. Nicholls

Condition monitoring of key elements in high speed textile machinery 142
J.M. Sharp

On the monitoring of a high speed mechanical feeder by means of an in-time windowing technique 148
E. D'Amato and G. D'Emilia

Condition-Based Maintenance Management

Condition based maintenance 155
G. Allenby

Cost effective computer software in maintenance management 162
P. Cross

Condition monitoring based maintenance in practice in a heavy chemical plant 169
J. Hensey and M.N.K. Nair

Condition monitoring at a continuous coal handling facility 176
R.J. McMahon

Condition-based maintenance in the FMMS environment 182
E.J. Morris

Predictive maintenance programme for GEN sets and reciprocating MUD pumps of oil fields 197
R. Pandian and B.V.A. Rao

A case study of the maintenance of Second World War military jeeps in Bhopal: A study in the resourcefulness of the local mechanics 203
M.A. Qureshi, Q. Ansari and A.F. Qureshi

Maintenance strategy in the '90s 207
P. Upshall

Industrial Risk and Waste Management

Risk and waste management in oil field 212
D.J. Barua and R. Pandian

Environmental Pollution Monitoring

Environmental pollution monitoring in oil exploration and exploitation 218
J. Chand

Quality and Reliability Management

Comparison of the digraph and FAULTFINDER methods of fault tree synthesis for nested control systems 225
J.D. Andrews and A.R. Khan

A study on the hierarchal quality control in CIMS 234
Zhixiang Chen and Changhua Hu

An experiment in automation of statistical process control using fuzzy logic 240
M.H. Lim et al.

Application of Intelligent Knowledge-Based Systems

Application of syntactic pattern recognition techniques to condition monitoring of machines 247
M.M. Ahmed and R.D. Pringle

An engineers's assistant for balancing of flexible rotors 253
R.J. Allwood, C.N. Cooper, G.D. Wale and M.H. Walton

On-line decision making for diagnostic and monitoring purposes 260
N. Ardjmand and Z.M. Ranic

Expert systems for quality management in glass container production 266
M. Gianotti, P. Molinari and A. Pozzetti

System-independent approaches to diagnostic knowledge reasoning and representation for mechanical and thermal systems 275
H.K. Ho, S.S.G. Lee, M.K. Lim and Y.W. Wong

Intellectualization of faults diagnosis of rotating machinery – expert system 281
X. Jiang and S. Wang

Amethyst: An expert system for the diagnosis of rotating machinery 287
Dr R. Milne

Integration of a PC based expert system shell and production line tracking system for failure diagnosis and rework definition 293
B.G. Oak

An expert system with circulating notice for emergency in underground coal mine 299
Y. Tominaga, K. Ohga and Z. Yang

Medical Monitoring and Diagnostics

Detection of periods of significant activity in noisy signals 305
A.J. Charles and A.S. Sehmi

Application of an expert system on the monitoring of human lungs 311
A.I. Khalil and A.K. Jawad

Intelligent classification in EMG decomposition 317
G.H. Loudon, A.S. Sehmi and N.B. Jones

Advances in Instrumentation and Control Technology

Octave band revisited – machine condition monitoring using octave band data collectors 326
J.H. Burrows

Tool wear monitoring sensors 343
K.H. Hale and B.E. Jones

System detection instrumentation 349
I.A. Henderson and J. McGhee

Real-time model and kinematic control of machine tools 355
C.Y. Li and C.R. Zhang

Digital capacitance displacement measuring instrument 361
Hongyan Li, Lishi Yang and Jun Li

Optimisation and diagnosis for numerically controlled machining 367
M.A. Rahbary and G.N. Blount

In-process measurement using an opto-electronic probe 373
I. Shams and C. Butler

A new three dimensdion piezoelectric transducer for diagnostic engineering 379
D.S. Shi, Y.T. Wang and S.H. Miao

Improving the prediction of condition using simple sensors 384
R.K. Stobart

A high accuracy test technology used for on-line measurement of abrasion wear in bearings 390
S.X. Zheng and W.R. Dai

Advances in Diagnostic Engineering

Machine diagnostics using the concepts of resolution ratio and automatic timing 396
D.M. Benenson, A. Soom, S.Y. Park and S.E. Wright

Condition assessment and forecasting form plant diagnostic data with pareto model 403
C. Cempel

Drift-fault diagnostics using fault-curves fitted to changes in dynamic response 407
P.R. Drake

Fundamentals of linear time-varying plant for diagnostic engineering 414
J. McGhee and I.A. Henderson

Computer Vision

An artificial intelligent vision system for inspection and monitoring 420
B.J. Griffiths

TEXYS: A real-time picture processing system for on-line quality control 426
M. Hajnal, I. Lovanyi and A. Nagy

Advances in NDT/NDE

Moire photography in engineering measurement 432
C. Forno

Photogrammetry in industry 437
R.A. Hunt

Energy Management

Energy management for conversion in oil rigs 442
D.J. Barua

Continuing Education

Continuing education/training on computerized automation 450
A. Yamamoto

Novel Techniques of COMADEM

Forecasting the change of operating condition for machinery via fuzzy classification 455
Y.D. Chen and L.S. Qu

Homosource ANC and its application to the machinery diagnosis 461
Xiao Li and L. Qu

The phase information in the rotor vibration behaviour analysis 468
Xiong Liu and Liangsheng Qu

In-situ measurement of cylindricity 476
Qiuming Xu, Jun Li and Hongxiang Yuan

Three-dimensional component inspection using a profile projector equipped with a fibre optic sensor 483
Q. Yang and C. Butler

Author Index 489

Introduction

Raj B.K.N. Rao
(COMADEM Centre, Faculty of Engineering and Computer Technology,
Birmingham Polytechnic, Perry Barr, Birmingham, UK)
Y.H.J. Au and B.J. Griffiths
(Department of Manufacturing and Engineering Systems,
Brunel University, Uxbridge, Middlesex, UK)

To create, to engineer, to err and to destroy are basic human traits. Human progress depends upon the balanced exploitation of some of these traits. Our quality and standards of living are closely intertwined with the effective management of some of these traits and the limited physical resources that are at our disposal. The only sensible approach to harmonize and blend these two spheres is to effectively monitor, diagnose and control our own actions and reactions, at every stage.

We are now privileged to live in an age of advanced technology which is significantly influencing the basic infrastructures of our societies, on which our lives now depend. The rapid rate at which the technological changes are spreading is causing much confusion and uncertainty in communities, governments and industries at different levels. This, in turn, is creating and widening gaps and mismatches of unknown dimensions in almost all human endeavours. Such undesirable tendencies must be monitored, diagnosed, prognosed and controlled with the utmost haste and urgency.

Many have expressed an understandable concern as to whether our industries can remain competitive in the 21st century. It is not superhuman to improve our quality and standard of living and the desired productivity levels, unless we aim to achieve an all-round efficiency, improve public morale and confidence in health and safety matters, and to inculcate the competitive spirits, at all levels. It is, therefore, obviously clear that today's solutions demand a holistic approach to many problems. Condition Monitoring and Diagnostic Engineering Management (COMADEM) provides the much needed international platform for multi-disciplinarians to meet, discuss and frankly exchange views, thoughts and information on any technological matters and issues that will affect our present and future civilizations.

We believe that through COMADEM a greater awareness, understanding and involvement between the management and the work-force could be firmly established. COMADEM is a money saver. Judicious applications of the principles and practices of COMADEM helps by creating the goods and services the customers want within time and within budget. COMADEM provides a positive disciplined attitude to cope with any undesirable changes with maximum efficiency and minimum delay.

The contents in these proceedings reveal the significant advances that are being made in a wide variety of engineering and other disciplines. The editors are indebted to all the authors who have spent considerable time, effort and energy in preparing and submitting their invaluable contributions to these proceedings at such short notice. Thanks are also due to Mark Hammond and his able staff for their patience and perseverence in producing these proceedings in a professional manner.

Condition monitoring today and tomorrow – an airline perspective

A.C.D. Cumming
Director of Engineering, British Airways, London, England

Abstract
The objectives, processes and potential for development of maintenance programmes applied to civil aircraft structures will be discussed.

In the area of engine condition monitoring, mention is made of the various techniques used in airlines to monitor engines while operating in service, and the methods employed to inspect them internally without disassembly.

From the avionics point of view, the experience with built in test equipment is discussed, together with the future prospects for improving its performance. Finally, the area of flight data recording and telemetry is referred to.
Keywords: Condition Monitoring, Aircraft Structure, Airworthiness, Safety, Boroscope, Engine, ACARS, BITE.

1 Introduction

The concept of Condition Monitoring is very important to my Engineering Department in British Airways, and indeed forms the very basis of the modern philosophy in aircraft maintenance that we have been instrumental in formulating, in conjunction with the industry and regulatory bodies, over the years.

The process of implementing, controlling and monitoring the Condition Monitoring programme is a very expensive one, involving considerable resources of manpower. It is, however one that I feel is cost effective and that provides the very necessary safeguards in terms of aircraft safety. I will speak today of three important areas with respect to Condition Monitoring from an airline's point of view - Aircraft Structure, Engines, and Avionics.

2 Aircraft Structure

Maintenance of the structure of an aircraft in the public transport category is predominantly On Condition. Defects are rectified as and when they are found by inspections carried out in accordance with a scheduled inspection programme.

Some structure, principally where high strength forgings are used, and where an economic inspection programme cannot offer adequate assurance against catastrophic failure, is subject to Hard Time maintenance, via either an overhaul programme or finite life replacement.

An element of Condition Monitoring exists, however, in that most operators, acting in accordance with the rules set out by their Airworthiness Authorities, will:

Analyse defect histories and adjust their maintenance programmes accordingly.

Report significant findings to their Authority and to the aircraft manufacturers. These reports will be reviewed together with those from other operators of the same type of aircraft. The operators will react to advice or bulletins which recommend a change to the maintenance programme or to the objectives issued by the Authority.

All operators who have a fleet of aircraft will carry out sampling programmes to identify the maintenance needs of structure where no economic requirement has been established by analysis, test or previous experience and where deterioration which is critical to airworthiness is improbable. If, for example, an operator has a fleet of 20 aircraft, he may be required after 5 years to inspect each such area on 2 aircraft each year until, after 15 years, the area has been inspected on all aircraft in his fleet. The process will then start again, possibly at a more intensive level and will continue unless and until significant defects are found. The operator will then be required to schedule regular inspections of that area in all aircraft in his fleet. Reporting back to the aircraft manufacturer is an integral feature of this type of programme in that knowledge is based on findings from all aircraft of the type worldwide.

The Condition Monitoring principle is particularly relevant to structure where the maintenance requirements:

Are highly dependent upon the type of operation.
Can only be quantified by service experience.

Airworthiness Directives currently being drafted will require operators to react to the findings of a baseline corrosion inspection and control programme and to work with their Authorities in developing and adjusting a schedule of maintenance that will limit corrosion to a level where the remaining sound structure is still capable of carrying all the design loads.

Condition Monitoring is a process which extends throughout the life of an aircraft and, in the case of structure, may be expected to become more significant as the effects of aging become more manifest and critical.

3 Engines

The most expensive functional entities on an aircraft are the engines. Therefore, in order to optimise their use and keep them operating as long as possible between overhaul and repair, while still maintaining the stringent safety standards which have become the hallmark of airline operations, it is necessary to monitor the condition of the engines while they are in service.

The first line of monitoring is done by the flight crew who monitor the engine instruments continually during flight, reporting on any abnormal indications. Following on from this there are the routine inspections carried out by the ground engineers consisting of visual inspections of the external engine parts. To monitor the internal parts we use various methods, the most important of which are:

The use of boroscopes, or as they are sometimes called, endoscopes. These devices are most commonly shown being used for internal medical examinations. We use them extensively for carrying out routine inspection of internal parts of the engine which are known to deteriorate with time, such as combustion chambers and turbine blades. This enables us to continue to operate the engines for longer periods without removing them for overhaul. Without this ability, the lives of the engine would have to be set much lower, determined by the worst engine. For example with the RB211 we are currently achieving lives of 14,000 flying hours which is equivalent to 3 years in service, but some engines due to the condition found by boroscoping only achieve 8,000 hours. Boroscopes are also very useful in the field of troubleshooting. If an engine is reported for being low on performance, a boroscope check on the rotating assemblies could determine that the problem is a blade failure, which would require an engine change, or perhaps that a bleed valve is open, which could be fixed insitu.

The analysis of the engine oil can reveal much important information. As an engine runs, minute particles of metal are produced from the oil wetted parts. These particles remain in suspension in the engine oil and are so small they are not trapped by the engine filters. With normal oil loss and replenishment the particles do not increase in quantity over the life of the engine. If on the other hand a part starts to wear abnormally, the particle count will increase. Spectrographic oil analysis is the tool we have which enables us to identify this increase in particles. The two most widely used methods in use in the airline industry are atomic absorption and spectrographic emission.

With the atomic absorption method, a small sample of oil is burnt, and the flame is analysed through a light source which is particular for each element. This is very accurate, giving readings as low as 0.1 parts per million (ppm) but the analysis is laborious unless the material being monitored is known.

Spectrographic emission is similar to atomic absorption in that a small quantity of the oil is burnt. It is not so accurate being only able to detect particle counts down to 1 ppm. It has the advantage that the quantities of all the elements can be read at one burn.

Which method should be used is very much a case of horses for courses. For example, the JT9D engine fitted to the Boeing 747 aircraft normally only produces ferrous particle counts of 0.2 to 0.4 ppm and an increase of 0.6 ppm is significant. Therefore atomic absorption is required. Conversely, the JT8D engine fitted to the Boeing 737 aircraft produces counts of 10 ppm with increases of 10 ppm being significant. We therefore use the emission method with the advantage of monitoring many more elements.

In addition to checking for metal contamination, oil is also checked for viscosity and acid content. Both these give an indication of the environment in which the oil is working; any change would be a warning that the environment is hotter than normal and therefore there is a defect such as an air seal allowing hot air into a bearing chamber for instance.

Coupled with the oil analysis is the use of magnetic chip detectors or MCDs. These are small, removable plugs fitted with a powerful permanent magnet and situated in convenient positions in the engine scavenge oil system, the main criteria being that all the scavenge oil must pass over an MCD. The disadvantage of MCDs is that they will only catch magnetic materials, but as most of the rotating parts are made from magnetic materials, the success rate is high. The function of the plugs is to catch any magnetic metal particles in the oil which have been generated by any of the oil wetted parts such as bearings, gears etc. By analysing the material it is possible to determine whether the metal catch is indicating the impending failure of a main line bearing, which would necessitate an immediate engine change, or wear on a gear or shaft which takes longer to fail and therefore only requires a planned change at some later time. We in British Airways are proud of the way we manage MCDs. The MCDs are removed on a routine basis and sent for laboratory analysis. Here the debris catch is microscopically examined by experts who determine by the shape and form of the metal pieces whether they are genuine failure indications. Any suspect particles are then analysed using our own X-ray Energy Dispersive equipment, which is capable of analysing the constituent elements in metal particles down to 0.010 inch in size. This has saved a large number of bearing failures occurring which are expensive and result in the inconvenience of an unscheduled engine change, which very often occurs away from base. It also reduces the secondary damage to a minimum. For example, the cost of an engine removal and repair for a successful "find" will be in the region of £20,000 whereas a missed bearing failure could be in excess of £500,000 and could be uncontained.

Similarly, we inspect the engine oil filter on some engines, although this is not so successful as the MCDs but does have the advantage of catching all the pieces, irrespective of the material.

Another tool that we use is Engine Performance Trend Analysis. This is where the engine parameters such as spool speeds, exhaust gas temperature, fuel flow and EPR (engine power) are taken under cruise conditions. These figures are fed into a computer which corrects them to standard day conditions. It then compares them with a set of figures which are the average engine parameters for a standard day. From this we can see any engine deterioration trends which is useful in anticipating when engines are coming to the end of their useful life and thus schedule their removal. This monitoring can also show up rapid changes in the parameters which are indicative of failures within the engine such as blade failure, or defects in the instrumentation system.

A further development of this is the Engine Module Performance Analysis. Most modern engines are designed to be taken apart in sections or modules such as the compressor or turbine. By recording additional parameters to those mentioned previously, such as pressure and temperatures through the engine, it is possible to determine on an engine with a deteriorated performance which module or modules are at fault. It is then only necessary to change those modules to restore the engine's performance, thus saving a great deal of time and energy by not overhauling serviceable modules.

Engine condition monitoring is extremely important in the operation of the modern jet engine. The cost of maintaining it is high - in British Airways we spend in the region of £250,000 per annum on oil analysis, MCD inspection and trend monitoring. This is easily offset by the advantages when, for example, there are savings of up to £500,000 by the early detection of an impending bearing failure. Last year on the RB211 engines alone there were 9 such cases. Increasing the life of an RB211 engine by 1,000 hours at an hourly cost of approximately £80 would save £80,000 per engine - and we have 265 RB211s in our fleet.

As for the future, we are presently installing an automatic data link system on our aircraft, known as ACARS. This will complement the on board Flight Data Recording System which continuously monitors engine parameters, by automatically sending the data directly to the Engineering base in real time. Thus, not only will the Development Engineers and Maintenance Engineers have the information ahead of the aircraft landing, they will also have the opportunity to ask the flight crew for further information under the exact operating conditions.

4 Avionics

The situation with regard to avionics these days is of course that the vast majority of the equipment is solid state, and therefore lends itself well to the Condition Monitoring concept. Great strides have been made in the recent past in the area of BITE (Built In Test Equipment). For instance, the first generation of digital avionics aircraft, including the Airbus A320 and the Boeing 757 and 767, had systems with a BITE efficiency design target of 95%, that is, the BITE system should identify correctly 95% of all faults. In practice, this figure has indeed been reached, but even so, of course it still means that 1 in 20 faults are not picked up by BITE. Of more immediate importance to the airline, with the requirement to get aircraft serviceable and on time, is the fact that the BITE circuits have been over-sensitive to transient conditions, either power spikes or sensor perturbations, that have caused the BITE circuit to erroneously annunciate a fault condition. This has caused a great deal of time and effort to be expended in tracing non-existent defects, but in addition has had the side effect of reducing the maintenance staff's confidence in the BITE systems - a case of crying "Wolf"!

To a certain extent, this problem has been the result of design engineers trying to get the BITE to pick up genuine fault conditions in as quick a time as possible - there is a trade off to be considered between this and false warnings. However, a number of modifications have been introduced with significant benefit, and it is true to say that now, the BITE systems are an indispensable feature of the avionics systems.

What of the future? Well, hopefully the airframe systems design engineers and the equipment design engineers will have reacted to the torrent of criticism (all constructive of course!) that has come their way from the airlines, and the next generation of avionics will not have the false triggering problem to the same extent. It would be a pious hope of course that it will have been completely eradicated - being under software control there is an inevitable element of an unforeseen bug either in the software specification or in the coding. In addition, the airlines will be looking for an increase in the design target from 95% to perhaps 99.5% so that the potential exists for a policy whereby under normal conditions if the BITE indicates that no problem exists in a particular black box then it remains on the aircraft. Ultimately, a time will come when BITE will be the sole arbiter of whether a unit should be removed for an operational malfunction.

Of course, all this BITE technology doesn't come free! In fact a significant proportion of the cost of the equipment is accounted for by the BITE function - probably between 30% and 40% - not because of any large hardware content but because of the large software content. It is interesting to note that the split of costs in a typical digital avionics system these days is around 80% for the software aspects and only 20% for the hardware. It is important therefore that we do not lose sight of the requirement for this BITE function to be cost effective over the life of the equipment. It will be a case of the law of diminishing return in the drive towards a fully effective BITE.

Another aspect of condition monitoring in the avionics systems that is presently being developed is that of Fault Tolerance. The next generation of aircraft, including the Airbus A330 and A340, and the Boeing 767X will have some fault tolerant avionics. The aim of this approach is for the systems to be designed so that any single fault will have no effect on the functional capability of the equipment, i.e. such a fault will be transparent as far as the flight crew are concerned. Furthermore, the probability of a second fault not occurring in the next 200 flying hours (the effect of which in combination with the first system may cause a functional degradation) will be 99%. This will mean that, an aircraft will not normally need any avionic line maintenance troubleshooting, and the aircraft can continue in service until it next has a base check. Once again, this concept will have an impact on the cost of the equipment and the airlines will need to monitor the efficacy of the concept in practice, but it certainly holds a promise of producing a worthwhile reduction in unscheduled unserviceability of aircraft.

Condition monitoring today and tomorrow – a manufacturing perspective

B.E. Jones
The Brunel Centre for Manufacturing Metrology, Brunel University, Uxbridge, England

Abstract
Improvements to condition monitoring and predictive maintenance of the process, and predictive metrology for control of product quality will continue as means to sustain a competitive edge in manufacturing. Development of sensing techniques and expert systems will be important.
Keywords: Manufacturing Metrology, Sensors, Predictive Maintenance.

1 Manufacturing

Manufacturing embraces all the production industries with the exception of energy and water supply. It includes metal and chemical production, mechanical and electrical engineering, and the manufacture of metal goods, vehicles, food, textiles, furniture, paper and plastics. As a consequence of the competitive need of industry to be highly efficient and quality conscious, manufacturing metrology is evolving from traditional engineering metrology dominated by the skills of quality inspectors at the end of production lines, to automatic inspection methods off–line, in–cycle and in–line, and utilising microelectronic, computer (hardware and software) and novel optical techniques (Jones 1987a, Kalpakjian and McKee 1987). Suitable sensing techniques, sensors and transducers are essential to this developing situation. It is important to maintain a realization of traceability in dimensional metrology (Beyer and Kunzmann 1989).

2 Condition monitoring

Condition monitoring is concerned with the analysis and interpretation of signals from sensors and transducers installed on operational machinery. The monitoring of a machine condition or 'health' is normally done by employing sensors positioned on the outside of the machine, often remote from the machine components being monitored. Analysis of the information provided by the sensor outputs is done using systems employing microcomputers. Some form of interpretation of the analysed output is then needed to establish what actions to take.

The justification for the large investment in developing condition monitoring methods and equipment is to avoid costly downtime and catastrophic failures. A condition monitoring system can also be a test and quality assurance system for continuous processes as well as discrete component manufacture. Thus techniques normally associated with automated condition monitoring, such as dynamic analysis, spectrum comparison and fault diagnostics can be used for product quality monitoring (Hughes 1988). Recent case studies report condition–based maintenance schemes (Brennan and Morris 1989, Thorpe 1989).

There are many condition monitoring systems/methods and diagnostic tools now available (Jones 1988). Condition monitoring and Diagnostic Engineering Management (COMADEM) is a multidisciplinary subject and key functional components of all COMADEM systems are the sensors and associated instrumentation.

3 Industrial metrology

The usage of sensors in manufacturing seems bound to increase, because of the need for automated and flexible processes (Jones 1987a). Thus while some post–process inspection is always likely to be employed, in–line inspection will increase, as will the use of robots. More sensors will be used in each stage of manufacture (pre–process inspection, machining/extruding, assembly, post–process inspection/testing) and in intelligent robots.

Industrial metrology is concerned with sensors to measure movement of machine tool parts and monitor tool wear (Hale and Jones 1990) and the dimensions of artefacts in machining centres, sensors for robots in flexible manufacturing systems, sensors to gauge mating parts for selective assembly or allowing for interchangeability, and sensors for inspection and testing of assembled or part–assembled products. Sensors are required in all the widely differing manufacturing fields. In general the dimensional shape and physical properties of functional parts need to be inspected.

4 Sensors

It is evident that to be a successful practitioner in the fields of condition monitoring and industrial metrology requires a good breadth of knowledge about sensing techniques, sensors and instrumentation. An awareness of technical trends is important, and useful texts are now available (Jones 1987b).

5 Examples of current developments

5.1 Improved wear–debris detection (Flanagan et al 1988)

The characteristics of wear–debris in the lubricant flow of a machine can indicate the nature and severity of wear in oil–wetted components and the greater the severity of the wear, the higher the concentration and size of the wear–debris particles. Most current wear–debris monitoring techniques operate off–line. Recently a prototype on–line wear–debris monitoring system has been developed which is able to identify the material and approximate size of metallic debris particles. An inductive sensor is used. Ranges of operation are

100–400 μm diameter for ferrous particles, and 200–500 μm diameter for non-ferrous particles. Work aimed at substantial improvements in the range of particle sizes which can be analysed is being undertaken.

5.2 Improved acoustic emission sensing (McBride et al 1990)

Acoustic emission (AE) has been shown to be a useful technique in monitoring the state of wear in machine tools. Conventionally, contacting piezoelectric transducers are used for AE detection, although operationally a non–contacting technique would be preferred. Recently a non–contact technique has been developed based on optical interferometry, using fibre optics, to realise a miniature and robust probe for detecting AE by measuring the small amplitude (~0.1 nm), high frequency (0.1 – 1 MHz) surface vibrations which it produces. The extended bandwidth should yield more complete information on the AE process than presently available.

5.3 Intelligent condition monitoring (Smallman 1988/89)

The intelligent condition monitoring of industrial plant is a rapidly expanding area of research. It combines two radically different principles: classical condition monitoring and intelligent knowledge–based systems (IKBS). IKBS has been defined as a syustem which emulates human problem solving behavioiur. The knowledge base is the heart of the system; it contains facts, heuristic rules and almost any piece of knowledge which an expert would use in the solution of a problem in the application area (domain) which the system covers. An IKBS system can carry out fault diagnosis, performance analysis and planning using intelligent pattern matching to recognise trends in the output of a condition monitoring system. Diagnosis can take place at three levels: alarm diagnosis, steady–state diagnosis (relatively certain diagnosis of problems, based on interaction with the operators, and historical and real–time data), and transient analysis (recommendations only are offered, as the time frame is too short).

5.4 COMADEM and a flexible machining system (Maksym 1988)

The Lucas Machine Division of Litton Industries in the USA has combined two of their boring mills, an automatic toolchanger, an AWGV, a tool preset area, and a CMM into a flexible machining system (FMS). The FMS is computer controlled, for example, the computer sequences the parts through the required operations including fixturing, machining, refixture, and inspection. The software system also includes a tool management system that manages the large tool inventory, calculates and tracks tool life, and directs the movement of tools between storage positions and maching operations. The CNC controller at each machine monitors the spindle drives to detect increases in horsepower that indicate tool wear and calls for tool change when the tool becomes dull. Potential problem areas such as coolant levels, motor temperatures, and violations of safety zones on the machine are also monitored.

Diagnostic messages for these and other problems provide early warning and trouble-shooting assistance to help increase system utilization. Lucas reports average productivity increase of 300 per cent over their previous conventional machining operations and consistently higher quality levels.

5.5 The application of condition monitoring techniques to product quality monitoring (Hughes 1988)

The demand for greater quality assurance points to 100% testing of the assembled product. There is a need for inspection techniques which combine the speed and accuracy of automated testing, with the flexibility and diagnostic capabilities of the skilled human inspector. The automated system needs to be given the sensors, analytical capability and accumultated 'experience' of the human. Techniques normally associated with automated condition monitoring, such as dynamic analysis, spectrum comparison and fault diagnostics may be used to create a sensitive inspection technique for mechanical products capable of detecting faults not easily found from performance tests. This is an inspection technique for the overall 'health' of a product. Instead of using the product to monitor the process, the technique is to monitor the process itself, and so control the quality of the product. By combining a knowledge of the acceptable range of data variations with closely controlled test conditions, and loadings, a high sensitive method emerges for verifying the quality of an assembled product.

6 Future trends

Proceedings of recent conferences indicate steady advances in condition monitoring and quality control of manufacturing plant (Rao and Hope 1988, 1989: ISMQC/IMEKO 1989: Beyer and Kunzmann 1989) Non-contacting optical devices are important (Whitehouse 1985) as are the use of expert systems (Au et al 1988). 'Deterministic metrology' looks promising, although there is a need to simplify the development phase using software based on real-time expert system architecture. Future developments are aimed at a significant reduction in the development effort. Condition monitoring and predictive maintenance of the process, and predictive metrology for control of the product quality must continue to increase in importance as the means to sustain that competitive edge in manufacturing (West 1989).

7 References

Au, Y.H.J., Griffiths, B.J. and Lister, P. (1988) A methodology for a manufacturing systems approach to process condition monitoring. Proc. COMADEM 88, September 1988, Birmingham (London, Kogan Page: eds. Raj B.K.N. Rao and A.D. Hope), 359–365.

Beyer, W. and Kunzmann, H. (1989) Today's realization of traceability in dimensional metrology. Proc. 3rd IMEKO Int. Symp. Dimensional Metrology in Production and Quality Control, September 1989, Aachen (Düsseldorf, VDI Verlag: VDI Berichte 761), 1–13.

Brannan, T. and Morris, E. (1989) Condition monitoring in the FMMS environment. **Maintenance**, 4 (1), 3–10.

Flanagan, I.M., Jordan, J.R. and Whittington, H.W. (1988) Wear–debris detection and analysis techniques for lubricant–based condition monitoring. **J. Phys.E: Sci.Instrum.**, 21, 1011–1016.

Hale, K.F. and Jones , B.E. (1990) Tool wear monitoring sensors. Proc. COMADEM 90, July 1990, Brunel University (London, Chapman and Hall), this volume.

Hughes, M.L. (1988) The application of condition monitoring techniques to product quality monitoring. **Meas. and Control**, 21 (4), 113–118.

ISMQC/IMEKO (1989) Metrology for Quality Control in Production. Proc. 2nd IMEKO TC 14 Int. Symp., May 1989, Beijing (Beijing, International Academic Publishers).

Jones, B.E. (1987a) Sensors in industrial metrology. **J. Phys.E: Sci.Instrum.**, 20 (9), 1113–1126.

Jones, B.E. (ed) (1987b) **Current Advances in Sensors** (Bristol, Adam Hilger).

Jones, B.E. (1988) Sensors and instrumentation for condition monitoring. Proc. COMADEM 88, September 1988, Birmingham (London, Kogan Page: eds. Raj B.K.N. Rao and A.D. Hope), 19–32.

Kalpakjian, S. and McKee, K.E. (1987) Automated inspection – an extended view. Proc. 8th Int. Conf. Automated Inspection and Product Control, June 1987, Chicago (Kempston, IFS), 13–23.

Maksym, E.J. (1988) Automated inspection for flexible machining systems. Report MTIAC SOAR–88–01, Manufacturing Technology Information Analysis Center, MTIAC Operations, Chicago.

McBride, R., Barton, J.S., Jones, J.D.C. and Borthwick, W.K.D. (1990) Fibre optic interferometry for acoustic emission sensing in machine tool wear monitoring. Conf. Electro–optics and Laser Int., March 1990, Birmingham (Richmond, Reed Exhibition Companies), 1–10.

Rao, Raj B.K.N. and Hope, A.D. (eds) (1988) COMADEM 88. Proc. Seminar, September 1988, Birmingham (London, Kogan Page).

Rao, Raj B.K.N. and Hope, A.D. (eds) (1989) COMADEM 89. Proc. 1st Int. Cong., September 1989, Birmingham (London, Kogan Page).

Smallman, C. (1988/89) Intelligent condition monitoring of gas compressor engines. **Meas. and Control**, 21 (Dec/Jan), 307–311.

Thorpe, P. (1989) Condition monitoring of machine tools. **Maintenance**, 4 (1), 17–23.

West, D.A.L. (1989) Condition monitoring in manufacture.
Proc. COMADEM 89,September 1989, Birmingham (London, Kogan Page: eds. Raj B.K.N. Rao and A.D. Hope), 25–28.
Whitehouse, D.J. (1985) Instrumentation for measuring finish, defects and gloss. **SPIE** Vol. 525, 106–123.

Dynamic performance of a hydrostatic thrust bearing with an ERFS and grooved lands

D. Ashman, BSc., PhD., C.Eng., M.I.Mech.E.
Birmingham Polytechnic, Birmingham, West Midlands, England
E.W. Parker, BSc., PhD., C.Eng., F.I.Mech.E.
Wolverhampton Polytechnic, Wolverhampton, West Midlands, England
A. Cowley, MSc., PhD.
UMIST, Manchester, England

Abstract
The work described in this paper deals with the design, construction and instrumentation of a high-speed multi-recessed hydrostatic thrust bearing test rig, which had the facilities of applying static and dynamic loading to the bearing arrangement together with varying the land and pocket geometry. The experimental dynamic loading results with an ERFS and grooved lands are then compared with those with a conventional plain pocket and flat lands.
Keywords: Bearing, Hydrostatic, Thrust, Pocket, Land.

1 Introduction

Developments in machine tool design have been considerable over recent years with the introduction of computer controlled machines and new cutting materials such as polycrystalline diamonds, boron nitrides and ceramics. To take full advantage of the new materials cutting speeds up to 20 m/s for steel and 60 m/s for aluminium are needed. Such cutting speeds place increased requirements on machine tool spindle and slideway systems if high reliability, large load-carrying capacity and long-life expectancy are to be achieved.

Recent research has concentrated on improving the high-speed performance characteristics of hydrostatic bearings as it was found that under high-speed conditions undesirable hydrodynamic and inertia pressure variations, together with large temperatures rises, alter considerably the quasi-static loading characteristics.

Much work has been undertaken by Mohsin et al. (1980), who re-designed the land and pocket geometry by introducing an external recess flow system using recess inserts (know as an ERFS) and orthogonal grooved lands. Further work by Ashman (1987) and Ashman et al. (1988, 1989) confirmed the preliminary theoretical investigation carried out on the new bearing design. It was found that an ERFS and grooved lands reduced considerably the frictional power consumption and harmful hydrodynamic pressure variations in both the lands and the recesses. Diagrams illustrating the configuration of a plain pocket and one which contains an ERFS are given in Fig.1.

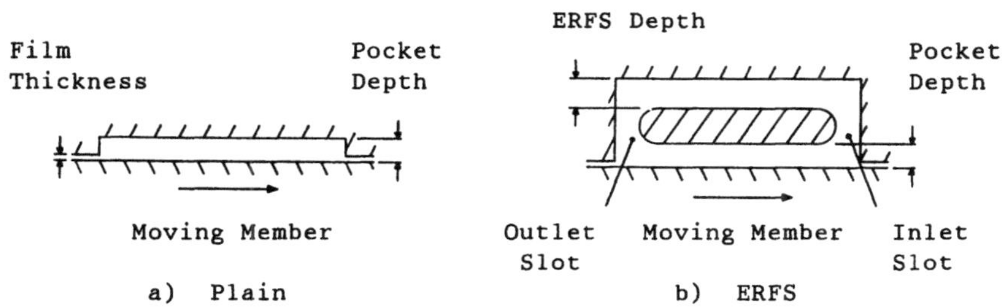

Fig.1 Geometry of Bearing Recesses

However, the quasi-static loading and high-speed performance of a machine tool is only a part of the characteristics required. An important feature is the resulting dynamic response of the bearing system, as this will often have a major influence upon the stable machining capacity of the machine.

The work presented in this paper describes an experimental investigation of the dynamic response of an annular multi-recessed hydrostatic thrust bearing with an ERFS using recess inserts and grooved lands. The work was carried out in order to determine whether the introduction of an ERFS and orthogonal grooved lands had any detrimental effects on the response of a conventional bearing with plain pockets and flat lands.

2 The Test Rig

A detailed description of the bearing is given in Ashman (1987). Figs. 1 - 3 illustrate the components of the test bearing. The bearing blocks (items 10 and 15), each had three recesses. Spacer and return flow inserts (items 26 and 27 respectively), which were placed in each pocket to change the recess geometry, were held in position by screws (item 25). The pocket depth and return flow channel could be varied between 1.5 - 4.0 mm in steps of 1.25 mm. For a given ERFS the inlet and outlet slots were equal to the depth of the return flow channel. Also, the oil inlet position to the recesses could be either in the pocket or return flow section. A diagram of the grooved land geometry used in the design is given in Fig.3.

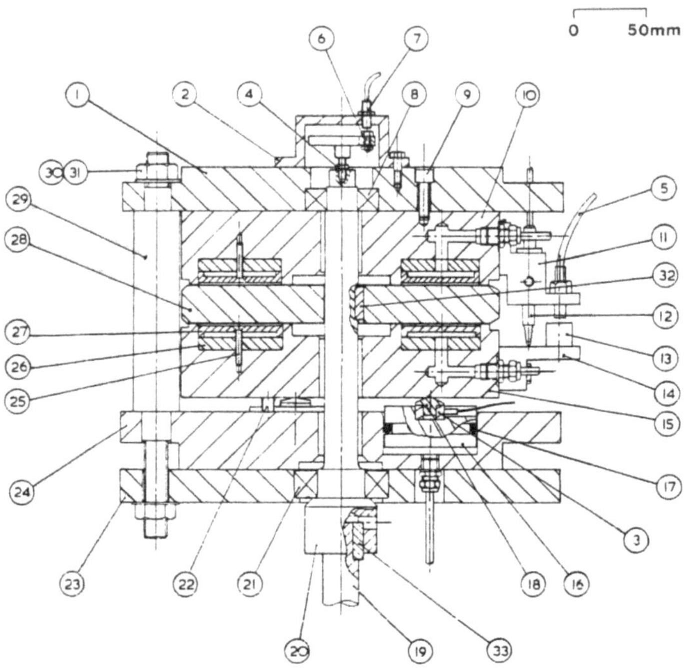

Fig.2 Cross-section through Test Bearing

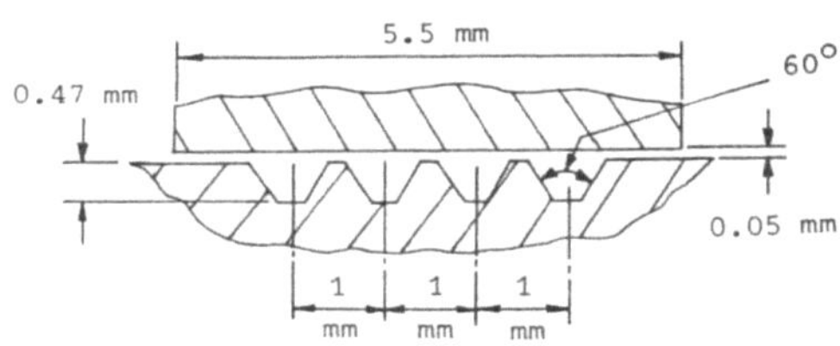

Fig.3 Grooved Land Geometry

For convenience the numerous bearing pocket and land geometry configurations were given an identification format of s/v/w/z where :

s = depth of recess (mm),
v = depth of return flow channel (mm),
w = land geometry. F for flat or G for grooved,
z = oil inlet position. C for in the return flow channel or R for in the recess.

The test bearings were supplied with oil via six variable restrictors (not shown), the latter being screwed directly into the bearing blocks to eliminate large pipe volumes between the compensating restrictors and the bearing recesses. The object of the hydraulic loading circuit was to provide a sinusoidal dynamic load superimposed on a static base load. A pressure control valve in the loading circuit controlled the static load, whilst the dynamic load was controlled by a Dowty servo value. Also, three variable restrictors were incorporated which allowed adjustment of the length of the connection to the loading rams (item 16). These restrictors were tuned to ensure in-phase loading between the rams over a range of frequencies between 0 - 70 Hz. This frequency range was a limitation placed upon the experimental programme by the hydraulic loading circuit. The lubricant used throughout the tests was Shell Tellus 27 Oil.

Other principal features were those concerned with the measurement of : -

(a) Static displacement by four equi-spaced Mercer probes [item 12],

(b) Dynamic force by three Kistler quartz load washers [item 3] placed in the three load rams (see Fig.2) in conjunction with a Solartron 1250 Frequency Analyzer and a four-channel oscilloscope,

(c) Dynamic displacement by four equi-spaced Bently Nevada non-contact displacement transducers [item 5], as shown in Fig.2.

(d) Static load by measuring the mean pressure in the rams,

(e) Frequency of oscillation and phase difference between the force and displacement measurements by a Solartron 1250 Frequency Analyzer.

3 Experimental Results

The quantities of force and displacement amplitudes were measured and then analyzed as follows :

$$\text{Column Stiffness} = \frac{\text{Force Amplitude}}{\text{Column Displacement Amplitude}}$$

Phase Angle = Phase Difference Between the Force and Displacement Amplitudes

It should be noted that the static stiffness is that which occurs at zero frequency. The experimental results of column stiffness and phase angle versus frequency with various pocket geometries, land types and oil inlet temperatures are plotted in Figs. 4 - 5. For all the tests the supply pressure to the restrictors, pressure in the recesses, film thickness and force amplitude were set to 10 bar, 5 bar, 0.05 mm and 640 N respectively.

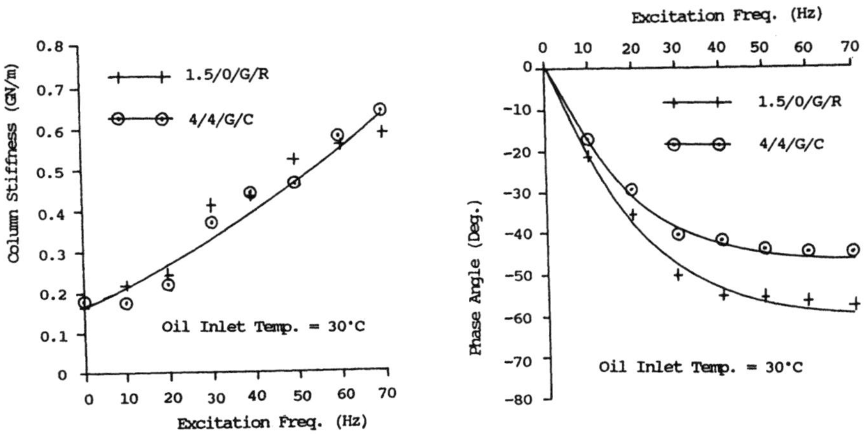

a) Column Stiffness v Freq. b) Phase Angle v Freq.

Fig.4 Dynamic Response - 1.5/0/G/R and 4/4/G/C

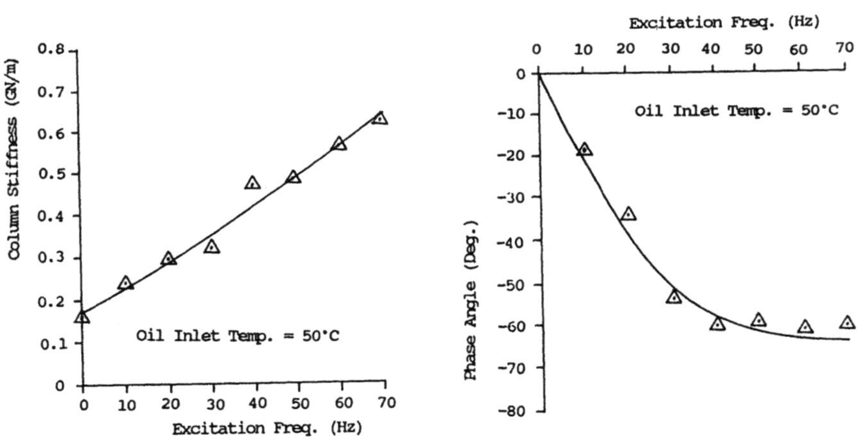

a) Column Stiffness v Freq. b) Phase Angle v Freq.

Fig.5 Dynamic Response - 1.5/0/F/R

4 Discussion

The dynamic responses of the bearing configurations 1.5/0/G/R and 4/4/G/C, operating at 30°C, are plotted in Fig.4. It was found that the increase in pocket volume did not affect the column stiffness significantly over the range of frequencies tested, however, a reduction of the phase angle was measured in the case of the 4/4/G/C configuration. This variation was attributed to oil compressibility effects in the larger pocket volume.

The results for the bearing configuration 1.5/0/F/R, operating at 50°C, are plotted in Fig.5. Comparison of the results with the bearing configuration 1.5/0/G/R operating at 30°C shows that the responses of the bearing arrangement were nearly identical. This was due to the fact that the flow resistance of the bearing arrangement, and therefore the dynamic constants (i.e. static stiffness and effective land and pocket damping), was similar in both cases.

5 Conclusions

The experimental investigation has shown that the introduction of an ERFS had only a marginal effect on the dynamic response and no effect on the static stiffness over the range of pocket inserts tested. Also, it was found that grooved lands did not affect the static and dynamic response when an identical hydraulic resistance was used.

6 References

Ashman, D. (1987) High-speed Performance of a Hydrostatic Thrust Bearing, Ph.D. Thesis, Wolverhampton Poly., England.

Ashman D., Parker E.W. and Cowley A., (1988) Behaviour of a High-speed Hydrostatic Bearing with Recess Inserts and Grooved Lands, 15th Leeds-Lyon Symposium on Tribology, Leeds University, England.

Mohsin, M.E. and Sharratt, A.H. (1980) The Behaviour of a Total Cross Flow Hydrostatic Thrust Bearing, Pro. 21th Int. M.T.D.R. Conf., Swansea, England, pp 449-459.

The use of stress wave sensors for the diagnosis of bearing faults

M.N.M. Badi, D.E. Johnson and G.J. Trmal
Bristol Polytechnic, England

Abstract

This paper investigates the condition of gearbox bearings using stress wave sensors which respond to very high frequencies. Accelerometer measurements were analysed in parallel to the stress wave sensors in order to provide comparison between the two methods.

The bearings were first examined individually on a special test rig, allowing an assessment in isolation away from the contaminating gearbox noise. Various 'intentional' faults were introduced and changes in the signature examined, using techniques based in the time and frequency domains, and thus enabling detection of such faults.

The bearings were then fitted into a gearbox and special detection techniques were used for extracting their signatures from the gearbox noise. All the 'intentional' faults were identified and fault diagnosis using the stress wave sensor method is hence assessed.
Keywords: Diagnosis, Bearing faults, Stress wave sensor.

1 Introduction

As most machines contain bearings, condition monitoring of rolling element bearings is important and has received much attention over the past years. The most common damage to rolling element bearings is spalling of the raceways and rolling elements. Rolling element contact with the fault generates impulses of a wide frequency range (Ratcliffe 1990) which can excite resonances in the bearing and machine from a few kilohertz to several megahertz (McFadden 1990).

Established techniques for bearing diagnosis can be classed in thetime and frequency domains. Time domain methods involve development of indices that are sensitive to the amount of impulsive vibration, eg RMS, crest factor, kurtosis and shock pulse counting. Frequency domain methods eg spectral analysis, synchronized averaging, cepstrum and high frequency resonance analysis involve the detection of impulse trains at some characteristic defect frequency (James Li 1989).

The objective of this paper is to present results on the use of stress wave sensors (SWS) for the diagnosis of rolling element bearing faults. These sensors operate by measuring the acoustic emissions (AE) caused by defects in materials. Unlike conventional AE techniques the SWS contain an onboard real time signal processer

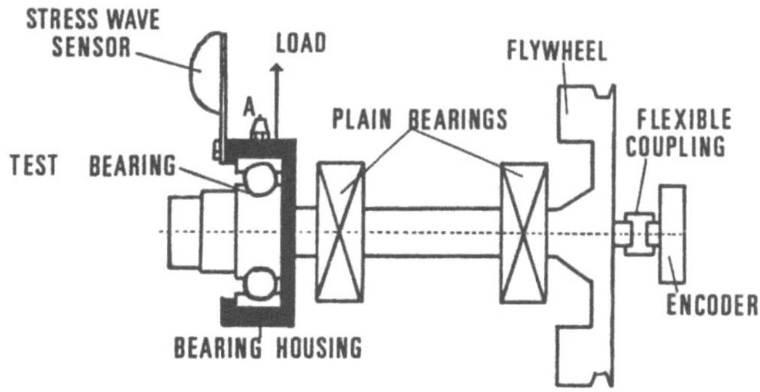

Figure 1 Bearing test rig (A = accelerometer)

which gives a usable 0-5V output signal (Holroyd 1990).

Two stress wave sensors were used with different operating frequencies; the high frequency SWS at 500KHz and the low frequency SWS at 60 KHz. The sensors were used to monitor bearings with simulated faults on a bearing testrig and in an automotive gearbox. Accelerometer measurements were also recorded and processed in a similar manner.

2 Experimental setup

2.1 Bearing test rig

A test rig was developed to operate with minimal noise and vibration, see figure 1. The subject bearing is mounted in a bearing housing which contains fixing points for both SWS and for two accelerometers. The accelerometers were positioned at right angles to each other in order to maximise pulse detection. Radial Loading is achieved by a cable attached to the bearing housing, a pulley and weights. The drive is from an electric motor via a belt to the flywheel. An encoder is fixed to the shaft to allow signal digitising with respect to shaft angular position. This will ensure minimal spectral smearing of peaks due to shaft speed fluctuations.

2.2 Data acquisition

Signals from the SWS (using the fast time constant 100 µs) and accelerometers are captured in the same manner. A CED1401 intelligent interface unit is used for digitising. Signals were low passed filtered and digitised using the encoder as a clock signal. Sixteen revolutions of the shaft are captured, 256 points per revolution. The CED1401 allows two 16 bit channels to be sampled simultaneously. Captured data is transferred to a Tandon PC for signal processing.

2.3 Signal processing

The captured signal is examined in the time and frequency (order) domains. Spectral analysis is performed by using a Fast Fourier Transform (FFT). With a time series of 16 shaft revolutions

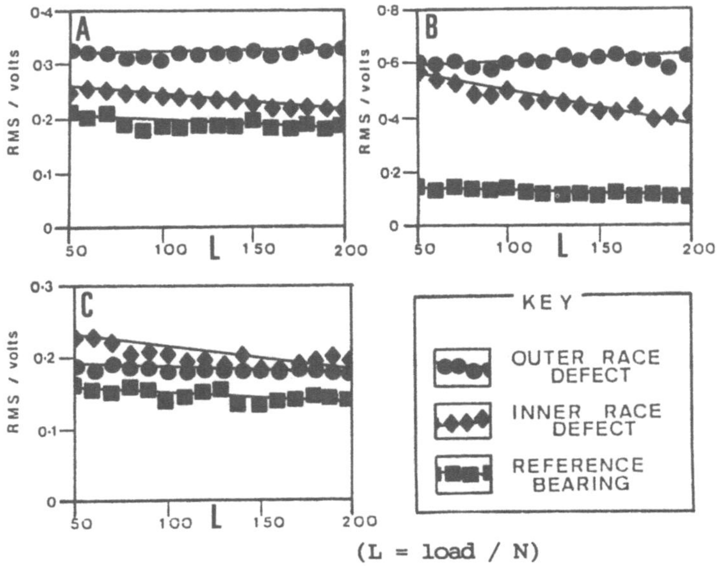

Figure 2 RMS level against load for three bearings using
A: High frequency stress wave sensor
B: Low frequency stress wave sensor
C: Accelerometer (resultant accelerometer signal)

(256 points per revolution) a resolution of 1/16 order is achieved in the order domain. This is necessary to allow precise identification of non-integer bearing fault frequencies. Application of the inverse FFT to the log power order spectrum generates the cepstrum. This is useful for examination of the "hidden" harmonic series in the frequency domain.

Time series evaluation involves calculation of the mean, RMS, and kurtosis. The resultant accelerometer signal is generated in the Tandon pc and is used for time series analysis. It is not used for spectral analysis as generation of the resultant causes frequency distortion.

2.4 Experimental method

Both SWS are digitised simultaneously at bearing loads of 50 to 200 N in 10 N increments. The two accelerometer signals are captured in the same manner. For time series analysis, 20 samples were recorded at each load setting and the average RMS, kurtosis etc values kept. Spectral analysis was performed on data captured at 100 N load. Once all three bearings were examined on the bearing test rig, they were transferred to a Ford Sierra automotive gearbox.

From the bearing dimensions the calculated fault frequencies (Sandy 1988) are:

Inner race fault = 4.94 orders
Outer race fault = 3.06 orders

3 Results and discussion

Figure 2 shows results of RMS from the SWS and resultant accelerometer signals against load. The figure shows that faulty bearings are separated from the reference (good) bearing in all cases. The low frequency SWS seperates more effectively while the accelerometer the least. With higher loads the bearing with the outer race defect shows a marginal increase in stress wave level, while the reference bearing and inner race defect bearing shows a marginal decrease (figure 2 A and B). The resultant accelerometer signal (fig 2 C) shows a slight decrease for all bearings with increasing loads. In general, a raised RMS level can indicate a fault independant of the applied load.

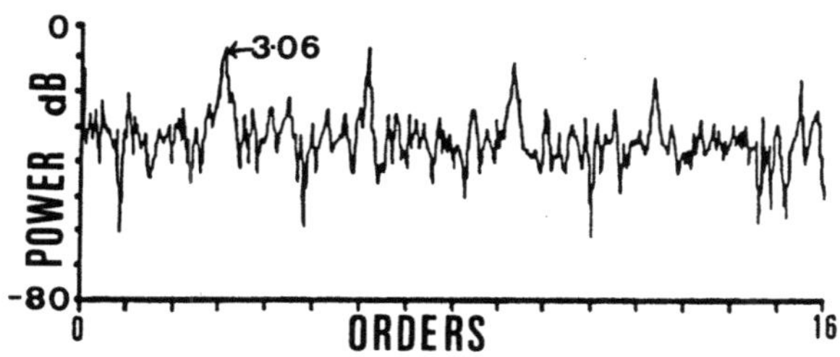

Figure 3 Spectrum for outer race defect using high frequency SWS

Figure 3 shows a typical power order spectrum for the outer race defect using the high frequency SWS. A series of harmonics can easily be seen, the fundamental at 3.06 being the calculated outer race defect frequency.

Figure 4 shows power order spectra for the inner race defect. The high frequency SWS (A) shows a peak at 4.94 orders. This is the calculated inner race fault frequency. Higher harmonics of this are present. The reference bearing (R) does not contain peaks at this frequency and has a flatter spectrum with peaks at 1 and 2 orders (due to loading of the shaft). The average level of the reference bearing is some 25 dB lower than the faulty bearing. For the low frequency SWS (figure 4, B) a peak is observed at the fault frequency but no higher harmonics. The reference bearing (R) is some 25 dB lower on average. The accelerometer order spectrum (figure 4, C) shows a totally different situation. The fault frequency at 4.94 orders is not present. Peaks exist at 4 and 6 orders in both the faulty and reference (R) bearing spectra, with the reference bearing no longer having a flat spectrum.

It is obvious that the order spectra for the SWS both separate the faulty from reference bearings and show peaks at the calculated bearing fault frequencies. The accelerometer however, show no peaks in the same spectral region, as the SWS, and so needs a different processing method.

Figure 5 shows the cepstrum calculated from the accelerometer order power spectrum shown in figure 4 (C). A peak at 0.204 in the cepstrum corresponds to 4.9 in the order domain. This is the frequency for the inner race defect.

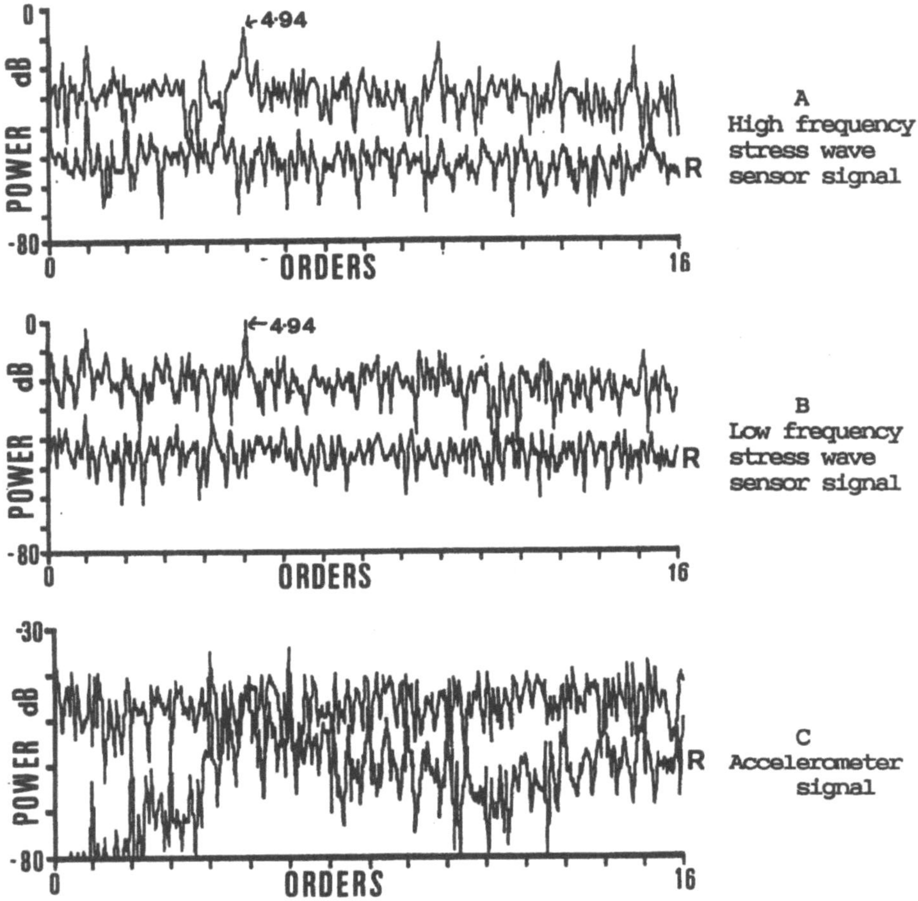

Figure 4 Order spectra for the bearing inner race defect
(R = reference bearing)

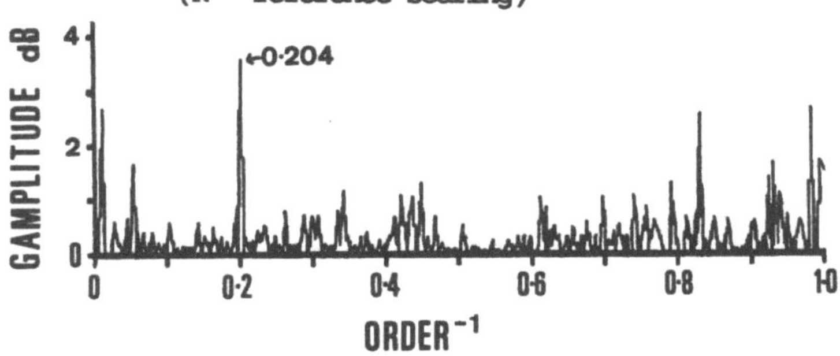

Figure 5 Cepstrum from accelerometer signal for inner race defect
(calculated from fig 4,C)

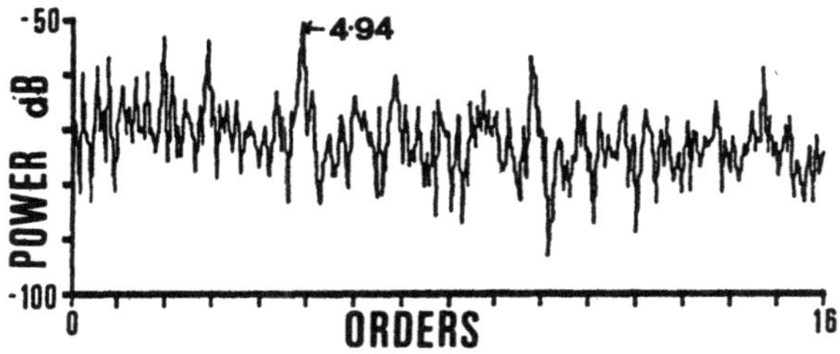

Figure 6
Order spectrum
for inner race
defect in gearbox
using high
frequency SWS

Figure 6 shows the order power spectrum calculated from the high frequency SWS signal for the inner race bearing fault mounted in a Ford sierra gearbox. A peak at 4.94 orders due to the inner race fault is present as well as higher harmonics. Accelerometers fixed to the bearing produced spectra similar to that shown in figure 4 C.

4 Conclusions

The rms value of the stress wave signal is a good indicator as it seperates faulty from good bearings well over a range of loads. The accelerometer can also be used for this purpose to a limited extent. With stress wave sensors it is possible to monitor the characteristic bearing defect frequencies in the order domain directly. An accelerometer signal, however, needs further processing (eg cepstrum analysis) to allow evaluation of the defect. Stress wave sensors detect bearing faults in the gearboxes used which are easily observed in the spectrum.

In general, the load does not appear to significantly affect neither the RMS level nor the characteristic fault frequency components of all the tested bearings.

The stress wave sensors are easy to install and need simple signal processing to evaluate bearing faults. Unlike most accelerometers, the stress wave sensors are not dependant on their mounted direction. The only draw back of these sensors are their slightly bulky size.

5 References

Holroyd, T.J, King, S.D. Randall, N. (1990). Machine condition monitoring via stress wave sensing **I Mech Eng.** Machine condition monitoring seminar, 45-48.

James Li, C. Wu, S.M. (1989). On line detection of localized defects in bearings by pattern recognition analysis. **ASME** Vol.111,331-336

McFadden, P.D. (1990). Condition monitoring of rolling element bearings by vibration analysis. **I Mech Eng.** Machine condition monitoring seminar, 49-53.

Ratcliffe, G.A. (1990). Condition monitoring of rolling element bearings using the enveloping technique. **I Mech Eng.** Machine condition monitoring seminar, 55-65.

Sandy, J. (1988). Monitoring and diagnosis for rolling element bearings. **Journal of Sound and Vibration.**

Monitoring of check valves in reciprocating pumps

T. Berther and P. Davies
Purdue University, R.W. Herrick Laboratory, West Lafayette, Indiana, IN 47904, USA

Abstract
Techniques are described for condition monitoring of check valves in reciprocating pumps that are used to charge pressure tanks to high pressures (310 bar). During a period of several months vibration measurements were made on cylinder heads of several pumps. The vibration signals typically showed transients that are caused by the closing impacts of the check valve plates. The frequency spectrum of the transients is complicated and, due to cross-talk from neighboring cylinders, it is not possible to draw conclusions from spectral information about a particular valve's condition. To isolate signatures from particular valves, time domain modeling and filtering techniques have been studied (e.g., Prony Series). One approach has been to model, in detail, the transient events that take place during each revolution of a pump. The goal of this analysis was to find parameters that can be extracted from the signals that show sensitivity to the changing condition of the valves. The results from different analysis techniques are discussed in the paper, and are correlated with the known condition of the valves in one of the pumps. Finally, suggestions for an implementation in a condition monitoring device are made.
Keywords: Reciprocating Pumps, Condition Monitoring, Prony Series, Vibration

1 INTRODUCTION

Techniques for monitoring the condition of check valves in reciprocating pumps are discussed in this paper. The aim of the research is to produce a measurement device that will rate a valve in a pump in such a way that the rating will correspond to a need for maintenance and repair of components. In reciprocating machines the valve impacts cause transient events in the vibration time history that cause difficulties when trying to analyze the signals using common FFT techniques. Other difficulties arise since impacts from neighboring valves appear in vibration signals taken close to a particular valve.

2 THE PUMPS AND THEIR OPERATION

The machines studied are a series of high pressure water pumps. The seven cylinders of each pump feed into a high pressure water system that is used to drive large presses. The pumps are driven by a synchronous motor at 300 rpm and load an accumulator system up to a pressure of 310 bar at a flow rate of 15 liters/second. Each cylinder (Fig. 1) consists of a spring-loaded suction and discharge valve. Typical failures on these parts are: worn or broken springs, broken valve plates, leakage between valve body and block and excessive leakage between plunger and stuffing box. Unlike leakage through the stuffing box, damage inside the valve cannot be detected easily.

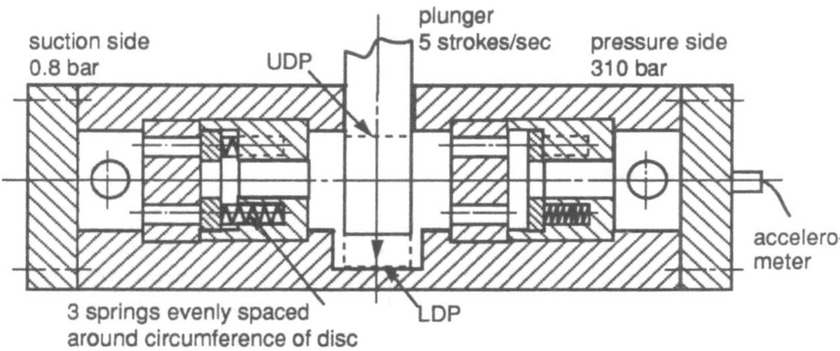

Fig.1 Schematics of cross section of pump cylinder

3 CURRENT MAINTENANCE TECHNIQUE

One indicator of a bad valve is the extra time taken by the pump to recharge the accumulator, but since the amount of the drawn water changes randomly, this parameter is imprecise.
A crude technique is to use a screwdriver to listen to what is happening inside the valve. The screwdriver is used to transmit the vibration from the machine to the human ear. For this, the pump operator needs a very "experienced ear" and the diagnosis is dependent on subjective judgement. A more sophisticated technique uses an ultrasonic measuring device similar to a stethoscope (UE Sys. 1987) to enable the listener to hear hissing sounds due to leakage. However, any decision regarding the valve's condition is still dependent upon the subjective judgement of the listener. Because of this, the pump's operating time is presently the key factor in determining whether it needs to be overhauled. This does not indicate which of the valves are causing the problems and so the pump must be fully overhauled to diagnose and repair the failed components. This approach is not cost effective, since the pump may run

with some defective valves for a long time before the overhaul, or some valves in good condition may be repaired needlessly .

4 MEASUREMENT SETUP FOR VIBRATION ANALYSIS

To gain a reliable, objective measure of the status of the valves, vibration signals were measured on-site with an accelerometer mounted on the front plate of each of the pump's cylinder valves. On one channel the vibration signal was recorded while the second channel was used to record the position of the crankshaft. The data was then transferred to a computer for further analysis.
It was found that low pass filtering the time histories at 6 kHz did not significantly affect the time history features that we were interested in tracking as the pump aged. The analysis therefore was performed in this frequency range.

5 ANALYSIS OF DATA

The features in the vibration signal could be split into three groups and related to different types of damage and wear: a) the signal could be treated as a **sequence of transients** so that the timing and the amplitude of the transients allows us to draw conclusions about pressure conditions, spring wear and delays due to sticking valve plates; b) the **shape** of the transients could be used to draw conclusions about the plate, the seats and the springs' condition; c) the **background** signal between the transients could be used to detect leakage. To build a reliable condition monitoring tool that can be used to make a detailed diagnosis of the valve condition, all three features must be analyzed. Long (1988) describes a commercially available package that employs pressure, vibration, ultrasound and ignition voltage measurements to detect faults in reciprocating machinery and suggests a diagnosis based on statistical analysis. However, the paper mainly addresses combustion engines and not pumps. Detailed study of the vibration signals from the pumps studied here indicated that analysis needs to be tailored to the working of the pump to allow detailed conclusions to be drawn about the valve's condition.

5.1 Timing (sequence of transients)

In fig. 2 is shown a measurement taken on a pump cylinder head after overhaul. The responses due to the closing of the valve plates can be seen clearly. The additional transients are caused by neighboring valves and are the reason why analysis of the whole signal causes problems.
Since the optimal timing of the valve closure is essential

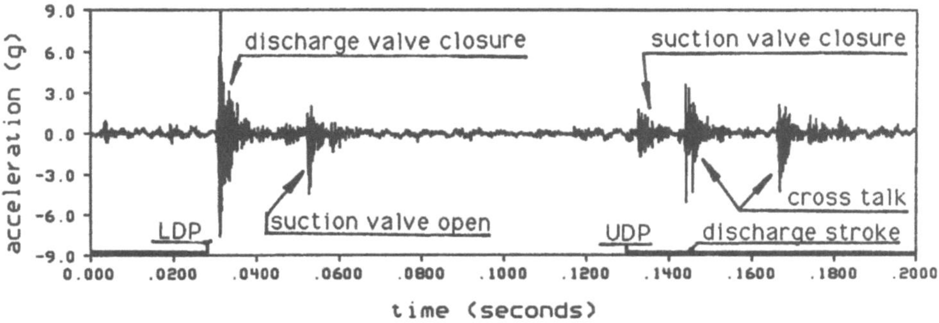

Fig.2 Vibration signal measured on a pump cylinder head during one revolution

for good efficiency, a program was developed to detect variations in timing and impact strength. For each of the pump's 14 valves the program analyzes: 1) the time delay of the valve plate closures relative to the corresponding dead point, 2) the standard deviation of the timing over several revolutions, 3) the amplitude of the impact response vibration signal, and 4) the standard deviation of the amplitudes over several revolutions. It was found that a newly overhauled pump will operate very regularly and that for example, the standard deviation of the timing is below 0.44 ms. Whenever one of the above parameters is outside a specified range, a value of 1 is added to a valve's rating. In the worst case a pump rating can add up to a total of 56 points (14 valves * 4 criteria). However, experience shows that a total rating of above 28 points is already too high: i.e.; the pump should be overhauled. While this scheme provides a good indicator of the overall pump condition it is still difficult to draw specific conclusions about individual valves using this technique.

5.2 Transients (Prony Series Model)

If the three supporting springs wear unevenly the resulting force is not aligned with the valve axis. Consequently the plate does not approach the seat straight and may produce a double impact. Berther (1990) applied the cepstrum to detect such changes in the valve.

Another approach is to parametrize the signal and track the parameter variation of different measurements. The Prony series can be used to model a signal as a sum of complex exponential function (sum of damped sine-waves). If $\tilde{y}(n)$ is an estimate of y(n), the signals can be modelled as:

$$\tilde{y}(n) = \sum_{i=1}^{M} A_i e^{s_i \Delta n}$$

where A_i and s_i are both complex. A_i is a function of the amplitude and phase, and s_i is a function of damping and natural frequency of the damped sine wave in the signal.

The Prony method can be used to compute A_i and s_i for a specified model order M. Davies (1989) describes the Prony algorithm among other time domain algorithms and lists references to further literature concerning the theory and their applications. Prony models were computed for transients caused by valve closures before and after a pump's overhaul. To locate the frequencies that are relevant to the system in both cases, 4 revolutions were analyzed (Braun 1987). Whenever the coefficients are close together it can be assumed that they represent the system properly. In fig.3 the difference between the response before and after overhaul can be seen. Some of the frequencies that were present shortly after overhaul disappeared (e.g.; 690, 1120, 3640 Hz) while other became more dominant (e.g.; 580, 920, 3050 Hz). It therefore can be concluded that the Prony coefficients allow one to detect changes in the pump cylinder. However, it has not yet been possible to relate specific frequencies to specific changes in the cylinder.

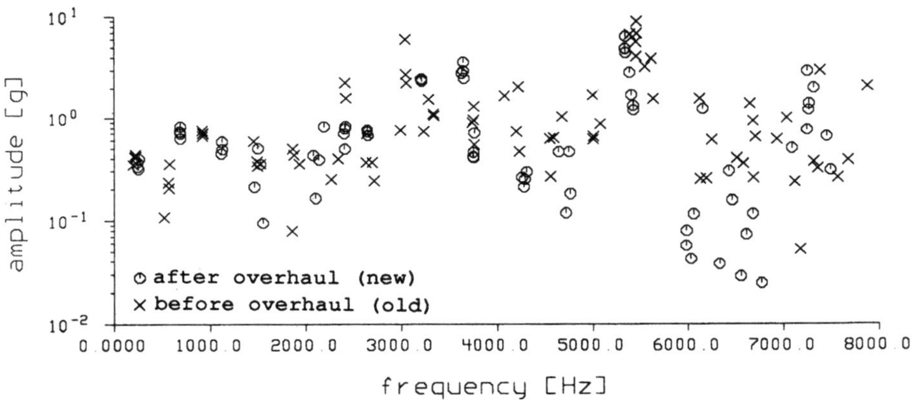

Fig.3 Prony series coefficients for valve plate closing impact responses (model order M=40)

5.3 Background Signal

During the discharge stroke; the suction-side valve in the cylinder has to seal against a very high discharge pressure. Very often this valve leaks. The additional vibration induced during the stroke can be detected as a hissing sound. However, it is difficult to identify such behavior by listening to the signal because each half stroke lasts only a tenth of a second and the impact responses often dominate the general sound impression. The same situation applies during the suction stroke when the discharge valve has to seal. If leakage is severe it can be detected because the hissing creates a broadband noise with a higher energy content than the usual background signal.

6 CONCLUSIONS

Valves that are not working correctly cause a significant drop in the efficiency of a pump. The mechanical behavior of the check valves results in complex vibration signals that contain several transient events in each cycle of the pump. The variation of these signals between different valves, even in newly overhauled pumps, are very large. This makes it difficult to identify and estimate parameters that indicate a particular valve's ageing. However, by dividing the signal into three categories, different types of damages can be recovered. Changes in the timing of the transients could be correlated with aging of the pump and efficiency decrease. By using the Prony series models, it was possible to detect dependence between the frequency coefficients and the valve's condition. Further analysis is required to relate these changes to the valves specific condition.

In a future step the findings discussed in this paper could be brought together in a PC-based condition monitoring package that would allow monitoring of reciprocating pumps. The rating procedure could then be automated and improved by using a neural network (Jakubowicz 1990). Whenever a pump is overhauled, the data obtained before overhaul together with the condition of the parts could be used to teach the network and adjust the parameters of the network. This network may then be able to pick up subtle effects that are difficult to define when using the present scheme. The pump rating scheme could be extended to include the Prony model coefficients, the cepstrum and inverse filtering results (Berther 1990). When installed, additional data could be gathered from this monitoring system which would improve the existing pump rating procedure.

REFERENCES

Berther, T. (1990) **CONDITION MONITORING OF CHECK VALVES IN RECIPROCATING PUMPS**, M.S.M.E. Thesis, Purdue University

Braun,S. and Ram, Y.M. (1987) DETERMINATION OF STRUCTURAL MODES VIA THE PRONY MODEL: SYSTEM ORDER AND NOISE INCLUDED POLES, **Journal of the Acoustical Society of America**, 81(5)

Davies, P. (1989) SOME TIME DOMAIN ALGORITHMS AND THEIR APPLICATION TO THE ANALYSIS OF MEASURED DATA, **from course notes Applied Digital Signal Processing**, Purdue University

Jakubowicz, O. and Ramanujam, S. (1990),**Proceedings of the International Joint Conference on Neural Networks, Washington DC**, January 1990

Long, B.R. and Schuh, D.N. (1988), FAULT DETECTION AND DIAGNOSIS IN RECIPROCATING MACHINERY, **Proceedings of tenth Machinery Dynamics Seminar**, The Palliser, Calgary, Alberta, 2627

UE Systems Inc (1987)., ULTRASONIC TESTING, **Plant Engineering**, June 18, p.82

Signature analysis of a gear box of turbine generator set

M.A. Rao
College of Engineering, Andhra University, Visakhapatnam, India
D.A. Rao
NSTL, Visakhapatnam, India
K.R.D. Roy
College of Engineering, Andhra University, Visakhapatnam, India

Abstract

This paper presents vibration signature analysis as a diagnostic and maintenance tool for a gear box of turbine generator set by following trends for individual components in the spectrum. Real time analysis has been carried out to pickup vibration as a function of acceleration. The data is obtained through a data acquisition system and condition monitoring amplification system. The data thus collected is analysed through FFT spectrum analyser.

1 Introduction

Vibration is a very harmful phenomena of the force transmission through a machine which provokes wear and result in breakdown. The dynamic behaviour of the machine parts will start changing when faults begin to develop, and some of the forces acting on this machine parts are also correspondingly changed resulting in the vibration level and influencing the shape of the vibration spectrum. Thus vibration signals carry much information about the running condition of the machine on the basis for using periodic vibration measurement and analysis as an indication of machine health and the need of maintenance.

All machines vibrate regardless of how well they are designed and assembled, and it has been found in industrial practice that good correlation exists between the characteristic vibration signatures of machines and their relative condition[1]. A practical method for judging vibration severity is to establish baseline signatures for a machine known to be in good condition and to monitor changes in these signatures with time[2]. In deciding upon the significant magnitude of change in a signature component an increase in vibration level is not significant unless it doubles. Therefore trend monitoring of vibration signatures is a more useful maintenance tool than a one-time survey of absolute magnitudes[3].

In this paper signature analysis of gear boxes of turbine generator set and propulsion plant are carried out to predict the present condition of the gear boxes by following trends for individual components in the spectrum.

2 Experimental setup

The turbine generator set gear box is a simple epicyclic one with an input pinion, 3 planetary gears and an annulus wheel. The details of the gear box of steam turbine propulsion plant are shown in fig.2. The details of the gear trains are indicated in table 1.

The monitoring (measuring) points of the gear boxes of gas turbine generator set and steam turbine propulsion plant are shown in fig.1 and 2.

The measuring and analysis instrumentation setup is shown in fig.3. Oscilloscope and measuring amplifier are used for conformation of better recording of vibration signature and Desk top computer is used for data processing and documentation.

3 Signature analysis

Amplitude and frequency modulations and the families of sidebands commonly found in gear box vibration spectra. These are often an indication of faults of various kinds. As the condition of gears deteriorate, increased side banding and modulation tends to occur. If a failure is present the harmonics of the gear mesh frequency together with side bands of both the gear meshing frequency and its harmonics will monitor these effects.

The spacing of these side bands contains very useful basic diagnostic information as to the source of a vibration problem, often tracing it to a particular gear in a complicated gear box. Even for gear boxes in good condition, the vibration spectra normally contains such side bands but at a level which remains constant with time. Changes in the number and strength of the side bands would generally indicate a deterioration in running condition.

3.1 Gear box of turbine generator set

A plot of vibration amplitude against frequency is known as the vibration signature of vibration spectrum of the machine which is obtained by frequency analysing the machine vibration signal. The amplitudes are measured in acceleration mode. Vibration spectrums are plotted between acceleration referred to 10-3 cm/Sec versus frequency for different monitoring points and both for good and bad condition turbine generator sets with gear boxes and are shown in fig.1. It is observed from the fig.4 that the vibration levels are high in the case of bad conditioned gear box when compared with levels of good conditioned gear box. In general the overall vibration levels should be around 60 dB. It is observed in both the gear boxes from the analysed data that the second harmonic peaks of output shaft. Rotational frequency increases more rapidly when compared to its fundamental frequency peaks.

It is also observed along with the output shaft rotational frequency and its harmonics that the presence of half the shaft rotational frequency peaks. This indicates the bad bearing operation.

The higher amplitude levels at harmonics of output shaft rotational frequency (fig.4c) indicates the misalignment of the output shaft of the gear box. Accordingly a check was carried out for misalignment. The measured level of alignment in this check was 0.18mm lateral horizontal, but the permissible level is 0.03mm. Therefore, realignment has been carried out. In the second gear box, though the levels are high at harmonics of output shaft rotational frequency, these are considered within the satisfactory limits. Similarly the peaks at half the output shaft rotational frequency are very low and considered within the satisfactory limits.

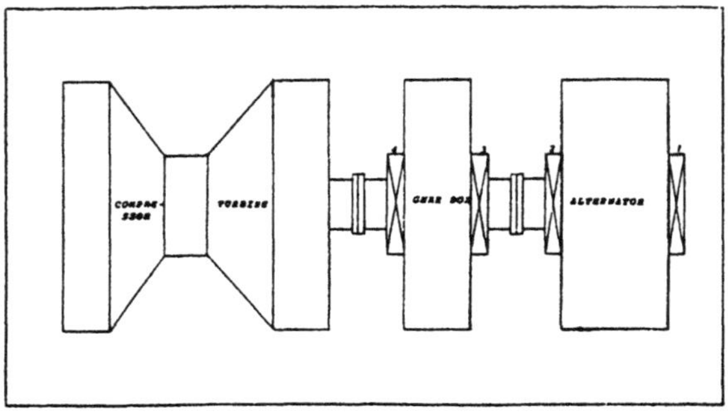

FIG -1. Turbine Generator Set

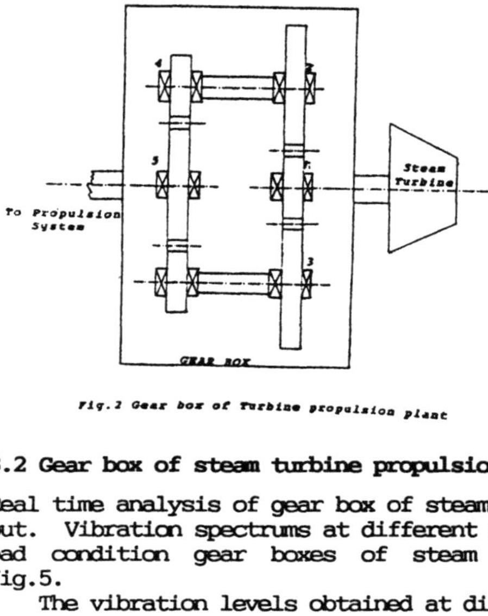

Fig.2 Gear box of Turbine propulsion plant

TABLE 1 DETAILS OF GEAR [illegible]

a) Gear train of Generator set •		
No. of teeth on input pinian, Z_1	:	28
No. of teeth on planetory gears Z_2	:	89
No. of teeth on annulus wheel, Z_3	:	235
Turbine rpm, N_1	:	12600
Alternator rpm N_2	:	1500
b) Gear train of propulsion [illegible]		
No.of teeth on input pinion, Z_1	:	33
No. of teeth on pimary gear, Z_2	:	170
No. of teeth on secondary pinion, Z_4	:	35
No. of teeth on secondary gear, Z_4	:	172
Turbine rpm, N_1	:	5746
Propeller rpm, N_4	:	227

3.2 Gear box of steam turbine propulsion plant

Real time analysis of gear box of steam turbine propulsion plant is carried out. Vibration spectrums at different monitoring points and both for good and bad condition gear boxes of steam turbine propulsion plant are indicated fig.5.

The vibration levels obtained at different points of these two gear boxes were compared with previous data available on gear boxes of the same type. The good conditioned gear box shows a reasonable satisfactory levels of vibration, while the bad continued gear box shows higher levels. These higher levels of vibration exists more prominently around primary and secondary meshing frequencies.

A phenomenon which is observed from the analysed data that the secondary mesh frequency peaks increases more rapidly compared to all other peaks (fig.5d). Later it is found that the reason for this occurrence due to tooth surface inaccuracies of secondary pinion and gear. The above reason can also be confirmed by the presence of series of upper and lower sidebands corresponding to secondary pinion shaft rotational frequency.

A second phenomenon observed in this is that the presence of second and third harmonics of secondary mesh frequency components. The prominent appearance of these harmonics reveals unacceptable secondary mesh teeth clearances and backlashes. In our checking it is found that the secondary mesh teeth clearance is above the acceptable limits.

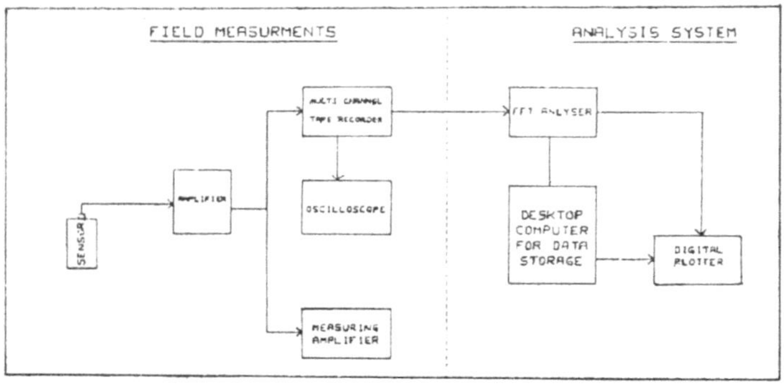

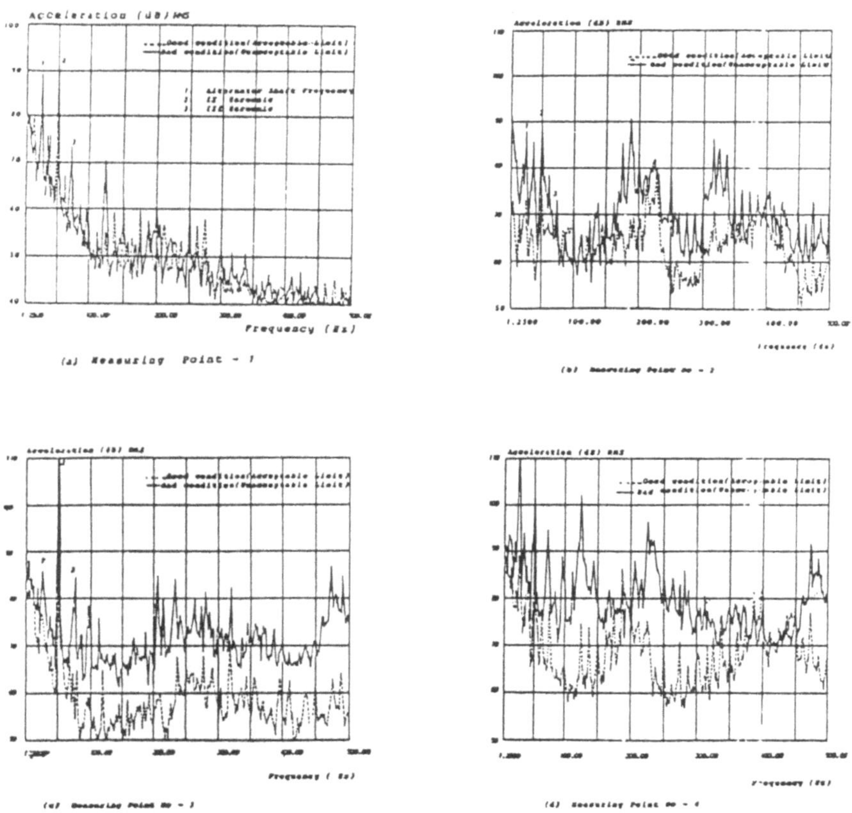

Fig.4 Vibration Signatures of Turbine Generator Set

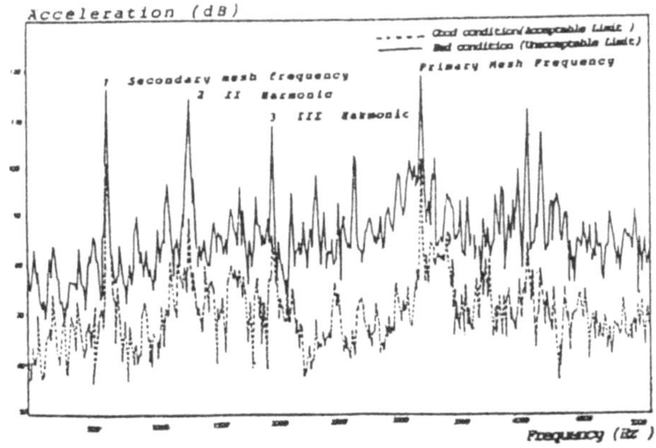

(a) Measuring Point No - 1

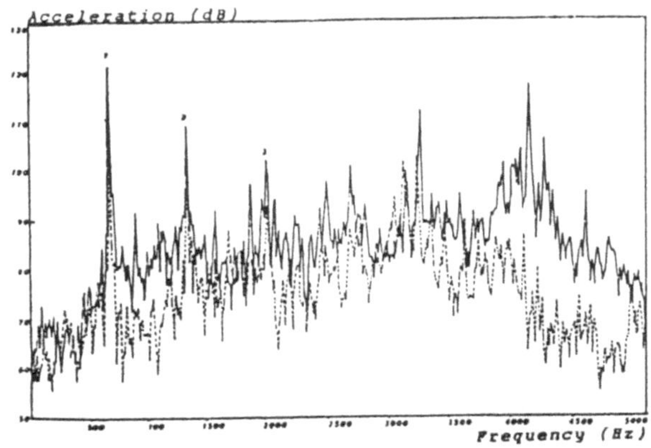

(b) Measuring Point - 2

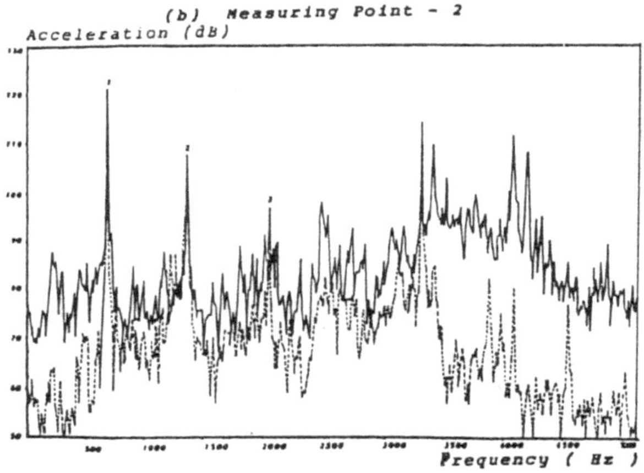

(c) Measuring Point No - 3

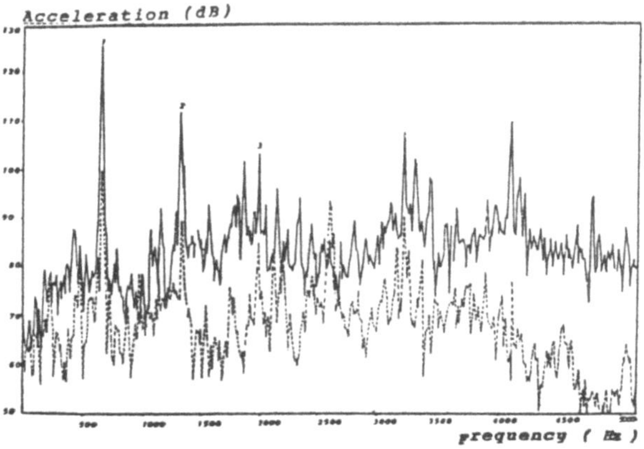

(d) Measuring Point No - 4

Fig.5 Vibration Signatures of Steam Turbine Propulsion Plant

4 Conclusions

The investigations outlined in these studies highlights the various possibilities of the faults in the gear boxes of turbine generator set and turbine propulsion plant through signature analysis. The periodic vibration measurements are very much required in order to have the information regarding the condition of the machines to obtain significant improvement in productivity and overall efficiency of plant.

5 References

Sankar, T.S. and Xistris,G.D(1971), 'Failure Prediction through the theory of stochastic excrusions of extreme vibration amplitudes', ASME Paper 71-VIBR-60.

Bowes, C.A.(1973), 'Shipboard vibration monitoring as diagnostic/maintainance tool' Int. Symp. Marine Engineering Society of Japan, November.

Glew, A.W. and Watson, D.C.(1968), 'Vibration analysis as a maintainance a tool' Trans.I.Mar.E.(Canadian Division) Supplement 32.

Condition monitoring of wear using acoustic emission techniques

R.J. Boness, S.L. McBride and M. Sobczyk
The Royal Military College of Canada, Kingston, Canada

Abstract
The paper describes measurements of acoustic emission RMS signals obtained from the wear testing of sliding metallic contacts. For the test conditions studied the direct empirical relationship between integrated RMS signal and the wear volume removed from the test ball, enables direct on line monitoring of the wear coefficients to be obtained.
Keywords: Acoustic Emission, Wear Testing, Wear Coefficient, Condition Monitoring, Dry and Lubricated Contacts.

1 Introduction

In spite of the considerable effort made over many years, a complete understanding of the wear mechanisms occurring between dry and lubricated contacts in relative motion still eludes us.

More than thirty years ago, following the classic work of Bowden and Tabor (1954), Archard (1953) derived an important wear model for unlubricated surfaces in the oversimplified form of:

$$\text{Wear rate} \quad \mathbf{V/L = K.W/H} \tag{1}$$

Expressed in this way wear rate, defined as volume of material removed (**V**) per unit sliding distance (**L**), is proportional to the normal load (**W**) and inversely proportional to the hardness of the softer material (**H**). The dimensionless parameter **K** of this equation is called the wear coefficient.

Although many observations do not follow the behaviour predicted by this model, it is now generally accepted as a suitable framework within which the quantitative aspects of this tribological subject can be discussed. The problem lies in the specific interpretation of **K**.

Experimental measurements of unlubricated wear, (see Archard (1980),book), show that wear rates, and consequently the wear coefficient **K**, vary over a considerable range; almost 10^5. This range is increased even further when lubrication and other wear reduction techniques are present.

It is known that **K** depends on the nature of the materials used, on the geometry and topography of the contacting surfaces, and on interfacial friction. The problem is further complicated by the presence of a lubricant film between the surfaces, whether it be a thick hydrodynamic film keeping the surfaces completely apart, an elastohydrodynamic film (1 µm thick), or even boundary lubricant effects which provide protective films by chemically reacting with the surfaces in contact.

At this stage it is important to realise that two engineering surfaces which are loaded together touch only over a very small part of their apparent area of contact. That is, the load is supported by a limited number of asperities on opposing surfaces, contacting each other.

The most common interpretation of the meaning of the wear coefficient **K** is that it represents the proportion of all asperity contacts which result in the production of a wear particle. The problem is to be able to relate the wear coefficient **K** to specified operating conditions.

Historically, wear measurements have been conducted on specialised test rigs which enable simple test pieces to be loaded and rubbed against each other, either dry or lubricated, and with a combination of rolling and sliding motions.

In order to obtain the wear coefficient for a particular test, normally requires that the test is stopped, while the test pieces are removed and examined under a microscope and the volume of material removed by the wear process is measured. In many cases as it is impossible to reassemble the test pieces in exactly the same position in the test rig new specimens have to be used.

This paper is concerned with the measurement and subsequent analysis of acoustic emission (AE) signals obtained during the wear testing of sliding metallic contacts, with the object of using such measurements for monitoring, in real time, the wear coefficient **K** as a function of time or sliding distance.

2 Experimental Apparatus and Procedure

These particular tests were carried out on a ball and cylinder test rig under conditions of pure sliding: a stationary ball being loaded against a rotating cylinder. Details of the test materials and lubricants are given in Table 1.

Table 1. Test Materials and Lubricants

	Diameter (mm)	Material	Hardness (Rc)	Surface Roughness (µm Ra)
Ball	12.7	52100 steel	64	0.2
Cylinder	49.0	8720 steel	61	0.56

Lubricants	Heavy Paraffin, Kinematic viscosity	67×10^{-6} m^2/s @ 37.8°C
	Light Paraffin, Kinematic viscosity	28×10^{-6} m^2/s @ 37.8°C
	SAE 30, Kinematic viscosity	68×10^{-6} m^2/s @ 37.8°C

Acoustic emission testing offers a novel method for studying the complex wear mechanisms that exist between loaded surfaces in relative motion. Traditionally, such techniques have been used to detect defects in materials and to monitor incipient fatigue growth cracks in critical structures. A complete bibliography of acoustic emission has been assembled by Drouillard (1979, book) and is periodically updated in the Journal of Acoustic Emission.

Acoustic emission may be defined as transient elastic stress waves, generated at a source, by the rapid release of strain energy within a material. These radiating stress waves are detected at the surface of the body by a suitable transducer. Subsequent signal processing producing results relating to the original AE source mechanisms which may arise from different phenomena such as asperity contact, micro crack initiation and growth, plastic deformation and flow etc. These are the mechanisms which are involved in the basic wear processes. Unfortunately, as any combination of these source mechanisms may be active at any time, the interpretation of AE signals can be difficult.

In a previous paper the authors (1989) have examined acoustic emission from rubbing contacts using a sophisticated data acquisition system which enabled number of events, peak amplitudes, rise rates, decay rates and pulse lengths, as a function of time, to be recorded. Because of the interference of the general signal level with the trigger settings in some of the tests, difficulty was experienced interpretating the vast amount of data collected. It was realized by Boness, McBride and Sobczyk (1990) that the simple measurement of the RMS of the AE signal contained considerable information concerning the wear mechanisms occurring in sliding contacts. In this paper AE signals were detected by a resonant type transducer (resonant frequency 750 KHz - band pass 500-1200 KHz) acoustically coupled with silicon grease to the ball housing, shown diagrammatically in figure 1. The transducers signals were amplified by an AE preamplifier, (60dB gain with frequency range of 300-1000 KHz), and channelled to a Hewlett-Packard model 3400A RMS meter providing an output proportional to the DC heating power of the input waveform. This signal was sampled at 0.1s intervals for the first 100s of testing and 2s interval thereafter. The resulting readings were recorded by a microprocessor and subsequently downloaded to a spreadsheet for analysis.

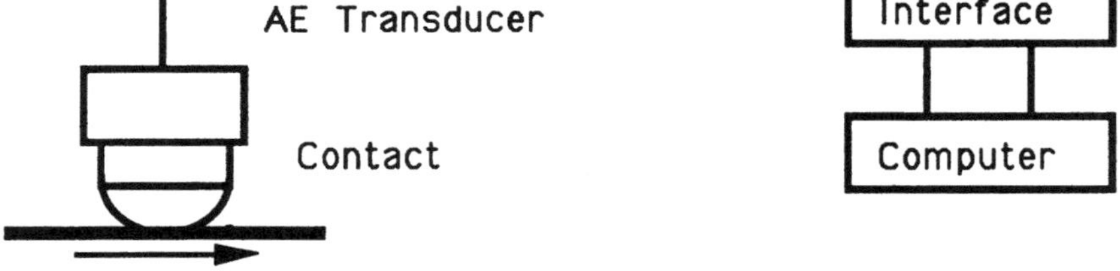

Figure 1. Schematic of Acoustic Emission Equipment

3 Discussion

A series of experiments were conducted where wear scar volume and AE RMS signal was measured as a function of running time. It was found to be more accurate to calculate the volume of material removed from the ball using direct measurements of wear scar area obtained from an optical microscope and an Omnimet image analyzer, than to base this calculation on the major and minor dimensions of the elliptical wear scar.

Typical wear scar volumes as a function of time, for both lubricated and unlubricated experiments obtained from the ball-on-cylinder test rig, are illustrated in figure 2. These results were obtained for a constant load of 10N and sliding speed of 0.57 m/s (225 rev/min). Based on elastohydrodynamic lubrication theory, the ratio of the film thicknesses between the ball and the cylinder and the RMS surface roughnesses, the λ value, was estimated to be .06 and .24 for the light and heavy paraffins respectively. Hence, substantial asperity contact was expected to occur for these tests.

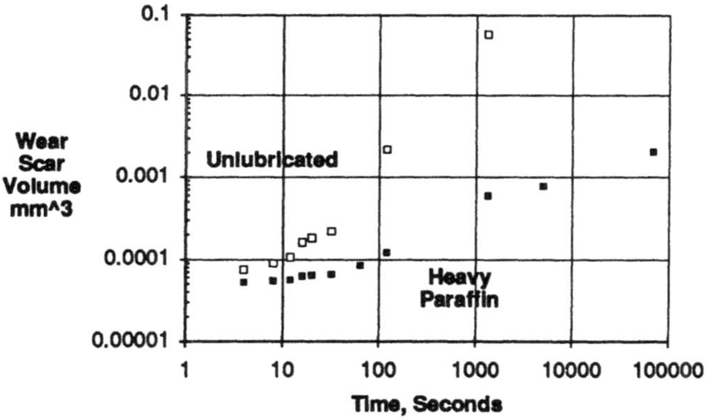

Figure 2. Wear Scar Volume vs Time

The resulting RMS acoustic emission signals, referred to the input of the preamplifier, are shown in figure 3. By plotting the integral of the RMS signal against wear scar volume, figures 4 , a relationship between these parameters became evident. In the spirit of simplification the existing data may be fitted by several power laws of the form:

Wear scar volume (mm^3) = constant x $\int(\text{RMS})^x \, dt$

Table 2 summarizes the empirical relationships between the integrated RMS data and the wear volume together with the wear mechanisms observed from scanning electron microscope examination of the rubbing surfaces. It is interesting to note that the power of the functions given in Table 2, i.e the slopes of figures 4 , show remarkable agreement for the mechanisms observed.

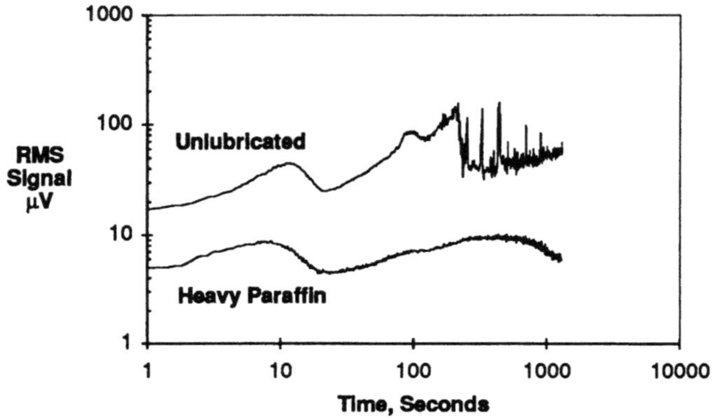

Figure 3. RMS Signal vs Time

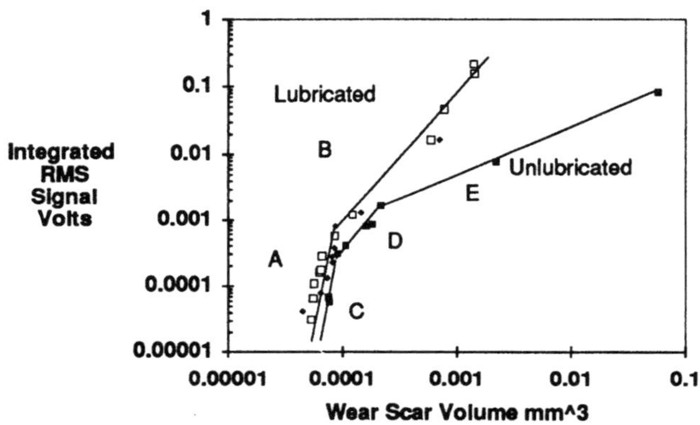

Figure 4. Integrated RMS Signal vs Wear Scar Volume

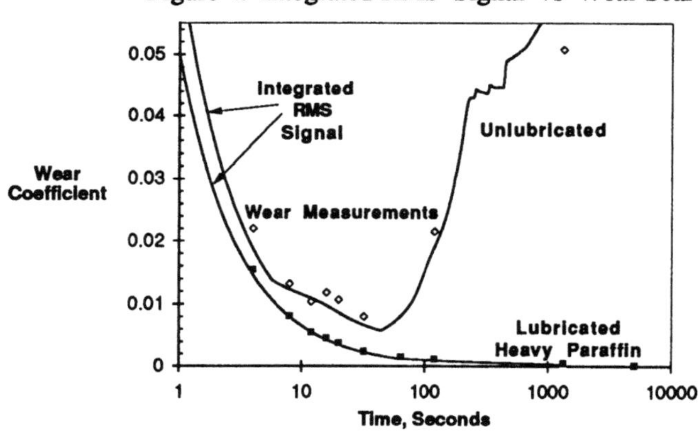

Figure 5. Wear Coefficient vs Time

Table 2. Integrated RMS Functions

Region	Function	Wear Mechanism
A	$0.000144\int(RMS)^{0.1}\,dt$	Initial wear (lubricated)
B	$0.00355\int(RMS)^{0.5}\,dt$	Adhesion (lubricated)
C	$0.00016\int(RMS)^{0.1}\,dt$	Initial wear (dry)
D	$0.00525\int(RMS)^{0.5}\,dt$	Adhesion (dry)
E	$3.162\int(RMS)^{1.5}dt$	Abrasion (dry)

Also of particular significance is that the basic empirical relationships between wear scar volume and integrated RMS signal is independent on the lubricant used. In order to obtain the wear coefficient **K** as a function of time the operating conditions load and speed together with the hardness of the softer material is substituted into equation (1). Figure 6 shows the wear coefficient **K** based on the integrated RMS AE signals together with the **K** value obtained directly from wear scar volume measurements. The remarkable agreement between these results indicate that RMS acoustic emission signals can provide direct on line monitoring of the wear coefficient **K**.

4 References

Archard, J.F. (1953) Contact and Rubbing of Flat Surfaces. **J. App. Phys.**, 24, 981-988.

Archard, J.F. (1980) Wear Theory and Mechanisms. **Wear Control Handbook ASME**, 35-80.

Boness, R.J., McBride, S.L. and Sobczyk, M. (1990) Wear Studies using Acoustic Emission Techniques. **Trib. Intl**, in press.

Bowden, F.P. and Tabor, D.(1954) **The Friction and Lubrication of Solids.** Clarendon Press, Oxford.

Drouillard, T.F. (1979) **Acoustic Emission - A Bibliography with Abstracts.** IFI/Plenum Data Company.

McBride, S.L., Boness, R.J., Sobczyk, M. and Viner, M.R. (1989) Acoustic Emission from Lubricated and Unlubricated Rubbing Surfaces. **J. Acous. Emis.**, 8, 1-2, 192-197.

The condition monitoring of machine tools

A. Davies and J.H. Williams
School of Electrical, Electronic and Systems Engineering,
University of Wales College of Cardiff, Cardiff

Abstract
The paper shows how the condition of mechanical characteristics of a machine tool can be inferred by the use of signal analysis of accessable servo currents. This approach does not degrade the reliability of the machine tool which can occur if additional sensors are used.

1 Introduction

Machine tool technology is now developing at a faster rate than at any time in its history. Despite these advances, developments in the strategy and techniques of machine tool maintenance have fallen behind in the general pattern of progress. This is of serious concern, since it can be argued that present maintenance techniques will be unable to cope with the new demands that will certainly be made upon them in the future. Consideration of this problem and some new condition monitoring techniques are presented in this paper

2 Condition Monitoring

To justify the application of CM the following criteria must be generally satisfied. (Harris, Williams, Davies 1989).

(i) the plant to which it is applied must be critical in terms of the effect that a breakdown has either on safety or lost revenue.

(ii)the plant must have observable characteristics from which trends in deterioration can be estimated.

The CM methods being applied to machine tools can be classified into two categories: 'Direct' methods in which data extracted from the machine directly relates to its condition and 'Inferential' methods in which data is processed by some form of model to yield a health assessment. An example of the latter approach is the Kearney & Treckers' remote Diagnostic Communication System (Hatschek 1982, Freeman 1980, Chamberlain 1980). This paper will outline newly developed techniques that also fall under this category.

3 CM Techniques

When setting up a CM scheme for a machine tool, Waterman (1981) suggests that a detailed analysis of all the components on the machine be undertaken to assess in each case the likelihood of failure; the effect of the failure when it occurs, and the time it takes to repair. Each component is ranked in order of criticality as far as CM is concerned. Birla (1980) has listed possible critical components: fluid and lubrication systems; work holding fixtures; rolling element bearings and gearboxes; pumps; slideways.

Vibration analysis is perhaps the most widely used and developed technique in the wider application of CM. Machine tools which have many rotating elements such as spindle bearings, gearboxes and drives could usefully adopt many of the vibration monitoring techniques. The most common of these are: r.m.s. and peak level measurements (Fox 1977); shock pulse measurements Kurtosis (Dyer 1978); low-frequency spectrum analysis (Claessens 1979); and signal averaging (Stewart 1979,). All these techniques rely on measurement and analysis of vibration produced by the equipment under test.

Stoferle (1976) describes a parametric adaptive modelling approach which is relevant to CM requirements. The system being diagnosed as modelled by a differential equation relating the equipment's performance (e.g. table position x(t) to an applied stimulus such as a control voltage y(t). This equation is written in its difference form of the general type:

$$x_k = f(x_{k-1}, x_{k-2} \dots y_{k-1}, y_{k-2}, \dots a, b, c, \dots) \quad \text{equ 1}$$

where k represents the sampling instant and a, b, c ... are coefficients related to the system's physical characteristics. Thus on line identification will furnish information on any physical changes in the system. It is convenient to use a recursive updating technique:

$$\begin{bmatrix} a_1 \\ a_r \\ a_n \end{bmatrix}_{k+1} = \begin{bmatrix} a_1 \\ a_r \\ a_n \end{bmatrix}_k + G_k \begin{bmatrix} x_{R+1} - M^T \end{bmatrix} \begin{bmatrix} a_1 \\ a_r \\ a_n \end{bmatrix}_k \quad \text{equ 2}$$

where a_i are the system's coefficients, G is the correction factor and M the system's coordinates. Acceptable bands are set within which the estimated system's coefficients can vary. While the coefficients remain within those limits the system is classified as healthy.

A practical modification of this modelling technique (Harris, Williams, Davies 1989) is to simulate the system in parallel with a digital model. This model is not allowed to free run (letting it proceed from a set of initial conditions unaltered except by inputs) but is forced to track the actual system behaviour within a prescribed error limit. This is achieved by resetting the initial conditions of the model whenever the error limit is consistently exceeded. The frequency of model re-sets giving an indication of healthy, degraded or catastrophic failure performance. The general scheme is shown in Fig. 1.

4 CM via servodrives

Fundamentally the 'heart' of the machine tool are the servo drives. Using a system approach the focus of the health of the machine is present in the dynamic responses of the servos. The gain and bias of the servo systems are usually available for adjustment to the maintenance engineer and in general as there are only two parameters concerned the operator can quickly learn the distinguishing features of each variable. Of more interest is the capability of observing mechanical degradation effects in the servo responses. Johansson (1976) maintains that the feed driver currents can indicate insufficient lubrication resistance to slide motion due to foreign body ingress between slides and slideways and motor condition. At UWCC some of the CM research has been focussed on this approach. Trials have shown that axis backlash (lost motion) can be monitored by measuring the current to the axis motor Hoh (1989). The machining centre used for the trails is a current model of a machining centre made by De Vlieg Machine Co., supplied under an SERC ACME Grant on 'Sensor Based Machine Tool Condition Monitoring.

The current sensor used is a standard built in device which is used to safeguard against overloads. This is attractive as no additional sensors are required and the overall reliability of the system is not impaired. Both the toothed drive belt and the thrust bearing were adjusted so as to produce lost motion up to 90 microns from a nominal value of 12 microns.

The adjustments to both belt and bearing are not easily controlled and the lost motion attained were of values that were considered reasonable. With the belt and bearing in their normal, acceptable condition the transient response of the motor current to a step demand in table position (x-axis) was recorded. This signal was used as a reference or nominal response. Subsequently the belt was adjusted so as to increase the lost motion from 12 microns to 26 and 50 microns. Transient responses were recorded for both settings. The procedure was repeated but with the thrust bearing adjusted to give lost motions of 40 and 90 microns. The deviations from nominal are shown in Fig. 2. It is seen that these mechanical effects are clearly observable in the motor current transients. To date the lost motion due to the thrust bearing, motor current interdependence is consistent and robust. The lost motion due to the belt, whilst observable in the motor transients, is not giving deviations which are scaled versions of the adjustments imposed.

5 Conclusions

The use of the servomotor current transients have been shown to offer a means of monitoring not only the servo system itself but mechanical changes in the load such as those producing dedregation of performance. Future work will develop a pattern recognition based diagnostic scheme (Williams), which will help to reduce down time of the machine tool.

6 References

Harris, C. G., Williams, J. H., Davies, A. 'Condition Monitoring of machine tools, Int. J. Prod. Res. 1989, Vol 27, No. 9, pp 1445-1464.

Hatschok, R. L., 'NC diagnostics, Special Report, American Mechanical, 1982, pp·161-168.

Freeman, G. C., 'Machine Tool diagnostics, Machine Tool Task Force Report Vol 4, Lawrence Livermore National Laboratory, Univ. of Cal., Livermore, Cal., USA, 1980, pp 7.17.1-7.17-16.

Chamberlain R.G., 'Diagnostics, Machine Tool Task Force Report, Vol 4, Lawrence, Livermore National Laboratory, Univ. of Cal., Livermore, Cal., USA, 1980, pp 7.16-1-7.16-8.

Waterman, N.A., 'Condition monitoring of machine tools, Proc. of 3rd National Conf. on Condition Monitoring, 1981, London, Confer. Secretariat London.

Birla, S. K., 'Sensors for adaptive control and machine diagnostics, Machine Tool Task Force Report, Vol. 4, Lawrence Livermore National Laboratory, Univ. of Cal., Livermore, Cal., USA, 1980, pp 7.12-1-7.12-70.

Board, D. B., 'Rotating Machinery Diagnostics Through Schock Pulse Monitoring (ISA), ASME, Winter Annual Meeting on Diagnosting Machinery Health Symposium, New York, 1978, pp 559-566.

Fox, R. L. 'Preventive maintenance of rotating machinery using vibration detection, Iron & Steel Eng. 1977, 54, pp 52-60.

Dyer, D. 'Detection of rolling element bearing damage by statistical vibration analysis, Jour. of Mech. Design, 1978, 100, pp 229-235.

Claessens, C. 'Condition monitoring of rotating equipment with vibration measurements, 'Proc. of the Eurotest Seminar on Plant Condition monitoring, Imp. Coll. London, 1979, pp 1-12.

Steward, R. M., 'The development of monitoring technology for rotating machinery. Proc. of 2nd Int. Conf. on Condition Monitoring, London, Brit. Coun. of Maintenance Associations, 1979, pp 3.1.1-3.1.19.

Stoferle, T. 'Automatische Uberwachung and Fehlerdiagnose an Werkzeumaschinen, Annals of C.I.R.P., 125, 1976, pp 369-374.

Johansson, K. E., 'Maintenance system for the Mech. Ind. including Automatic Condition Monitoring System and Administrative System for Maintenance, Rep. LiTH-IKP-R-071, Sin Doping Institute of Technology, 1976.

Hoh, S. M., Condition Monitoring for Machine Tool mechanical axis drive servo-unit using a built-in-axis motor current sensor', U.W.C.C., Cardiff, Tech. Note No. 171, 1989.

Williams, J. H., Transfer Function Techniques and Fault Location, Res. Mono. Eng. Dynamics and Control Series, Research Studies Press, John Wiley & Sons Ltd., 1986.

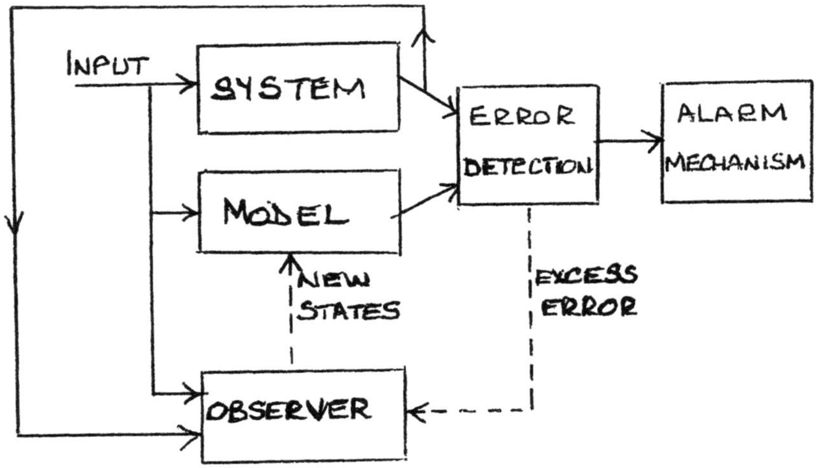

FIG. 1 TRACKING RATE ANALYSIS SYSTEM

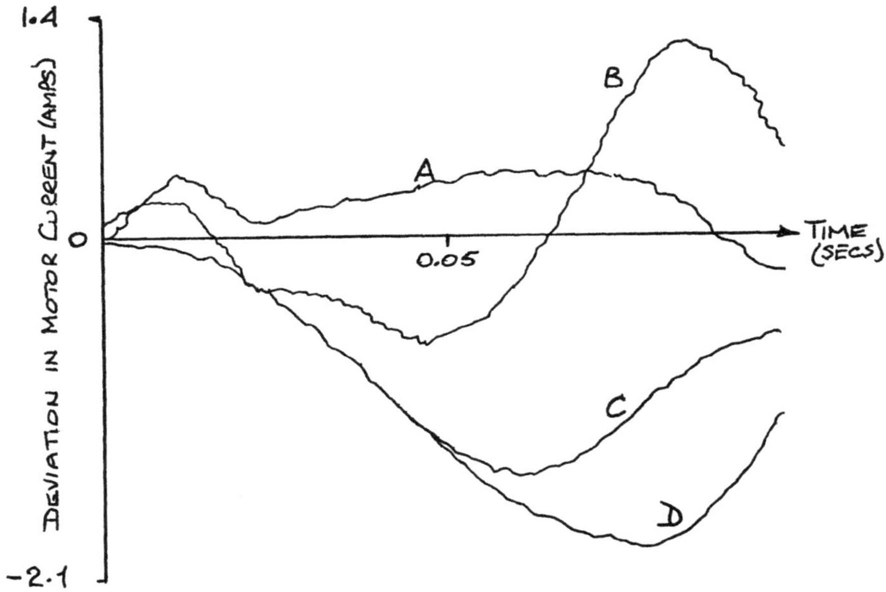

A	BACKLASH	OF	26	MICRONS	DUE	TO	BELT	
B	"	"	50	"	"	"	"	
C	"	"	40	"	"	"	THRUST	B'R'G.
D	"	"	90	"	"	"	"	"

FIG. 2. SERVO RESPONSE DEVIATIONS DUE TO LOST MOTION

Condition monitoring systems for machining applications

M. Raghunandan and R. Krishnamurthy
Manufacturing Engineering Section, Indian Institute of Technology, Madras, India

Abstract
Efficient production machining needs heavier chip loads to be used, because of which, probability of damage (to tool, workpiece or machine) has increased. It is important, particularly for untended manufacturing, to detect inciepient damage, before it can accumulate and lead to catastrophic consequences. Adaptive control and condition monitoring systems have been proposed for detecting chip congestion, tool wear and breakage.

Many signals like tool forces, tool tip temperature, torque, vibrations, acoustic emission (AE), etc., have been used in condition monitoring of machining.Of these, vibrations and AE signals look prospective because of the ease of measurement and ruggedness and sensitivity of sensor.

Comparitive analysis of the results obtained by researchers, in monitoring the machining process, using these two signals, indicates AE to be more sensitive to the actual cutting process. Moreover, AE signals seem to be affected by process variables only and are not dependent on the structural rigidity of machine tool, as do vibration signals. Hence, real-time determination of structural modes is needed for an effective vibration based monitoring. Studies with vibration and AE signals, are on, at present, to investigate the effect of process variables and develop efficient signal processing methodologies.
Keywords: Tool Condition Monitoring, Vibration Signals, Acoustic Emission, Pattern Recognition.

1.0 INTRODUCTION

The Need to operate at optimal efficiency, when products of higher accuracies and complex geometries are produced, under conditions of low product life-cycle and variable demand, has lead manufacturing towards minimally manned or unmanned production. On the other hand, to obtain optimal efficiency, machines have to operate under conditions which make it difficult to predict completely the effect of process variables on the workpiece, tool or machine. Hence it is important that any damage (to tool, workpiece or machine) be determined at the incepient stage, before it can lead to catastrophic failure.

Condition monitoring and adaptive control have been proposed to resolve these problems, particularly in the context of stochastic nature of the manufacturing system. Conditon monitoring systems have been proposed for detecting chip congestion, tool wear and breakage. Adaptive control systems have been advocated for compensating variations in tool and work materials and surrounding environment. Many technical and economic aspects relating to use of condition monitoring systems are dealt with in some of the published literature [Weck 1983, Tonhoff 1988].

Lack of fundamental process knowledge - relating to use of techniques to wide range of materials and machining conditions [Lister 1986], and lack of rugged sensors and inadequate sensing methodologies [Dornfeld 1988] seem to be the main obstacles in successful implementation of condition monitoring systems. Hence, research has been directed towards applying artificial intelligence concepts, such as, pattern recognition [Emel 1988], fuzzy logic [Li 1988] and parallel processing and distributed computing [Chryssolouris 1988, Dornfeld 1988] to manufacturing.

2.0 TOOL CONDITION MONITORING

Monitoring methods can, in general, be classified as direct or indirect. Direct methods deal with measurements of the volumetric loss of cutting tool material; indirect methods with changes in cutting parameters resulting from deterioration of tool. The main advantage of indirect methods is that they are on-line and real-time techniques. A variety of signals like tool forces, tool tip temperature, torque, power, motor current/voltage, vibrations and acoustic emission (AE) have been attempted for on-line measurement of tool condition. Discussion of certain specific sensors and their ability to perform various monitoring tasks [Tlusty 1983], of various direct and indirect methods [Lister 1986], and comparison of commercial systems [Osipova 1987] can be found in published literature.

Vibration signals have long been used for monitoring machinery health and AE analysis is an established NDT technique. This paper will discuss applicability of these two signals for determining tool and machining conditions.

3.0 VIBRATION SIGNAL BASED MONITORING

Vibration signal analysis is attractive because of the ruggedness of sensor and the ease of measurement. According to [Martin 1986], vibration signals are much less sensitive to tool condition, than AE or tool forces. This is mainly because vibrations during machining are produced as a secondary effect of tool wear and breakage. Vibrations are produced due to rubbing action at the work flank interface [Weller 1969], formation of built-up edge, waviness of work surface [Shaw 1962], or dynamic changes in tool forces, apart from those due to associated gear contacts. Moreover the signal reaches the sensor after damping and attenuation at various levels.

Table 1 gives a comparison of the work done by various researchers

Table 1 Vibration based monitoring - comparison of results

Reference to published work	Process	Variables	Freq. KHz	Signal Processing
1 Weller 1969	Turn	Tool overhang, Tool & work matl., Work size, Machine, Depth, Speed, Feed	4 to 8	Total power content
2 Martin 1976	Turn	Speed.Feed,Tool position w.r.t workpiece end	2 to 3	Power Spectral Density (PSD)
3 Pandit 1982	Turn	Speed,Feed,	4 to 5	Data Dependent systems (DDS)

in using vibration signals. A linear model has been fitted, to relate PSD to volume of metal removed and tool wear [Martin 1976]. But [Pandit 1982] indicates that actual power of the signal at, principal mode of tool holder, decreases to a minimum at 0.25...0.36 mm of flank wear, after which, an increasing trend is noted with increasing wear. It has also been established that the significant feature of the vibration signal, which is related to tool wear is, the first principal mode of the tool holder system [Weller 1969, Martin 1976, Pandit 1982].

4.0 AE SIGNAL MONITORING

Research over last several years has dealt with AE based sensing methods, both for detection of tool wear and fracture and for analysis of cutting process. These have given useful insights on the signal processing required and the generating mechanisms involved. A review on use AE signals in machining is found in [Dornfeld 1988, Lee 1987].

4.1 AE FROM METAL CUTTING

AE is defined as the transient elastic energy spontaneously released in materials undergoing deformation, fracture, or both. Deformation mechanisms, fracture, decohesion of inclusions, realignment or growth of magnetic domains and phase transformations have been proposed as sources of AE [Dornfeld 1980]. Two types of AE are observed in metal cutting - burst AE and continuous AE [Kannatey-Aisbu 1982].

In metal cutting, burst AE has been associated with chip and tool breakage and continuous AE with shearing at primary and secondary zones and wear on tool face and flank [Emel 1988]. Fig.1 gives various zones which act as sources of AE. A detailed analysis of generating mechanisms of AE in metal cutting can be found in [Dornfeld 1988].

It has also been indicated that the signals measured at workpiece and tool side are different - mainly due to the high acoustic impedence of various shear zones [Uehara 1984]. They put forth the hypothesis that AE on tool side will contain information about the contact at tool-chip and tool-work piece interfaces and on tool chipping (hence on tool wear and breakage), while the signal on workpiece side will contain information on plastic deformation, shearing and contact on tool-workpiece interface. Further, it was found that the power spectrum of different work materials is different and signal measured on tool side is effected by tool wear.

Thus, unlike vibration monitoring, a strong dependency of AE generated on process variables can be anticipated because AE is generated by fundamental process events. Analysis of time-based statistics (count and count rate) and energy (V_{rms}) of AE signals has been attempted. A review of various methods therein can be found in [Teti 1988]. Table 2 gives details of the affect of process variables on AE signals generated. Apart from speed, feed, depth of cut and tool angles, material dependent properties like morphology of chip formed and hardness, chip-tool contact length and geometry of contact affect AE signal intensity widely. Moreover events like chip breakage, collision, and lubricat-ion produce AE signals which may over-shadow those from the cutting process [Lee 1988].

4.2 AE BASED TOOL CONDITION MONITORING

A variety of techniques have been attempted for monitoring of tool wear and breakage using AE signals. Table 3 gives a summary of the techniques reported by various researchers.

Analysis based on count, count rate, statistical properties of amplitude distribution (Skew, Kurtosis, B functions, Mode) have given only limited success due to their dependency on process variables. Pattern recognition [Emel 1988] of AE signals using power spectral components as features, has shown good results in determining tool wear and breakage and chip noise.

The reliability of pattern recognition algorithms is dependent on the features selected. The main disadvantage of the pattern recognition based on spectral components is that large number of orthogonal components are generated, necessitating the use of feature selection algorithms, which may not select the optimum features. But, low order Auto-Regressive (AR) models have been shown to be giving good fit,to the normalized (DC component filtered) time data. The coefficients of

Table 2 Effect of process variables on AE generated

AE parameter	Process variables affecting the AE parameter
1 V_{RMS}	Strain rate, Bulk volume of matl. undergoing Deformation [Teti 1988] Chip contact length [Kannatey-Aisbu 1982]
2 Average AE intensity	Depth, Rake angle, Speed, Noise sources (Chip breakage & collision, Lubrication) [Lee 1988]
3 Count, Count rate	Feed, Speed, Depth [Teti 1989]

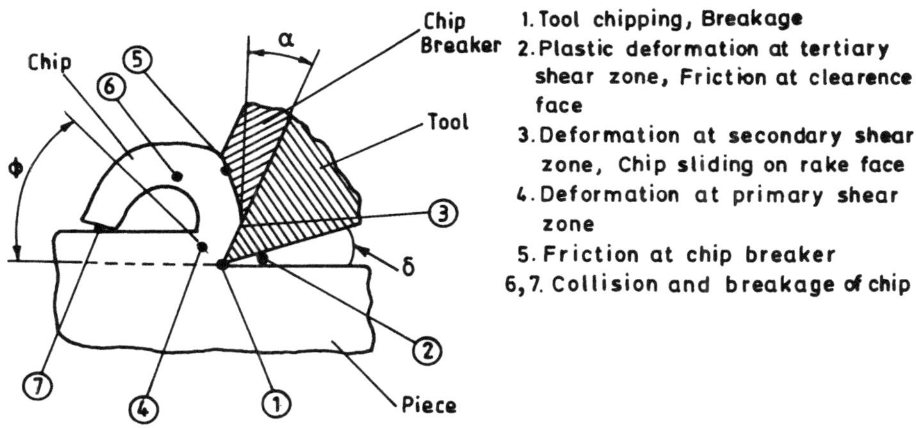

Fig.1 Different AE sources in machining [after Roget 1988]

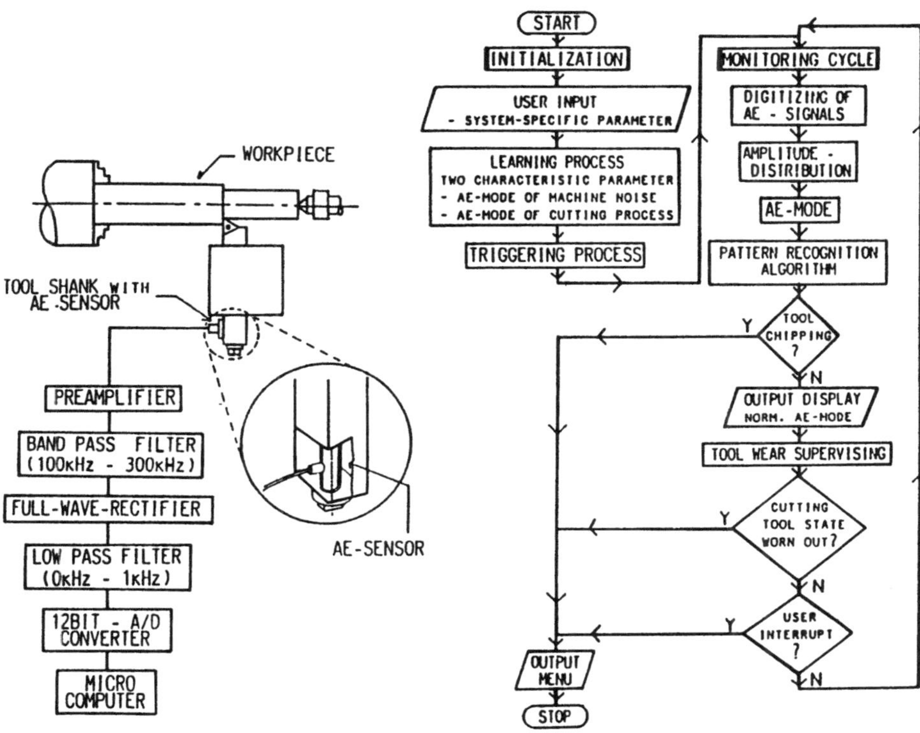

Fig.2 AE experimental set-up and signal processing flow chart. [after Inasaki 1988]

Table 3 AE based monitoring - comparison of results

Reference	Condition	Process	Signal processing details
1 Kannatey-Aisbu* 1982	Wear	Turn	Skew,Kurtosis,B function of Amplitude distriibution of RMS
2 Emel 1988	Breakage Wear	Turn	Pattern recognition using power spectral components
3 Inasaki 1988	Breakage	Turn	Mode of Amplitude distribution
5 Liang 1989	Wear	Turn	Auto-regeressive model of RMS
6 Teti 1989	Wear	Turn	Count, Count rate

this AR model can be used as feature vector for a pattern classifier algorithm, eliminating the need for feature selection [Liang 1989].

Study is in progress to develop pattern recognition techniques suitable for AE study. Figure 2 shows typical experimental setup used for collecting AE signals and for subsequent analysis. Using suitable filters, rectifiers and A/D converters it is possible to collect AE signals in a micro computer for fuller analysis. Figure 2 also illustrates a typical flow chart for pattern recognition in AE monitoring. Thus, by using AE technique, it is possible to accurately recognise state of machining.

5.0 CONCLUSIONS

1. Since vibration signals are dependent on the tool holder dynamics, real-time determination of natural frequencies is required, for an effective vibration based monitoring system. Moreover, due to the presence of noise from machine and other signals from several contacting surfaces,such a system might be limited in actual usage, owing to said constraints for proper inference.

2. In spite of definate indications available on the effectiveness of AE in providing information on the cutting process and tool deterioration, much conflict exists with regards to the signal processing methodologies necessary for extracting this information.

3. Differences in fracture toughness of materials, which imply different velocities of propagation of cracks, should cause AE whose frequencies are charecteristic of material. Different mechanisms of production of AE - like fracture and dislocation movements - should produce differing frequencies. It remains to be investigated, if spectral components are sufficient for determining tool wear and fracture, or information on amplitude levels is also necessary.

REFERENCES

Chryssolouris, G. Guillot, M. & Domroese, M. (1988) A decision making approach to machine control. **ASME J. Engg. Ind.**,110, 397-398

Dornfeld, D.A. and Kannatey-Aisbu (1980) Acoustic emissoin during orthogonal metal cutting. Int. J. Mech. Sci., 22, 285-296

Dornfeld, D.A. (1988) Monitoring of the cutting process by means of AE sensor. **3rd Int. Machine tool Engineers Conf. Tokyo, Japan**

Emel, E and Kannatey-Aisbu, E (1988) "Tool failure monitoring in turning by pattern recognition analysis of AE signals" **ASME J. Engg. Ind.**, 110, 137-145

Inasaki, I. Blum, T. and Suzuki, I. (1988) Development of a condition monitoring system for cutting tools using AE sensor. **Bull. JSPE**, 22, 301-308

Kannatey-Aisbu, E. and Dornfeld, D.A. (1982) A study of tool wear using statistical analysis of metal cutting AE. **Wear**, 76, 247-261

Lee, M Thomas, C.E. and Wildes, D.G. (1987) Prospects for in-process diagnosis of metal cutting by monitoring vibration signals. **J. of Met. Sci.**, 22, 3821-3830

Lee, M Wildes, D.G. Hayashi, S.R. and Kermati, B (1988) Effect of tool geometry on AE intensity" **CIRP**, 37, 57-60

Li, P.G. and Wu, S.M. (1988) Monitoring drilling states by a fuzzy pattern recognition technique. **ASME J. Engg. Ind.**,110, 297-300

Liang, S.Y. and Dornfeld, D.A. (1989) Tool wear detection using time series analysis of acoustic emission. **ASME J. Engg. Ind.**, 111, 235-242

Lister, P.M. and Barrow, G. (1986) Tool condition monitoring systems. **Int. Machine Tool Des. Res. Conf.**, 271-288

Martin,P. Mutel,B. and Drapier,J.P. (1976) Influance of lathe tool wear on the vibrations sustained in cutting. **Int. MTDR Conf.**, 251-257

Martin, K.F. Brandon, J.A. Grosevenor, R.L. and Owen, A. (1986) A comparison of in-prcess tool wear measurement methods in turning. **Int. MTDR Conf.**, 289-296

Pandit, S.M. and Kashou, S. (1982) A Data Dependent Systems strategy of on line tool wear sensing. **ASME J. Engg. Ind.**, 104, 217-233

Osipova, S.S. and Shraibman, J.S. (1987) Tool monitors for use in a cutting process. **Sov. Engg. Res.**, 7, 77-78

Roget, J. Souquet, P. Gsib, N. (1988) Application of acoustic emission to the automatic monitoring of tool condition during machining. **Materials Evaluation**,225-229

Shaw, M.C. and Sanghani, S.R. (1962) On the origin of cutting vibrations. **CIRP**, 11, 59-66

Teti, R. and Dornfeld, D.A. (1988) Modelling and experimental analysis of acoustic emission from metal cutting. **ASME J. Engg. Ind.**, 111, 229-237

Teti, R. (1989) Tool wear monitoring through AE. **CIRP**, 38, 99-102

Tlusty, J. and Andrews, D.G. (1983) A critical review of sensors for unmanned machining. **CIRP**, 32, 563-581

Tonhoff, H.K. Wulfsberg, J.P. Kals, H.J.J. Konig, W. and Van Lutterveld, C.A. (1988) Developments and trends in monitoring and control of machining process. **CIRP**, 37, 611-622

Uehara, K. Kanad, Y. (1984) Identification of chip formation mechanisms through AE measurements. **CIRP**, 33, 71-74

Weck, M. and Vorsteher, D. (1983) Monitoring and diagnosis systems for numerically controlled machine tools. **Int. MTDR Conf.**, 229-237

Weller, E.J. Schrier, and H.M. Weichbrodt, B. (1969) What sound can be expected from a worn tool? **ASME J. Engg. Ind.**, 91, 525-534

Tool wear monitoring by the analysis of surface roughness

W.P. Dong and W.Y. Wang
Beijing Institute of Technology, Beijing, China

Abstract
In this paper the trend of surface roughness change thereby the trend of tool wear state change are discussed from machine tool dynamics point of view. An intelligent strategy to monitor tool wear by the analysis of surface roughness of works is proposed.
Keywords: Monitoring, Tool Wear, Surface Roughness, Machine Tool Vibration.

1 Introduction

The on-line tool wear monitoring is acknowledged to be an important need for unmanned manufacturing. Various on- or off-line techniques proposed to measure different aspects of machining process have been developed for sensing tool failure either directly or indirectly. Some papers have discussed tool wear monitoring by the analysis of surface roughness, it is agreed that an inherent relation between surface roughness and tool wear exists. Thus researches have proposed some contact and non-contact methods for measuring surface roughness and then for determining tool condition. However, it is observed that surface roughness is not a solely increasing function with tool wear, in some cases, it would decrease as tool wears severely.

In reality, besides the effects of geometrical and physical aspects, vibration of machine tools is another important aspect which affects surface roughness. Hence the rule of surface roughness change caused by vibration of machine tools, which is related to tool wear, must be considered carefully in the monitoring.

In this paper, tool wear monitoring is addressed by the analysis of surface roughness. The trend of surface roughness change is discussed from machine tool dynamics point of view. Experiments are conducted by using an on-line contact sensor and an intelligent strategy to monitor tool wear by the analysis of surface roughness is proposed.

2 Discussion of surface roughness change

A commonly recognized change of surface roughness versus time is from A to C shown in Fig. 1, where two stages exist, the first stage A to B corresponds to initial use of a new tool, since the cutting edge becomes acute after some hard grits were worn out, Ra becomes smaller. In the second stage B to C, tool wears gradually, since the tool corner radius becomes larger and larger, Ra increases versus time, in other words, Ra increases versus tool wear.

Accouding to traditional metal cutting theory, the cause of Ra increasing with tool wear is because that the waveform on works is copied by the shape of tool tip. In our opinion, however, the enhancement of vibration versus tool wear is another main cause which increases Ra. Moreever the effect of vibration on Ra not only plays an important role in second stage, but also it produces another stage C to D (see Fig. 1) which corresponds to rapid wear stage of a tool. At this stage, after Ra gets a maximum value, then it decreases. This phenomenon is observed in many of our experiments, it can not be explained by the traditional theory, but it can be explained by machine tool vibration theory.

If the tool is continually used after dramatic wear, Ra gets to the fouth stage. In this stage, because of very blunt of the tool, it is not in normal cutting but in extrusion, so Ra decreases.

3 Effect of machine tool vibration on surface roughness

For illustrating the effect of machine tool vibration on surface roughness, a famous machine tool dynamics model proposed by Merritt is used. The transfer function of the closed loop machining system in which the Fourier transform of nominal cutting thickness $a_0(t)$ is as input and the Fourier transform of vibration $y(t)$ is as output is obtained,

$$H(\omega,\zeta,k_c)=\frac{Y(\omega,\zeta,k_c)}{A_0(\omega)}=\frac{k_c b u W(\omega,\zeta)/k}{1+(1-\mu e^{-j\omega\tau})k_c b u W(\omega,\zeta)/k} \quad (1)$$

$$W(\omega,\zeta)=1/(1+j\zeta\omega/\omega_0-\omega^2/\omega_0^2) \quad (2)$$

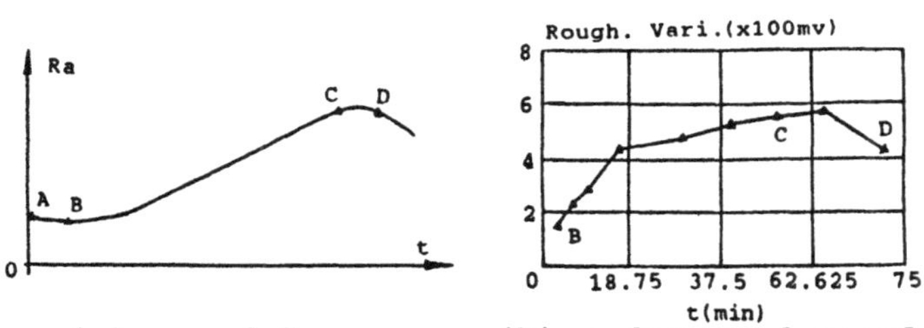

(a) a model (b) a detected result

Fig.1. Change of surface roughness versus time

where m and k represent the mass and stiffness of the system respectively, c includes system damping coefficient and cutting damping coefficient, k_c stands for cutting stiffness, u denotes direction coefficient, b is cutting width, μ is overlap coefficient and τ is time difference between two sequential cuts.

It is obvious that the transfer function of the system $H(\omega,\zeta,k_c)$ is the function of damping ratio and cutting stiffness. Rewrite equation (1) as,

$$Y(\omega,\zeta,k_c)=|H(\omega,\zeta,k_c)A_0(\omega)| \tag{3}$$

It can be seen that the vibration $Y(\omega,\zeta,k_c)$ is strongly dependent on the system dynamics characteristic, therefore it should be examined how the $Y(\omega,\zeta, k_c)$ as well as $H(\omega,\zeta,k_c)$ change with ζ and k_c.

Suppose that full overlap cutting be conducted, that is μ=1, multiply $H^*(\omega,\zeta, k_c)$ ("*" means conjugation) in the two sides of equation (1),

$$|H|^2=\frac{k_c^2p^2|W|^2}{1+2k_cpq+2k_c^2p^2(1-\cos\omega\tau)|W|^2} \tag{4}$$

$$p=bu/k \qquad q=Rel[(1-\cos\omega\tau+j\sin\omega\tau)W] \tag{5}$$

Derive the partial derivative of $|H|^2$ with respect to ζ and take its sign into account,

$$sign\left\{\frac{\partial|H|^2}{\partial\zeta}\right\}=sign\left\{\frac{\partial|W|}{\partial\zeta}+k_cp\left(2q\frac{\partial|W|}{\partial\zeta}-r|W|\right)\right] \tag{6}$$

where

$$r=Rel[(1-\cos\omega\tau+j\sin\omega\tau)\partial W/\partial\zeta] \tag{7}$$

Substitute equation (2), (5), (7) into equation (6), it is obtained,

$$sign\left\{\frac{\partial|H|^2}{\partial\zeta}\right\}=sign\left\{-\left(4\zeta\frac{\omega^2}{\omega_0^2}+2k_cp\frac{\omega}{\omega_0}\sin\omega\tau\right)\right\} \tag{8}$$

It can be proven as the vibration frequency ω is approximately equal to the system inherent frequency ω_0, which corresponds to an induced chatter, the part in braces of equation (8) is larger than zero. Therefore, the sign of $\partial|H|^2/\partial\zeta$ is negative, it means the amplitude of |H| would decreases as ζ increases, that is to say, as tool wear induces an enhancement of ζ, the value of |H| would decreases, and hence the amplitude of vibration $|Y|=|HA_0|$ decreases, thus reducing surface roughness. Similarily, the sign of $|H|^2$ change versus cutting stiffness is also derived

$$sign\left\{\frac{\partial|H|^2}{\partial k_c}\right\}=sign\left\{\left(1-\frac{\omega^2}{\omega_0^2}\right)^2+4\zeta^2\frac{\omega^2}{\omega_0^2}+k_cp(1-\cos\omega\tau)\left(1-\frac{\omega^2}{\omega_0^2}\right)+2\zeta\frac{\omega}{\omega_0^2}\sin\omega\tau\right\} \tag{9}$$

It can be proven that the sign of $\partial|H|^2/\partial k_c$ is positive as ω is approximately equal to ω_0. It means that the amplitude of |H| would increase with cutting stiffness k_c, thus increasing the amplitude of vibration $|Y|=|HA_0|$ and the waveform left on work surface.

It has been pointed out by Tlusty and others that cutting stiffness and cutting damping ratio would increase with tool wear, but in normal wear stage of a tool the increase rate of k_c is greater than that of ζ, however, in severe wear stage of the tool, it is vice versa.

According to this conclusion, besides the effects of geometrical and physical aspects, surface roughness would increase or decrease with cutting stiffness and cutting damping coefficient. In the stage B to C (Fig. 1) which corresponds to normal wear stage of a tool, the effect of k_c on $|H|$ is larger than that of ζ on $|H|$, therefore the global effects is increasing $|H|$, and hence increasing surface foughness.

As in the stage C to D (Fig.1), tool wear happens severely, the influence of ζ on $|H|$ exceeds the influence of k_c on $|H|$, so this causes a reverse change of surface roughness.

4 On-line measurement of surface roughness

For measuring surface roughness, a measurement system (Fig. 2) which is composed of an on-line profilometer and other instruments including an eddy current sensor, a low pass filter and a microcomputer etc. is built.

As the tool carrige moves along the axial direction slowly, the stylus moves along the contour of the work, and the induced sheet travels along the radius direction because of action of the double sheet springs. The displacement y, then, is sensored by the eddy current sensor, and next the signal detected passes through a preamplifier and a low pass filter and is sampled by an A/D converter. Finally, the microcomputer processes the sampled data, gives Ra, ΔRa or other desired parameters, and makes a judgement about the tool condition.

Because of the restriction of material and radius of the stylus, the travel speed of the carrige and the stiffness of the springs should be selected carefully. It has been proven that the system can measure such a surface that corresponds to the N7 surface of ISO-302 1978 standard.

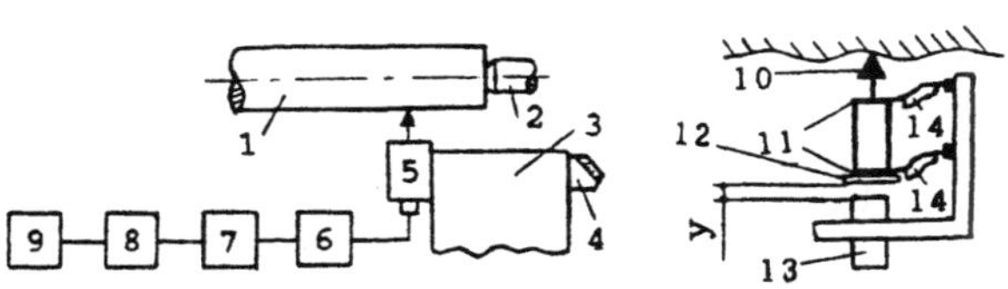

Fig.2. Measurement system and profilometer
1-work, 2-centre, 3-tool holder, 4-tool, 5-profilometer, 6-preamplifier, 7-low pass filter, 8-A/D converter, 9-microcomputer, 10-stylus, 11-sheet spring, 12-induced sheet, 13-eddy current sensor, 14-damping meterial

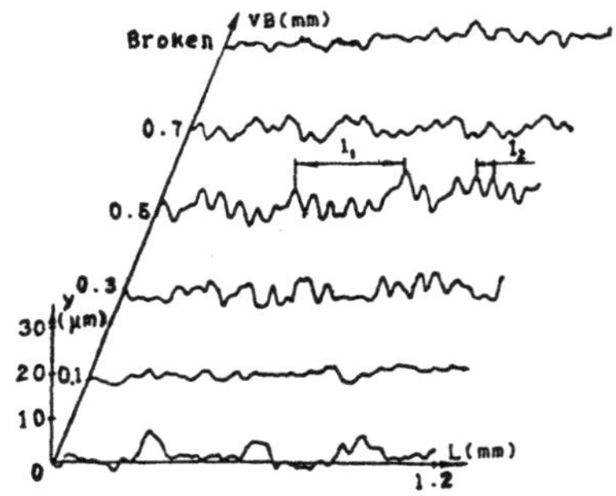

Fig.3. Measured waveform on a turned surface
v=128 m/min, f=0.107 mm/rev, a =0.5 mm
Work: S45C Tool: YT15, $\gamma_0=12°, \alpha_0=8°, \kappa_r=90°, \kappa_r'=10°$

Fig. 3 shows a series waveform measured on a turned surface. The low freqrency wave whose wave length is l_1 is caused by vibration and the high frequency wave whose wave length is l_2 is caused by the cutting of minor edge. Obviously, vibration has higher effect on the magnitude of surface roughness than physical and geometrical aspects do. It can be noted that the geometrical effect has relation with vibration effect, the larger the vibration effect is, the larger the geometrical effect is. Moverever, the trend of the waveform change with tool wear coincides with what we discussed above, hence tool condition can be monitored by the state of surface roughness.

5 An intelligent strategy for tool wear monitoring

As we know, surface condition is also dependent on cutting parameters (cutting speed, feed, and cutting depth), if a work which has complex configuration is machined, various cutting parameters are adopted at different parts of the work, hence, it results in different surface states.

For excluding the effects of cutting parameters on judgement, an intelligent strategy can be adopted. Since production in unmanned machining is a batch process, surface roughness can be measured at the same part of each work. In doing so, all cutting conditions except for tool condition are the same at these parts, the difference of surface roughness between different works is only caused by tool wear not by cutting parameters, hence tool condition can be determined by the change of surface roughness.

Generally, Ra is analyzed in the monitoring. But since the Ra has a very strict definition and it should be measured and calculated carefully, it is very complex to obtain real Ra in situ, hence another two parameters are used instead, one is the variance of surface waveform σ^2, the other is its change $\Delta\sigma^2$ These two parameters reflect statistical features of work surface, they can be easily calculated by,

$$\sigma_m^2 = \frac{1}{N}\sum_{i=1}^{N}\sigma_i^2 = \frac{1}{nN}\sum_{i=1}^{N}\sum_{j=1}^{n}(y_{ij}-\bar{y}_i)^2 \qquad \Delta\sigma_m^2 = (\sigma_m^2 - \sigma_{m-1}^2) \tag{10}$$

where m denotes work number, σ_i^2 is the variance of the i-th sample section, $\bar{y}_i$, y_{ij} represent the mean value and the j-th sample value of the section respectively, n is the number of data sampled in one section and N is the number of sampled sections in one appraisal length. In this strategy, two criteria are set, one is absolute criterion, the other is relative criterion,they are expressed by,

$$\sigma_m^2 > \sigma_s^2 \tag{11}$$

$$(\Delta\sigma_{m-1}^2 > 0) \wedge (\Delta\sigma_{m-1}^2 \Delta\sigma_m^2 < 0) \tag{12}$$

where σ_s^2 is a restricted maximum variance of waveform measured on surface of a standard work. Equation (11) ensures that the surface roughness is within design requirement and equation (12) ensures that tool wear is detectable as it gets into dramatic wear zone although the surface roughness is not beyond requirement. If one of these two criteria is satisfied, the tool is regarded as failure. According to this strategy, some experiments are conducted, one of detected results is shown in (Fig.1 (b)).

6 Conclusions

The change of surface roughness with tool wear is discussed in the paper. It stresses a less noticed fact that the surface roughness may become smaller as a tool gets into dramatic wear zone. The phenomenon is analyzed and illustrated from machine tool dynamics point of view, and it is pointed out that vibration is one of main resources which produce waveform on work surfaces. According to the rule of surface roughness change, an intelligent strategy which is for excluding the effects of cutting conditions on on-line tool wear monitoring and for analyzing surface condition conveniently can be adopted.

7 References

Dong, W.P. (1987) Research of tool failure and machining condition monitoring in unmanned machining. Ph.D. thesis

Merritt, H.E. (1965) Theory of self-excited machine tool chatter. Trans. ASME, Series B, 4, 447-454

Masaki, M. (1985) Variation of machined surface with progressive of tool wear, Bull JSPE, 3, 216-217

Tlusty, J. (1963) The stability of the machine tool against self-excited vibration in machining. Int. Res. in Prod. Eng. ASME, 465-474

Characterization of the acoustic emission behaviour during quasi static punching tests

P. Souquet, C. Bouhelier, Dr E. Grosset, A. Maillard and L. Li
CETIM Senlis, France
Dr J. Au and A. Mardapittas
Brunel University, UK

Abstract
This work was carried out in the framework of a BRITE program. The aim is to develop an expert system for tool wear monitoring in milling, drilling and blanking using multisensor systems. In blanking, the first step consisted in working on a tensile machine to characterize the process of press punching without the spurious noise of the machine and to evaluate the influence of the main punching parameters on the acoustic emission behaviour. According to these results, the possibilities of punching monitoring were clearly demonstrated in quasi static tests although they must be confirmed in industrial environment.
Keywords : Acoustic Emission, Tool Wear Monitoring, Blanking, Brite Project.

1. Introduction : European Brite Project

This project has begun since January 1989.

Eight partners are involved :

BRUNEL UNIVERSITY (England)
CENTRE TECHNIQUE DE L'INDUSTRIE DU DECOLLETAGE (France)
COMPUTER TECHNOLOGY COMPANY (Greece)
NICOLAS CORREA S.A. (Spain)
INSTITUTO DE AUTOMATICA INDUSTRIAL (Spain)
TEKNIKER (Spain)
HERIOT WATT UNIVERSITY (England)
CENTRE TECHNIQUE DES INDUSTRIES MECANIQUES (France)

The task of monitoring the blanking operation comprises three main phases :

a) Quasi static tests
b) Development of test stations
c) Progressive wear tests

The present results concern the first stage which consisted of studying the blanking process on a tensile machine.

2. Experimental procedures

Different blanking parameters were tested :

a) Various materials : mild steel, stainless steel, aluminium alloy and copper alloy
b) Various thickness 1, 2 and 3 mm
c) Various radial clearance 10 %, 5%, 3,3 %
d) New and worn tools
e) Various diameters 5 and 8 mm

The worn tool was a tool of 8 mm diameter provided by CETIM with natural wear /1/.
The comparison of new tool and worn tool has been done by changing the punch only.
The AE signal was analyzed as follows :

a) Amplitude distribution of the AE in the rupture zone
b) Global parameters of AE versus displacement

3. Experimental results

3.1. Influence of the radial clearance

When a radial clearance is changed from 5 % to 3,3 % for a stainless steel sheet 3 mm thick, peak amplitude decreases and the duration of the AE event in the rupture zone increases.

The same behaviour is observed when the radial clearance changes from 5 % to 10 % for a stainless steel sheet 1 mm thick.

3.2. Influence of the sheet material

Two families of materials can be separated by the analysis of the AE behaviour independently of the thickness :

a) Mild steel, copper alloy and stainless steel, which give an intense discrete type acoustic emission at the rupture. Moreover, the material can be arranged in ascending order according to the AE peak amplitude and the corresponding displacement of the punch at which the rupture AE occurs.

b) Aluminium does not give an intense acoustic emission in the rupture zone but presents a low continuous type acoustic emission around the yield point observed for mild steel and copper alloy.

3.3. Influence of the sheet thickness

General AE behaviour is not modified by thickness. However, the study of the maximum peak amplitude of the intense discrete type AE in the rupture phase shows that there is no clear evolution between AE amplitude and thickness. In this case, only the displacement corresponding to the maximum peak amplitude increases with thickness (Table 1).

Table 1. Thickness inflence on the AE Signal for stainless steel

Thickness	Peak value of AE in rupture area	Displacement corresponding to intense AE
10^{-3} m	dB	10^{-6} m
1	53	856
2	48,7	1 510
3	56	2 150

3.4. Influence of the punch diameter

These tests were conducted on a C.Frame press at Brunel University by using mild steel. The AE amplitude distribution becomes narrower and more peaky for the 5 mm punch than for the 8 mm punch. That indicates a decrease in the AE activity at the rupture between the 5 mm punch and the 8 mm punch.

3.5. Influence of wear

The results should be analyzed according to materials : materials with AE in the rupture zone (mild steel, copper alloy, stainless steel) and materials without AE in the rupture zone (aluminium alloy).

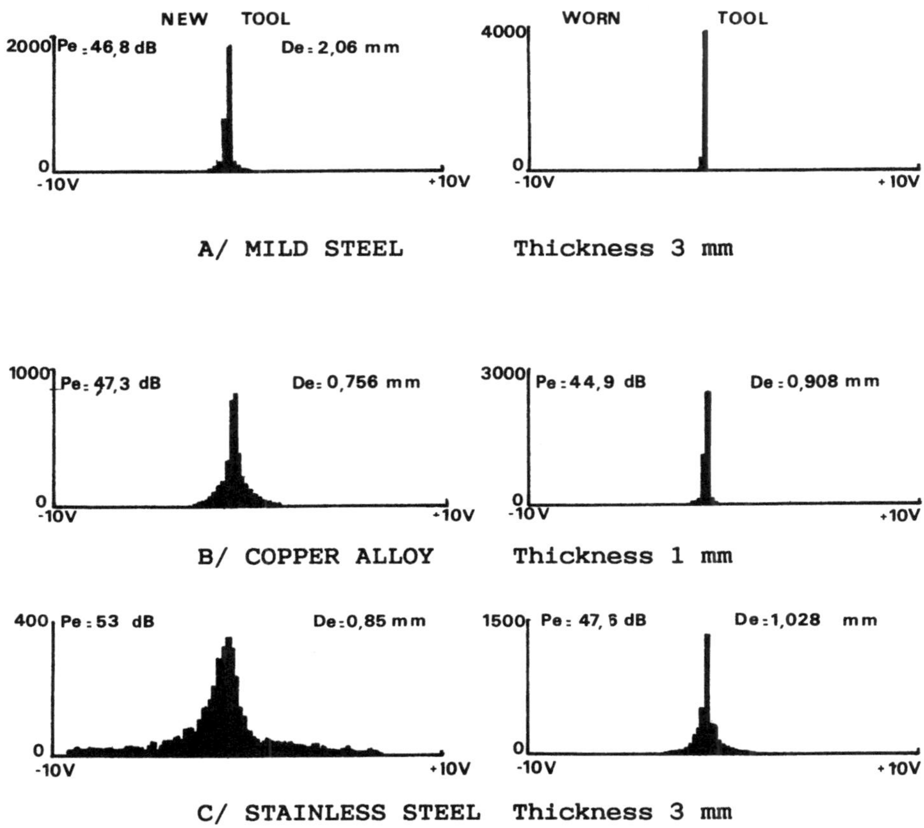

AMPLITUDE DISTRIBUTION

Pe = Maximum AE peak amplitude in the rupture zone
De = Duration of AE event in the rupture zone

Figure 1 : Differences on AE signals between a new tool and a worn one

a) For mild steel, copper alloy and stainless steel, the maximum peak amplitude of the intense AE in the rupture zone is lower when a worn tool is used than with a new one. At the same time, the displacement corresponding to the maximum AE amplitude increases (Figure 1).

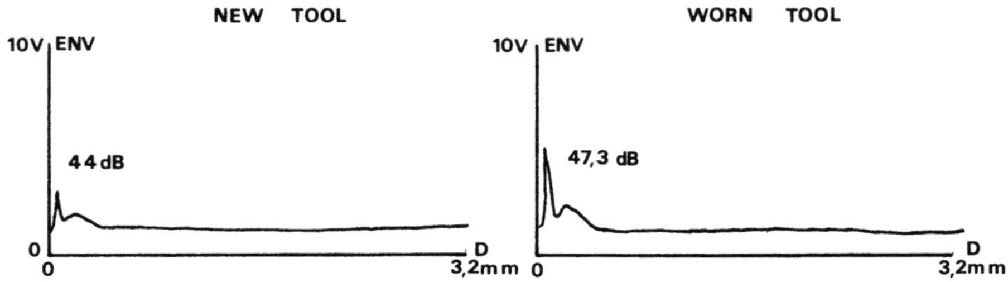

ALUMINIUM Thickness 2 mm

Figure 2 : Differences on AE signal between a new tool and a worn one

b) For aluminium alloy, the AE around the yield point is higher when a worn tool is used than with a new one (Figure 2).

4. Blanking monitoring by acoustic emission : overview of problem

In a blanking operation, material plays a predominant role on the AE behaviour. Two classes of materials can be distinguished : mild steel, copper alloy and stainless steel give an intense AE in the rupture phase and aluminium gives only a low continuous acoustic emission around the yield point. The general AE behaviour is not influenced by the thickness. A change in the radial clearance induces a change in the AE behaviour, and a change of the tool diameter mainly affects the AE amplitude.

However, the conclusions of the punch diameter influence cannot be final because only two values were tested and the geometry of the industrial punching tool can also be very complex.

In order to distinguish a worn tool from a new one, it is absolutely necessary to distinguish the two families of materials at present time. However, the use of new signal processing techniques developed in this project or the results obtained on an industrial blanking machine with a higher punching speed could allow to define a common criterion to monitor wear for all materials.

The main difficulty of the acoustic emission measurements should be the scattering of the results which will probably require use of averaging techniques.

Among the acoustic emission measurements used for these tests, the most profitable are :

a) The analysis of the raw acoustic emission by the maximum peak amplitude and the statistical parameters such as standard deviation and kurtosis.
b) The analysis of the envelope of the signal.

The other parameters as burst count and RMS with large time-constants are not in accordance with the high speed phenomenon studied in blanking.
For the ring down count, it is possible to use it during the next laboratory study, but this parameter strongly depends on the operating conditions as threshold and gain, it seems difficult to use it in industrial surroundings.

5. Acknowledgements

The authors wish to acknowledge the support provided by the European Commission, project partners and industrial collaborators.

6. References

(1) E. GROSSET, P. PEYRE, P. CHERRY, J. GASNIER, C. TOURNIER
Les dépôts PVD et CVD en poinçonnage et relevage des collerettes - CETIM Info N° 111 - July 1989

The determination of cutting tool condition using vibration signals

T.N. Moore
Queen's University, Kingston, Canada
Z.F. Reif
University of Windsor, Windsor, Canada

Abstract
This paper presents the results of studies undertaken to determine if vibration signals can be used to detect failure in two important classes of metal removal tools: the twist drill, and the multi-insert face mill. Twist dills of several sizes were operated to failure and vibration was measured by means of accelerometers. The output signal was analysed for several descriptors in order to determine basic relationships between the deteriorating condition of the tool and the resulting changes in vibration. The results indicate that vibration is very good in detecting and, to a degree, predicting tool damage. Measurement of wear is less effective, but it can be improved by more complex monitoring procedures. In the case of the face mill, cutting inserts with known amounts of edge fracture or wear were used, in various combinations, during the milling process. The resulting vibration signals were analysed to determine both the type and severity of insert failure that can be recognized using various signal analysis methods. It is shown that both edge fracture and wear can be detected, albeit with varying degrees of confidence.
Keywords: Drill, Face Mill, Diagnostics, Tool, Failure, Breakage, Wear, Vibration.

1 Introduction

An important element of the automated process control function is the real-time detection of cutting tool failure, including both wear and fracture mechanisms. The ability to detect such failures "on-line" would allow remedial action to be undertaken in a timely fashion, thus ensuring consistently high product quality and preventing potential damage to the process machinery.

Of all the metal removal processes employed in industry, by far the most prevelant are those of drilling and milling.

The objective of the present study is to investigate the possibility of using vibration signals generated during the drilling and face milling processes to detect both progressive (wear) and catastrophic (breakage) tool failure.

2 Drills

2.1 Experimental technique

To study the basic relationships between the progressing tool wear and the resulting changes of vibration, a large number of representative tools was operated until failure, as described in Moore and Reif (1985) and Reif and Lau (1989). Because of the lack of space in the machining area and the very hostile working environment, accelerometers were preferred as the vibration transducers. The location of their attachment was restricted by practical considerations to a suitable stationary surface, as near to the rotating tool as possible. In most cases, the upper surface of the spindle casing satisfied these requirements. The advantage of this location is that the signal from the same accelerometer can also be used for monitoring the condition of bearings and for some other diagnostic functions.

During the cutting sequences, the vibration signal was recorded on magnetic tape and in the digitized form on computer diskettes. Subsequently, the recordings were analyzed for several different descriptors. The progressing wear was measured along the cutting edges by means of a travelling microscope.

2.2 Discussion of results

The origins of the measured vibration are the forces which are produced at the drill tip by the cutting action. Because of the different dynamic properties of the transmission paths, significant changes in the vibration response may exist for different machines using the same tools. Additional variation is caused by workpiece materials. Even castings of identical components supplied by different foundries can introduce significant changes in the measured vibration response. Consequently, unique vibration signatures for particular faults do not exist and perhaps the most effective monitoring procedure is to determine a descriptor which in a particular application exhibits sufficiently large changes from baseline values to reliably indicate, without significant masking by other sources of vibration, a developing fault.

Allowing for a number of practical operation aspects, vibration acceleration was found to provide best overall results. The simplest approach, in terms of the required hardware and data treatment, is to use maximum values of the vibration acceleration. This approach gives satisfactory results for detecting tool breakages or severe damage. From Fig. 1 it is evident that drill breakage produces sufficiently large changes in the vibration response to be reliably recognized by the monitoring software. On the other hand, this method is unsatisfactory for determining wear and predicting developing failure of tools. Fig. 2 shows vibration acceleration changes for a particular drill with several levels of wear. It can be seen that maximum values of acceleration alone would not provide a good measure of tool wear and that the density of the signal is much better in that respect. It follows that a vibration descriptor, which allows for both the amplitude and density, such as, for example, the rms value of the acceleration signal, would produce better results. This was confirmed by the obtained results. Another procedure with reasonably

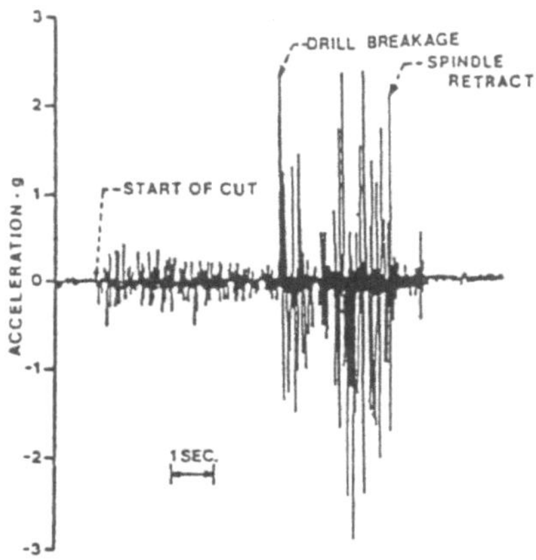

Fig. 1 Acceleration levels for drill breakage.

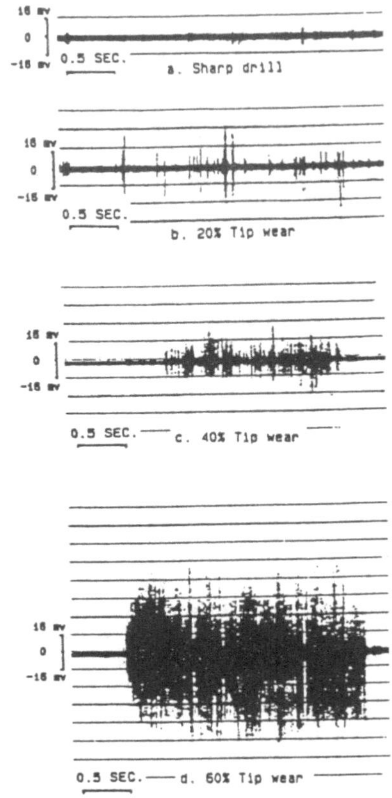

Fig. 2 Acceleration levels for progressive stages of drill wear.

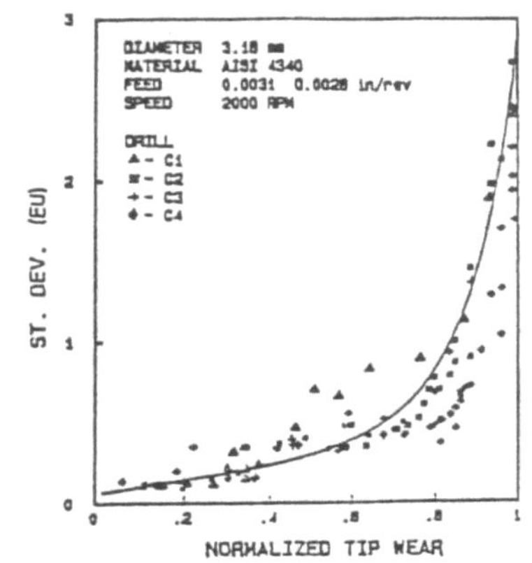

Fig. 3 Variation of standard deviation with wear for four drills.

good application potential is the counting of spikes, which exceed a certain specified level. This function can be performed relatively easily by electronic methods. Other descriptors, such as crest factor, cepstrum, power levels of selected frequency spectrum bands produced generally acceptable, if not outstanding, results.

Several statistical properties can also be extracted from the vibration signal. In this particular investigation, best results were obtained with standard deviation and variance. A typical set of results is shown in Fig. 3, where standard deviation is plotted against normalized wear. The latter is obtained by dividing a wear value by its magnitude at failure for the same drill. This was necessary in order to eliminate the significant variations in endurance of nominally identical new drills. It can be seen from Fig. 3, that, as wear progresses, a well defined trend exists with this descriptor and that, when normalized, the curves for the four drills are practically coincident.

3 Face mills

3.1 Experimental technique

The experimental studies were carried out using a 3hp vertical milling machine. The cutting tool was a 3.81 cm diameter, three insert, face-mill employing Carboloy TPC-322E grade 370 tungsten carbide cutting inserts. The "standard" workpiece was a mild steel plate with a length of 30.5 cm, a height of 15.2 cm, and a width of 1.3 cm. While cutting, the mill traversed the length of the workpiece performing an interrupted, symmetric cut. The vibration generated during milling was measured on the workpiece clamp. The vibration signals were tape recorded for later analysis in the laboratory.

The first series of tests consisted of milling cuts made using various combinations of worn and sharp inserts in the three-insert cutter. The worn inserts were obtained by machining a "wear block" under very light, and non-interrupted, machining conditions. Once the desired flank wear had been obtained, the insert was put aside to be used subsequently in the actual interrupted face-milling tests performed on the standard workpiece.

In the second series of tests milling cuts were made using various combinations of "fractured" and sharp inserts in the three-insert cutter. Due to the difficulty in obtaining controlled amounts of edge fracture during normal machining operations, it was decided to simulate fracture of the insert using electrical disharge machining to remove a controlled amount of the insert cutting edge.

Inserts with various magnitudes of wear and fracture (ranging from 0.13 mm to 0.78 mm) were used in the experiments. See Pei and Moore (1990) for a detailed description of the combinations utilized.

3.2 Discussion of results

3.2.1 Fractured inserts

Fig. 4 shows typical acceleration level versus time histories.

Fig. 4(a) is for the case of three sharp inserts. Note that the engagement of each insert in the workpiece is clearly evident and that all engagements share similar characteristics, although they are by no means identical.

Fig. 4(b) shows the resultant acceleration signal for the combination of two sharp inserts and one insert with a 0.39 mm fracture. The sharp inserts produce signals consistent with those shown in Fig. 4(a) while the fractured insert produces a significantly different output. The reduced output level for the fractured insert is a result of the much smaller depth of cut associated with this insert.

It would seem from the time domain data available, that the use of either an "envelope detection" or a "threshold crossing" scheme would provide the ability to automate the detection of tool fracture in a a multi-insert milling operation.

Fig. 5 shows some typical frequency spectra for various tool conditions. All frequency domain analyses discussed in this paper were obtained using eight ensemble averages and were calculated over a bandwidth of 0 to 20,000 Hz using 800 lines of resolution.

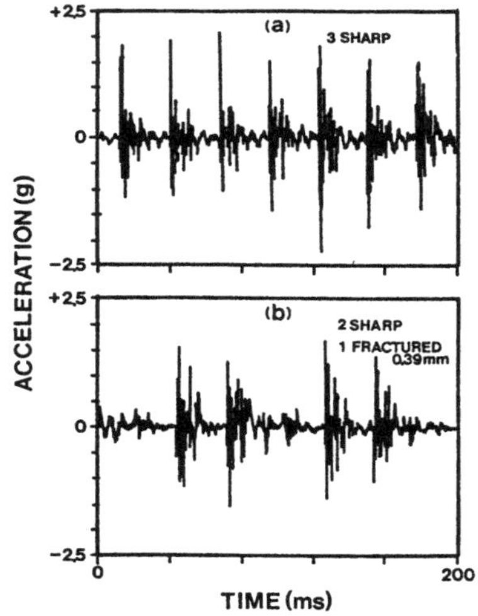

Fig. 4 Time domain data - insert fracture

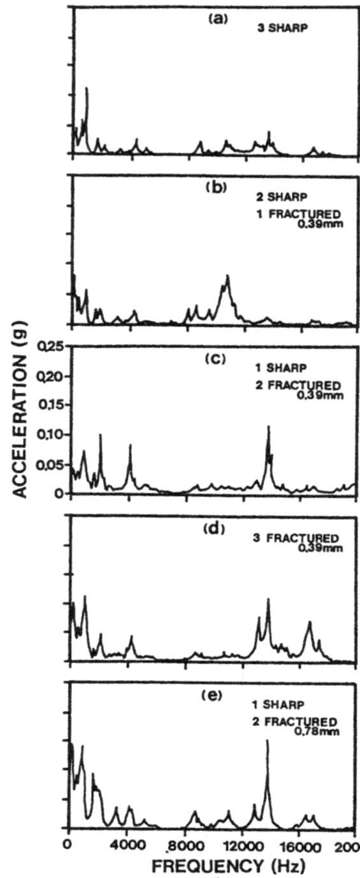

Fig. 5 Frequency domain data - insert fracture

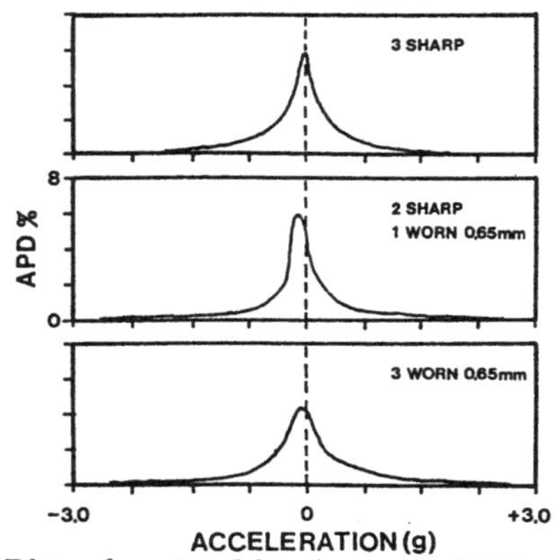

Fig. 6 Amplitude probability density - insert wear

It is immediately apparent that, in general, fracture phenomena is indicated by an increase in the level of spectral components within the range of 10,000 to 17,000 Hz. A comparison of Fig. 5(a) and Fig. 5(b) indicates a noticeable increase in the spectra around 11 kHz when a single insert fracture of 0.39 mm is present. For two fractured inserts, Fig. 5(c), the peak shifts to around 13.5 kHz. For the case of three fractured inserts, Fig. 5(d), both the 13.5 kHz peak and an additional peak at about 17 kHz are apparent.

Comparing Fig. 5(e) and 5(c) it is seen that, in general, increasing the size of the insert fracture results in an increase in the spectral peak associated with that failure condition.

At the present time the eight ensembles used to obtain the spectral data are not synchronously averaged. That is, the data acquisition for the FFT analysis does not necessarily start at the same point on the cutter rotational cycle. Since the "data window" (the time required to obtain one set of data points) is significantly shorter than the time to complete one cutter revolution, it is possible that, in many instances, the data samples used for the FFT analyses do not contain information from the "failed" inserts. It may, therefore, be possible to improve the performance of the spectral data by utilizing a combination of synchronous averaging and delayed triggering to ensure that data representative of each insert in the cutter is obtained and processed.

3.2.2 Worn inserts

The acceleration-time histories for the worn inserts do not, in general, produce the noticeably different engagement signals evident in the case of fracture (see Fig. 4(b)). However, by processing the data in a slightly different manner it is possible to detect evidence of tool wear.

Fig. 6 shows the amplitude probability density (APD) for several tool conditions. This distribution shows the percent of time a given level was present within the sample period. The data shown is for eight ensemble averages. It thus seems possible that insert wear could be detected using such features as the location of the peak in APD, the magnitude of the peak, and the area under specific segments of the distribution.

As with fracture, the presence of insert wear resulted in a significant increase in the spectral components within the 10 kHz to 13 kHz band. Although this would seem to indicate that the presence of flank wear could be detected using simple spectral analysis, it is not yet clear if this method would be sufficiently discriminating to permit reliable determination of the magnitude of flank wear.

4 Conclusions

This paper has presented results from studies undertaken to determine if vibration signals generated during the machining process could be used to detect the failure of drills and face mill inserts.

It was shown that, for both types of tool, vibration acceleration signals can, if appropriately analyzed, give indications of both wear and fracture, albeit with varying degrees of confidence.

5 References

Moore, T. and Reif, Z. (1985) Detection of tool breakage using vibration data, **SME Manufacturing Transactions**, pp. 45-50.

Pei, J. and Moore, T. (1990) Detection of face mill tool failure using vibration data, **CSME Mechanical Engineering Forum**, Toronto.

Reif, Z. and Lau, H.K. (1989) Monitoring the condition of drills by means of vibration, **Proc. 7th Intl. Modal Analysis Conf.**, pp. 1022-1027.

Monitoring tool wear during the turning process

S. Taibi, J.E.T. Penny and J.D. Maiden
Department of Mechanical and Production Engineering, Aston University, UK
M. Bennouna
Department Genie Mecanique, Ecole Mohammadia d'Ingenieurs, Morocco

Abstract
Automatic in-process detection of tool wear is an important requirement in unmanned machining. Cutting tool wear has been investigated previously by applying frequency analysis to the tool vibration and cutting force signals. When applied to turning operations this technique provided a successful means for assessing changes in the cutting conditions. This paper shows how the force and vibration signatures change due to the tool wear and how they can be used to monitor on-line cutting tool wear.
Keywords: Vibration, Turning, Tool wear, Monitoring

1 Introduction

Where real time monitoring techniques are not available, the cutting tool condition has to be inspected off-line. This increases the non-productive machine time and reduces the efficiency of the metal cutting operations. Tool failure between scheduled inspections may result in workpiece and even machine tool damage. An automated on-line tool and cutting condition monitoring system is a necessary prerequisite for ensuring successful operation of unmanned machining processes.

For this purpose, various methods have been proposed since Taylor established the tool life equation. Most of these methods are classified into two groups:

(1) Direct methods: These methods generally involve taking measurements associated with the volumetric loss of cutting tool material, Jelty S.(1984).
(2) Indirect methods: These methods correlate the changes in such characteristics of the cutting process as force, vibration and sound, to the cutting tool conditions; for example see, Tarn J.H. et al.(1989), Danai K.(1987), Rao S.B. (1986) Rotberg et al.(1987) and Weller E. J. et al. (1969).

Frequency analysis techniques have been used to monitor tool wear in turning, drilling and milling. It has been found that the frequency spectra obtained by monitoring cutting forces and/or machine or tool vibration vary as the tool wears, and a high frequency component occurs which increases in magnitude as the tool approaches the end of its life. The purpose of the present work is to report some results obtained in the monitoring of tool wear condition through the analysis of the spectra of tool vibrations and cutting forces. A series of tests have been performed under standard conditions with increasing degrees of wear. The interrelation between the different spectra and flank wear has been investigated. Initial results give a good correlation with the tool wear. Further work, incorporating different cutting conditions, work material, tool geometry and carbide tools, will need to be done to discover whether the results obtained to date are repeatable and whether they can be extended.

2 Experimentation

Tool wear monitoring must be on-line and insensitive to variations of the cutting parameters. In this first approach, off-line analysis was performed and the cutting conditions were held as constant as possible. The test data collection consisted of dry cylindrical turning of mild steel at a cutting speed of 32m/min, a depth of cut of 1.5mm and a feed rate of 0.32mm/rev. The cutting tools were all high speed steel and had a nominal relief angle of 6º, a rake angle of 0º and an overhang of 30mm. At the beginning of each test the tool was sharp and the test continued until it reached a preset wear limit. The tool wear was measured off-line after each cutting operation. Small variations in the alignment, in the angle between the tool and the workpiece, and in the tool overhang have been shown not to be sufficient to invalidate results from these tests, see Pulak et al.(1986).

Tool vibrations were sensed with two accelerometers, one mounted on the top of the tool post and the other mounted horizontally on the back of the tool post while tangential and feed forces were measured using a Kistler three-component measuring platform dynamometer. The signals were recorded on a four channel tape recorder. The experimental setup is shown in Fig.1.

3 Experimental results

Tests were originally made over a frequency range of 0-25kHz and showed that the field of interest lay in the range from zero to 6kHz. To determine the free vibration frequencies of the machine tool, the transient vibrations were obtained by

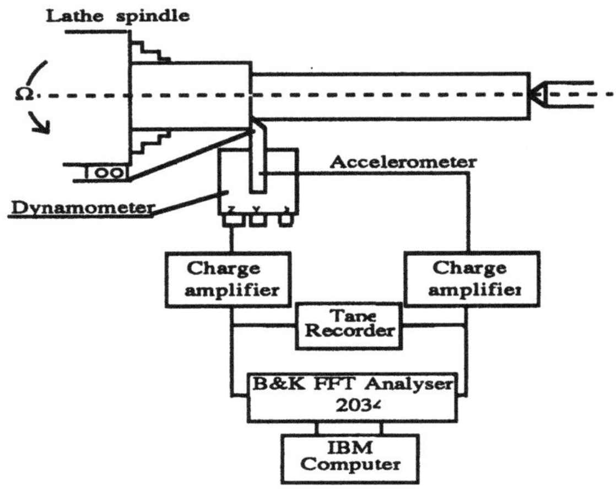

Fig.1: Schematics of experimental setup.

impacting the toolholder-dynamometer system with a hammer blow to the tool tip and measuring the response with an accelerometer mounted at different positions on the tool and on the tool post. It was found that the principal natural frequency of the tested system was at about 3.3kHz. Tool vibration and induced forces with the lathe idling were also recorded in order to single out forcing frequencies introduced by the driving device. Fig.2 and Fig.3 depict typical spectra of the force and the vibration signals in the tangential direction, when running at 32m/min, in the 0-6.4kHz frequency range, as the tool flank wear grows to 0.387mm.

The first point of interest is that the force amplitude, in the first moments of the tool life, decreases slightly due to the built-up edge formation. Moreover, the vibration spectra were in evident contrast to the trends shown in the spectra obtained as the tool wear increased. This suggested that, for further tests, recording should be deferred until a just noticeable level of wear had occurred.

The main cutting force amplitude, Fig.2, decreases as the frequency varies from 0 to 6.4kHz. Most of the power signal is then contained at low frequencies. All force spectra exhibit a trough between 3.2 and 4kHz. This force drop-off is usually observed at resonant frequencies owing to the decrease in the apparent mass of the system. Moreover, the frequency where this trough is observed decreases as the tool wears and the peak observed in the frequency response depicted in Fig.4 is also shifted to a low frequency, implying that the system comprising the workpiece, tool and tool support becomes more flexible in contrast to what would be expected. Initially, the cutting force remains nearly constant with wear. As the wear increases, the cutting force becomes more sensitive to wear and increases in the full frequency range. This increase is marked up to 4.6kHz after which it is less evident.

The acceleration spectra, Fig.3, show that low frequencies dominate when the tool is new. However, a broad peak is generated within the 2.8-4kHz frequency band. As machining progresses, two peaks become visible. Clearly, the acceleration signal power contained under 2.7kHz is much smaller than that contained in the upper frequencies. It is also noticeable that at various harmonics of workpiece rotation frequency the narrow band noise increases with wear more than the total power. This is particularly evident in the spectra up to 2kHz, see Fig.3. As the tool wears, a peak becomes evident at 1677Hz which appears to be sensitive to wear and would appear to signal the moment at which the cutting conditions become unstable.

This evidence leads one to consider the inertance depicted in Fig.4 as a function to monitor the tool wear. This frequency response does not show any useful information for frequencies under 2.8kHz and from 4.2 to 6.4kHz. The analysis is therefore concentrated in the 4.2-6.4kHz range. In this frequency range, an important increase is observed at the beginning of machining. The rate of increase decreases as the tool wears. When the tool is new, the contact between the tool and the workpiece is restricted to an almost line contact that offers a small amount of resistance to the oscillations. In general, low frequencies dominate when the tool is new. As the tool wear develops, the larger contact area increases the amount of workpiece material elastically deformed. This increase in the frictional area causes the vibration sensed on the tool post to increase more than the force as measured by the dynamometer in the vertical direction.

4 Conclusion

The above discussion can be summarised in the following points:

(1) The cutting forces and the tool vibrations during machining were recorded with a band up to 25kHz. The frequency band of interest was found to be in the 0-6kHz.
(2) The power contained in the range up to 2.5kHz represents a very small percentage of the total power of the acceleration signal. Low frequencies dominate when the tool is new and as the tool wears a broad peak becomes more and more important in the range of 2.8kHz to 4kHz.
(3) The dynamic cutting force spectra exhibit a significant decrease at the resonant frequency of the tool post system which is influenced by the tool conditions and reduces as the wear increases. The dynamic system comprising the workpiece, tool and tool support

becomes more flexible due to wear.

It was shown that the acceleration sensitivity is more related to the cutting conditions, whereas the force sensitivity is more related to the dynamometer-tool holder system dynamic characteristics. The inertance exhibits an important variation in the 2.8-4kHz range which should be more useful for tool wear determination. The maximum sensitivity in variation of the inertance spectra occurs in a frequency range depending on the dynamic characteristics of the tool-machine tool system; nevertheless the sensitivity found in the range specified seems to be satisfactory and appear to be promising at least for certain class of dynamometers and machine tools.

The next steps in the development of this research will be as follows:

(1) To extend the above results to a wide variety of materials. No tests have been conducted thus far on materials other than mild steel.
(2) To model the cutting process mathematically in order to explain the observed variation in different spectra.
(3) To develop a system for the automatic recognition of the tool failure.
(4) To transform the method from off-line to on-line with improved computational speed.

5 References

Danai K. and Ulsoy A. G. (1987), A Dynamic State Model for On-Line Tool Wear Estimation in Turning. **ASME, Journal of Engineering for Industry**, 109, 396-399.
Jetley S. (1984), Measuring Tool Wear On-line: Some practical considerations. **Manuf. Eng.**, 93, 1, 55-60.
Pulak B., Evers M. G., Ren H. and Wu S. M. (1986) A Feasibility Study of On-Line Drill Wear Monitoring by DDS Methodology. **Int. J. Mach. Tool Des. Res.** 26 3 245.
Rao S. B. (1986), Tool Wear Monitoring Through the Dynamics of Stable Turning. **ASME, Journal of Engineering for Industry**, vol.108, p183-190.
Rotberg J., Lenz E., and Braun S., (1987), Mechanical Signature Analysis In Interrupted Cutting. **Annals of the CIRP.**, vol.36, p249-252.
Tarn J. H., and Tomizuka M., August (1989), On-Line Monitoring of tool and Cutting Conditions in Milling. **ASME, Journal of Engineering for Industry**, Vol.111, 206-212.
Weller E. J., Schrier H. M., Weichbrodt B., (1969), What Sound Can be Expected from a Worn Tool?. **ASME, Journal of Engineering for Industry**, 525-534.

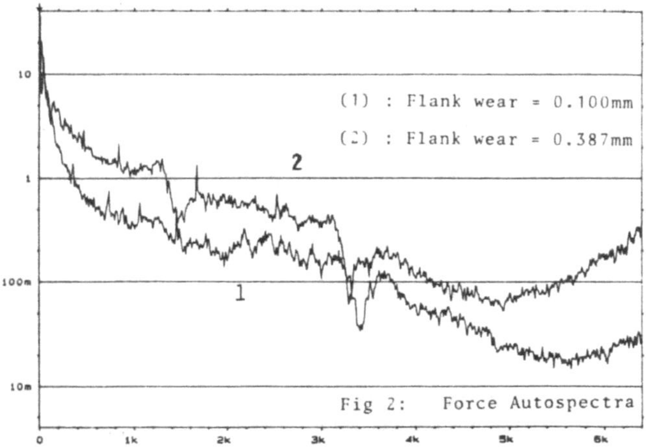

Fig 2: Force Autospectra

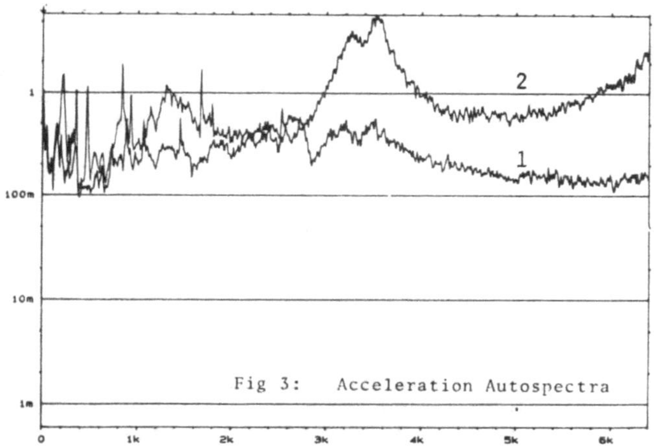

Fig 3: Acceleration Autospectra

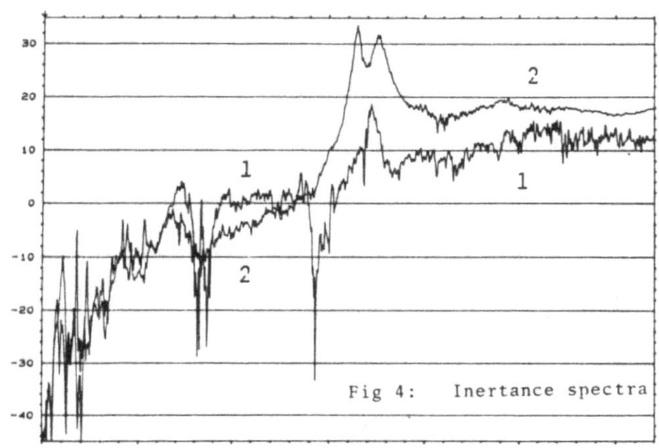

Fig 4: Inertance spectra

Monitoring of a production grinding process

G.J. Trmal
Engineering, Bristol Polytechnic, Bristol, England
C.B. Zhu
Engineering, North China University of Technology, Beijing, China
P.S. Midha
Engineering, Bristol Polytechnic, Bristol, England

Abstract
The introduction of CNC technology minimises worker's participation in operating the grinding machines and thus makes it difficult to learn by experience. Yet, a combination of CNC technology and monitoring techniques offers considerable potential for optimising the grinding process. Monitoring of power used in grinding is simple and it does not interfere with the normal grinding operation. Evaluation of recorded power can provide important information about wheel sharpness and grinding stiffness. This information can be used to optimise for a burn free grinding.
Keywords: Grinding, Process Monitoring, Power, Stiffness, Specific Energy

1 Introduction

Traditionally, grinding has been treated more as an 'art' than science and therefore long experience has been the pre-requisite for the selection of correct grinding conditions for a particular application. Research in recent years has attempted to provide a basis for the prediction of grinding conditions but validity of proposed methods has only been tested in limited experimental work under simplified conditions. Therefore, although these methods provide a useful guide, the interpretation of predicted data to specific applications reqiures judgement based on exprience.

The paper proposes a method for optimising grinding conditions based on the determination of specific energy and grinding stiffness. The specific energy can be related to thermal damage in workpiece material and thus, those grinding conditions that lead to possible thermal damage can be avoided.

Determination of specific energy has been based on the power monitored during the grinding operation. It is envisaged that with a combination of suitable expert system and monitoring devices, the CNC machines can guide the operator to use optimum griding conditions. In addition, these machines can be made to learn from monitoring the process and update the existing knowledge.

2 Power Monitoring

It is important that transducers used for monitoring under production conditions are sufficiently robust so that they can be installed in rather hostile environment of abrasive particles and coolant mist. In addition, they should not obstruct the working area and make the loading/unloading of components and resetting of the machines difficult.

In this investigation power was measured by a robust transducer connected in an electrical circuit of the wheel motor drive. A typical trace of grinding power in a cylindrical grinding operation is shown in fig. 1. Before the advancing wheel makes the first contact with a rotating workpiece the 'idle' power is consumed to cover losses in the bearings and drive.

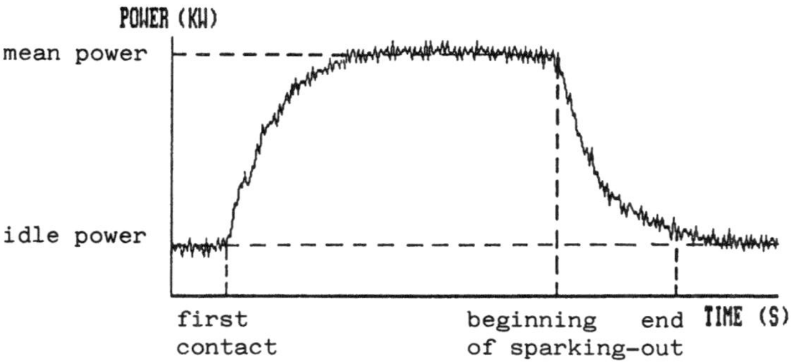

Fig.1. Power-time recording of a cylindrical grinding cycle.

At the first contact between the wheel and workpiece the grinding force starts to develop and deflects the work-wheel system. After the deflection of the system reaches the maximum value determined by the normal grinding force, the material is removed at a steady rate. At the begining of the sparking-out period the infeed stops but the wheel and workpiece continue to move while the system deflection is being eliminated. The wheel is withdrawn at the end of the sparking out period and the power returns to its 'idle' level. The power trace contains information about the machine stiffness which can be derived from the rate of decay of power during the sparking out period. It also contains information about the cutting ability of the wheel that can be expressed in terms of specific energy.

3 Specific Energy as a Measure of Efficiency of the Grinding Process

The specific energy was determined by integrating the power above the idle value and relating it to the volume removed as calculated from measured workpiece diameters before and after grinding. Specific energy determines the heat generated during the grinding process. If it is excessive, the heat can cause thermal damage to the component.

To analyse the process, Malkin (1974) used the theory of a moving heat source developed by Jaeger. He assumed a uniform power flux through the wheel- work interface moving across the surface of aworkpiece and derived a formula for calculation of the surface temperature. The temperature depends on specific energy, rate of material removal, workpiece surface speed and the equivalent diameter, which provides a measure of conformity of the wheel and workpiece. It also depend on the thermal properties of the workpiece material.

Using experimental results, Malkin established that burn marks appear on the surface of a carbon steel component if the temperature, as determined by his formula, exceeds 575 degrees centigrade. On this basis he determined the maximum value of the specific energy (Eb) for burn free grinding. The value depends on the removal rate, work speed and equivalent diameter, as indicated in Equation (1):

$$Eb = 7.17 \frac{Deq^{0.25} \times vw^{0.25}}{Z'^{0.75}} + 6.21 \quad (J/mm^3,\ mm,\ mm/s,\ mm^2/s) \qquad (1)$$

where:
- Eb = specific energy on the onset of burns
- Deq = equivalent diameter = Ds×dw/(Ds+dw)
- Ds = wheel diameter
- dw = work diameter
- vw = work surface speed
- Z' = specific rate of material removal

Correct combination of the wheel type, dressing and coolantmust reduce the specific energy below this value.

4 Evaluation of Stiffness of the Grinding System

The stiffness of the grinding system is an important parameter as it affects the grinding performance and optimum selection of conditions. Yet, very little attention is paid to it in practice. The nominal infeed rate might never be achieved because the grinding system is too flexible and the required stock is removed before the steady state can be reached. Such a situation is shown by curve 'b' in Figure 2. Curve 'a' represents the situation for a relatively stiff grinding system.

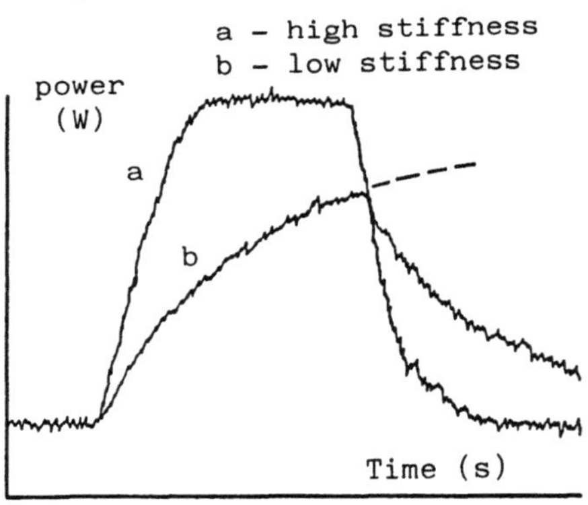

Fig.2. Grinding power with high and low grinding stiffness.

In this investigation, stiffness of the grinding system has been determined from the rate of decay of power during the sparking- out period. It has been shown by Trmal (1979) that under simple assumptions of a linear wheel- machine- workpiece system, constant specific energy during the whole cycle and constant ratio of tangential to normal force the decay in grinding power during the sparking out period can be determined as follows:

$$P = Po \times e^{-A \times t} \qquad (2)$$

Where:
P = instantaneous power
Po = power at the begining of the period
t = time

$$A = \text{constant} = k \times \frac{vs \times u}{dw \times E \times B} \qquad (3)$$

k = grinding stiffness
vs = wheel surface speed
u = tang. to normal force ratio
dw = work diameter
B = work with
E = specific energy

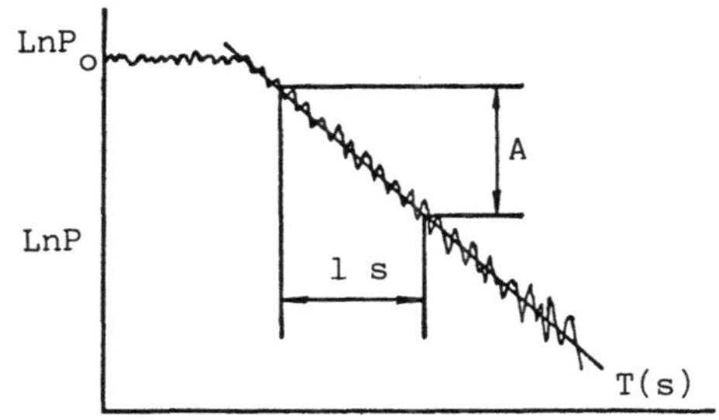

Fig.3. Linear power decay during the sparking out period.

When ln power is plotted in the time domain as shown in fig.3, then the constant A can be determined as the slope of a linear power decay curve during the sparking out period. From that constant the grinding stiffness can be easily calculated because all the other parameters are known.

5 Experimental confirmation

A series of tests have been carried out to determine the validity of the assumptions and the accuracy and repeatibility of determining the specific energy and the stiffness of the grinding system. Two workpieces of different stiffness were used in the tests. The stiffness of the workpieces was assessed by a simple deflection test prior to grinding. The workpiece dimensions as well as the measured stiffness values (k1) are shown in fig. 4.

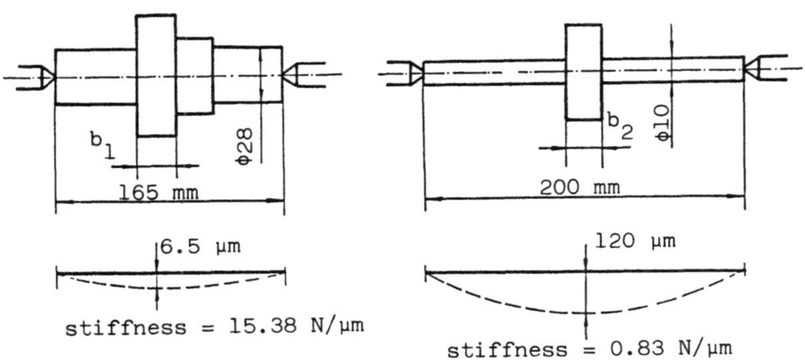

Fig. 4. Components used in tests.

The stiffness (k2) of the spindle and the infeed mechanism was not determined but it was estimated from the previous work on a similar machine (Trmal 1979)) that the combined infeed- spindle stiffness is about 20 N/um. This would result in total grinding stiffness (k) of 8.69 N/um in case 1 and 0.796 N/um in case 2 (k =k1 x k2/(k1+k2)).Both components were plunge- cut ground in several experiments under conditions quoted in Table 1.

Machine: Jones & Shipman 1070 Wheel type: WA60K Speed: 26.9 m/s
Workpiece material: En 31 Dressing: diamond blade

	Case 1: (Work speed,Vw=0.393 m/s)				Case 2: (Vw=0.387 m/s)		
Test No:	1.1	1.2	1.3	1.4	2.1	2.2	2.3
Dress lead(mm/rev)	0.05	0.05	0.05	0.3	0.05	0.1	0.3
Feed speed (mm/s)	0.007	0.015	0.025	0.03	0.007	0.015	0.025
Removal rate(mm2/s)	1.05	2.09	3.46	4.12	1.29	2.58	4.28

Table 1. Details of Test conditions

Grinding power was recorded during the tests and the specific energy was determined by integrating the power record and relating it to the stock removed. Selected power records can be seen in figure 5. Corresponding plots of ln Power against time during the spark-out period are shown in fig. 6. The slope of decay in power in this graph was used to determine the stiffness of the machine using the technique described in section 4. The force ratio u was assumed to be 0.4.

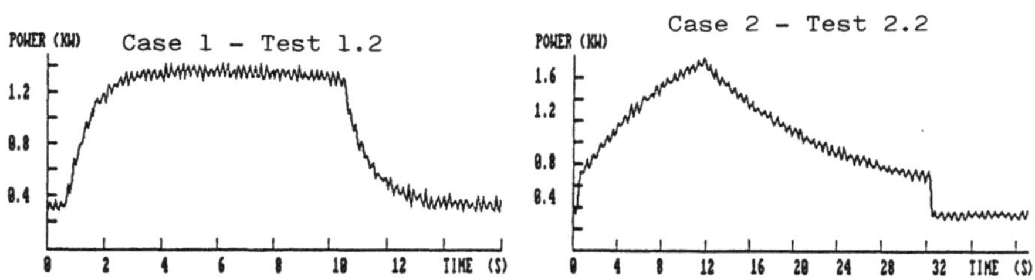

Fig. 5. Typical records of power.

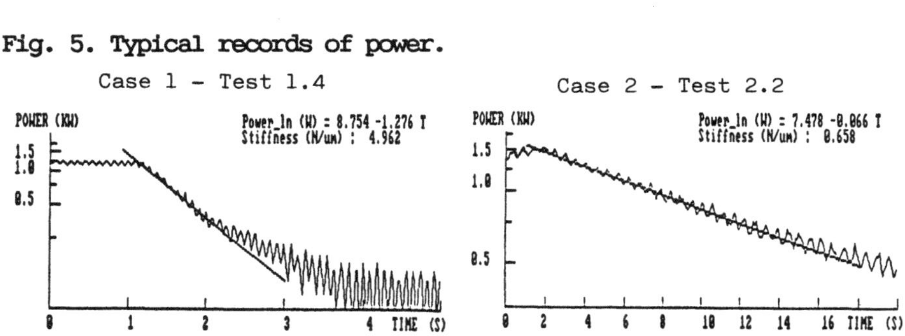

Fig. 6. Graph of power against time in the spark-out period.

The values of stiffness and specific energy derived from the monitoring of power can be seen in table 2.

	Case 1				Case 2		
Test No:	1.1	1.2	1.3	1.4	2.1	2.2	2.3
Specific energy (J/mm3)	55	45	42	26	65	63	38
Grinding stiffness (N/um)	6.3	5.8	7.5	4.9	0.66	0.65	---

Table 2. Summary of test results.

As cen be seen from the table, the values of stiffness are in a reasonable agreement with the expected values quoted earlier.

6 Thermal damage

The temperature determined from Malkin's theory should have resulted in surface burns. However, no damage was found under visual inspection. It is quite possible that the burns generated during the grinding cycle could have been removed during the sparking out period when the grinding rate was diminishing. The result depends on grinding stiffness which determines the rate of reduction in the removal rate and also the depth of the material removed during this period. A complete theoretical analysis of this problem is difficult and probably only of a limited practical importance. However, a company which monitors the grinding power and obtains information on the specific energy as well as as grinding stiffness for its grinding operations can easily relate the acceptable levels of specific energy to its own limits of thermal damage, optimise its grinding conditions and produce good quality components in minimum time.

7 Conclusion

Grinding power is an useful parameter which can be easily monitored in a production environment.

Specific energy,which is a measure of wheel sharpness, and grinding stiffness, affecting the grinding performance, can be derived from the monitored power.

Danger of surface burns can be assessed from the specific energy and grinding stiffness and the process can be optimised.

8 References

Malkin, S. (1974) Thermal aspects of grinding, part 1 and 2 J. of Eng. for Ind. Trans ASME V 96 p 1177-1191

Trmal, G.J. (1979) In process control of size of a ground component. Proc. 20th MTDR Conf. p 405-411

The technique of mesh obscuration for debris monitoring – the COULTER LCM II

M.P. Cowan and R.A. Wenman
Coulter Electronics Ltd, Luton, England

Abstract
Debris monitoring of fluid power systems can now be performed by the technique of mesh obscuration. The contaminated fluid is passed through a fully characterised filter mesh under constant, laminar flow conditions. Particles larger than the effective pore size of the mesh are retained on its surface and reduce the open area available for flow. The upstream pressure rises in proportion to the number of particles retained, because of increasing degree of mesh obscuration. If several different meshes are used, a particle count distribution may be obtained at Industry Standard size levels. This simple principle ensures that the results are independent of colour or opacity of the fluid. Contamination such as water does not effect the results.

The COULTER® LCM II is a fully automatic instrument which may be operated throughout a wide temperature range. Particle counts are obtained at size levels of 5, 15 and 25μm, with fluid viscosity and temperature also reported. The COULTER LCM II has low pressure on-line capability, together with fully programmable remote control and data transmission via a standard computer interface. Programmable count thresholds for all particle size levels and viscosity may be used to indicate, and/or act upon set required sample conditions. This rugged and portable instrument is also suitable for batch sampling either in the Laboratory or at remote sites.

Performance results are presented for Industry Standard calibration materials, and for typical debris monitoring applications, including linearity, accuracy and repeatability data.

The flexibility of the COULTER LCM II and the simplicity of the mesh obscuration principle will enable this highly innovative instrument to be easily integrated into any debris monitoring program.
Keywords: Mesh obscuration, debris monitoring, contamination, BS 5540 part 2.

1 Introduction

Modern hydraulic systems increasingly consist of components with fine finishes and close tolerances between moving parts. Particles of debris are the major cause of wear leading to a deterioration in performance and ultimately catastrophic failure. Debris monitoring can be of assistance in several ways; it can be used to determine the degree and efficiency of filtration required for a specific application and guide the selection of the most cost effective filtration system. It can also be used for trend analysis to predict component wear and potential failure.

The COULTER LCM II is a fully portable instrument capable of counting the solid particulate contamination present in most lubricating oils, between a kinematic viscosity of 10 and 500 cSt. The analysis period is typically less than five minutes, counts are given greater than 5 and 15µm according to BS 5540 part 2 (ISO 4402), and greater than 25µm. The principal advantage of mesh obscuration is its independence of oil colour and opacity. Water in oil emulsions can be characterised for solid particle content, and small numbers of air bubbles do not significantly alter the results.

2 Theory

The patented technique of mesh obscuration has been previously described by Hunt, Tilley and Rice in 1984 and by Wenman et al in 1988, these contain a complete description of the theory. Briefly, lubricating oil is passed through a fully characterised filter mesh under constant, laminar flow conditions without dilution or concentration. Particles larger than the effective pore size of the mesh are retained on its surface and reduce the open area available for flow. The upstream pressure rises in proportion to the number of particles retained because of the increasing degree of mesh obscuration. If several different meshes are used, a particle count distribution may be obtained at Industry Standard size levels. This simple principle ensures that the results are independent of colour or opacity of the fluid. Other contamination, such as water droplets, does not affect the results.

The COULTER LCM II contains a DC motor and fixed displacement hydraulic pump assembly. When analysing a sample the processor-controlled motor runs at an automatically selected constant speed intelligently based on three parameters: 1) the volume of sample available, and 2) the contamination level, and 3) the viscosity. The optimum measuring conditions are estimated at the start of each sample analysis by an expert system.

The filter mesh holder assembly contains 5, 15 and 25µm filter meshes, each in an easily removable holder. A hydraulic slide changeover valve ensures continuous

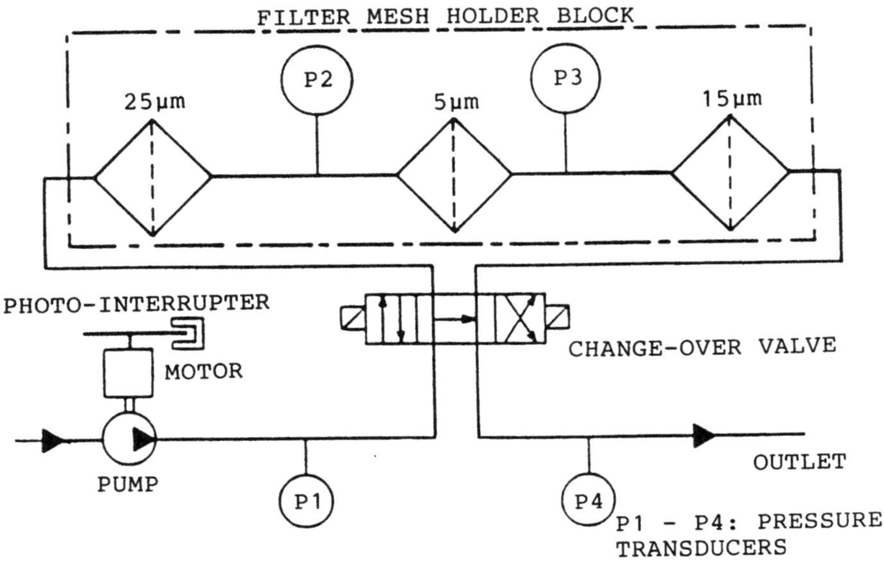

Fig. 1 LCM II Schematic

operation by routing the flow in either direction. In the first direction the 15µm and 5µm meshes are used, and in the other the 25µm and 5µm meshes. The whole assembly is designed to minimise particle entrapment and reduce dead volume, long term continuous operation is possible by flushing the trapped particles from the surface of the meshes; again this is automatically done during the normal analysis run. Four pressure transducers measure the pressure drop across the meshes from which the count is calculated.

3 Performance

In this study, three production instruments were used in order to establish the typical performance of the product. The instruments were evaluated using mineral hydraulic oil which had been deliberately contaminated with A.C. Fine Test Dust (ACFTD). This is a well characterised material used as an Industry Standard and calibration material, the reference distribution of which can be found in BS 5540 part 2.

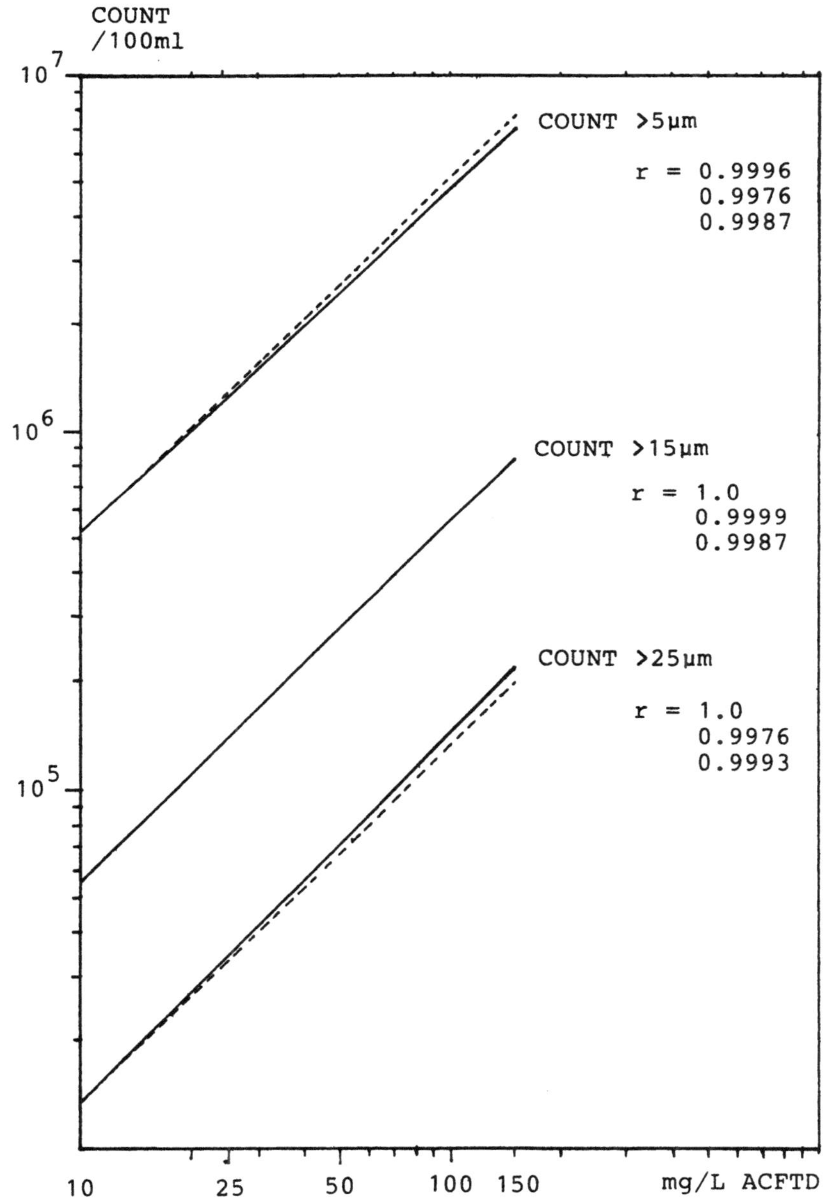

Fig. 2.

COUNTS/100ml Vs. MASS of ACFTD/L for nominal 250cSt oil

3.1 Linearity

Figure 2 shows the essentially linear response of the LCM II to increasing mass for each of the three filter meshes up to 150mg/L of ACFTD. This is equivalent to an ISO code of 24, the highest level of contamination for which there is a code. In the figure the dashed line shows the ideal and the solid line the actual response. Linear regression analysis of the data shows an excellent fit to a straight line with little scatter in the data. Oil viscosity was approximately 250cSt in this example, results of similar quality were obtained using other oils with viscosities between 10 and 500 cSt.

3.2 Reproducibility

Table 1 shows typical within run precision for five consecutive counts at varying masses of ACFTD (at each size level filter). The coefficient of variation was typically less than 3% over the instrument's range.

Table 2 shows that the LCM II measurement of viscosity was independent of the contamination level and was exceptionally reproducible.

Table 1. Within run precision for nominal 250cSt oil

Concentration	Coefficient of Variation (%)		
(mg/L)	Count >25μm	Count >15μm	Count >5μm
10	1.9	1.7	2.1
25	3.3	1.7	0.7
50	2.2	0.5	0.7
100	2.4	2.0	2.4
150	1.2	6.0	2.5

Table 2. Viscosity measurement for nominal 100cSt oil

Concentration (mg/L)	Viscosity cSt	Coefficient of Variation
clean	99.7	0.7
10	102	1.1
25	99.7	1.1
50	100	0.6
100	100	0.3

4 Other features

In addition to the basic features already indicated, **(guaranteed performance)** the LCM II has a number of enhancement features. All LCM II instruments leave the factory calibrated to ACFTD, according to BS 5540. If other Standards are used, a sample single point calibration for each size level can be performed.

The in-built expert system makes the analysis fully automatic with both discrete and on-line sampling capability. Control and data storage can be remote through the computer interface, enabling full integration with process management. Alarms can be triggered when preset ISO codes or viscosities are reached, a valuable feature if monitoring for example, a filtration system. Once the required cleanliness has been reached, the filtration can be stopped thereby optimising the use of plant, energy and time.

If the instrument has low confidence in any measurement, for whatever reason, the result will be flagged to alert the operator. In-built software continually monitors the integrity of the pressure transducers and, if a failure occurs, will prevent further analysis whilst indicating the error.

5 Conclusion

The LCM II measures particulate contamination and has been shown to respond linearly with the mass of debris. Measurement of viscosity is shown to be independent of contamination and is very reproducible. Its principal advantages over competitive techniques are that it is independent of colour or opacity of the fluid, the refractive index of the particles, water contamination, and the material composition of the debris. It can analyse debris in emulsions of up to 40% water in oil.

The flexibility of the COULTER® LCM II and simplicity of the mesh obscuration principle will enable this highly innovative instrument to be integrated into any debris monitoring program.

6 References

BS 5540: part 2 (1978) (ISO 4402) Method of defining levels of contamination (solid contaminant code).

Hunt, T.M. Tilley, D.G. and Rice, C.G. (1984) Techniques for the assessment of contamination in hydraulic oils. Proc. IMechE Conf. C244/84, 57-63.

Wenman, R.A. Bunker, R. and Miller, B.V. (1988) Optimization of the mesh obscuration technique for the measurement of particulate contamination in lubricating liquids, in Particle Size Analysis (ed P.J. Lloyd), John Wiley and Sons Ltd., Chichester, 63-73.

Finding what matters – health monitoring by FCM

T.M. Hunt
Associate Consultant, Lindley Flowtech Ltd, Bradford, West Yorkshire, UK

Abstract
Too often health monitoring of machines and systems has been spoilt in concept because of inadequate monitoring devices. Why do we monitor? Basically it is because we want to know any change from near-perfect health. We want to detect those early signs of malfunction. We want to be warned of impending failure.

On the other hand we do not want to be switching off our process at every whim and fancy of the monitor, only to find that a small gremlin of absolutely no consequence is passing through. We do not want to be told to shut down when there are still thousands of hours work left in the machine.

It is in this detection of what really matters that the Fluid Condition Monitor (FCM) is becoming more and more accepted. From an analysis of the machine, and possible serious changes in the fluid, the FCM is able to show, at the right time, when that condition is reached where action must be taken.

After a discussion of machine health and the associated fluid changes, the paper describes the FCM's capabilities and gives examples of it in use where its output is related to real machine condition.
Keywords: Filter Blockage technique, Fluid Condition Monitor, Lindley Flowtech, Machine, Off-line, Oil, On-line, Reliability, Wear Debris.

1 Introduction

What is the purpose in fitting a condition monitor? It seems almost an irrelevant question - but far from it! The history of condition monitoring is by no means straightforward, and many ingenious approaches have been taken to sell monitoring products which unhappily finish their short lives as paper weights or even less.

If you buy a monitor, it must be with the objective of having a device which gives you information relating to the condition of your equipment. This information should be of such content that it enables you to take corrective action before serious consequential damage occurs. It is no use having a monitor which tells you when the system has already broken down; what is needed is advance warning.

But what is it that makes a monitor so prophetic?

2 Early Signs

The predictive ability of a monitor is that feature of it which enables it to

DETECT THE FIRST SIGNS OF DECAY, RELIABLY

One example of unreliability is that of the early chip plugs used by US Army personnel reported by Beerbower (1975). If I remember rightly something like 30% of the indicated engine 'faults' were spurious - it being the chip plug which was at fault! It took many years for this to be forgotten and forgiven.

Several examples of the inability to detect the early signs were observed in work undertaken at the Fluid Power Centre in Bath, and reported by the author, Hunt (1986). Figure 1 shows the results.

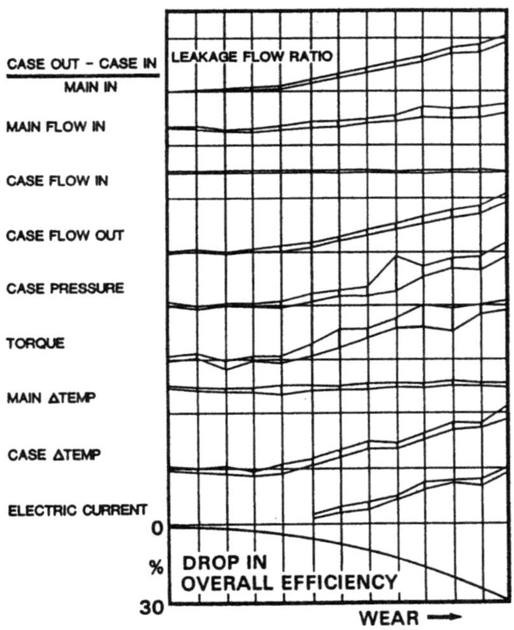

Fig. 1. Change in transducer output with wear (Hunt 1986)

Some 15 different transducers and techniques were used on an axial piston pump as it deteriorated to a degree of almost total uselessness. Few of these were able to spot any change at all in the early stages of wear (up to 5% loss in overall efficiency), and most of the others, whilst noting a change, gave such scattered readings that confidence in them was limited. (The width of the lines gives an indication of the variableness of the measured values at each stage of wear). Basic vibration analysis, including FFT, produced no clear results, but Cepstrum analysis did indicate differences with wear, but the sensitivity was low and the repeatability unknown.

The detection of the early signs is what is going to sell a monitor to management. It makes economical sense to save a machine before catastrophic failure occurs and to repair it at a convenient time

3 Fluid changes

One aspect which was not covered by Hunt (1986) was the debris in the oil analysis. This would not have been a true test because the wear was accelerated, and no attempt was made to monitor during the wearing process but rather after each stage of decay. Debris Analysis is, however, a technique ideal for use in the real situation at the time that wearing is taking place. Even more, where a device like the Fluid Condition Monitor is used, it can show up the early stages par excellence as we shall see below.

The fluid being pumped around a machine, regularly carries the evidence of fault, if present. That is just why our blood is so seriously monitored and checked at regular intervals.

> Lubrication oil and hydraulic fluids carry solid particulate contaminant, which can cause failure - and that is one major reason for fluid monitoring.
>
> Another reason for monitoring is that the fluid carries debris from components which are wearing or collapsing.

Because of the variable nature of these particles, see Hunt (1988),-

Fig.2.Shapes of common solid debris in engineering fluids

the shape, the size, metallic, non-metallic, magnetic, non-magnetic - most detectors are somewhat limited in their ability to detect them, particularly if the fluid is chemically contaminated as well.

But the evidence is there, the early evidence waiting to be spotted.

4 The Fluid Condition Monitor

The reason why the Lindley Flowtech Fluid Condition Monitor (FCM) is so suitable for monitoring fluid debris is that it can detect all debris in whatever fluid. It does not matter whether the fluid is coloured, or mixed with water or air, particles down to 5 micron, or even less, are detectable, accurately.

The technique of Filter Blockage is used. The technique involves the measurement of pressure drop as a filter blocks. In practice what happens is the debris is trapped in a refined mesh as the fluid passes through it. The pressure drop across this mesh thus rises quickly if there is more debris present, more slowly if there is little there.

In the FCM the times and pressures are measured so precisely, that an actual count of particles greater in size to the mesh is achieved. This has been described in detail by Raw and Hunt (1987). By automatic reversing of the flow, the mesh is back flushed and continuous operation occurs.

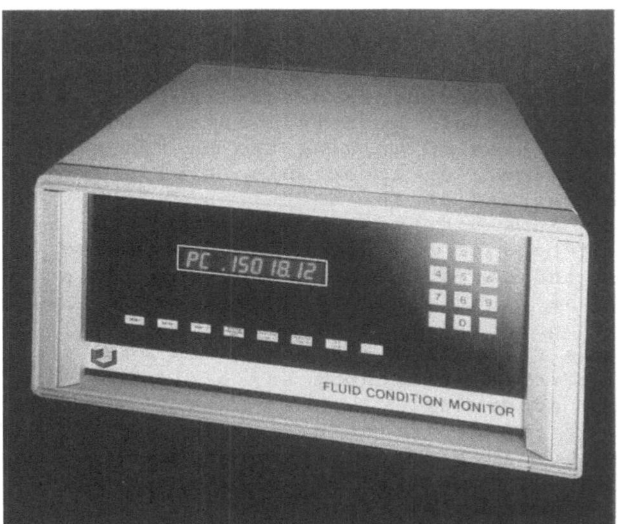

Fig.3. The Fluid Condition Monitor (FCM)

Another valuable feature is the on-line fitting. The FCM can be permanently attached to, or plugged into a sample point on, a high pressure line to give continuous on-line monitoring of the fluid as critical machinery is operated. The only extra to the FCM is a fluid interface to provide a fixed flow at low pressure from the high pressure line (this is supplied by Lindley Flowtech). Alternatively, the FCM can be used off-line, by means of its inbuilt pump, connected to a reservoir or sample container.

5 Examples of immediate fault detection

Fig.4 The Rover 200

A very recent example of fault detection in good time, could be taken from the Rover Group use of the FCM. The FCM functions as a monitor of the polyol ester fluid involved in the robots used in welding the Metros and Rover 200's at Longbridge. An acceptable level of particulate has been agreed at the NAS Class 7 cleanliness level. Some nine months after the FCM was put into service, instead of the usual NAS 7 or below, NAS 11 was suddenly indicated. The system was stopped and examination of the fluid indicated, even by eye, many large aluminium and non-metallic particles present. Corrective action was taken, and the robot was able to be restarted after a short delay without any serious consequential damage occurring.

Another example of the detection efficiency of the FCM was given by the author in a paper presented last year at COMADEM 89, Hunt (1989). In that illustration a valve was found leaking on a flushing rig.

6 Conclusion

Although it is appreciated that there are many very exciting monitors and monitoring techniques available, it is felt that the only really viable monitors, as far as management is concerned, are those which detect the early stages of decay.

It is suggested that one such suitable monitor is that which looks at all debris in the fluid, as exampled by the Fluid Condition Monitor.

7 Acknowledgements

The author would like to thank the directors of the Rover Group for permission to mention the data relating to their use of the FCM; he would also like to thank Mr Paul Wilkes of their Materials Laboratory, for his helpful advice. Additional help has also been given by Mr Gary France and the staff of Lindley Flowtech Limited, which is much appreciated.

8 References

Beerbower, A. (1975) Mechanical failure prognosis through oil debris monitoring, in the US Army Air Mobility Research and Development Laboratory report TR-74-100,p.72

Hunt, T.M (1986) Condition monitoring of hydraulic systems, in the 2nd Naval Hydraulics Conference at Manadon, pp. 19/1-19/16.

Hunt, T.M. (1988) Particle Counting by size - which method is right? in the IMPARTECH '88 Conference organised by the Filtration Society at Imperial College, London.

Hunt, T.M. (1989) Debris counting for consistent monitoring, in the COMADEM 89 International Congress (eds R.B.K.N.Rao and A.D.Hope) at Birmingham Polytechnic, Kogan Page, pp.348-352.

Raw, I.and Hunt, T.M. (1987) A particle size analyzer based on filter blockage, in the Condition Monitoring '87 International Conference (ed. M.H. Jones) at the University College of Swansea, Pineridge Press, pp. 875-894.

Grease – a suitable case for condition monitoring

M.H. Jones
Department of Mechanical Engineering, University College, Swansea, UK

Abstract
A method of wear debris analysis of grease by emission spectroscopy (rotrode) is described. A sampling procedure and sample preparation technique are included. A case history of damage in a highly loaded thrust bearing with resulting redesign is detailed.
Keywords: Condition Monitoring, Grease, Emission Spectroscopy

1 Introduction

Wear debris analysis of liquid based lubricants has been carried out for more than forty years, through spectrometry since 1945, and subsequently ferrography and related techniques. Semi-liquid lubricants such as grease have, however, been relatively ignored despite the majority of bearings in industrial applications being grease lubricated. The US Army have initiated, during the last few years, a program of grease analysis by ferrography. This requires breaking down the grease structure by the use of solvents. Applications of this technique have been restricted to the swash plate bearing of one of the army helicopters. Other applications have been reported by the ferrograph manufacturers (1). Ferrography, which entails the magnetic separation of debris, i.e. ferrous particles, has to utilise an alternative syphon system to ensure large particle detection. A burette feeding the solvent and grease mixture over the ferrogram slide has been found to be adequate; an alternative method has been described in the reference (2).

A disadvantage of the above methods is that only ferrous debris is separated for diagnosis. Further techniques have to be used to identify whether the particles are mild steel, alloy steel, cast iron, etc. The use of X-ray analysis would be required for more precise identification. A further disadvantage is that time taken for these procedures makes the analysis very expensive and therefore restrictive.

An ideal technique for grease analysis should enable the detection of all the elements likely to be encountered in typical applications, e.g. component materials, lubricant additives and contaminants, minimum sample preparation and low cost. This paper proposes such a technique.

2 Sampling procedure and analysis technique

Taking samples of grease has always been perceived to be a problem as to whether the sample is representative of the bearing condition. Experience over more than two years has shown that within various set procedures, the grease sample may be truly representative and the results reproduceable.

The method proposed entails the use of nylon tubing attached to a 'Vampire' sampling gun inserted into the loaded region of the bearing; several strokes of the pump (sampling bottle attached), should extract two to three inches (6 cms) of grease. The nylon tube is cut and the ends sealed for transit.

Preparation of the sample entails the following:

a. Force a weighed amount of grease from the tube into a beaker.

b. Add to the grease three-times its weight of base oil (Conostan 75 clean base oil standard has been found most suitable for dilution). Thorough mixing results in a semi-liquid sample suitable for analysis.

c. Pour the mixed sample into a specimen cup for emission spectrographic analysis. The pre-sparking on the rotrode ensures an even coating of the graphite disc, prior to analysis.

Table 1 lists the analysis of three successive sparks from the same spectro sample cup.

3 Case Study

With increasing demands made on the manufacturing plant by new and more difficult products, a strip mill which has been committed to PCM (plant condition monitoring), had extensive experience of vibration monitoring, spectrographic oil analysis, wear debris analysis and numerous non-destructive techniques, however, no experience had been gained on grease applications. The taper roller thrust bearings used in the mill, which were grease lubricated, was an application for which a monitoring technique was required. If one of these bearings fails to respond to its control, or responds too late, the effect on the product being rolled is often to cause asymmetrical cross section, the strip will veer to one side of the mill and jam i.e. 'cobble'; both loss of product and mill damage ensues. The robust nature of the mill, and the fact that other faults may also cause 'cobbles', often results in excessive bearing wear or damage proceeding unnoticed for a considerable time. The environmental conditions are extremely hostile with the bearings subject to high pressure water and oxides.

Condition monitoring these thrust bearings has proved extremely difficult, the slow speed of rotation makes vibration analysis impossible, and as there had been no flow of lubricant, wear debris analysis was considered inappropriate.

Fig.1 illustrates the general arrangement of the taper roller thrust bearing which is 19.5 ins. outside diameter and 5.75 ins. high with 18 rollers. When rotating, the bearing moves at a speed of 2 rev. per min. See Appendix for further details.

Initially, the cartridges locating the bearings are grease and oil lubricated to ensure a lubricant film protects the contact areas between the rollers and thrust faces at the time of installation. Thereafter the bearings are grease-lubricated on an autopulse system - supplemented by hand feeding at the time of repair shifts. The grease in this application was of number 2 consistency and lithium based.

Fig.1 shows the application point for the autopulse grease system at the cartridge top and the drain hole and plug at its base. Grease samples were extracted by first removing the plug and then inserting the nylon tube; suction through the tube draws off sufficient grease for analysis. Extra grease was drawn off into a sample bottle when possible. This grease was diluted with solvent and the magnetic particles extracted for further examination and confirmation of the spectrographic results obtained as described above. The extracted debris was 'fixed' to sellotape and attached to cards for future reference in a similar manner to that described by airline operators using magnetic plugs.

Samples were taken at intervals of two weeks. An example of grease analysis which clearly identifies possible bearing damage is shown in Fig.2 (8th February 1988) The critical elements in this particular case have been found to be nickel and chromium which are alloying elements in this bearing. The first analysis shows 11 ppm chromium and 23 ppm nickel. The bearing specification has 1.5 per cent chromium and 3.5 per cent nickel, i.e. the same ratio as that detected in the grease. Debris separated from the grease sample was prepared for electron microscopy. Fig.3 shows the extent of the wear debris and X-ray analysis identifies the source as the high alloy steel of the bearing.

Confirmation of the level and analysis of the debris was given in the subsequent grease sample(4th March 1988): values having increased to 31ppm chromium 67ppm nickel.

The bearing cartridge was removed for inspection at the next downshift. Grease samples were extracted for further analysis as shown in Fig. 2 (10th March 1988). Figs. 4 - 6 show the extent of the damage to the bearing. The taper rollers had suffered excessive damage to the smaller diameter end, whilst the thrust face had cracked in numerous places, these cracks extending through the thrust plate.

Further monitoring of the bearings detected damage at this location on 25th April, and at another location on 24th May which are shown in Figs. 7 and 8. Over the nine month period, five bearings were removed as a result of the grease analysis, inspection confirming bearing damage to have occurred in each case.

During the monitoring programme the detection of excessive wear in numerous bearings resulted in a redesign of the bearing which will be modified to utilise a higher capacity bearing.

4 Conclusion

A simple method has been developed using existing spectrographic equipment for monitoring grease lubricated bearings.

During the monitoring programme, areas of excessive wear have been highlighted with subsequent plant redesign.

Experience gained by monitoring grease used in bearings has resulted in diversified applications such as crane slew ring bearings on offshore oil rigs and helicopter gearboxes.

5 Appendix

Details of the taper roller thrust bearing:
The taper roller thrust bearing, housed in a cartridge, transmits the screw force. The taper roller thrust bearings are 19.5 ins. outside diameter by 5.75 ins. high with 18 rollers. All component material is SAE E3312 Static thrust capacity 1672 tonnes. The thrust faces of the clamp rings securing the bearing to the screws for upward movement are aluminium bronze coated.

6 References

Bowen, E.R., Bowen, J.P. and Anderson, D.P. (1978) Application of Ferrography to grease lubricated systems, **46th Annual Meeting NLGI**.

Jones, M. H., (1983) Techniques used to enhance the use of Ferrography, **Wear**, 84 (1), pp. 111-113.

Lubricant : Army Spec. Grease

Component : Swashplate
Model : not provided

Sample Date Sample No. Lubricant Hours/mileage Equipment Hours/mileage Makeup Lubricant		28 jun 88 n/p n/p n/p	28 jun 88 n/p n/p n/p	28 jun 88 n/p n/p n/p	
Wear Elements	Iron	36	30	30	
	Chromium	0	0	0	
	Aluminium	26	21	22	
	Copper	8	7	8	
	Lead	3	0	0	
	Tin	0	0	0	
	Nickel	1	0	0	
	Manganese	1	0	1	
	Titaniium	0	0	0	
	Silver	0	0	0	
	Molybdenum	0	0	0	
Additive Elements	Zinc	2	1	2	
	Phosphorus	30	25	28	
	Calcium	101	89	87	
	Barium	0	0	0	
	Magnesium	250	230	236	
Contaminant Elements	Silicon	1102	1013	1044	
	Sodium	266	250	235	
	Boron	1	0	0	
	Vanadium	0	0	0	

Table 1 : The analysis of three successive sparks from the same spectro sample cup

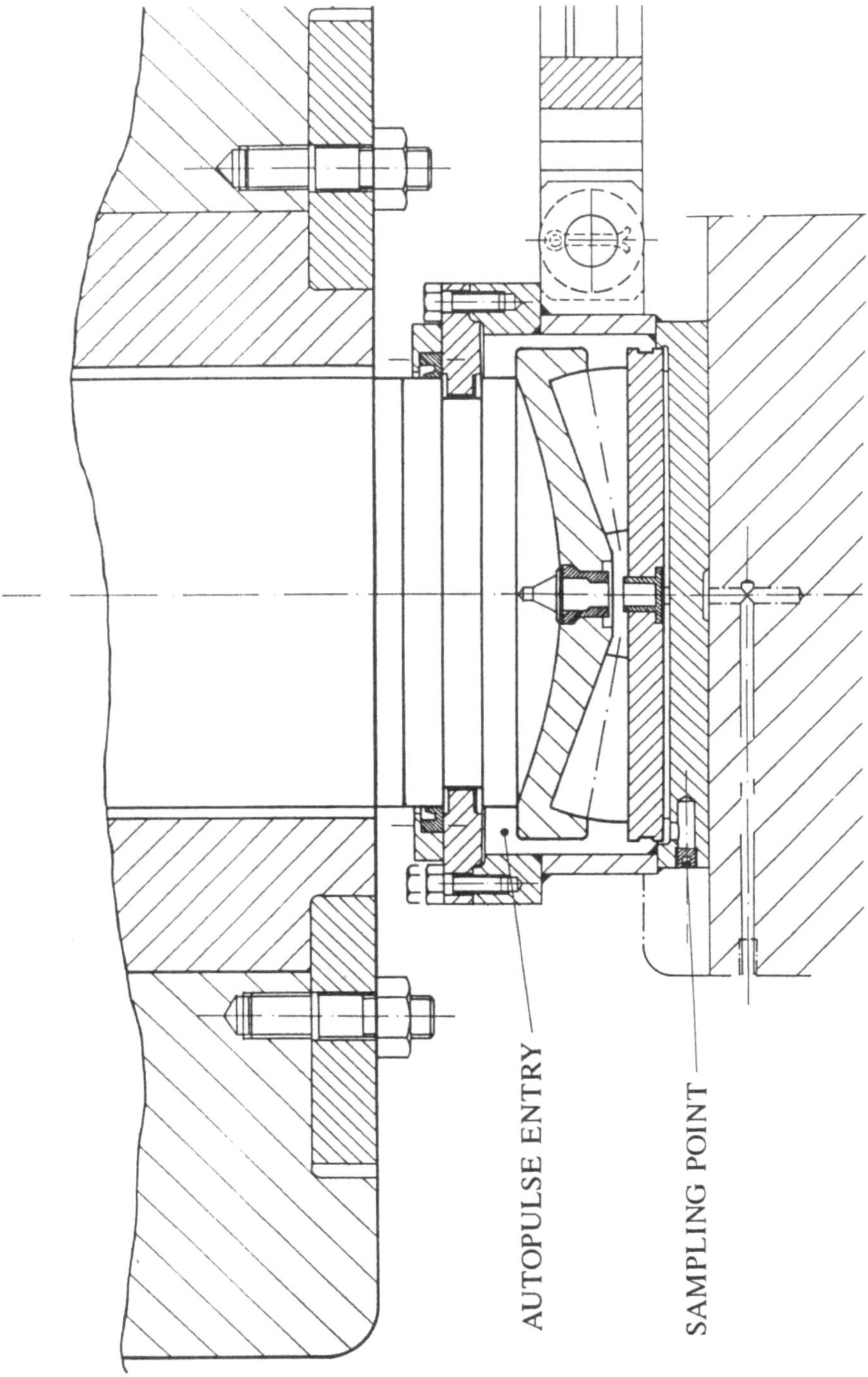

Fig. 1 Thrust Bearing Assembly

Lubricant : No.2. Lithium Base

Component : Screwdown Bearing
Model : Conical Thrust

		8 feb 88	4 mar 88	10 mar 88	14 mar 88	29 mar 88
Sample Date		8 feb 88	4 mar 88	10 mar 88	14 mar 88	29 mar 88
Sample No.						
Lubricant Hours/mileage		n/p	n/p	n/p	n/p	n/p
Equipment Hours/mileage		n/p	n/p	n/p	n/p	n/p
Makeup Lubricant		n/p	n/p	n/p	n/p	n/p
Wear Elements	Iron	>1000**	>1000**	>1000**	192	383
	Chromium	11**	31***	18***	1	2
	Aluminium	2	2	4	1	16
	Copper	5	71**	56**	45	25
	Lead	1105	1503	1097	>1000	>1000
	Tin	3	6	2	6	2
	Nickel	23**	67***	39***	1	2
	Manganese	19	47**	28	3	7
	Titaniium	0	0	0	0	2
	Silver	0	0	0	0	0
	Molybdenum	0	0	0	6	1
Additive Elements	Zinc	9	14	12	13	27
	Phosphorus	3	6	3	12	31
	Calcium	2	4	4	20	311
	Barium	0	0	0	1	2
	Magnesium	0	1	1	1	14
Contaminant Elements	Silicon	0	2	0	6	60**
	Sodium	2	3	4	7	42
	Boron	5	13	10	25	12
	Vanadium	0	0	0	1	1

Figure 2 : Spectrographic Analysis of grease samples showing taper roller thrust bearing

Fig. 4 Thrust face damage

Fig. 5 Thrust face damage

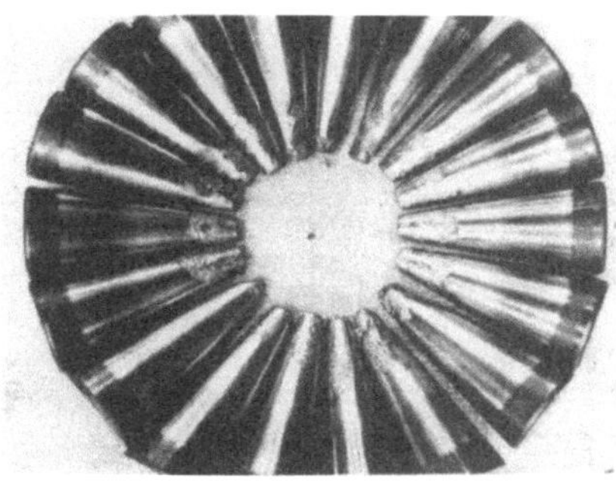

Fig. 6 Rolling Element damage

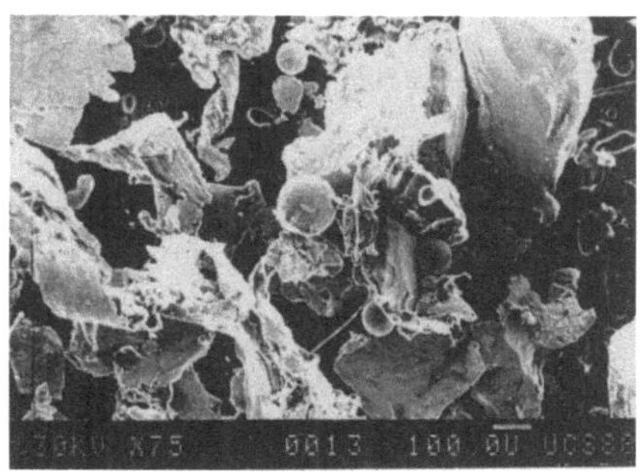

Fig. 7 Debris extracted from grease

Lubricant : No.2. Lithium Base

Component : Screwdown Bearing
Model : Conical Thrust

Sample Date		14 mar 88	29 mar 88	12 apr 88	25 apr 88	7 may 88
Sample No.						
Lubricant Hours/mileage		n/p	n/p	n/p	n/p	n/p
Equipment Hours/mileage		n/p	n/p	n/p	n/p	n/p
Makeup Lubricant		n/p	n/p	n/p	n/p	n/p
Wear Elements	Iron	192	383	345	1380***	74
	Chromium	1	2	2	12**	1
	Aluminium	1	16	2	5	2
	Copper	45	25	26	38	15
	Lead	>1000	>1000	1542	598	1264
	Tin	6	2	3	2	2
	Nickel	1	2	4	23**	1
	Manganese	3	7	6	18	1
	Titaniium	0	2	0	0	0
	Silver	0	0	0	0	0
	Molybdenum	6	1	1	1	1
Additive Elements	Zinc	13	27	7	6	4
	Phosphorus	12	31	7	4	5
	Calcium	20	311	7	37	3
	Barium	1	2	1	0	0
	Magnesium	1	14	1	2	1
Contaminant Elements	Silicon	6	60**	2	11	2
	Sodium	7	42	2	4	2
	Boron	25	12	8	9	1
	Vanadium	1	1	0	0	1

Figure 7 : Spectrographic Analysis of grease showing evidence of damaged bearing

Lubricant : No.2. Lithium Base

Component : Screwdown Bearing
Model : Conical Thrust

		25 apr 88	7 may 88	24 may 88	3 jun 88	9 jun 88
Sample Date		25 apr 88	7 may 88	24 may 88	3 jun 88	9 jun 88
Sample No.						
Lubricant Hours/mileage		n/p	n/p	n/p	n/p	n/p
Equipment Hours/mileage		n/p	n/p	n/p	n/p	n/p
Makeup Lubricant		n/p	n/p	n/p	n/p	n/p
Wear Elements	Iron	148	74	407**	43	32
	Chromium	1	1	5**	0	0
	Aluminium	3	2	8	2	2
	Copper	32	15	15	12	34
	Lead	>1000	1264	>1000	>1000	>1000
	Tin	1	2	1	0	0
	Nickel	1	1	9**	1	1
	Manganese	3	1	6	1	1
	Titaniium	0	0	1	0	0
	Silver	0	0	0	0	0
	Molybdenum	0	1	56	2	13
Additive Elements	Zinc	8	4	16	12	17
	Phosphorus	1	5	0	0	4
	Calcium	6	3	73	26	13
	Barium	0	0	2	1	0
	Magnesium	1	1	3	0	1
Contaminant Elements	Silicon	2	2	18	4	4
	Sodium	5	2	16	5	8
	Boron	1	1	43	54	11
	Vanadium	0	1	2	0	0

Figure 8 : Spectrographic Analysis of grease showing evidence of damaged bearing

Engine monitors for efficient take-off performance monitors

R. Khatwa
Department of Aerospace Engineering, University of Bristol, England

Abstract
The Take-Off Performance Monitor (TOPM) reviewed in this paper has the capacity to detect and declare performance deficiencies during the ground roll by considering critical take-off lengths, current achieved acceleration and engine health. The Mass Momentum Method (MMM) and the Simplified Gross Thrust Method (SGTM) are proposed as suitable methods to compute in-flight gross thrust.
Keywords: Take-Off Monitor, Flight Safety, Rejected Take-Off, Engine Monitor, Turbofan, Isentropic Flow, Thrust Measurement.

1 Introduction

The report of the Air Florida take-off accident from Washington National Airport in January 1982 recommended the development of a TOPM as existing airworthiness requirements contained several inadequacies. Currently, the scheduled accelerate-stop distance may be exceeded during a rejected take-off (RTO) even before the decision point (V_1) is reached, as various factors such as runway contamination, degraded engine performance and tyre failures can adversely affect the airplane acceleration and are not adequately considered during type certification. Furthermore, for aircraft certified in accordance with US Federal Aviation Regulations the accelerate-stop performance for contaminated surface operation is based on a clean, dry runway.

An efficient TOPM would assist the pilot in keeping the progress of the take-off constantly in view, so as to make it easier to decide if a take-off can safely be continued or to support the decision to abandon it, even before he reaches V_1. Despite the recommendation noted, only limited progress has been made in producing a reliable system. In contrast to TOPM systems developed to date, which do not encompass in-flight thrust measurements, this paper focuses on the role and development of an efficient Engine Performance Monitor (EPM) to enhance the pilot's go-no-go judgement.

2 General principles for a display

Fundamental to the objectives of this research is the desire to produce a display with reliable predictive capacities which will provide warnings of take-off dangers earlier than would otherwise be the case. Numerical integration of the set of equations, to which filter outputs contribute values, will provide at least

(a) The current position of the aircraft on the runway (l_c), a continually lengthening bar growing within a simple runway outline.
(b) The position along the runway at which a decision speed will be reached (V_{Rc}) and indicating that the take-off 'can go'. This could be represented by a narrow bar across the runway strip, and preferably coloured and visible as the 'next marker' along the strip.
(c) The runway position (V_{Rm}), calculated **backwards** from the screen or stopway, at which the same decision speed must be reached in order to satisfy the minimum safe take-off requirements and at which the take-off 'must go'. This is the ground roll limit position.
(d) A prediction of stopping distance for the current speed (S_c).

A simple form of the display would be as shown in Fig. 1.

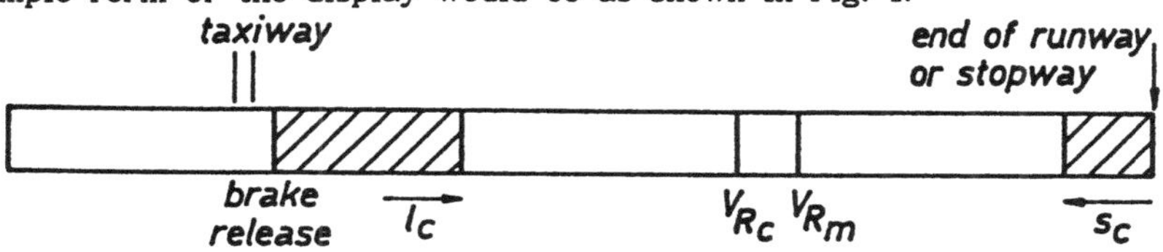

Fig. 1. Simple TOPM display.

All four variables will be re-evaluated regularly (Khatwa 1990), subject to data being collected from

- Inertial Reference System (accelerometers)
- Air Data Computer (ADC) (wind/velocity sensors)
- Tyre health monitors (eg Tyre Pressure Indicating System)
- Engine parameters in order to monitor engine health
- An algorithm that estimates runway rolling friction coefficient

The relative positions of the two warning bars 'can go' and 'must go' will provide predictive assistance to the pilot. These bars would migrate if the take-off parameters (eg engine health, tyre conditions, runway contamination) changed during the ground roll and if the order of the two bars is reversed the pilot can forsee problems before they arise. Display formats which incorporate critical lengths such as the stopping distance from V_1 (S_{V1}) are considered in (Khatwa 1990).

3 Engine Performance Monitoring

Improper power setting or sub-standard thrust development would not only affect the acceleration to the V_1 speed adversely, but also limit the climb performance as in the Air Florida incident. It could be that a very long runway would allow for eventual safe approach to V_1 and take-off, but the climb performance thereafter would be insufficient and unsafe. Engine related problems also reduce the level of reverse thrust available during stopping. A means of monitoring the engine health would warn the crew of any anomalies and enable the pilot to initiate a safe RTO. Realistic predictions of stopping performance are a necessary prerequisite to aid the pilot in making the correct go-no-go judgement. The EPM must communicate with stopping performance algorithms so that the extent of the engine failures can be accounted for.

3.1 In-flight thrust measurement

One particular TOPM presently under development (Srivatsan 1987) monitors engine health by comparing achieved Engine Pressure Ratio (EPR) with a predetermined value for the current conditions and recommended throttle setting. However, this method cannot detect anomalies independent of EPR. Two 'gas path' methods, namely the MMM and the SGTM (Hughes 1981) are suitable for real-time thrust computing, the latter method having been developed to overcome the difficulties imposed by complex traditional techniques.

The general concept of these methods is to relate real nozzle performance to that of an ideal nozzle through use of empirically established coefficients. These coefficients account for factors such as three-dimensional, friction and instrumentation effects.

A mathematical model of the Boeing 747 aircraft and of a two-spool turbofan engine with a thrust rating similar to the Rolls Royce RB-211 were employed to investigate both methods. The hot and cold flows in the engine are considered as separated propulsive jets. A schematic diagram of the engine is given below.

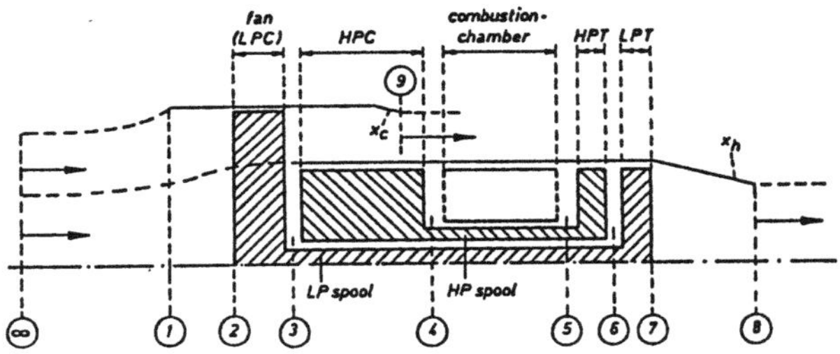

Fig.2. Two-spool turbofan engine.

3.2 Mass Momentum Method

This method essentially determines the force caused by the change in momentum of the fluids passing through the engine and a derivation of the equations is included in (Khatwa 1990). The nozzle is considered unchoked whenever

$$\frac{P_E}{p_\infty} < \left(\frac{\gamma - 1}{2}\right)^{\frac{\gamma}{\gamma-1}}$$

and the basic thrust equation is

$$F_G = \frac{2A_E p_\infty \gamma C}{\gamma - 1}\left[\left(\frac{P_E}{p_\infty}\right)^{(\gamma-1)/\gamma} -1\right] . \tag{1}$$

For sonic nozzle velocities

$$\frac{P_E}{p_\infty} \geq \left(\frac{\gamma - 1}{2}\right)^{\frac{\gamma}{\gamma-1}}$$

and the appropriate thrust equation is

$$F_G = CA_E\left[\left(\frac{2}{(\gamma + 1)}\right)^{\gamma/(\gamma-1)}(\gamma + 1)P_E - p_\infty\right] . \tag{2}$$

The contributions to the total thrust from both core and fan streams must be established. Assuming isentropic flow, then the total pressure at the nozzle exit is equal to the gas total pressure at the nozzle entrance. Thus measurements of the turbine discharge total pressure (P_7), the total pressure in the by-pass duct (eg P_3, but preferably at a point near the exit) and the ambient static pressure fulfil the requirements of (1) and (2).

The gross thrust coefficient is defined as the ratio of the actual thrust measured on a thrust stand to ideal thrust obtained from an ideal nozzle at a given pressure ratio, ie

$$C = \frac{F_M}{F_I} . \tag{3}$$

In practice the data is collected up to the maximum available nozzle pressure ratio in the ground test. For higher nozzle pressure ratios that are available in flight, the data are either extrapolated or further testing is required in an altitude facility.

3.3 Simplified Gross Thrust Method

This method was originally developed to calculate gross thrust of an engine incorporating afterburners. The equations modified for the nonafterburning engine are derived in (Hughes 1981). For an unchoked nozzle

$$\frac{P_x}{p_\infty} < \left(\frac{\gamma_x - 1}{2}\right)^{\frac{\gamma_x}{\gamma_x-1}}$$

$$F_G = \frac{2\gamma_x}{\gamma_x - 1} A_x p_x \left(\frac{P_x}{p_x}\right)^{\frac{\gamma_x-1}{2\gamma_x}} \sqrt{\left(\frac{P_x}{p_x}\right)^{\frac{\gamma_x-1}{\gamma_x}} - 1}\ \sqrt{1 - \left(\frac{p_\infty}{P_x}\right)^{\frac{\gamma_x-1}{\gamma_x}}} . \tag{4}$$

For the choked nozzle

$$F_G = \frac{2\gamma_x}{\gamma_x - 1} A_x p_x \sqrt{\left[\left(\frac{P_x}{p_x}\right)^{\frac{\gamma_x-1}{\gamma_x}} - 1\right]} \left(\frac{P_x}{p_x}\right)^{\frac{\gamma_x-1}{2\gamma_x}} \left\{\sqrt{\frac{\gamma_x - 1}{\gamma_x + 1}} + \frac{\sqrt{\frac{\gamma_x - 1}{2}}}{\gamma_x}\left[\left(\frac{2}{\gamma_x + 1}\right)^{\frac{\gamma_x}{\gamma_x-1}} - \frac{p_\infty}{P_x}\right]\left(\frac{\gamma_x + 1}{2}\right)^{\frac{\gamma_x + 1}{2(\gamma_x-1)}}\right\} . \tag{5}$$

For both the above cases the value of P_x is calculated as

$$P_x = P_I\left\{1 - \left(\frac{\gamma_x}{\gamma_x - 1}\right)K\left[\left(\frac{P_I}{p_x}\right)^{\frac{\gamma_x - 1}{\gamma_x}} - 1\right]\right\} \tag{6}$$

The nozzle discharge static pressure measurement, p_x, is made at a point 'x' inside the nozzle. See Fig. 2. Thus in addition to all measurements required for the MMM, the SGTM demands knowledge of the pressure p_x. The empirical parameter that adjusts the calculated gross thrust to equal the measured gross thrust is K. Ground thrust stand/altitude facility tests are conducted to obtain plots of p_x and P_I. The factor K in (6) is iterated until a match is obtained between calculated and measured gross thrust for values of p_x/P_I.

Ideally the ratio of specific heats, γ, should be obtained from a table-look-up as a function of the exhaust gas temperature, ie

$$\gamma = f(T_E) \quad . \tag{7}$$

Thus additional instrumentation is required but if the temperature probes are not present then a constant γ can be adopted and its effect corrected by the nozzle coefficient (Hughes 1981).

3.4 Reference gross thrust

An engine performance deficiency is annunciated if the difference between the reference gross thrust for a healthy engine and actual thrust developed exceeds a preselected threshold. The reference gross thrust must be indicative of the aircraft flight speed and ambient conditions for the recommended power setting and is obtained from an empirical model of the engine by virtue of several table-look-ups, ie

$$F_R = f(M, T_\infty, p_\infty, \delta) \quad . \tag{8}$$

Inputs such as Mach number and ambient air total temperature and static pressure are extracted from the ADC. A decision making algorithm considers factors such as number of engines failed, stopping distance remaining, and the relative positions of the 'can go' and 'must go' bars prior to generating 'go' or 'abort' warnings.

4 Results and discussion

The engine developed for this study was operated at numerous conditions to establish the equivalent theoretical thrust stand data outlined above. A comparison is made of calculated and measured gross thrust for both the MMM and the SGTM at full power in Fig. 3 for standard sea level conditions. Fig. 4 illustrates reference and measured gross thrust for an engine experiencing a reduction in intake efficiency as might occur in the case of icing or foreign object ingestion. Fig. 3 indicates good agreement between measured and actual thrust, and Fig. 4 confirms that the EPM has the capacity to detect and annunciate engine malfunctions, including those independent of EPR. The response of the TOPM algorithms to a rather more severe engine malfunction is illustrated in Fig. 5. Lengths such as V_{Rc}, S_{v1} and S_c are lengthened.

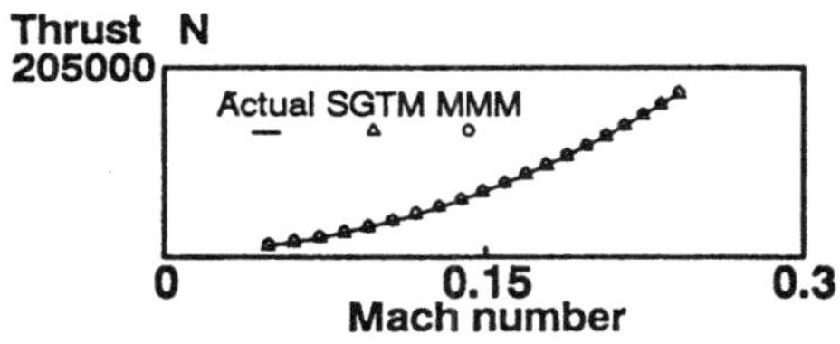

Fig.3 In-flight thrust measurement

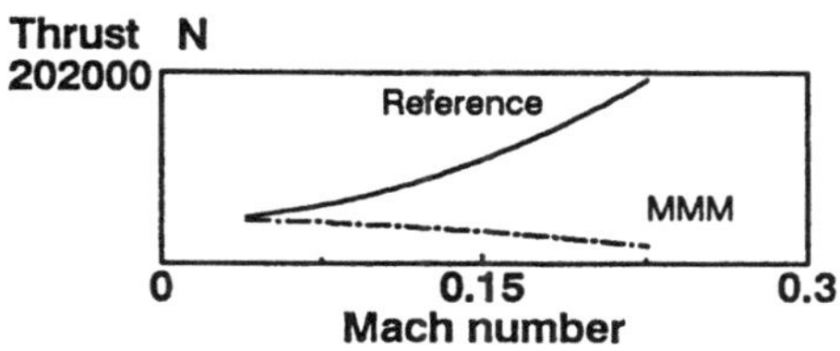

Fig.4 Engine malfunction detection

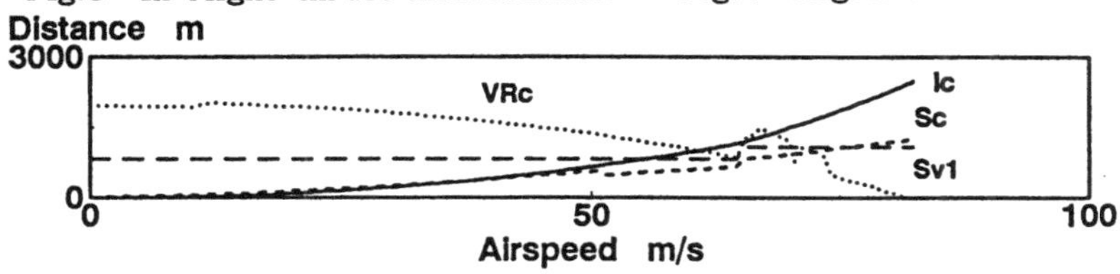

Fig.5 TOPM response to engine failure

These techniques of assessing the take-off progress would warn the crew that the airplane performance is at a lower-than-normal rate.

5 Conclusion

Detection of any shortfall in the performance early in the take-off run would thus not only prevent a high speed overrun, but would aid the pilot in judging the point at which an undisturbed take-off were both possible and safe. These analytic studies and extensive flight tests by NASA/Computing Devices (Canada) Ltd confirm that in-flight thrust measurement is feasible and consequently the EPM can enhance the role of the TOPM.

6 References

Hughes, D.L. (1981) **Comparison of Three Thrust Calculation Methods Using In-Flight Thrust Data,** NASA TM-81360.

Khatwa, R. (1990) **The Development of a Take-Off Performance Monitor,** PhD Thesis (in preparation), University of Bristol.

Srivatsan, R. et al. (1987) Development of a takeoff performance monitoring system, **J. Guidance and Control,** 10, 433-440.

7 Appendix A. Nomenclature

A	area	m^2	P	total pressure	N/m^2
C,K	nozzle flow coefficient		T	total temperature	K
M	Mach number		γ	ratio of specific heats	
F	thrust	N	δ	throttle position	deg
p	static pressure	N/m^2			

Subscripts

E	nozzle exit	1-7	engine stations
I	nozzle inlet	R	reference condition
M	thrust stand measurement	∞	ambient conditions

Condition/performance monitoring of internal combustion engine by continuous measurements of crankshaft twist angle

J.T. Lenaghan and A.I. Khalil
Humberside College of Higher Education, Hull, UK

Abstract
This paper describes a new technique to be used for the condition/performance monitoring of internal combustion engines. It relies on successive measurements of the crankshaft twist angle using optical sensors in conjunction with a microcomputer.

Initial results obtained from a rotating shaft driven by an induction motor and loaded by a dc generator were comparable with those of the theoretical model. Similar measurements from a crankshaft are being carried out to derive information relating to engine health.
Keywords: Crankshaft Monitoring, Twist Measurement, Torsion, Engine Health.

Notation

- α angular acceleration
- θ crank angle
- Θ angle of measurement zone
- Φ initial angular displacement between the strips of film
- φ twist angle
- Ω a shaft uniform angular velocity
- ω angular velocity of (crank)shaft
- N number of zones per revolution
- t_1 time for a rotation of Θ
- t_2 time interval between measuring points across shaft

1 Introduction

A great deal of interest has been shown in recent years by many diesel engine operators in utilising condition/performance monitoring for predictive maintenance activities. In response, many academic researchers and instrumentation manufacturers have undertaken to search for new ideas and develop advanced measuring systems. The most significant development has been made through the advent of high speed microprocessors. Vibration monitoring of machinery is probably the main beneficiary of this advancement, yet application of this technique to internal combustion engines remains troublesome. This is due to the complexity of the vibration spectra and the numerous exciting forces that exist in an operating engine.

The present work is an attempt to ease the problem of engine vibration monitoring by employing a different sensing device from that conventionally used in this field. The new technique is based on the successive measurements of the crankshaft twist angle with the help of a microcomputer. One immediate result of the analysis of these measurements would be directly related to the state of the crankshaft. The torque, which is a measure of engine performance, transmitted to the crankshaft would also be monitored this way.

This method has so far been examined on an induction motor drive shaft loaded by a dc generator. The results obtained were found to be comparable to those of the theoretical model of the shaft. This finding, the difference in the values of stiffness and other related factors were used to assess the feasibility of applying the technique to an engine crankshaft.

This proposed technique has the advantages of being relatively inexpensive, non-intrusive and tolerant to the engine's operating environment.

2 Instrumentation

2.1 Theory

This section includes a brief list of the fundamental relations on which this technique is based.

The measurement of twist angle is calculated from the ratio of the time between similar events at each end of the shaft (t_2) and the time (t_1) for a predetermined angle of rotation (θ) as illustrated in Figure 1. This can be represented as follows:

$$\varphi = (t_2/t_1)\theta. \qquad (1)$$

The above equation was derived from the truncation of a series to its first term. The inaccuracy in twist angle due to this mathematical

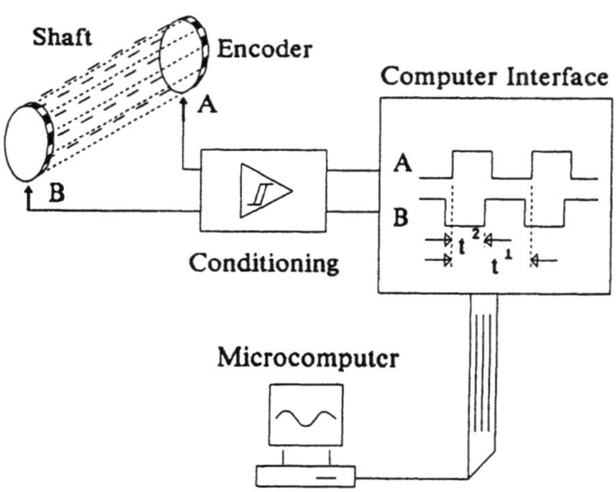

Fig. 1: Transfer of Data from Shaft

approximation is given by

$$\varepsilon = (1/8)\,|\alpha/\omega^3|\,\Theta^3(1/t^1). \quad (2)$$

The error arising from this inaccuracy becomes significant only under severe transient effects.

2.2 The Transducer

The transducer comprises two strips of photographic film, developed with an integral number (20 in the case of the development rig) of clear and opaque zones, mounted on a reflective backing and wrapped round the shaft circumference. These strips are illuminated by infra-red LEDs appropriately situated. A reflective type opto-electronic switch was used to generate an electrical signal relating to the instantaneous position of the shaft. The switch was held by an adjustable bracket close to a film strip. A similar arrangement was made at the other end of the shaft.

The signals produced from the opto-switches were transmitted to a remote box containing the conditioning circuitry linked to the microcomputer interface, see Figure 1. The conditioning circuitry is required to convert the analogue signals received to TTL voltage levels. The interface consists of the microprocessor interrupt circuitry, a 2MHz crystal oscillator and two 16-bit counters.

2.3 Signal Processing

The signals acquired were processed in order to obtain information related to the twist angle. This was carried out in the software according to the following reasoning.

Consider a shaft in uniform angular velocity, Ω. The time for one revolution, T is given by

$$\sum_{i=1}^{N}(t_1)_i = T. \quad (3)$$

Hence the intervals measured are on average, T/N. The strip misalignment will cause these intervals to differ by a factor k_1 so that

$$(k_1)_i\cdot(t_1)_i = T/N. \quad (4)$$

To determine the twist angle it is necessary to find the initial angular displacement between the two strips, Φ, which is derived from

$$\Sigma(t_2)_i = \Phi T/\Theta. \quad (5)$$

The time interval, t_2 will also vary by a factor, k_2 from the ideal such that,

$$(k_2)_i\cdot(t_2)_i = \Phi T/\Theta N. \quad (6)$$

Equations (3)-(6) are used in the calibration of the

instrumentation. This process involves the calculation of the values of the constants, $(k_1)_i$ and $(k_2)_i$, and the initial angular displacement of the strips, Φ.

The angular displacement, including that due to twist effect, between the two strips is calculated from equation (1) using the appropriate substitutions. The difference between the calculated value and the initial displacement would represent the twist angle of the shaft. However, the inaccuracy in the measurement of the time intervals which in turn causes a corresponding error in the value of the measured twist angle is estimated by

$$\delta\varphi = (\sqrt{5})\Theta(\delta t/t^1). \quad (7)$$

The signal processing has been implemented in the software which is also used for the acquisition and analysis of the results.

2.4 The Test Rig

Figure 2 shows a schematic diagram of the rig. It comprises, in addition to the elements described previously, a 20mm diameter bright draw mild steel drive shaft, a 4kW squirrel cage induction motor and a separately excited field dc generator which is coupled by belts and pulleys to the shaft. This arrangement was made in order to examine the twist angle of the shaft under different loading conditions which were set by computer controlled thyristor switching. In constructing the rig provision was made to ensure that the effects of mechanical and electrical interference with the measuring instrumentation were reduced as much as possible.

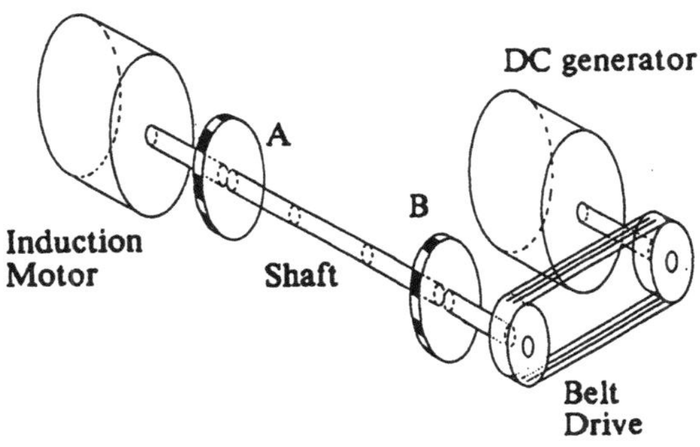

Fig. 2: Development Rig Schematic

3 Results and Discussion

As mentioned before, the technique of successive measurement of the twist angle was examined using the rig described in section 2.4.

Prior to starting the test, a calibration procedure was conducted by running the shaft without load at a constant speed and using the method of section 2.3.

The twist angle of the shaft was measured under both transient and steady state running conditions. Figure 3 shows results obtained under induction motor direct-on-line starting. A maximum value of 12 degrees, indicating a severe transient effect, can be seen at the beginning of the test. This is followed by fluctuating lower readings of the twist angle which eventually depict a steady state condition. Under full load the value of this angle is found to be about 2 degrees. Figure 4 shows results obtained from the theoretical model based on previous work, Jones(1967) and Cowburn(1984). The graph compared favourably with that of Figure 3 plotted using results obtained by the successive measurement of the twist angle. Similar graphs were obtained under different loading conditions. It is therefore reasonable to suggest that the technique was used to monitor both the behaviour of shaft and the running conditions under which it was driven.

For this application, the instrumentation was adequately used with an absolute resolution of 0.1 degree twist angle. Considering the full load steady state condition this gives a measuring dynamic range of 26dB. This value was used as a basis for verifying the feasibility of applying the technique to an engine crankshaft. To do this the first step was to estimate the maximum angle of twist exhibited by the shaft. A theoretical model from Carter(1928) was used for this purpose and a value of 0.5 degree twist angle was estimated. The second step was to use the measuring dynamic range obtained previously to estimate the new absolute value of resolution

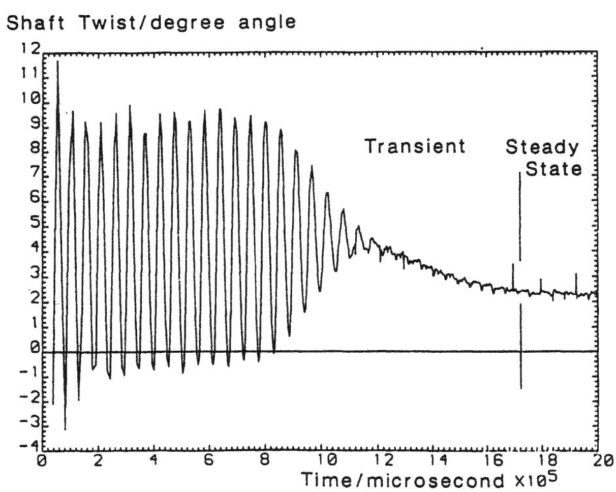

Fig. 3: Direct on line start up. Measured twist.

required for engines' application. Accordingly, a four-fold reduction was needed in order to allow for the much higher stiffness of the crankshaft chosen for the next step of the test program. In fact, a ten-fold reduction was achieved and arrangements are being made to apply the successive measurements technique on a 2300cc petrol engine available in the College.

4 Conclusion

A new method of monitoring the condition/performance of internal combustion engines was presented. Results obtained using the new technique on a specially designed rig were discussed. It is concluded that the application of this technique to engine crankshafts is feasible.

5 References

Carter, B. C. (1928) Empirical Formula for the Stiffness of Crankshafts in Torsion. **Engineering,** July, 36-39.

Jones, C. V. (1967) **The Unified Theory of Electrical Machines.** Butterworth, London, UK.

Cowburn, D. (1984) **The Mechanical Performance of Automotive V-belts.** PhD Thesis, University of Bradford, UK.

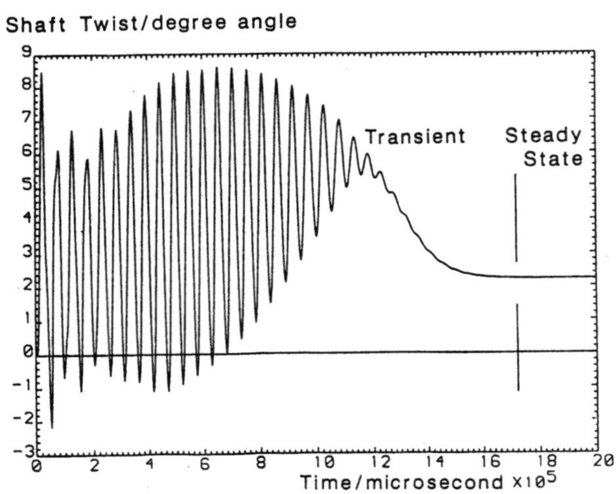

Fig. 4: Direct on line start up. Modelled twist.

Rotating machinery noise and vibration study

G.H. Lim
School of Mechanical and Production Engineering, Nanyang Technological Institute, Singapore 2263

Abstract
A project has been set up and implemented at the School of Mechanical and Production Engineering for the study of rotating machinery noise and vibration. The aim of the project is to allow students to analyse the vibratory displacement, velocity and acceleration spectra when the machine is running at certain speeds, to compare and correlate the noise and vibration signals and to determine the characteristic frequencies of various rotating machine components.

This paper describes the experimental setup, the kind of results that have been obtained and the conclusions drawn. It is found that the frequency spectra for the machine running at 240 rev/min and 480 rev/min contain many characteristic component frequencies which centre at these two machine running frequencies. These are due to the unbalance and misalignment of the shaft. The frequency spectrum due to noise at the same machine frequencies did not produce identical spectrum as that due to vibration. However, it is observed that when the vibration level is increased, the noise level also increases.
Keywords: Rotating Machinery, Dynamic Analyser, Spectrum Analysis, Vibration, Noise.

1 Introduction

Vibration and noise are invariably produced when a rotating machine is running. In recent years, there have been considerable interest in the maintenance techniques based on the monitoring and analysis of vibration

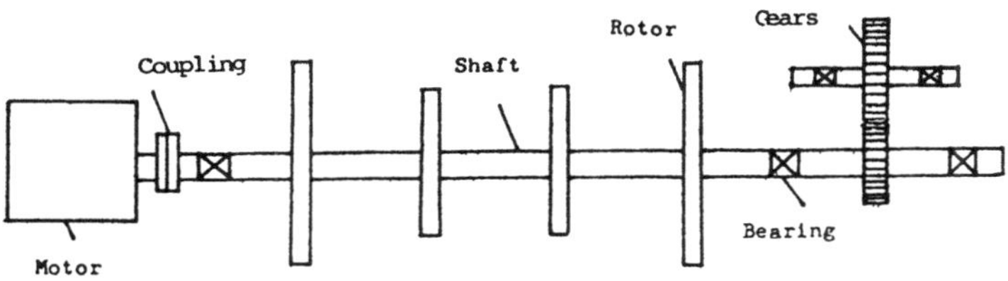

Fig.1. The rotating machine.

characteristics generated by machines. Although the condition monitoring and analysis of a machine will increase the operational safety level, it will not be able to prevent failures. What it will do is to enable the progressive deterioration of a machine to be detected and for appropriate remedial action to be initiated. (Collacott, 1977)

Recognizing the importance of introducing engineering students to the science of rotating machinery condition monitoring, a project has been setup and implemented in the laboratory. This paper describes the experimental setup and presents the main results that have been obtained.

2 Experimental setup

The experimental rig was designed to simulate an industrial plant. A typical industrial plant may consist of a generator, a turbine, a shaft, a reduction gearbox and bearings. Figure 1 shows the experimental rig fitted with the following key components: an electric motor, a shaft supported by three bearings and connected to the motor via a coupling, four rotor disks placed at fixed intervals along the shaft between two bearings, and a reduction gear attached at the end of the shaft.

Some damaged parts are also available for the experimental study. These include a gear with a damaged tooth, a worn bearing and a cracked shaft.

3 Experimental measurements

3.1 Instrumentation

The following instruments are available for use in the tests:

- a Spectral Dynamics Dynamic Analyser (SD 375). This is a microprocessor based dual channel frequency domain dynamic analyser
- accelerometers and signal processing units
- proximity probes and signal processing units
- a sound level meter
- a vibration meter
- a tachometer
- a 1/3 octave band pass filter
- an oscilloscope
- a x-y plotter

3.2 Procedures

The rotating machine is initially set running at say 240 rev/min. Using the sound level meter, which is connected to the Spectral Analyser, the noise spectrum on real time and on 1/3 octave band filter is produced. The vibration level meter is then used, set to internal filter. The rms values of displacement, velocity and acceleration are recorded. The vibration level meter is then set to external filter and using the 1/3 octave band-pass filter, the rms values of displacement, velocity and acceleration are also recorded. Subsequently, the displacement of the shaft is measured using the proximity probes.
The whole procedure is then repeated with the machine running at say 480 rev/min.

4 Results and discussion

4.1 Correlation between noise and vibration signals

It is generally accepted that noise is produced as a result of the vibration of components. However, this does not necessarily mean that for each peak that is displayed in the spectrum for noise, a corresponding peak in the spectrum for vibration will be observed.

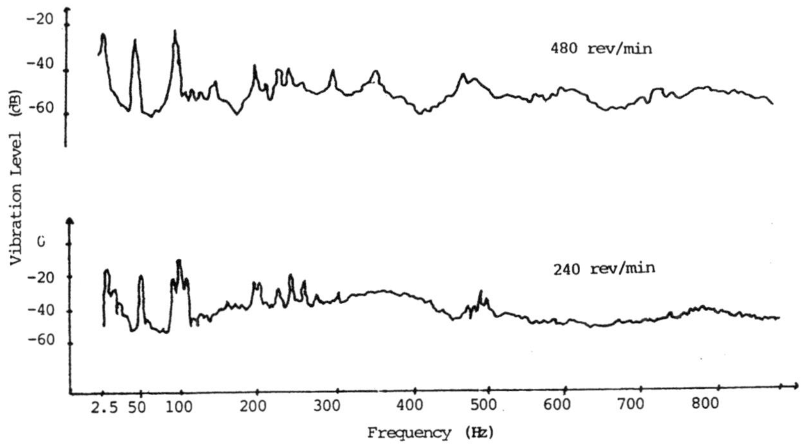

Fig. 2. Vibration spectrum in terms of velocity.

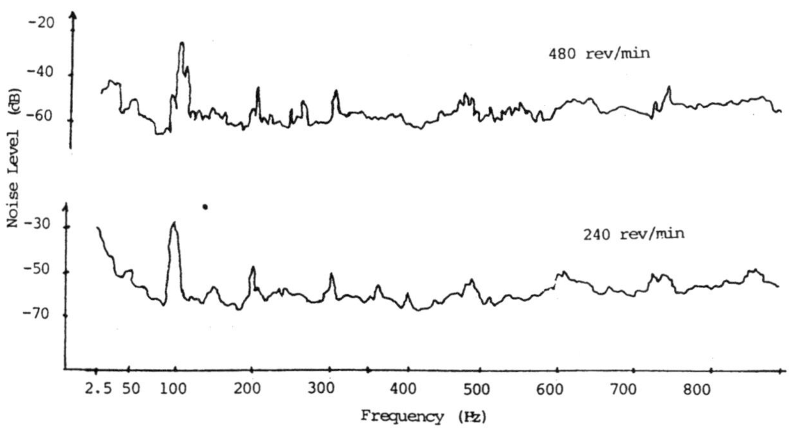

Fig. 3. Frequency spectrum in terms of noise.

It is apparent that given two vibration sources of the same amplitude and frequency, the noise produced by these sources need not be of the same form when analysed using the spectrum analyser. This is because different materials radiate noise differently even though they may be similarly excited. The spectrums for vibration and noise, as shown in Figure 2 and Figure 3 respectively, show that there is no direct correlation at the motor speed of 240 rev/min. In fact, superimposing the two spectrums shows that at low frequencies, where there were sharp peaks in the vibration spectrum, the noise spectrum yielded broad peaks. The intensity of the peaks for the vibration spectrum at 2.5, 50 and 100 Hz showed relatively similar values whereas for the noise spectrum, this was found to be varying.

4.2 Vibratory displacement, velocity and acceleration spectra

It is known that the vibratory velocity leads displacement by a phase angle of 90° and acceleration leads velocity also by a phase angle of 90°. If the phase angle is to be neglected, then the velocity can be obtained by dividing the acceleration signal by a factor proportional to frequency; the displacement obtained by dividing the acceleration signal by a factor proportional to the square of the frequency. (Broch, 1984)

The velocity spectrum in Figure 2, shows peaks throughout the frequency range of 2.5 to 880 Hz. In fact, the velocity spectrum is considered to be an ideal choice for frequencies ranging from 10 to 1000 Hz. For this reason, the vibratory velocity spectrum was chosen to be representative of the variations in vibration when compared with the noise spectrum, as discussed under section (4.1).

4.3 Characteristic frequencies of components of machine

Table 1 shows the characteristic frequencies of some of the components of the rotating machine. It can be seen that most of the frequencies are centred around the low frequency region.

The characteristic frequency due to the gears can be readily traced on the frequency spectrum at 240 Hz. Another peak which could be seen at 480 Hz is due to the harmonics of the characteristic frequency of the gears.

The velocity spectrum in Figure 2 registered many peaks in the middle frequency range. The first peak at 50 Hz could be due to the mains supply frequency. Thus, electrical installations not properly shielded would affect the spectrum in this way. The rest of the peaks were due to harmonics of the mains frequency.

Table 1. Characteristic frequencies of components at 240 rev/min

No.	Nature of fault	Frequency of dominant vibration (Hz)
1	Rotating members out of balance	4
2	Misalignment and bent shaft	4, 8
3	Mechanical looseness	4, 6, 8, 10
4	Unbalanced reciprocating forces & couples	4, 8
5	Electrically induced vibration	4
6	Damaged or worn gears	240, 480
7	Damaged rolling element bearings	More than 1000

5 Conclusion

An in-house experimental rig, built for the study of rotating machinery noise and vibration, has been described. It is shown that spectrum analysis displays identifiable frequencies for the rotating elements and therefore changes at a given frequency can be pinpointed directly to the rotating component concerned.

Furthermore, this study highlights the need for careful fault diagnosis to avoid unnecessary and costly maintenance shutdowns in industrial plants.

6 References

Broch, J.T. (1984) Mechanical Vibration and Shock Measurement. Brüel and Kjaer, 2nd edition.

Collacott, R.A. (1977) Mechanical Fault Diagnosis. Chapman & Hall, London.

Vibration analysis as tool for computerised machine monitoring

G. Luelf and Dr R. Vogel
Thyssen Stahl AG, Duisburg, Germany

Abstact
Machine condition monitoring under the aspect of production quality and installation characteristics will be an increasing fact for the view of profitability. From todays point of view condition monitoring is only practical by using certain measurement methods. The registration of machine vibration seems to be very helpful, because with this kind of measurement it is possible to register the immediate condition of the system and even the developement of abrasion during running time. Examples from the iron and steel industries demonstrate, how easy it is to recognise the installation characteristics and estimate the condition of the system.
Keywords: Condition Monitoring, Installation Characteristics, Frequency Domain Fast Fourier Transformation, Spectrum, Cepstrum, Correlation, Trending

1. Introduction

The continuous co-operation of different manufacturing plants and processes is characteristic for production flow in the iron and steel industries. The undisturbed production is an essential feature for productivity, quality of the products and profitability. Therefore maintenance of the very high standard systems took more importance during the last 10 to 15 years. Higher availability of the plant equipment is the keyword for the nearest future. A way to reach this aim is a change in the maintenance philosophy, a change from the preventive to the predictive maintenance. Basis of such a procedure are guarded informations on:

- the actual condition of the plant including the knowledge of recived loads,
- the calculation of influence of maximum load and load cycles
  - the behaviour of demage by known defects

  With the help of suitable software it's possible to give a certain prognose about the plant capability. Two solutions are leading to this aim:

1. The use of nonparametric simulation models of the pruduction process and the plant enables, with a provided model adaption (accommodation to the real system), the theoretical calculation of loads, which are nessesary to run the process. These results will later be the basis of calculation determined statements about the

responsibility of the plant.
2. The observation of plant and process by measuring and recording the effektive load and the conversion into load collectives or into trendings.

The analysis of this measurements is the basis of further calculations with regard to possible wear or start of failures in plant parts. A few universities are trying to solve the methods and it is to belief, that getting a reliable system to view the production line will need a few more years. Meanwhile some measuring- and analysis techniques have been developed, in order to realize with a high reliability the application of condition monitoring in the plant.

2. Ascertainment of plant load

For ascertainment of the plant load, there are available, as a rule, a multitude of physical data from control and measuring systems (e.g. current, speed, temperature. etc.), as well as from special ratings (e.g. torque, acceleration etc.). Their informational content, however, is highly varying as is apparent in **Fig. 1** from the recorded signal curves.

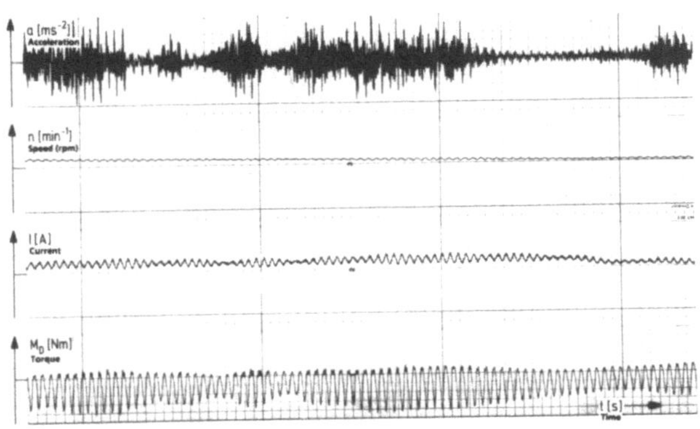

Fig. 1: Signal-time curve of torque, current, rpm and linear acceleration on roller bearing

- Time curves for torque, current, and rpm are characterized by a sinusoidal base frequency which, in the acceleration time curve, is recognized only in the area of very low amplitudes.
- Amplitudes of the torque are greater than the static threshold, slackness effects at the load arrest point are visible.
- The acceleration-time curve shows a superposition of high-frequency amplitudes by low-frequency components.

This causes a certain "modulation" of the signal curve which is also found again in the signals of torque and current.

No direct connection or conclusion as to the cause of vibrations can be perceived from these recordings; the frequency of visible sinusoidal vibrations (f=45Hz) is far above the number of revolutions of the drive (f=10Hz); it cannot result from the load either since a constant load is lifted. This dynamic force which is very high for a mechanical drive leads to a constant lifting and abutting of the gear tooth flanks, the so called 'hammering' which is an audible action.

Experienced engineers and operators know such phenomena and are able to make statements regarding the plant condition by touching or by listening to running noises. Man as an analyser of vibrations is, however, much too subjektive in his evaluation and easily fooled by interference effects.

3. Analysis in the frequency domain

The signal analysis in the frequency range is a highly meaningful possibility for an objektive evaluation of measurements; however, it cannot be interpreted detached from the plant.

The signal analysis within the scope of plant control is now no longer done in the time range but rather, as already shown, in the more meaningful frequency range. By means of an FFT (Fast Fourier Transformation), all time signals are tranformed into frequency or power density spectrums **(Fig. 2)**.

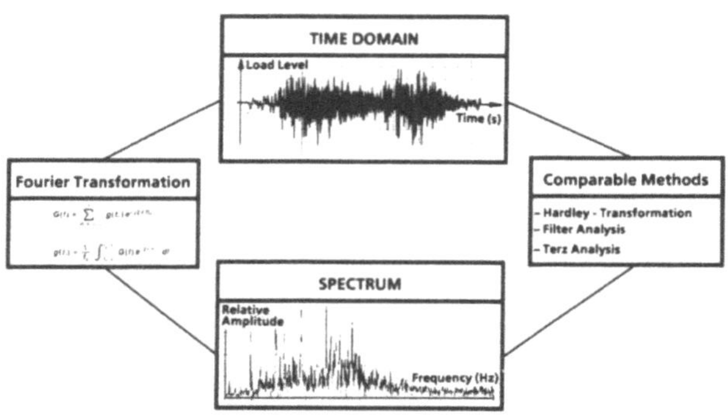

Fig. 2 : From Time to Frequency Domain

Related to the signal-time curves of **Fig. 1**, this means four power density spectrums according to **Fig. 3**

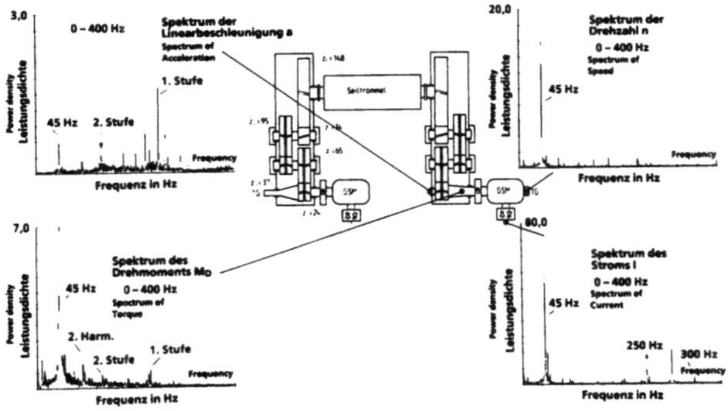

Fig. 3: Power Density Spectrums of the Signal Time Curves of **Fig. 1**

- At 45 Hz, all spectrums have a dominant amplitude value.
- In ranges above that, only the current (at 250 and 300 Hz) and the acceleration measuring point showrelevant amplitude values.
- At the acceleration measuring point, the highest amplitudes are between 200 and 300 Hz.

4. Computerized analysis in the frequency domain

Dependencies are not directly visible from these spectrums; direct assignment of influencing variables from the mechanical drive and the control and measuring systems is not possible. The fast fourier transformation and analysis techniques, setting up on the results of the FFT, have been generally accepted now. The use of computerised measuring analysis methods leads to the possibility, to take different procedures into account and to bring different views and informations into the analysis. **Fig. 4** shows the widespread procedures used for the frequency analysis.

5. Examples from the iron and steel mill area

The applicability of the presented procedure for monitoring the actual plant condition by means of a vibration analysis is shown by way of presenting examples from the iron and steel mill area. The use of this monitoring techniques in our woks occurs essentially on plants with anusual faults in time. An example shows a directly driven ventilator with a floating fan. Reason of the temporary damage was the break down of roller bearings on the shaft and in the driving engine.

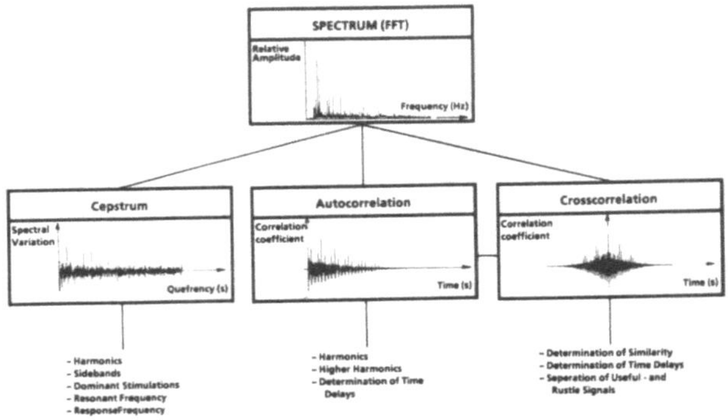

Fig. 4: Analysis Techniques in the Frequency Domain

The measurement of acceleration with linear accelerometers (15 kHz linear range) took place near to rotorshaft bearings and to the drive engine bearing. the analysis results are shown in **Fig. 5:**

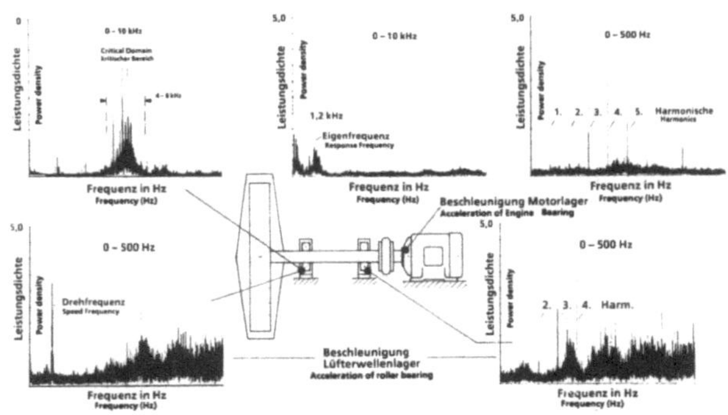

Fig. 5: Ventilator with Floating Fan

Current, torque and rpm where nessesary for reference.

- The bearing beside to the fan has the highest amplitudes in the range of 6 kHz. High amplitudes in this area means defects of roller races due to pittings.
- in the range of drives rotating frequencies is only the speed frequency extremly high, that means an unbalanced fan.
- The measurements of the shaft rooler bearings give no hits about failures on the bearings, but the exsistance of a single peak at

the speed frequency and their higer harmonics points out a typical sign of failure in adjustment of fanshaft and drive motor.
Using only one measurement this example shows three main failures. The following example will show one failure with three different analysis methods - spectrum, - cepstrum, - autocorrelation **(Fig. 6)**.

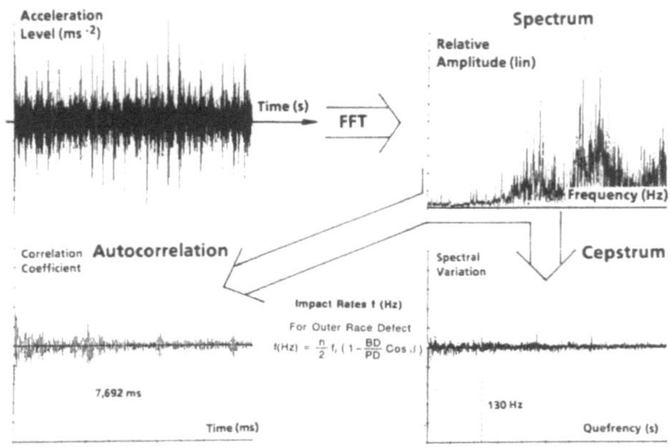

Fig. 6: Methods of Analysation

The figure shows at the top left the acceleration time signal with a dominate 130 Hz frequency, which is not found in the 10 kHz spectrum. Here are the highest amplitudes in the range between 6 and 7 kHz. In contrast to the spectrum the cepstrum and autocorrelation obvious marks the 130 Hz frequency. Using the expression for the impact rate, the failure is to relate to the bearings outer race.

The damage of roller bearings by material abrasion leads to another phenomenum in the spectrum. Increasing abraison means the developement of pittings on inner and outer race. The overrolling ball will stimulate the inherent frequencies of the bearing, increasing amplitudes in this range of the spectrum mark the increasing abraison in the bearing **(Fig. 6, top right)**. The inherent frequency ranges of roller bearings are between appr. 2 and 20 kHz in relation to the size of bearing.

Next step in the use of the methods of machine condition monitoring will be the installation of these techniques with a new process-technique in the hot strip mill area, which will take place very soon.

References:

Stearns,S.D (1979) Digitale Verarbeitung analoger Signale, Oldenbourg Verlag, München,Wien

Bruel & Kjaer (1988) Structural Testing, Part 1; Mechanical Mobility Measurement, DK BR 0458 - 12

Bruel & Kjaer (1985) Maschinenüberwachung, DK 0382-11

Using a PC based system for machinery transient analysis

C. Nicholls
Nikat Associates, Chester, England

A transient event is the unsteady-state response of a dynamic system to a changing excitation. Mechanical transients can occur during run-up or run-down of a machine, or when the machine's regular operation is disturbed by external or internal events. Mechanical Transient Analysis is the procedure by which the mechanical response of a machine to changing excitation is captured, displayed and examined.

For some machines, transients such as run-ups, run-downs and disturbances may be critical events. They require on-line monitoring and, if problems are detected, necessitate immediate decision making. The performance of vibration analysis on a critical machine during these transient events is a common practice in some industries and has several purposes:

1 Determine Machine Condition Before and After its Overhaul

Before overhaul, evaluating a machine's condition will uncover critical faults that can be fixed during the overhaul period. After overhaul, the machine run-up transient analysis is important for identifying faults that have not been corrected or that were created during the overhaul period.

Transient analysis can detect many rotating machinery faults including unbalance, mechanical looseness, structural resonances, cracks, rubs and instabilities. Transient analysis may detect the severity of those faults before start-up, thus avoiding possible catastrophic failures.

2 Determine the Effect of Disturbances on Normal Operation

During normal machine operation, sudden process changes may cause a machine's vibrations to change radically. Such changes can be captured and analyzed as transient events. Valve position, damper position, back pressure, flow rate and load are just few examples of the disturbances that may occur during normal operation that may cause signature changes as vibrations are excited near the natural frequency of the structure or the critical speed of the shaft. Transient analysis can help in locating the source of disturbance and once found, operational or mechanical parameters can be modified before damage occurs. Transient analysis can also determine the severity of the incident before a shut-down is ordered. Finally, transient analysis permits analysis of the trend in the frequency domain to help determine if the degradation is permanent or temporary.

3 Verify Machine Design and Specification

Analytical methods are usually used to assure that a new machine has a stable design. However, most of these methods are theoretical and, as such, are subject to inaccuracies in their interpret ation of how a machine might perform during actual operation. Transient analysis of a new machine, particularly during commissioning, is highly desirable (and sometimes required) in order to verify actual design parameters such as bearing and gear stability, shaft critical and structural resonance.

Transient analysis may detect many of the following rotating machinery faults.

3.1 Unbalance

Unbalance occurs when the mass of a rotor is not equally distributed about its rotating centreline. Although there may be many sources of unbalance in a rotor the resultant is centralised at a point called the heavy spot. This heavy spot causes vibration at 1X the running speed and is directly proportional to the amount of unbalance. As the running speed increases, the magnitude of vibration 1XRPM increases proportionally to the square of the speed.

3.2 Mechanical Looseness

Mechanical looseness is the result of loose mounting bolts, excessive bearing clearance or a cracked structure and can result in a large number of harmonics. As the running speed increases, the magnitude of vibration of the harmonics also increases but not uniformly.

3.3 Resonances

Resonances are excited when the running speed reaches the natural frequencies of either the shaft, the housing or the structure of the machine. The resonant frequencies of a machine will not change when the running speed changes. Resonance frequencies can be determined using Nyquist, Bode, or cascade plots.

3.4 Rubs

When an impeller, vane or blade rub exists, vibration occurs at the blade pass frequency or one of its multiples. If a shaft or seal rub exists, vibration occurs at half harmonics. The frequency of the rubs increases as running speed increases.

3.5 **Cracks**
From its initiation through propagation to failure, cracks are a difficult type of failure to predict. Changes in operating speed or load can affect the vibration levels associated with a crack. As the crack increases in size, the natural frequency of the rotor decreases. Shifts in rotor critical frequencies can correspond to crack growth.

3.6 **Instabilities**
Transient analysis focuses on the trend of the vibrational signature as the running speed increases or lapsed time changes. Bode and Nyquist plots will reveal the trend of the frequency signature and assist in predicting the stability of the system.

CRITICAL ELEMENTS OF A TRANSIENT SYSTEM

Several factors distinguish a well-designed system. Using a simple flow diagram, Figure 1 represents the flow of data from its pre-acquisition stage through processing and ending with its post-processing and diagnostic stage.

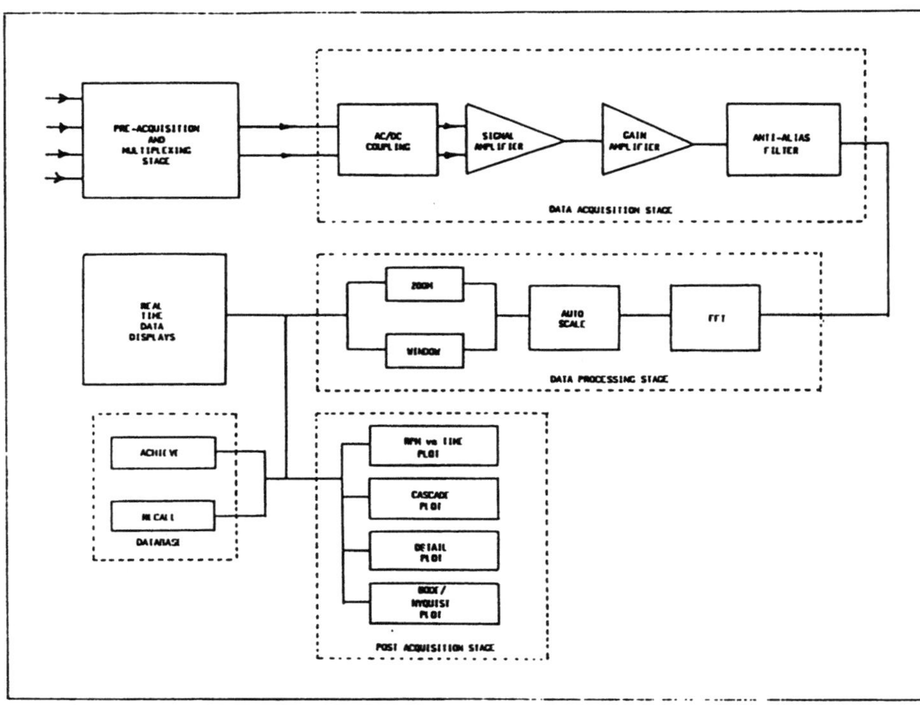

Figure 1

1 Pre-acquisision and Multiplexing Stage

Most transient measurements require the collection of data at more than two points during the event. This requirement dictates that the system either scan many parallel data channels or rapidly scan data in a sequential manner. The use of a multiplexer makes it possible to collect data on as many points as desired, at minimal system cost and with minimum interruption. More importantly, a multiplexer organises data collection procedures. Without multiplexing capability, the user is either limited to the number of channels resident in the physical transient system, or is confronted with developing an extensive (and probably expensive) custom multi-channel data acquisition system.

2 Data Acquisition Stage

The data acquisition phase of a transient event has to be fast and accurate. It is a fact that the accuracy in most of today's data acquisition systems far exceeds that required for analysis. Therefore, overall analysis is usually limited by the accuracy of the transducers being used.

Due to the nature of transient events, the speed of the acquisition system is always important. Speed is especially critical, however, when dealing with certain very brief and rapidly changing transient events. As such, top priority in the design of the system must be given to speed of data acquisition. While useful in subsequent analysis, other features such as graphic displays, signal con-ditioning and signal processing must not hamper data acquisition rates.

Input signals must be amplified during the data acquisition stage in order to reject noise signals from the input stage. In addition, gain may be required in order to amplify the signal before its digital conversion. Signal offset causes superimposition of low dynamic signals at large offsets. AC/DC coupling can be used to eliminate signal offsets. Finally, anti-alias filtering is an absolutely essential requirement for input signal conditioning. Without it, frequency components that are above the Nyquist frequency will not be removed and will 'wrap around' into desirable signal data. Also at this stage, signal conditioning is required in order to transform the input signal voltage to some other user defined engineering unit. Calibration factors should be carefully chosen to eliminate any significant overall error.

Triggering is the acquisition of data based on some significant event. This can be a signal in one of the channels being analysed or some event such as TTL poise or key phasor that trips the trigger. The user should have the option to start or stop the data collection on the trigger, or to choose the time trigger mode that will allow the triggering event to happen in the middle of some data window. Additional, helpful triggering features include the ability to specify the triggering level and the slope.

3 Data Processing Stage

One of the advantages of a PC-based transient analysis system is its ability to use internal memory to store and retrieve collected data, eliminating the need for a tape recorder. PC-based systems are also highly desirable because of their speed, flexibility and accessibility.

Once data has been retrieved from its database, the user may wish to display time waveforms or frequency spectra for diagnostic purposes. A PC-based system can be used to display such data, or other diagnostics that the user may prefer or need. For example, using a digital FFT, data can be transformed into the frequency domain. In this domain, each peak may represent one or more time domain component.

The system must be able to perform digital integration and differentiation. This capability allows conversion of data collected with any type of transducer into the most convenient or appropriate user-specified units.

The user should also have the option to choose the scaling units for both axes. The presence of an autoscaling feature eases the process of displaying the overall broadbank picture. In some instances, however, the user will need higher resolution over a smaller part of the graph. This can be accomplished through use of a zooming and windowing feature. Several other options should be made available in order to give the user extensive flexibility in generating and manipulating screen plots and displays. These include, but are not limited to, complete cursor capabilities, on-screen magnifying and on-screen reducing.

4 Real Time Data Display

Real-time displays are important because machines may experience excessive vibrations during a transient event. It is worth having the capability to view some characteristics of these vibrations. In addition, real-time displays assure the validity of the data. However, data acquisition speed should not be sacrificed for real-time display capability. It should be possible to display, among other things, synchronous magnitudes and phase of each channel, but again without interrupting data collection.

5 Data Base of Test Results

A database should be developed solely for storing and manipulating data collected during a transient event. Data files should be organized into categories representing date, time, machine, the plane and the points from which transient data was collected. Each category may represent data for a point, a machine, or a group of points that have the same characteristics. The user should then be able to place the data in order of priority. Such division of the database will ease and speed operations when storing and retrieving data.

6 Post processing and Diagnostic Display

Certain displays are essential for proper diagnosis of transient data:

a. RPM vs Time plot,
b. Detail plot,
b. Waterfall display,
d. Bode-Nyquist plots.

6.1 RPM vs Time plot

The plot in Figure 2 is a simple representation of machine speed in RPM, as captured by the key phasor. It is plotted relative to the start of data collection time. The graph represents the change of machine speed with respect to real time.

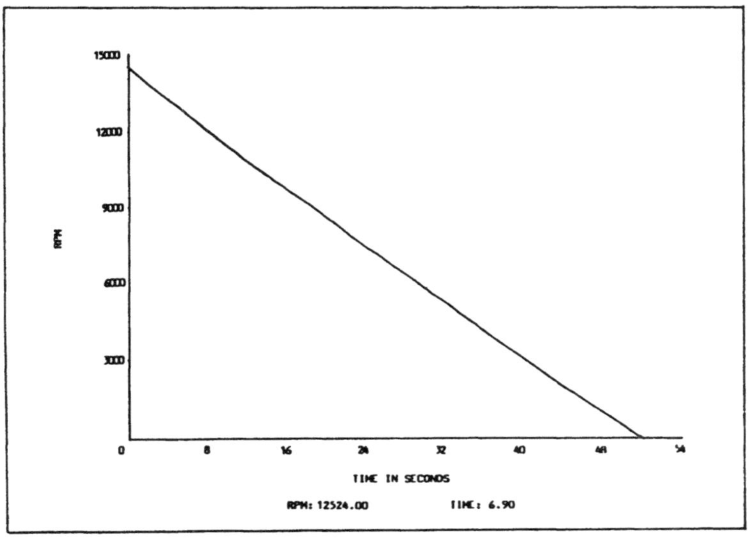

Figure 2

6.2 Detail plot

A detail plot is composite time plot. It allows the user to select a window from the whole waveform plot and to view it as an independent plot in the time domain, frequency domain or both. To get the most from a detail plot, it should be displayed on the screen adjacent to the entire waveform. This technique is illustrated in Figure 3. Detailed plot windows, selected by the user, show a more detailed plot with higher resolution. For two-channel transient systems, a two waveforms/two windows screen is preferred. The bottom window may be used to show a portion of the time domain data and/or an instantaneous or averaged frequency spectrum. As before the user should have full control over the partitioning of these screen displays.

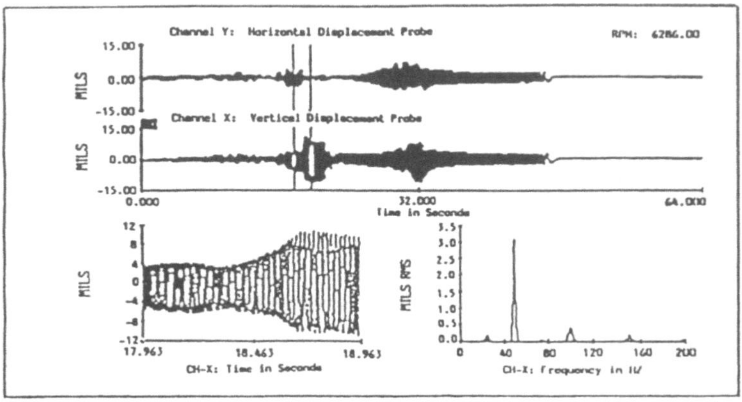

Figure 3. Detail time plot

6.3 **Waterfall Display**

The waterfall display, or spectrum map, gives the analyst a view of the machine characteristics relative to its speed or actual running time. The operator should have the option to plot as many spectra as needed on the same screen (again with full control over the plotting parameters). This is shown in Figure 4. Some of the control parameters that aid diagnosis include the number of lines in the spectra, the starting and ending RPM, starting and ending time, delta time or RPM between spectra, plot staggering and baseline suppression. Certain cursor capabilities, if available, speed up and ease the analyst's job. These include zooming, expanding, contracting, selection of any single spectrum from the waterfall display to be plotted alone and re-plot of the waterfall display.

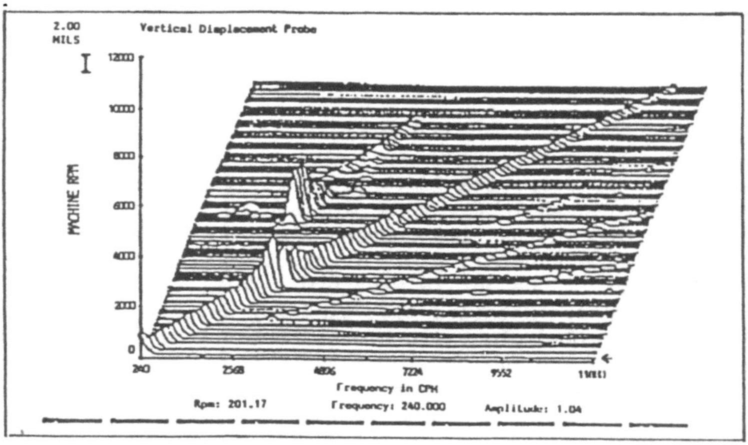

Figure 4. Waterfall display

6.4 **Bode/Nyquist plots**

Bode and Nyquist plots are important tools for analysing transient events. These plots display the syn-chronous amplitude and the synchronouse phase values of the transient relative to the machine speed (Figure 5). The number of points used in the plots is user specified, as is the selection of the harmonic of interest and the number of machine cycles necessary to compute each point. Additional desirable features are the ability to subtract the runout component from the amplitude and phase plots and the ability to toggle back and forth between Bode and Nyquist plots.

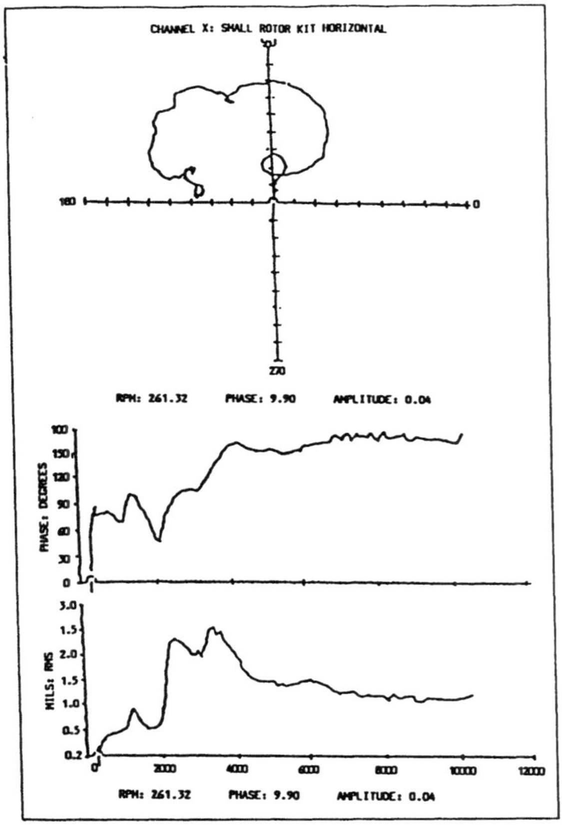

Figure 5

ADVANTAGES OF A PC-BASED TRANSIENT SYSTEM

Today's stand-alone ditigal transient systems have become highly sophisticated and somewhat multifunctional. However, they need the flexibility, practicality and power of a computer, along with special application software, in order to become an integrated part of an efficient problem solving system. PC-based transient systems reach or exceed the sophistication and functionality of stand-alone systems.

The PC-based transient systems are part of the unlimited expansion environment of the personal computer with all of its wide range of use and application. They have the potential to become a major tool for helping with engineering decisions and for providing engineering solutions. Several related systems can be linked together in order to establish a network between management, engineering and maintenance personnel.

1 Speed

The speed of transient analysis operation on any PC-based system is only limited by the speed of the computer used. In many industrial environments there are computer resources available that are already properly configured for this application; often, only memory modifications are required. A typical system would be XT or AT compatible and would have 20-40MB of hard disk and 12.5 M Hz clock. A maths co-processor is also required.

2 Memory

In capturing a long transient event, the amount of data that can be collected becomes critical. As a rule, most transient events require the storage of large amounts of data. Stand-alone systems are usually connected to a tape recorder in order to store the data collected. Personal computers are capable of storing megabytes of data directly to the data base, eliminating the cost and the complexity of transferring data from a multi-channel tape recorder.

3 User Interface

Some PC-based systems are database-oriented to facilitate the user's manipulation of the data collection procedure. Only that data which meets the user's specified requirements, specifications and conditions is stored in such a configuration. A system thus designed saves space, time and money for the user.

4 Price

In today's competitive PC market, mass production has resulted in a price decrease for true IBM-compatible computers. In addition, the amount of hardware required for PC-bases transient systems is moderate. Of course, many other applications are available for use with a properly configured PC. Thus considered, significant savings can be expected when purchasing a PC, as opposed to a single-function standalone spectrum analyser, for transient analysis.

References

Bitar, E., Pardue, E.F. and Sidlowski, D.G. (1988) Machinery Transient Analysis is Simplified when using a PC-Based Analysis System: Vibration Institute's 12th Annual Meeting, Session XI.

Doebelin, E.O. (1980) System Modeling and Response Theoretical and Experimental Approaches: Wiley & Sons, Inc.

Rothschild, R.S. (1987) The Second Revolution in Real-Time Spectrum Analyzers: Sound and Vibration, V.21,No.3,14-22.

Smith,Strether and McGinn, Sean (1987) the Design of High-Performance Data Acquisition Systems: Sound and Vibration, Vol.23,No.2,18-24.

Talmadge, R.D. and Banaszak, D.L. (November 1985) New Signal Analysis Package for Microcomputers: Sound and Vibration, 6-8.

Thanos, S.N. (September 1987) No Disassembly Required: Mechanical Engineering, 86-40.

Thompson, W.T. (1981) Theory of Vibration with Applications: Second Edition Prentice-Hall, Inc.

Condition monitoring of key elements in high speed textile machinery

J.M. Sharp
Department of Aeronautical and Mechanical Engineering, University of Salford, Salford, UK

Abstract
Knitting machines convert yarn into fabric using a latch-needle-cam system. The latch needles, cams and yarn are the key elements. The forces that arise in this system are critical in limiting machine speed and fabric quality due to either needle or yarn breakage. Typical high speed commercial knitting machines have between 2000 and 4000 needles of approximately 0.45mm in thickness. Just one damaged needle will severely affect the fabric quality therefore knowledge of needle condition is vital to maximise machine productivity.

Existing instruments can only detect broken needles. Several novel sensitive transducers were designed and developed to measure cam-needle and yarn-needle forces which were capable of identifying damaged needles. The transducers were controlled by microcomputer and along with other ancilliary instrumentation formed the basis of an on-line condition monitoring system of the key elements in high speed knitting machines.

The transducers used in the research were for one particular machine but can be further developed to form a portable condition monitoring system for use as a portable diagnostic tool on a textile production floor containing many knitting machines.
Keywords: Knitting, Textile, Forces, Transducer, Condition Monitoring, On-line Monitoring.

1 Introduction

The two principal methods of producing textile fabrics are weaving and knitting. The latter has a significant market share particularly in apparel applications, but also in furnishing, surgical, industrial and geotextiles (Dutton, 1981). The basic technologies are old in most cases such that a better understanding is necessary in order to be able to improve the knitting process. A research programme was undertaken to determine, both theoretically and experimentally, the dynamic forces in high speed weft knitting.

This paper describes how the transducers used in this programme can be used to monitor the condition of the needles and be encorporated into an on-line condition monitoring system.

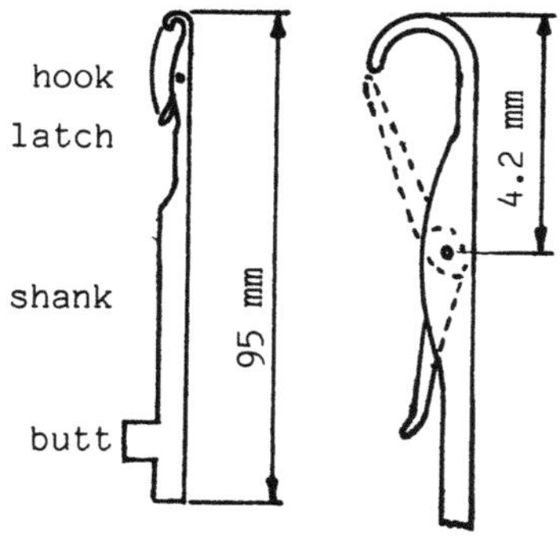

Fig 1 The latch needle

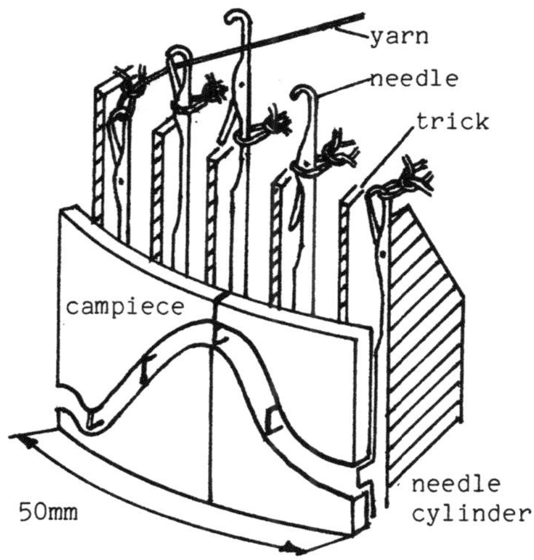

Fig 2 Circular knitting machine

1.1 The basic mechanism

The vast majority of weft knitting machines convert yarn into fabric using a latch-needle-cam system. A latch needle, shown in Fig 1, comprises of a hook, a protruding pivoted latch which is free to revolve on the circumference shown, and a butt which protrudes from the shank. Fig 2 shows the arrangement of needles in a basic circular machine. The needles are located in a needle bed in equally spaced vertical grooved slots known as tricks. The needles are not generally free to move but are restricted by trick-wall pressure and sometimes a band spring (not shown). The needle butts protrude from the needle bed and may be moved by the exertion of force.

To achieve the motion necessary for loop formation, relative motion must occur between the needle butts and the cam profiles. In the configuration illustrated, the cams are rigidly mounted and stationary and the needle cylinder revolves. Yarn is delivered to the needles via the yarn guide. As each needle rises, the existing loop opens and clears the latch and the hook engages the yarn. As the needle descends, a new loop is pulled through the existing loop, producing a continuous tube of fabric interlocked vertically and horizontally. The mechanism is therefore self-acting in terms of loop formation. A number of auxiliary elements may be used to aid this process, but these are not considered.

Cam tracks are made up of individual pieces as shown in Fig 2. These facilitate the control of fabric characterisitcs and structure. Many different combinations of cams and needles are possible, but all employ the same basic principles (Smirfitt, 1975). The commercial machine used in this work was of the cylinder-and-dial type. A dial is a second set of needles in a radial plate perpendicular to the needle cylinder which allows a greater variety of structures to be produced.

1.2 Theoretical models

The forces that arise in the system are critical in limiting machine speed, determining overall system performance and contribute to vibrations, stresses, noises, wear and needle failure. Theoretical models were derived to predict forces in the knitting zone (Sharp, 1986). The needle cam interaction was fully analysed, including cam design techniques (MacCarthy, 1986). Sensitive transducers were designed, manufactured and developed to measure the knitting forces in order to verify the theoretical predictions (Sharp, 1986)

2. The transducers

Several transducers were designed

(1) Cam force Transducers: these measured the reactive force for a single needle butt on a cam profile on both cylinder and dial
(2) Yarn Force Transducer: this measured the yarn input tension immediately prior to entering the loop formation zone
(3) Fabric Force Transducer: this measured the takedown tension of the knitted fabric

This paper concentrates on the cam force transducer for the condition monitoring of the needles.

2.1 Cam-force transducers

Two essentil requirements were that the transducer should measure force associated with a single needle, and the principle should be applicable to any campiece at any instant of time. As the needle descent phase is critical in loop formation and associated with high force levels, the campiece that controls the needle in this phase was chosen for experimentation.

The principle of measurement was to suspend a section of cam as an end-mass from the free end of a square-cross-section cantilever beam, as shown in Fig 3. Strain was induced in the beam when the needle butt is driven against the cam profile. Strain gauges are positioned on the vertical and horizontal faces close to the support, where maximum strain occurs. By calibration with known forces, the output from the strain gauges is transformed into the horizontal and vertical components of the cam-needle butt reactive force.

The cam force transducer is illustrated in Fig 4 in position in a cylinder cam box. Fig 4 also illustrates the method used to isolate the measuring needle. Two identical cam pieces are used. Over half of their thickness is removed from the rear face. A hole is cut through the cam section housing behind the cam piece location, sufficient for the cantilever beam. One of the cam pieces is fastened to the beam and the other is fastened in its standard position to the cambox housing. This positioning ensures that the butts of the standard machine needles contact the rigidly supported cam without contacting the transducer cam.

A longer needle butt is used to contact the rear cam transducer.

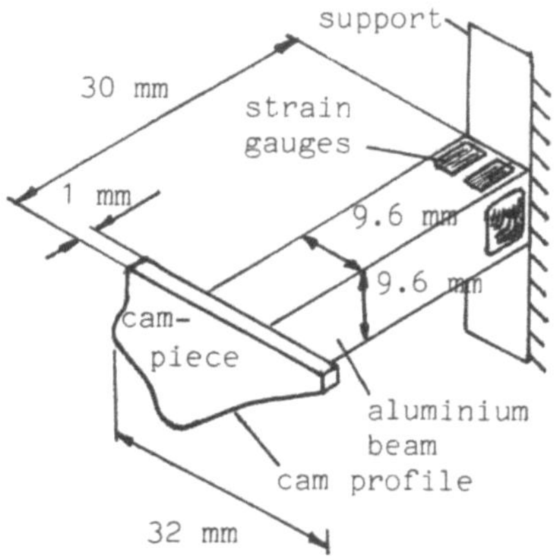

Fig 3 Cantilever beam transducer

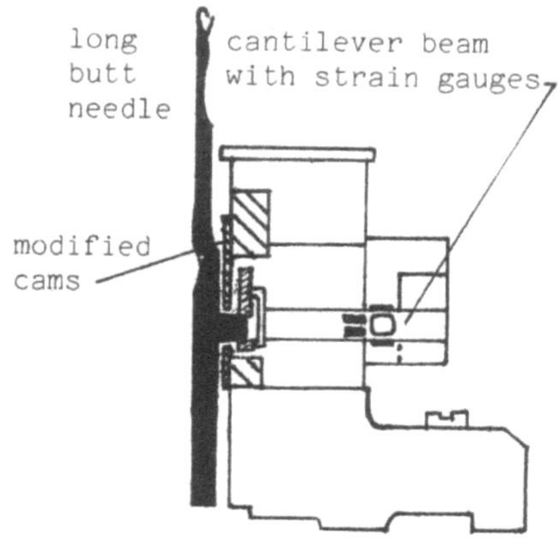

Fig 4 Cam force transducer

If the rear cam is offset vertically by a small amount, then the path of a single long butt needle in a bank of standard needles, is isolated on the transducer cam. In order to ensure that the path of the hook of the long butt needle is precisely the same as the standard needle, it is necessary to reduce its butt height by an amount equal to the vertical offset of the rear cam.

The same principle was applied on the dial, In addition, the transducer was designed to be easily removable from the section for calibration or to use different profiled cams.

2.2 Instrumentation

When mounted and operational, the transducers were interfaced with timing and speed monitoring instrumentation and a microcomputer-based

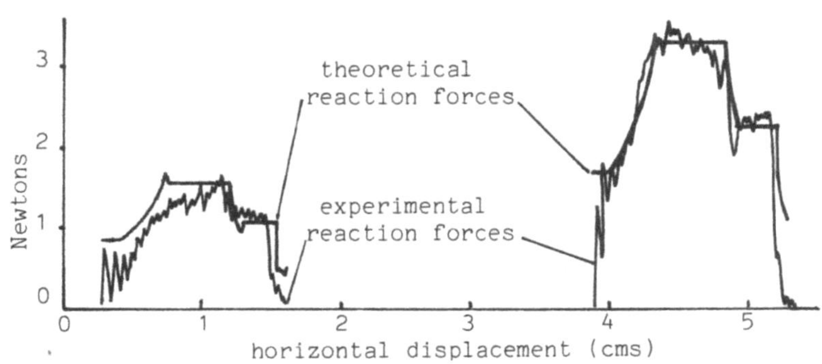

Fig 5 Cam-needle reaction forces

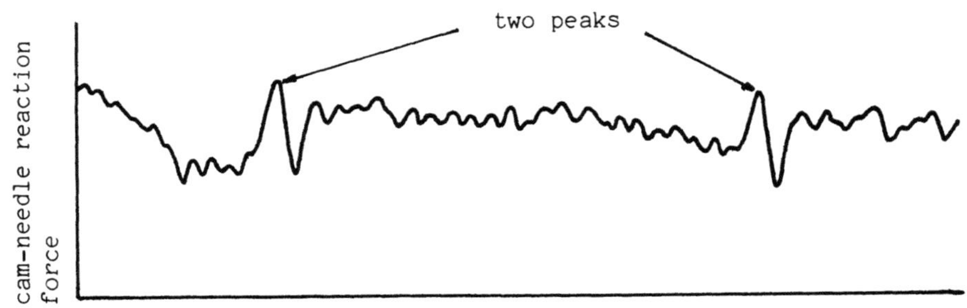

Fig 6 Condition monitoring of needles using cam force transducers

data logging system. Each needle follows the campiece profile (attached to the transducer) for approximately 0.02 seconds at maximum machine speed which requires a fast sampling rate. The conversion time was 22 μs per reading. The recorded signals could be aligned accurately to the position of a needle in the cycle. The transducer readings were downloaded from the microcomputer to a mainframe computer to produce detailed plots of experimental and theoretical forces.

The instrumentation was developed around a Bentley RJ48E which is a typical large diameter (0.76 metre) double jersey fabric machine. It had two sets of 1700 needles in the cylinder and the dial, all 0.45 mm thick. The microcomputer was a Research Machines Limited (RML) 380Z, whilst the computer was a Cyber CDC 170-720 machine. The instrumentation and transducers have previously been explained in detail (Sharp et al, 1989)

3. Condition monitoring of needles

The transducers and instrumentation were used in two extensive sets of experimental investigations (MacCarthy 1986; Sharp, 1986). The effects of many variables were studied. The variable of interest in this paper is that of cam-needle reaction force as it can be used to monitor aspects of needle condition. The left hand traces in Fig 5 shows the theoretically predicted and experimentally recorded (by the transducer) cam-needle reaction forces for a standard needle on a conventional arc/tangent cam profile. The right hand traces relates to precisely the same condition except that the needle had a much larger trick needle resistance. This plot shows the sensitivity of the transducers, whilst highlighting the effect of trick resistance on cam-needle reaction force. The trick-needle resistance was determined by measuring the force required to initially move a needle in its trick (Sharp, 1986). Over 90% of needles had an average value of 0.55 Newtons, whereas a few had trick-needle resistance of upto 1.5 Newtons.

The key defects are: (1) Bent needle, (2) Damaged trick, (3) Faulty latch, (4) Broken hook, (5) broken butt. Needles with defects number (4) and sometimes (3) and (5) can be detected by commercially available sensors such as Circscan by Meinersdel Ltd, Stockport, Cheshire. Numbers (1), (2) and (3) will result in significantly increased trick resistance/cam-needle reaction force, whilst for (4) it will be low and for (5) it will be virtually zero. These variations in force can all be measured by the transducers thereby identifying a defective needle.

As discussed in Section 2.1 the transducer measured the forces of long butt needles set in banks of short butt needles. For effective condition monitoring all needles should be measured thus the transducer was attached to a campiece which all the needles contacted. Two needles with high values of trick needle resistance (0.7N) were set 25 needles apart. To ensure identification the needle immediately behind each of the high resistance needles was removed. Fig 6 shows the vertical component of cam-needle reaction force at normal machine speed. Each peak is a needle impacting the campiece.

The two high resistance needles (and the two missing needles) are clearly identified showing that these transducers can be used for the condition monitoring of needles. A reduction in campiece width and an increase in transducer beam natural frequency are required to improve this measuring system. The microcomputer software could be programmed to monitor the cam-needle reaction force with pre set alarm conditions for minimum and maximum acceptable values.

4. Conclusions

The transducers and instrumentation can be used to monitor the condition of needles in commercial high speed weft knitting machines. These transducers operated effectively under realistic operating conditions. Such transducers and instrumentation discussed in this paper are portable and if the campieces were modified for different machines to enable the transducers to be fitted then it could be used as a portable condition monitoring system, with diagnostic capabilities, on a textile production floor.

The author is very grateful for the support provided by the SERC (GR/B/92935) and Bentley Engineering, Leicester, UK.

5. References

Dutton, W.A. (1981) Progress in weft knitting. Textile, 10(2).30.

MacCarthy, B.L. (1986) A kinematic and dynamic analysis of latch needle cam systems. PhD thesis, Uni of Bradford, Yorkshire.

Sharp, J.M. (1986) Dynamic forces in high-speed weft knitting machinery. PhD thesis, Uni Bradford, Yorkshire.

Sharp, J.M. MacCarthy, B.L. & Burns N.D. (1989) Transducers for the Measurement of forces in textile machinery. Trans Inst MC, VII, N3.

Smirfitt, J.A. (1975) Introduction to weft knitting, Merrow, Technical Library, London.

On the monitoring of a high speed mechanical feeder by means of an in-time windowing technique

Enrico D'Amato and Giulio D'Emilia
Dipartimento di Energetica, University of L'Aquila, Italy

Abstract
The setting up of a technique (instrumentation and data processing procedures) to diagnose a high speed mechanical feeder, as an example of a complex mechanism, in an efficient and economical way is described.

The procedure, which is computer based, takes into account the information deriving from the knowledge of the mechanical subsystems motion laws, with the aim of defining a limited number of useful and significant time windows in the mechanism cycle.

The measuring technique is based on the use of piezoeletric accelerometers, due to the requested frequency response in diagnosing the considered class of damages. The aspects related to the definition and optimization of the windows, with reference to the kinematic and dynamic characteristics of the studied mechanism, are also discussed.

Experimental results, referring to tests where artificial damages were imposed in the studied mechanism, prove the capability of the technique to be useful in identifying some defects in the mechanism.
Keywords: monitoring, mechanism, data processing techniques, time windows, instrumentation

1 INTRODUCTION

To perform an efficient diagnostic activity on a complex mechanism it's not only necessary to detect any possible out of order condition, but also to identify the relevant damaged components.

The application of the most popular techniques appears efficient only when the mechanical system's abnormalities are related to components (rolling bearings, gears,...) whose operating conditions can be easily characterized by typical frequencies, Tranter (1989), Alfredson and Mathew (1987), Mathew (1987), Angelo (1987), Lyon (1987).

When, on the contrary, the defects introduce broadband vibration signals (shock or transients), the natural modes of the mechanical system are excited with consequent masking of the signals to be used in diagnostic procedure, Bowen and Lyon (1989), Seth (1989). In these conditions, moreover, the analyses are difficult because of the interaction of the signals related to different damages, and due to the complexity of the transmission paths which reduces the capability of signature recovery procedures, Braun and Seth(1977), Braun (1989).

These topics result of great significance when the detection of a particular damaged component is to be considered or when it's necessary to optimize the experimental procedure and the signal processing techniques.

In fact, the aim is to separately provide the best signal to noise ratio for every damaged component; in this case a very big subject is opened both on the parameter to monitor, and on the measurement techniques (selection and positioning of the sensors) and on the signal processing techniques. These aspects appear more evident in cases of on line industrial monitoring applications where simple and reliable equipment and time saving signal processing procedures are requested.

This paper describes a novel diagnostic procedure and a first application on a mechanism composed by several elements linked to constitute a kinematic system. The proposed methodology deals with the above mentioned difficulties, starting from a theoretical examination of the system's kinematic behaviour, with the aim of selecting critical operating conditions, which could result useful to detect the different damages in the mechanism.

The information which seems to be interesting is mainly referred to time intervals where the acceleration changes the sign; the related inversion of the inertial loads causes impulsive excitations when the links are characterized by large clearances: this can be due to design or machining errors or, more easily, to in service wear.

With reference to this information, some zones of the mechanism cycle have been focused, opening time windows.

The following topics have to be considered for the application of the windowing technique :

- windowing criteria;
- influence of the sensor position;
- propagation of vibration due to each defect in the mechanism;
- methods of identifying the contribution of the single damages.

In order to get an experimental validation of the proposed methodology, some tests were performed on a prototype of high speed mechanical feeder characterized by alternate and rotating motion of several elements. Some known defects were introduced to simulate out of order conditions.

2 THE PROPOSED METHOD

The processing of the data referred to the whole cycle, is often time expensive; moreover it could emphasize the effects of structural and working phenomena (low damped structural resonances, system input motion disturbances) in comparison with transient signal, often the ones carrying the requested information.

In the author's opinion, a first important aid, relevant with the selection of the cycle's time intervals to be focused on for diagnostic purposes, derives from a kinematic theoretical study of the mechanism. On the other hand the dynamic information concerning the kinematic behaviour of the system is generally available from the design step.

For the class of the mechanical systems object of this study the most important parameter to be considered is the acceleration which, in combination with mass distribution, determines the changes of the inertial loads. The knowledge of these phenomena allows to define the starting time (in many cases depending on each subsystem) of the occurring abnormalities related to sign changes of the acceleration for the mechanism's subsystems. Using this theoretical information, it's possible to pinpoint a component's group whose behaviour is more critical and has to be monitored with particular attention.

Even though the actual behaviour can be different from the theoretical one, this procedure has the advantage of allowing a quick and really inexpensive analysis of the mechanism in the whole without instrumenting all the mechanical components; this is particularly useful when the mounting of the sensor on all the mechanical components is difficult or quite impossible.

The problems to be solved for the correct application of the proposed methodology are:

- windowing parameters optimization (width, time positioning in relation to the focused phenomena)
- influence of the sensor positioning on the identification of the single defect contribution to the global vibration level.
- damages identification procedures.

Some tests performed on a complex mechanism confirmed the importance of these topics.

The mechanism considered in the experimental activity is the prototype of a high speed mechanical feeder, Cavagna and Giordana (1984), which realizes a linear alternate motion and dwells of the final element (a platform) by means of dead center points accumulation of two four-bar linkages. The links

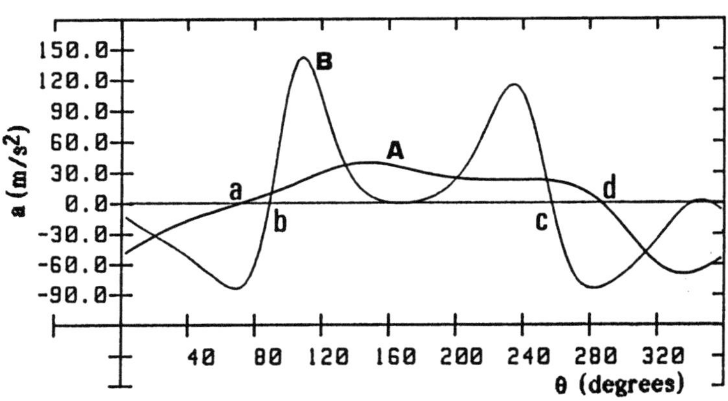

Fig. 1 Theoric tangential acceleration behaviour during a cycle, for the accelerometers A and B.

between the mechanism's components are obtained by means of pins and journal boxes (grease lubricated because of their alternate motion). The theoretical analysis showed the different time instants of acceleration inversion for the two four bar linkages (named A and B in the following), (points a, b, c, d in Fig. 1). Fig. 2 shows the experimental acceleration behaviour (average on 32 cycles) measured on the mechanism links to be monitored; piezoeletric accelerometers were used. The comparison shows a satisfactory agreement between the theoretical and experimental acceleration trend and in particular between the predicted and measured number and position during the cycle of the zero crossings.

As a first attempt, two windows have been defined at inversion

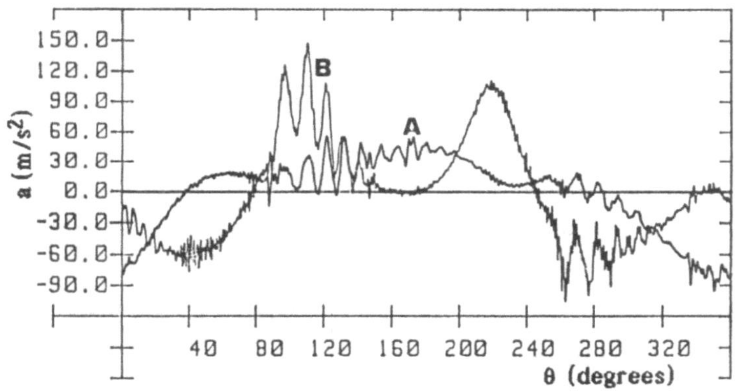

Fig.2 Experimental acceleration behaviour during a cycle (average on 16 cycles), for the accelerometer A, and accelerometer B. Mechanism in good health conditions; period of a cycle 0.292 s.

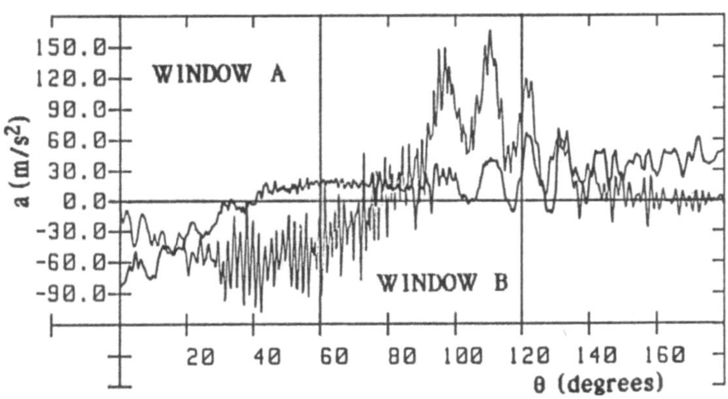

Fig. 3 Particular of the windowed period intervals. Same working conditions as in Fig. 2.

points a and b because they showed a better time resolution in the experimental diagram (see Fig. 2). The window was symmetrical around the zero crossing of the acceleration diagram and the width has been defined so as to not include the zero crossing of the other subsystem (Fig. 3). The data have been smoothed by an Hanning window and frequency analysed by FFT standard procedures. A first accelerometer has been placed on the head of the connecting rod in the A mechanical subsystem, the second accelerometer on a similar point of the B.

Starting from a mechanism in good health conditions, damages have been introduced in links of subsystems A and B, substituting journal boxes having project clearances with same type boxes in which wearing conditions have been simulated. Tab.1 shows the trend of the vibration energy in two different frequency ranges (where the vibration energy level was higher) in window A. Tab. 2 contains similar data for window B.

Table 1. Power ratio (db) for the A-windowed signals.

FREQUENCY RANGE (Hz)	SENSOR POSITION	MECHANISM HEALT CONDITIONS		
		Good	Damaged A	Damaged B
400÷600	ACA	0.0	8.6	1.0
	ACB	0.0	17.7	2.0
900÷1200	ACA	0.0	6.8	2.0
	ACB	0.0	14.0	6.0

The criterion used for identifying the damage was to compare the power contents of the accelerometer frequency spectra in different frequency ranges: it is assumed that the damage is related to the link whose acceleration inversion falls in the window where a higher energy variation occurs.

Table 2. Power ratio (db) for the B-windowed signals.

FREQUENCY RANGE (Hz)	SENSOR POSITION	MECHANISM HEALT CONDITIONS		
		Good	Damaged A	Damaged B
400÷600	ACA	0.0	5.5	10.0
	ACB	0.0	3.5	10.0
900÷1200	ACA	0.0	5.0	13.0
	ACB	0.0	6.0	13.0

Some considerations are possible:

- the power spectrum variations with reference to the good health conditions are similar in the two considered frequency ranges;
- the uncertainity on the measured data can be estimated in ±4 dB (average on 16 cycles).
- the presence of the damage occurring in the A subsystem can be

located by comparing the energy levels detected by ACA accelerometer in the chosen windows. Likewise, the presence in the B submechanism can be pointed out, also if less clearly, by comparing the data of the presented Tables: it can be noted that the differences in Table 2 for the damage B are greater than those in Table 1.

These results showed the efficiency of the proposed method, even though further developments are requested, to completely deal with the above discussed problems.

3 CONCLUSIONS

This paper describes a theoric and experimental methodology to identify the fault causes in a complex mechanism having more subsystems.

A preliminary theoric analysis of the kinematic motion laws of the different subsystems has been used to focus the attention on cycle time intervals, that seemed particularly important from a diagnostic point of view. In fact, the occurrency of specific kinematic conditions (acceleration sign changes only in a part of the kinematic chain, etc.) has been examined, in order to identify working conditions where a specific link defect could be more important than the other defects in the mechanism.

The already performed tests concerned the identification in a mechanism prototype of single and known defects, introduced for method validation purposes.

The experimental results showed that the proposed method can supply useful information to individuate the specific defects, even though simple and rough data processing techniques are used (evaluation of the vibration energy in specific frequency ranges).

In future work other data processing techniques will be used (envelope analysis of the structural resonances signals, cepstrum analysis, etc.), in order to increase the method sensitivity in single defects discriminating.

Particular attention will be paid to the following topics:

- sensitivity evaluation of the proposed methodology in order to diagnose the defect occurrency and its worsening;
- individuation of frequency ranges which are the most significant for diagnosing purposes;
- study of the sensor positioning influence, with the aim of optimizing the utilized instrumentation, from a point of view both functional and economical.

4 BIBLIOGRAPHY

Tranter, J., (1989) The fundamentals of, and the application of the computers to, condition monitoring and predictive maintenance. Proc. of 1st IMMDC, Las Vegas, NE.

Alfredson, R.J., Mathew, J.,An evaluation of some recently developed techniques for machine condition monitoring. Research

Paper, Monash University, Australia.
Mathew, J.(1986) Machine condition monitoring using vibration analyses. Acoustics, Australia, Volume 14 n. 1.
Angelo, M., (1987) Vibration monitoring of machines. Bruel & Kjaer Technical Review, n.1. .1.
Lyon, R.H, (1987) Machinery noise and diagnostics. Butterworths, Boston, NJ.
Bowen, D. L. and Lyon, R. H., (1989) Recovery of diagnostic signals for machine transients". Proc. of 1st IMMDC, Las Vegas, NE.
Seth, B. B., (1989) Isolation of abnormalities in rotating machines using signature analysis. Proc. of 1st IMMDC, Las Vegas, NE.
Braun, S.G., (1989) An overview: MSA - Mechanical Signature Analysis and diagnostic applications. Proc. of 1st IMMDC, Las Vegas, NE.
Braun, S.G., Seth, B.B. (1977) Signature analysis methods and applications for rotating machines". ASME Paper n. 77-WA/Aut-5.
Cavagna, C. and Giordana, F.,(1983) Una metodologia per la verifica sperimentale della legge di moto di un azionamento meccanico, Ingegneria, n. 7-8.

Condition based maintenance

G. Allenby
Geoff Allenby – Marketing Manager, SPM Instrument UK Ltd, Bury, England

Abstract
This paper deals with the **philosophy and commitment of condition based maintenance** and the reasons why, whats in it for me ?.

What are the costs ?, is there a pay back ?.
What type of equipment is available to the engineer and what does it do ?.
A full and candid appreciation of the **Shock Pulse Method.**
What does the **Shock Pulse Method** do and why ?.
Why is **Shock Pulse** so unique.
Keywords: Phillosophy, commitment, why, cost, pay back, **Shock Pulse Method.**

Introduction.

Condition based maintenance is a discipline that is gaining momentum within industry, but not always with a a full appreciation of the consequences and implications.

There is a basic need to understand and accept the **philosophy** of condition based maintenance, i.e. that the discipline, correctly applied, will show a return on investment and that every person, from the managing director down, understands that this is an essential discipline in to days modern industry.

Once the **philosophy** has been accepted, the next stage is the **commitment.**

Condition based maintenance does cost money to implement. The most suitable equipment has to be carefully selected and purchased, systems and routines have to be set up and organised along lines to suit individual needs.(see flow chart, fig.1.) The setting up of routines and personnel is a critical component of a successful operation and should be treated accordingly.

Pay back, if the **philosophy** and the **commitment,** by every one has been accepted in a professional manner, then there will be (not maybe) a **pay back.**

CONDITION MONITORING – SYSTEM ENGINEERING

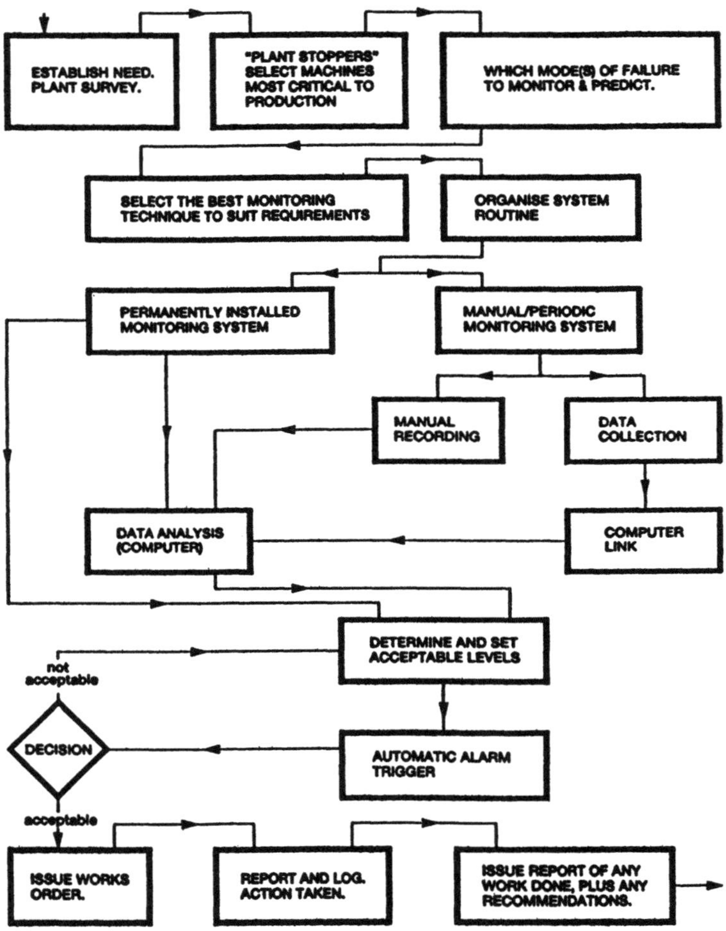

fig.1.

For obvious reasons, it is virtually impossible to put a monetary value to this statement, because every manufacturing or service industry has varying parameters to contend with. However, there is a simple statistical example of the benefits of condition monitoring, see table a).

Table a). Electric Motors

motor size	costs p.a.	cost 50 p.a.	2% saving /h.p./p.a.
300 h.p.+	7.5 pds.	112,500 pds.	2250 pds
150 to 300	16.5 pds.	185,625 pds.	3712 pds
up to 150	36.75 pds.	137,812 pds.	2756 pds
		Total Savings =	8718 pds.

This is a very simple and conservative example, but is indicitive of the benefits to be obtained.

There are a number of other areas that operations can relate to and place a definate cost against savings achieved through condition based maintenance, these are as follows ;

1) manufacturing and/or process plant downtime. Probably the most significant cost to a company, because in todays industry, 24 hours, 7 days a week, 360 days a year are quite common. Consequently any lost production cannot be recovered and is lost forever. Therefor the value of predictive maintenance in planning repair or rectification work at a convenient time, must be a considerable improvement on a "repair by breakdown" policy, which will inevitably incur consequential damage to machinery.

"Time based" maintenance, although an improvement over breakdown maintenance, still causes problems, i.e. unexpected machine breakdowns will still occur, but also, maximum machine utilisation will not be achieved due to the conservitivness of the life time calculations with the inherent built in safety factors.

2) Significant reductions in running costs of the maintenance department, due to planned replacements being carried out quickly, more efficiently and at a convenient time (not 3 o'clock in the morning) with all the right equipment for the job, available.

3) Reduction in stock holding, this becomes possible because with the implementation of the condition based maintenance programme, it is practical to plan the replacement and consequently, the require components well in advance.

Condition monitoring equipment. As mentioned earlier, it is vital that careful consideration to the equipment be given and that the equipment purchased will provide the most effective **pay back** in the shortest possible time. One other important factor, is that the equipment purchased can be readily interpreted and understood by all the people involved in this commitment.

There are a number of valid and effective techniques available to the engineer, some of which are described follows;

a) **Shock Pulse Method**, a unique method monitoring the " true operating condition " of a rolling element bearing. (see full appreciation of the technique). Can be used in either portable or installed mode, with software.

b) Vibration monitoring, a well established method for determining the physical movement of a machine or structure, due to out of balance, mounting or alignment. This equipment can be obtained as simple, easy to use and understand, or as sophisticated real time analyser. Normally a " finger print " or similar is required to trend a problem, which if used in conjunction with BS or ISO standards, enables related problems to be determined. This type of equipment is available in both portable and installed modes, with software.

c) Alignment. Shaft alignment or mis-alignment, constitutes of the order of 50% of machinery problems.

There are portable machinery alignment computers available to industry, which provide accurate, repeatable and consistently good alignments. These instruments normally, not only provide shimming details of front foot and back foot information, but also offset and angularity readings, thus enabling good alignment to be achieved. N.B. It is advisable to use good quality pre-cut shims to carry out the movements indicated by the instrument.

d) Thermography. Although this is a rapidly developing technique, it provides, via colour camera's and video's, clear indications of heat losses, hot spots, cold spots, such as switch gear or any piece of plant or production process, where temperature or its effect is important. It can be used both as a maintenance tool or a quality assurance tool.

e) Debri analysis. This technique is well proven in all types of industry and works on the principle of taking a known quantity sample from, for example a gearbox, then analysing the amount and type of foreign particles present in the sample. This will show such problems as gear wear, if the sample detects particles of gear material present.

f) Oil analysis. Oil analysis differs from debri analysis in so far as this technique allows an assessment of the actual condition of the oil in use. i.e. is the oil quality good enough for the application after a period of use, is it " burnt " or exceeded its useful life.

Shock Pulse Method - Appreciation.

The **Shock Pulse Method** is a unique technique,invented some twenty years ago, for monitoring the true operational condition of a rolling element bearing by measuring the pressure wave generated by an instantaneous mechanical impact.

This pressure wave or pulse, is a function of velocity not mass, as opposed to vibration i.e. physical movement of the machine or component.

By providing information to the measuring instrument (portable or installed), such as r.p.m., mean diameter and the bearing type, then measuring the the shocks generated by the bearing it is possible to carry out a full analysis of the bearing including not only early damage, if present, but most important, the vital oil film thickness, (elasto hydrodynamic wedge or EHD.) Evaluation of the bearing to this degree enables the engineer to be aware of a problem before it becomes a problem, because if the EHD can be established and maintained, it is possible to double the theoretical L-10 life. (see fig.2).

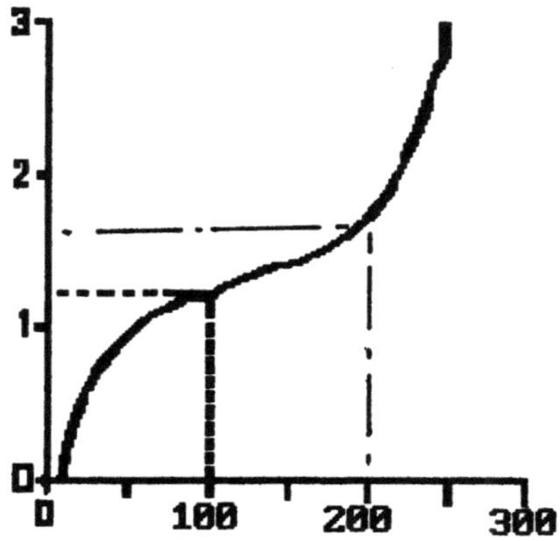

% of catalogue service life -L10

$$\frac{\text{lubrication film thickness}}{\text{composite surface roughness}} = \lambda$$

fig.2.

One of the most significant features of using the **Shock Pulse Method** is, because the input data and measuring data is absolute; **NO FINGER PRINT** of a bearing is required, regardless of wether the bearing is brand new or ten years old. The analysed reading clearly presents the actual operating condition and will not be influenced by other machine conditions, such as excessive vibrations due to other causes.

In practice the two techniques are complimentary, i.e. one will tell you what the other cannot, consequently evaluation of a problem is simplified.

There is still no other technique capable of analysing a rolling element bearing to the degree that **Shock Pulse** can, despite the numerous claims such as measuring high frequency bands.

The **Shock Pulse Method** operates on the basis that any instantaneous mechanical impact can be detected at a high frequency, consequently the technique works on a very discreet high frequency and by mechanically tuning the **Shock Pulse** transducer, thus making it very responsive, then by filtering the signal the absolute measurement of the bearing is possible, it is not a compromise.

Continued research and development by **SPM** engineers has ensured the continued international success of the method by improving a unique technique beyond all other "pretenders"!.

The importance of measuring the EHD can be seen from fig.3.

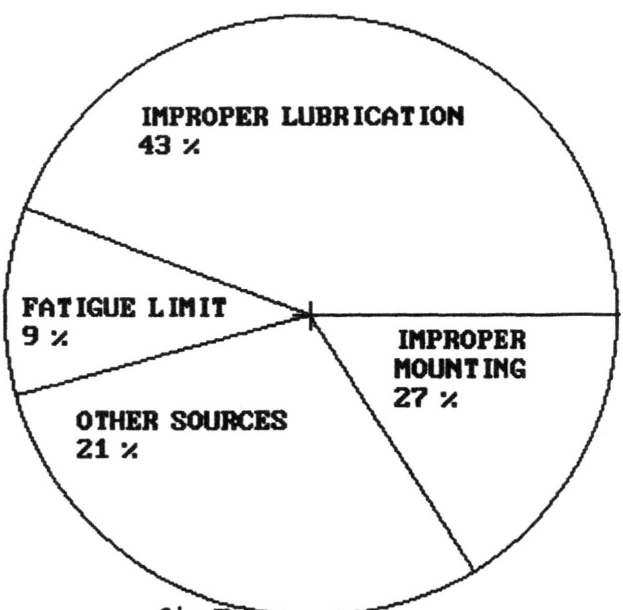

fig.3.

This pie chart clearly shows that approximately 70% of bearing failures are due either directly or indirectly to lubrication, to much, to little, wrong type, badly installed or mis-alignment. Only 9% actually achieve the fatigue life. This means that, because the bearing is the heart of any rotating machine, the useful working or production life is dramatically reduced, consequently any improvement in the operational life of the bearing automatically results in an increase in the availability of the machine.

Cost effective computer software in maintenance management

P. Cross
COMADEM Centre, Faculty of Engineering and Computer Technology, Birmingham, England

Abstract
This paper outlines the development of a maintenance management software package that was designed for use at the Atomic Energy Authority Laboratory, Harwell, in Oxfordshire. Although the software includes many of the characteristics of the more expensive commercially available systems, this package was created at minimum cost, and yet provided a 'tailor made' environment for recording planned, and unplanned maintenance, and condition monitoring, for a wide range of plant and services throughout the site.
Keywords: Maintenance Management, Database, Asset Register, Plant History, Condition Monitoring, Calendar Maintenance, Utilisation Maintenance, Spreadsheet, Network.

1 Introduction

Maintenance management software packages have now been available for over a decade, and a considerable variety of packages have been produced, varying in price from around £1,000 to £50,000.

During my employment with the Atomic Energy Authority at Harwell in Oxfordshire, I was asked to review and implement a suitable maintenance management package for use within the Authority site, which would centralise the work of several different maintenance departments, and correlate data into some useable form. The package was to embrace calendar and utilization based maintenance, and condition monitoring, and be capable of supporting several thousand assets.

Stringent monetary restraints were imposed on the project, and a detailed justification report for any expenditure was required.

Following a detailed examination of the various software packages available, and visits to review other Authority sites, it was finally decided that none of the then available packages could provide the facilities we needed, within the finances available to the project. It was therefore decided to create, from first principles, our own maintenance management package, and a team of three people were committed to the task, none of whom had any extensive prior experience of computing. The system thus created was subsequently networked to five workstations on the site.

Further development of the system is being conducted at the Birmingham Polytechnic and will encompass many of the features of computer maintenance management packages currently available.

2 The Maintenance Management System

2.1 System Development

The task of designing and implementing the maintenance system was allocated to a small department, not contained within the line management of the maintenance management structure.

The scope of plant and services to be included in the scheme was limited to the work of eight departments within the Engineering Division, which were roughly divided according to discipline, ie. Electrical (distribution), Electrical (services), Refrigeration and Air Conditioning, Fitting Shop, etc. Each department was supervised by a senior foreman, and each operated its own work recording and maintenance control systems, independently of the other maintainance departments.

In determining the parameters around which the computer system would be required to operate, the following factors had to be taken into consideration:-

(a) Existing practices
(b) Labour agreements
(c) Cost effectiveness
(d) Work study appraisal
(e) User friendliness
(f) Training needs
(g) Technical support
(h) Security

Existing planned maintenance and condition monitoring schemes, which already had clearly laid down procedures and maintenance periodicities, were to be absorbed into the overall maintenance plan, without recourse to unnecessary changes. Where no such schemes existed, or where updating or modification of schemes was required, then this work was to be undertaken first.

One example of this was the boilerhouse, where the eight boilers and ancillary equipment were maintained on a calendar basis, with maintenance tasks scheduled throughout a twelve month cycle. This scheme was revised, a utilisation based maintenance programme was introduced, and the maintenance periodicities redefined. Hour counters were fitted at strategic points within the plant operating systems, and readings taken at predefined intervals, so that maintenance tasks could be anticipated, and then performed as required.

2.2 System Specification

Detailed proposals of what would be required of the overall maintenance plan were defined, as a product of discussions and meetings at both craftsman and management levels. The frequency of these meetings, and ready exchange of ideas was an essential contribution to the introduction of the maintenance plan. Requirements were:-

(a) Comprehensive plant history of all plant and machinery listed in the assets register.
(b) Breakdown of costs against any nominated cost centre.
(c) Accumulated man/hour totals, by grade, against any nominated plant item, type, or cost centre, over any given timescale.
(d) Recurring faults, or defective item analysis.
(e) Costing.
(f) Planned maintenance job scheduling.
(g) Unplanned maintenance work recording.
(h) Work in progress.
(i) Condition monitoring.
(j) Selective report writing for individual system users, ie. Progress reports, Financial statements etc.

From this, it was decided that three integrated software packages would be required to operate the system; a database, a spreadsheet and a graphics package. Those purchased were dBase III+, Quattro, and Freelance, at a total cost of around £900, and an IBM PC was made available for the project.

2.3 Database Management

The database was used to provide the file management of the Assets Register and the Plant History, to provide a mechanism for operating the planned maintenance job scheduling system, for costing planned maintenance work when performed by external contract staff, and for creating the reports and listings that the system would require.

Using dBase 111+ to create the files we needed proved less complex than was at first anticipated, and a couple of days spent with the manual enabled us to proceed with relative confidence. All the files were created using the dBase 111+ 'Assistant'; a system of pull down menus that enable an unskilled user to create and manipulate files, without recourse to programming.

The Asset Register, which initially ran to over 2 thousand items, was compiled first, and here the advantages of creating a system unique to the user comes into its own, since field definitions, sizes and types can be of any number, or configeration desired, within the scope offered by dBase 111+. Similarly, screen formats (ie. the manner in which the record is displayed on the screen for adding or changing data) may be designed to resemble existing documentation. Data listed against each asset on the Asset Register included a unique plant identification number, (essential for searching, locating or filtering records,) its description, location, cost code, etc. and maintenance details ie. periodicity, maintenance instructions identifier, week number, month and year the task is due, and the maintenance cost. From this it was possible to list the maintenance due within any period, the costs required against any cost centre, the total time required, etc.

Next, a Plant History file, and associated screen format, was created, and by using dBase 111+ locate and search commands a flexible data retrieval system was designed, ie. planned and breakdown maintenance details on any plant type or item could be listed over any desired time scale. A complete summary of all work carried out in any Division by any maintenance department, can be shown. Accumulated man hours devoted to a given task can be summed, and the costs accrued to date calculated, etc.

Some modification of existing documentation was required, so that imformation could readily be entered into the database fields in the computer records. Two new types of maintenance card were designed, printed, and issued to the appropriate departments, so that planned maintenance tasks could be carried out, and the work done recorded and certified in a standard format. A modified job card, used for all tasks other than planned maintenance, was designed and issued, to replace an existing form. The Work Study department produced flow diagrams illustrating the paths along which the data could be validated prior to entering it onto the computer program.

2.4 Condition Monitoring

Condition monitoring was to be employed in the following work areas:-

(a) Calorifiers and plant rooms
(b) Gas mains and valves
(c) Compressed air valves and traps
(d) Steam main valves and line traps
(e) Overhead electric crane rails
(f) Ducts, and duct covers
(g) Water reservoirs
(h) Cooling towers

(i) Extract fans etc.

Owing to the vast number of these items throughout the site, a means of systematic inspection, testing, and data recording was needed. To achieve this, the site was divided into several work areas, and the assets of each type and in each work area listed. A series of computer generated worksheets were then designed, using the dBase 111+ report format. These provided a detailed print-out of location, identity number etc. of the asset, and the measurements to be taken and recorded during the inspection. This data was to be recorded onto the worksheet by the craftsman carrying out the inspection, and would then be entered onto the spreadsheet. From this, subsequent inspections would ultimately produce an adequate volume of data so that a graph could be generated for trends analysis.

Data recorded included GIA (gas in air) readings for gas valves and mains, cooling water chemical analysis readings for cooling towers, crane rail securing bolt torque settings for overhead electric cranes, etc.

It would, perhaps, be fair to note that while this may appear a fairly torturous route to achieve the desired result, the object of the exercise is not so much to reduce downtime, as may be paramount in a production environment, but to provide a safe and reliable system of performance monitoring best suited to a research environment.

2.5 The Spreadsheet

The spreadsheet provided the storage of condition monitoring data, as already described, and the file management of the utilisation based maintenance control system.

Each asset that was listed in the master Asset Register within dBase 111+ as requiring utilisation based maintenance, was also listed in the spreadsheet. Units of utilisation were quantified as either hours or operations. Four columns of data were created on the spreadsheet; hours/ops. total, hours/ops. since last maintained, hours/ops. remaining to next maintenance, and hours/ops. at which the next maintenance was due. These figures could be updated daily by entering the current meter readings from the various assets. Once an asset meter reading came within 10% of its maintenance period, then the appropriate maintenance card could be issued to the department concerned, and the date of issue noted on the spreadsheet. This proved to be a highly effective, yet simple system of utilisation maintenance monitoring.

2.6 The Graphics Package

This provided a simple means of converting tabular information from the spreadsheet into graphs, pie charts or bar charts. The package proved equally effective for producing forms, to be used for collecting maintenance data, and, for enhancing written reports to managers and users of the maintenance system.

2.7 Additional Features

A later addition to our software that enhanced considerably our database capability was a dBase 111+ add-on, 'Report Writer'. This enabled job and maintenance cards to be computer generated, thus reducing hand written work to virtually zero.

Also, computer generated memos to other departments, which acted as reminders that a piece of plant may be out of action whilest maintenance was taking place, could be automatically produced.

The report writing facility within dBase 111+ was very limited, so this additional software provided the extra support needed to make maximum use of the data we were storing.

2.8 Networking

The system as described worked well enough, but certain inherent flaws in the basic concept needed attention.

Primarily, work, other than planned maintenance, could only be recorded onto the computer retrospectively, ie. once the work had been done, the job card completed, and the data added to the Plant History file. Work in progress could only be determined by a physical check on the job or maintenance cards that were in circulation at any one time. (A simple system had been employed in each department since the inception of the project, whereby job and maintenance cards were stored in racks of three columns width. The first column indecated work awaiting attention, the second, work in progress, and the third, work completed.)

However, it was decided at this point to network the system to all users, and a further four PC's were purchased and installed. Users were limited to accessing only the Plant History files relating to their own department, for adding and reading records only. Now, work could be entered onto the users computer as it arose, and therefore all work outstanding or in progress could be identified and monitored.

2.9 Later Developments

Further improvements have since been made to the basic system, to enhanced user friendliness with simpler pull down menus, and a redesigned Assets Register will make better use of available disk space.

3 Conclusion

A maintenance control system, that has been derived by the

user, has certain advantages over a commercially available package; not only should there be a considerable saving in software costs, but the system will have been created around the specific needs of the user. It can grow and develop as experience of the system increases, and this familiarity will greatly improve the fault finding capabilities of the user.
Extra facilities may be incorporated into the system at no extra cost, and networking need not be a problem.

Condition monitoring based maintenance in practice in a heavy chemical plant

J. Hensey and M.N.K. Nair
Auginish Alumina Limited, Askeaton, Co. Limerick, Ireland

1 Introduction/Abstract

At the outset, let me thank you on behalf of our company, John and myself for this opportunity. As the title suggests we are going to talk about the practical aspects of condition monitoring in the overall maintenance management of the largest heavy process plant in Ireland.

2 Company background, products etc.

Aughinish Alumina is based on the Aughinish Island near Limerick on the Shannon Estuary. It is a refinery designed to extract alumina (aluminium oxide). Alumina (Al203), a fine white granular powder is extracted from Bauxite ore and is the material from which the metal, aluminium is smelted.

The plant which is among the most modern in the world can produce over 900,000 M.T. of alumina a year from 1.8 million tonnes of bauxite from Africa.

The process area is located on 360 acres in the Northern end of the 1,000-acre site and consists of the plant itself, a marine terminal through which basic raw materials are imported and finished product exported and a steam generating plant consisting of three boilers. The plant keeps more than 56 million gallons of process solution circulating through tanks, pressure vessels and pipes. The process is continuous, 365 days a year. Design of the plant incorporated extensive environmental safeguards.

The plant has its own workshop, warehouse and service facilities including medical and fire services. Tens of thousands of trees were planted in different parts of the Island and more than 100 varieties of birds live in harmony in a well preserved wildbird sanctuary established by the company.

Construction began in the 1970's and was completed in 1983. During construction, more than 6500 people were employed and for a time, it was the biggest construction project in Europe.

Alumina production began in September 1983. Capital invested in the project was in excess of 1.25 billion US Dollars.

The Company is owned by ALCAN of Canada (65%) and BILLITON INTERNATIONAL METALS (35%) part of the Royal Dutch Shell Group. It employs approximately 650 people.

The plant was grown out of vision based on low costs, high prices, and handsome prosperity for all. The company operates in the global market-place, importing most of the raw materials and selling the product in the international market.

However, Aughinish Alumina does not have the advantage of many of its competitors who have built their plants close to the bauxite mines. Hence, remaining competitive in the league is of utmost importance.

3 Maintenance Management Philosophy

The mission of AAL is :

"TO PRODUCE, SAFELY AND HARMONIOUSLY THE REQUIRED QUANTITY OF QUALITY ALUMINA IN AN ENVIRONMENTALLY ACCEPTABLE MANNER AND AT A COST THAT IS NO HIGHER THAN INDUSTRY AVERAGE"

AAL is a single product, mature technology "ore processing' business. Hence the focus is and will be on 'Manufacturing excellence'. Management of maintenance plays a vital role in this highly capital-intensive industry operating under severe conditions. At AAL, Maintenance is an activity of major scope and primary importance. Nearly 40% of the workforce (~300 employees) are engaged directly in the maintenance function. This consists of Managers, Engineers, Technologists, Supervisors, Planners & highly skilled craftsmen. The workforce is young and ambitious. The technology is most modern with very high level of automation built-in.

The equipment in this high pressure (900 psig), high temperature, heavy chemical plant is subjected to corrosion, erosion and caustic stresses. Maintenance Costs are about 12% of the total operating costs and about 39% of the controllable costs. There are over £5 million worth of inventory in the stores. Huge amount of data is collected and processed. Well developed computer systems are extensively used in the maintenance of the plant.

Fulfilment of AAL's mission of excellence therefore requires a substantial contribution from maintenance.

The maintenance management strategy at AAL is derived from the belief that : **"Maintenance needs to and must be professionally managed"** i.e. : optimising the four resources, Men, Machines, Materials and Money and ensuring the health of the assets.

This is enshrined in the published 'Maintenance Mission/Philosophy/ Principles' Document.

The strategical approach is grouped under 6 major categories :

a) Organisation & administration
b) Maintenance Concepts/principles
c) Maintenance engineering
d) Training - maintenance management and technical
e) Audits
F) Data Processing Systems

Due to time constraints, I cannot go into details of all these. However, let me elaborate on the maintenance concepts/principles which is the most pertinent to this conference. These consists of the following; P.M., Planning & Scheduling, Work order systems,

Performance Indices and Management Controls. Again, let me elaborate only on the P.M. concepts in practice in the plant.

Our definition of P.M. is "Routine periodic inspections designed to detect faults or activities designed for the upkeep of running equipment, i.e. lubrication, minor adjustments etc. P.M. should be a combination of overlapping maintenance and operations functions in order to develop a sense of ownership and caring of equipment"

Preventive Maintenance thus encompasses all the routine P.M. Schedules, **Condition Monitoring,** Reliability Centred Maintenance concepts and non destructive examination, i.e. all activities which prevent failures and ensure or increase the equipment life.

The objectives of my talk so far were to give you a 'flavour' of the overall context in which the **condition based maintenance** functions at the Aughinish Alumina Plant. Mr. John Hensey, Senior Maintenance Engineer responsible for steering Condition Monitoring in the plant will now go into the details of the subject.

I shall conclude later, time permitting, the maintenance achievements and the future plans.

4 Condition Monitoring in practice at AAL

A central Condition Monitoring group consisting of 3 fitters and 1 electrician under a supervisor and steered by a mechanical engineer was set up in late 1987. The objective was to provide a thrust to benefit from the new technology and to develop further initiatives in preventive maintenance.

The condition monitoring team provides information and analysis to each of the production sections who in turn integrate the information and analysis with their own knowledge and act accordingly. Each section takes its own oil samples for analysis by condition monitoring.

Condition monitoring recognises that mis-alignment and unbalance account for the majority of equipment problems. These areas are engaging the most attention.

4.1 Objectives of Condition Monitoring

'To maximise the performance of the company's assets by monitoring their condition and ensuring that they are installed and maintained correctly'

Strategies :

1. Oil condition analysis and oil debris analysis
2. Vibration monitoring and analysis
3. Alignment checking
4. Balancing and balance checking
5. Any other appropriate method.

Lubrication Oil Analysis

This is divided into - "oil condition analysis" and "oil debris analysis"

a) Oil Condition based on (1) viscosity and (2) water content. No other analysis performed, as engine oils are not included

b) Oil debris analysis tracks levels of
(1) metal content, (2) contaminant content, (3) total solids, (4) particle size distribution

2. Vibration Monitoring and Analysis

Vibration analysis has two components "vibration monitoring and analysis" and "vibration levels checking"

a)vibration monitoring assesses installed equipment condition over time. Deterioration are only reported in **exception** reports.

b) vibration level checking examines new and overhauled equipment for quality of alignment, balance and assembly.

3. Alignment

Alignment is carried out using the "reverse checking method" automated using optical transducers and computers. Alignment should be carried out cognizant of any distortion flanges may be causing and thermal expansion corrections.

4. Balancing

Balancing shall be carried out on all rotors passing through the workshops. They shall be balanced according to the international standard ISO 1940 initially and later to 'one grade better', if economical.

4.2 Longer Term Developments Planned

The following developments are currently being pursed

. mobile equipment lube oil analysis
. motor winding condition monitoring
. application of 'expert systems' to vibration analysis
. field balancing techniques
. thermography
. improved information integration

In addition, a demonstration rig is being built to show the effects of misalignment and unbalance on vibration. This will also be an alignment training rig.

4.3 Benefits/Problems.

The major benefit has been the overall improvement in the equipment availability and reduced costs as you would hear from Kutty. However, I would like to share some of the problems we faced and still need addressing.

. feeling of being behind the scenes, as opposed to sharp end maintenance. Also perceived as "checkers:
. Being a centralised group and of craftsmen status, the role is difficult.
. Reduced potential earnings
. Need for more training, appropriate training not readily world-wide.

CONDITION MONITORING

STRATEGY	EQUIPMENT/METHOD	BENEFITS	PROBLEMS	DEVELOPMENTS
1(a) Oil Condition Analysis	. Condition analysis contract with local laboratory	To be quantified	. Cost of analysis . Sample taking not representative sampling	. Perform analysis in-house . Improve methods of
(b) Oil Debris Analysis	. Debris analysis contract with local laboratory		. Production's understanding and perceived use of analysis.	. Internal Education programme of workshop
			. Analysis of results	. Build database of materials of construction
2(a) Vibration Monitoring Analysis	. IRD 890	Number of problems found decreasing	. In depth analysis of vibration weak	. Advanced training for team
	. Issuing exception reports	Found decreasing	. Little accumulated experience	. Encourage stability in team.
		Reduced mtce. Cost	. World wide available training costly and scarce.	. Reassurance testing and elimination
(b) Vibration Level Checking	. IRD 890	Quality Control	. Accepting need for baseline	. Internal education programme
	. Issuing "Baseline" Reports	Increased Equipment Availability	. vendor checking not done	. specs included in purchase orders.

CONDITION MONITORING

STRATEGY	EQUIPMENT/METHOD	BENEFITS	PROBLEMS	DEVELOPMENTS
3. Alignment	. Laser alignment using "Optalign"	Faster Benefits	. lack of training programme in-house	. regular retraining
	. Reverse checking method	Seen in reduced vibration problems	. Flange distortion of casings & alignment.	. Policy of repairing flange misalignment
			. Poor thermal expansion compensation.	. Develop database of corrections
4. Balancing	. IRD B50/350 balancing machine	Seen in reduced vibration problems	. Time taken operation	. Streamline balancing
			. Machine set-up course inaccuracy	. modify to improve it
			. new components not balanced orders.	. balance spec's to be included in purchase
			. expertise level low	advanced training
5. Others	. Strobes . Shock pulse monitoring . Temperature . Visual inspect			

5 Maintenance Achievements

During the past 2 years with good team effort and in keeping with maintenance mission and philosophy, major achievements were realised in the following areas :

a) production increased and the plant is operating at 114% of the rated capacity
b) Lost time went down by nearly 55%
c) Safety improvements have been substantial with zero lost-time accidents
d) Routine maintenance costs went down by 27% (i.e. about $3 million)
e) Maintenance productivity increased considerably
f) No contract labour employed for routine maintenance jobs
g) Savings of 20% achieved on contract equipment hire.
g) Stores inventory brought down from about 10M to slightly over £5M
i) productivity improvements in plant shut down and major turnaround jobs have been remarkable

Perhaps one of the most important benefits due to this approach to the management of maintenance to the company has been to **become competitive** in the league of world producers of alumina in terms of cost, quality, safety and consistency, not to speak of the improved overall morale and confidence of everyone.

6 Future

We feel that the foundation has been laid and awareness created for improved management of maintenance. Of course this has resulted in substantial gains. But, it is realised that considerable potentials still exist.

Without losing any of the gains made, and with emphasis on training of the most important asset, viz : human-ware and assisted by user-friendly computer systems, future thrusts will be directed towards improving the equipment reliability (zero breakdowns), designing out maintenance, multiskilling, reducing costs and ensuring safety and environment standards, thereby making Aughinish Alumina Limited a safe, healthy and pleasant place to work. Condition based maintenance, you can be sure, will be playing a key role in this.

Thank you.

Condition monitoring at a continuous coal handling facility

R.J. McMahon
Port of Tyne Authority, Newcastle Upon Tyne, England

Abstract
Continuous coal handling facilities operate worldwide and are used for stocking of and movement of coal at power stations and loading/unloading of ships. Typical systems cost millions of pounds to install and operate, any failure due to poor or ineffective maintenance can result in severe financial penalties for operators and carriers. The Port of Tyne Authority has such a facility which was installed in 1985 and has relied on a planned preventative maintenance program, one of the major requirements of this system requires items of equipment on the plant to be shutdown at regular intervals to allow for inspection and maintenance work to be undertaken, the frequency of which are based on recommendations by original manufacturers of the equipment which may or may not be justified by the condition of the machinery. One of the first areas to be analysed and investigated over a period of 1 year was debris analysis.

1 Introduction

Thirty years ago the handling, distribution and storage of coal was labour intensive now it is machine intensive. The Port of Tyne facility handles between 80,000 - 100,000 tonnes of coal in a week. Storing, handling and loading 6 days a week, 24 hours per day. The maintenance of the plant operates by use of a planned preventative maintenance program based on a IBM personal computer on a program called (Comac).

During the operation of plant, difficulties occurred during the initial period, involving control systems and general design parameters based on theoretical assumptions. During the past 2 years the plant has had some of its' major problems eliminated and has become more reliable, maintenance of the plant is a vital aspect and to this end it was felt that a different type of maintenance program may be more advantageous, i.e. condition based monitoring.

2 General

2.1 Description of Plant

The plant consists of an automatic rail offloading facility a coal stacker rated at 2,800 tonnes per hour, a coal reclaimer rated at an average of 2,000 tonnes per hours, a one line conveyor belt system rated at 2,800 t.p.h. and 2,000 t.p.h. with an average belt speed of 5 m/s, the lengths of the conveyor system vary from 100M to 450M in length, an automatic sampling system is also incorporated as well as automatic belt weighing facilities, a shiploader also is incorporated in the plant facilities on a berth capable of loading deep water ships. All operations are controlled by a centralised control room through electronic control systems PLC operated, (Programmable Logic Controllers).

The plant has handled a total of approximately 8 million tonnes to date and it is envisaged that yearly totals range from 3.5-4 million tonnes of coal, since the plant was originally designed at an optimum of 16 hours per day and 1.8 million tonnes per year the maintenance becomes a vital aspect, a stoppage on either incoming or outgoing facility stops the plant until a repair is effected, it is therefore crucial that if a breakdown does occur that it is repaired immediately and the minimum of downtime incurred.

2.2 Definition of Condition Monitoring

Condition monitoring is the technique of monitoring the operating characteristics of the plant, equipment or systems in such a manner that changes in monitored characterisation can be used to predict the need for maintenance before serious deterioration or breakdown occurs.

An evaluation of the various methods of achieving condition monitoring on this plant was then undertaken and two ares were identified which it was felt could be applied successfully. These being:

Vibration Analysis
Debris Analysis

This paper deals with Debris Analysis whose results are complete after one years' operation. Vibration Analysis continues to be analysed.

2.3 Debris Analysis

Most systems of built-in condition monitoring require the installation of expensive apparatus as well as the time of skilled technicians to oversee and maintain such apparatus and most importantly to interpret the information it produces. These factors are significant among the reasons why manufacturers are not eager to install condition monitoring, despite its proven benefits such as increased productivity, machine utilisation time and longer life cycle.

Several lubricant manufacturers carry out systems to analyse lubricants and an initial survey suggested such companies as Shell, Century Oils, B.P., Esso and Mobil Oil Company.

It was decided to analyse a service offered by Century Oils called Cent. Controlled Engineering Tribological service.

Century Oils was chosen for the following reasons:

2.3.1. Results were presented in a clear and concise manner and could be easily interpreted.

2.3.2. There was a large number of elements within the oil analysed.
2.3.3. Company showed greater awareness of problems in similar environments.
2.3.4. The period of time involved in the taking of samples to receiving the results and the actions advised.
2.3.5. The comments and actions recommended presented very clearly.
2.3.6. Cost per sample.

It is believed that lubricants in use provide a tangible medium not only for measuring wear debris and contamination but can be used as part of a new technological approach to analyse, compare, diagnose and predict what is happening within a system with a degree of accuracy and specific identification which will provide a powerful management tool in the maintenance of such equipment.

Basically the technology is similar to that used in preventative medicine whereby data banks are created over a period of time for patients and as a result of examinations, discussions, tests and analysis of blood samples, they are then used to monitor and direct ongoing use or abuse of the body.

3 The Cent Service

The Cent Service relies on development and detailed research between the supplier and the customers engineering and maintenance staff.

The objectives of the service are to determine the condition of the lubricant and the assessment of the condition of the equipment from which the sample of oil is taken. This part of the service is obtained by interpretation of the results obtained from a series of predetermined analytical tests carried out by regular oil samples.

From the oil samples a comprehensive series of tests are undertaken and the results are interpreted by competent personnel qualified in both engineering and metallurgical disciplines. These results are then assessed over a period of time, changes which occurred in both the condition of the lubricant and also equipment from which the sample of oil was taken are observed.

Rapid increases in wear elements which occur may be easily identified and possibly used to predict a failure.

This information can be provided in numerical tabular form, however to examine such a mass of figures and validate any remarks could be impracticable.

A Go-No-Go system has been instituted which combines with a graphical trend analysis which displays those salient points appertaining to comments on the report.

This simplified the interpretation of the results enabling the necessary action to be taken.

Machine components and metallurgical data files have been compiled together with the trends relating to wear metal generation during service conditions.

From comprehensive data bases, limits on wear metals and all other elements within the analysis can be established.

In some cases, this enables not only a prediction of imminent machine failure to be made but also identified the specific components involved and the operating conditions that the machine is experiencing.

This information assists in maintaining a high degree of machine reliability and identifies problem areas, it allows the user to maximise the use of the lubricant and extend the life by giving positive indications, it also indicates if filtration systems are operating by the levels of contamination, i.e. dirt, dust, water and also any additives in the oil can be identified.

4 Description of Equipment

For the purpose of evaluation gearboxes were chosen in all the vital areas which could cause major breakdowns and delays.

All gearboxes chosen are manufactured by Flender and Baur.

The Flender gearboxes being Redurex double reduction bevel-helical gear units, foot mounted and torque area mounted which are linked to a oil filled flexible coupling directly driven by an AC motor.

Baur gearboxes being three phase spiral bevel multi mounting geared motor/gearboxes.

All gearboxes on the plant are listed below for this evaluation.

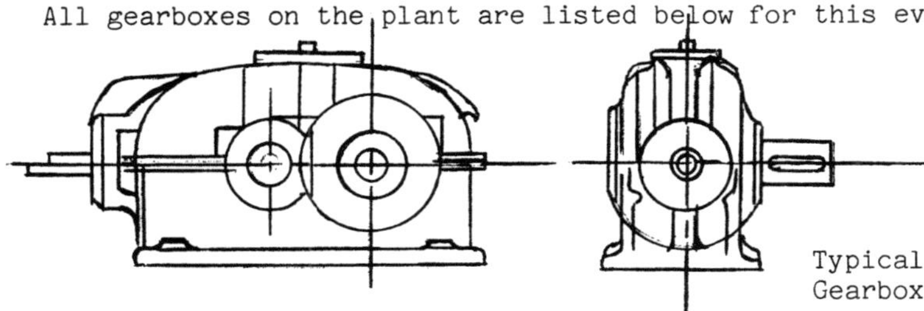

Typical view C3 Gearbox

Five gearboxes were chosen initially, C3 main drive, two on stacker main drive, reclaimer bucket wheel and C2 main drive. At a later date a further 11 gearboxes were added to the evaluation, these being C1, C4, C5, C6 and C7, Stacker Long Travel x 4, Reclaimer Slew x 2.

Initially to establish a data base 4 samples of each gearbox were taken over a fairly short period, i.e. 2-3 weeks and then an assessment made of any trends were undertaken. After this a normal sampling period was established.

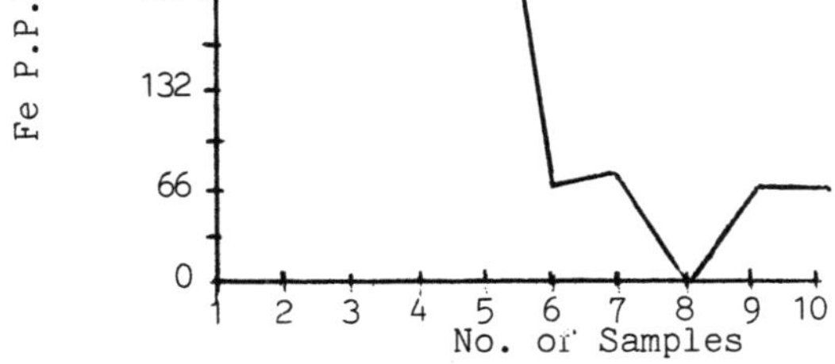

Graph Showing:
Fe/No of Samples
C3 Gearbox

After the initial samples had been analysed it was decided with consultation of Century Oils that 15 elements were to be monitored, these giving the necessary information to enable a positive indication of the condition of the gearbox, (see below).

All measurements in P.P.M. except water content %WT Kinemetric Viscosity also being part of the elemental analysis.

The Elements Monitored were as follows:

		Typical Source
Fe	- Ferrous iron	Cylinders, gears, rust crankshaft
Zn	- Zinc	Additives, bearings platings
Al	- Aluminium	Pistons, bearings, dirt, additives
Cr	- Chromium	Cylinders, rings coolant gears
Ni	- Nickel	Shafts, gears, rings
Mo	- Molybdenum	Rings, additives, steels
Pb	- Lead	Bearings, fuel, paint additives
Cu	- Copper	Bearings, bronze
Sw	- Sodium	Coolant additives
Ca	- Calcium	Additives, water
P	- Phosphorous	Additives, coolants, gears
Si	- Silicon	Dirt, lubricants
Mg	- Magnesium	Bearings additives, supercharges, dirt.

5 Results Over Period of 1 Year

Table Gearbox	KV40 (PPM) Below	Wear/Contaim FE	ZN	21	Additives P	Water	Condition
C1	300	50	10	5	200		Satisfac.
C2	333	30	5	5	-		"
C3	350	396	-	-	900		Investigate
C4	336	46	-	-	200		Satisfac.
C5	340	37	10	5	200		"
C6	330	20	20	20	220		"
C7	330	50	50	50	180		"
Stacker No.1	330	100	-	-	-		"
" No.2	330	239	-	-	-		Oil Changed
" L/T No.1	230	93	-	-			Satisfac.
2	390 Peak	1571	-	-		4.3%	Oil Chaged
3	220	156	-	-	300		Satisfac.
4	230	160	-	-			Oil Changed
Reclaimer Slew 1	388 Peak	1551	-	-	-		Investigate
" " 2	220	1793	-	-	-	1.66	"
Bucket Wheel 1	220	60	-	-	300	-	Satisfac.

5.1 Summary of Results
10 Gearboxes satisfactory, 3 gearboxes oil changed and 3 gearboxes required further investigation. C3 Reclaimer Slew 1 and 2.

5.2 C3 Gearbox
This gearbox was taken out of service and examined visually. Gears showed fretting corrosion and the hollow shaft showed signs of substantial wear. The gears, bearings and hollow shaft were all replaced and the gearbox put back into service, sample taken after repair showed condition satisfactory.

Possible reason for wear overloading of conveyor.

5.3 Reclaimer - Slew Gearboxes
The oil was changed and then resampled, the results showed a slight decrease.. Investigations showed the final drive pinions to have considerable wear on both gearboxes. These pinions have been temporarily repaired, and the manufacturers contacted to assist in either repairs or replacement. These two gearboxes are still in service and remain a problem.

6 Discussions/Conclusions
As can be seen from the results out of 16 gearboxes sampled, 62% condition was satisfactory, 19% required an oil change and a further 19% demanded action in order to resolve problems. The evaluation has proved considerably worthwhile giving the user a complete picture of the various gearboxes on the plant without having to take the gearboxes out of service and allowing a planned removal for decisions. A trend has been established on all the gearboxes considered and a history of any failure recorded. Problems which occurred were sometimes simply lost in transit, the samples have to be taken cleanly and correctly, a planned program of sampling has to be established and the cost has to be maintained at a low level.

This system is at present being further expanded to include all the gearboxes at Tyne Coal Terminal. Debris Analysis has proved to be a successful medium but has to be used in conjunction with other factors i.e. vibration, temperature, engineering knowledge and evaluation of the environment and duty of the particular machines being monitored. The Cent Service has proved that provided you evaluate the results on a trend basis and not as one sample in isolation of all the factors that the system is worthwhile and gives the engineer a much clearer picture of the condition of the plant without the need to strip down and examine costly process plant at pre-determined intervals.

Condition-based maintenance in the FMMS environment

E.J. Morris, BSc.
Resources Development Manager, British Aerospace, Brough, UK

SYNOPSIS

Following the installation of a Flexible Milling Manufacturing System at British Aerospace, Brough, it became obvious that new maintenance policies would have to be devised. Using the information gained on similar machines over the previous three years, combined with the results of a year's trials on proprietary monitoring equipment, it was decided to adopt a mixed policy of preventative and condition based maintenance.

To cope with the F.M.M.S. environment a device was installed capable of unmanned, round the clock monitoring. The device is synchronised with the logic controller of the milling machine and capable of performing most of the currently available monitoring techniques through different software modules.

This paper will concentrate on the practical steps which should be undertaken to achieve a mixed maintenance policy integrated with production and commercial aspects, thus achieving a truly integrated manufacturing environment.

CONDITION MONITORING IN THE F.M.M.S. ENVIRONMENT

Background

At the British Aerospace factory at Brough, North Humberside a decision was taken in 1984 to embark on a programme to convert a number of existing Kearney and Trecker Max-e-Trace Routing Machines for use in a Flexible Milling Manufacturing System (F.M.M.S.). Projected investment costs were in the order of £5m.

These six existing machines, which were to form the core of the F.M.M.S. facility, were of varying vintage from 5 to 15 years old. They were a mix of two and three spindle machines, all were three axis vertical milling and routing machines with Allen Bradley C.N.C. control systems. All except one used hydraulic powered axis drives, the exception being D.C. driven. Spindle drive in all cases was provided by rotary frequency changers.

The initial F.M.M.S. conversion package included automatic robot arm controlled tool changing, automatic pallet loading, high speed brushless D.C. router heads, adaptive controlled spindle speeds, D.N.C. control, solid state 50 h.p. static invertor spindle drives, "at source" swarf extraction and various other "state of the art" sophistications. The basic machines, i.e. axis drives, gearboxes, etc., although refurbished as part of the conversion, were retained.

These converted machines came on line between 1987 and 1989. They, of course, represent a Maintenance Manager's nightmare in that the core machines are ageing veterans, whilst the conversion largely consists of unproven development systems using prototype equipment. Needless to say, these machines were expected to produce at maximum capacity from Day 1.

Optimistic performance figures projected increased output per machine by a factor of five. This was to be achieved not just by improved logistic efficiency in terms of tool changing and loading, etc., but also by increasing the speeds and feeds of cutting.

Ultimate intentions are to achieve an unmanned, lights out, round-the-clock operation and production programming already reflects these intentions. The problem facing Maintenance Management was of maximising the availability and reliability of these hybrid machines.

Once the F.M.M.S. installation had been approved it became apparent that the previous policy of breakdown maintenance was untenable in terms of disruption to the production cycle. Conversely, a fully planned maintenance schedule was equally unacceptable due to the increased machine downtime involved. So it was decided to pursue a policy of mixed preventative and condition based maintenance in order to reduce total lost time to a minimum.

The mechanical components for the core of the F.M.M.S. machines were common to the previous generation of N.C. machines, so it was possible to analyse the predicted failure modes of the machine before installation.

Analysis of failure data showed where a preventative maintenance scheme would be cost effective, and such a scheme was set up as each machine was converted to F.M.M.S. Analysis also showed that with many faults the spread about the mean time between failures was large so that preventing a failure by preventative maintenance would not be cost effective. Indeed, in some cases, a policy of breakdown maintenance was determined to be more effective. It was decided that, where possible, maintenance would be based on the observed condition of a machine component. In order to establish a workable monitoring programme, it was realised that the environment which had been created would have to be analysed.

THE F.M.M.S. ENVIRONMENT

1. The Working Environment

The most obvious difference with traditional machining areas is that the machines are now remote from the operators, with a large number of automatic systems liable to move without warning. The unmanned area is large and surrounded by an eight foot high fence with interlocks. (See fig. 1). This means that the best possible condition monitor, an experienced operator, has been removed. On previous machines the operator would report a "funny noise", or "a bit of a roughness", which could be investigated by maintenance crews in planned downtime. An experienced operator will also adjust a machine by altering speeds and feeds in order to continue with the current job until something can be done. It was quickly realised that remote monitoring was needed in order to provide adequate and continuous cover. In many ways, the remoteness is similar to some process industries where access is either impossible or restricted during production. The idea of "dummy runs" for monitoring during downtime was rejected as being too time consuming when that downtime was needed for repair. Nor would this method provide information fast enough; by the time readings have been taken and analysed the planned downtime would be over.

2. The Planning Function

The other similarity between process industries and the F.M.M.S. environment is that to the production planning function the system brings about an almost continuous flow of work which can be carefully scheduled ahead. The cost of unscheduled downtime is, therefore, comparable with the process industries and the savings available through monitoring are of the same order. As the production function is able to rapidly alter schedules to meet changing conditions, so the maintenance function must be capable of using similar methods to schedule the work to be done during planned downtime, or even unplanned breakdown time should a breakdown occur.

3. The Machine

Whilst at the scheduling level, the F.M.M.S. environment can be viewed as being highly planned and continuous. When the actual process of cutting is considered, the parameters of the process are constantly changing. Speeds, feeds and other conditions are changing rapidly in response to the part programme. If a monitoring programme is to have validity, then any reading must be taken in the context of the conditions prevalent at that time. Indeed, monitoring the conditions of a reading is as important as actually taking the reading. To achieve knowledge of the conditions it is necessary to link into the machine and continually access the demanded speeds, feeds, etc. This should be backed up, where necessary, by speed sensors which can, as a by-product, give indications of machine condition through observing the response of a machine to a change in the demanded conditions.

This policy of linking into the machine leads logically to the concept of a machine in which the condition measuring functions are performed by modules designed directly into the machine, with 'on board' data acquisition, analysis and prognosis. The advent of cheap micro-processing facilities must eventually make such a system standard on many machine controllers.

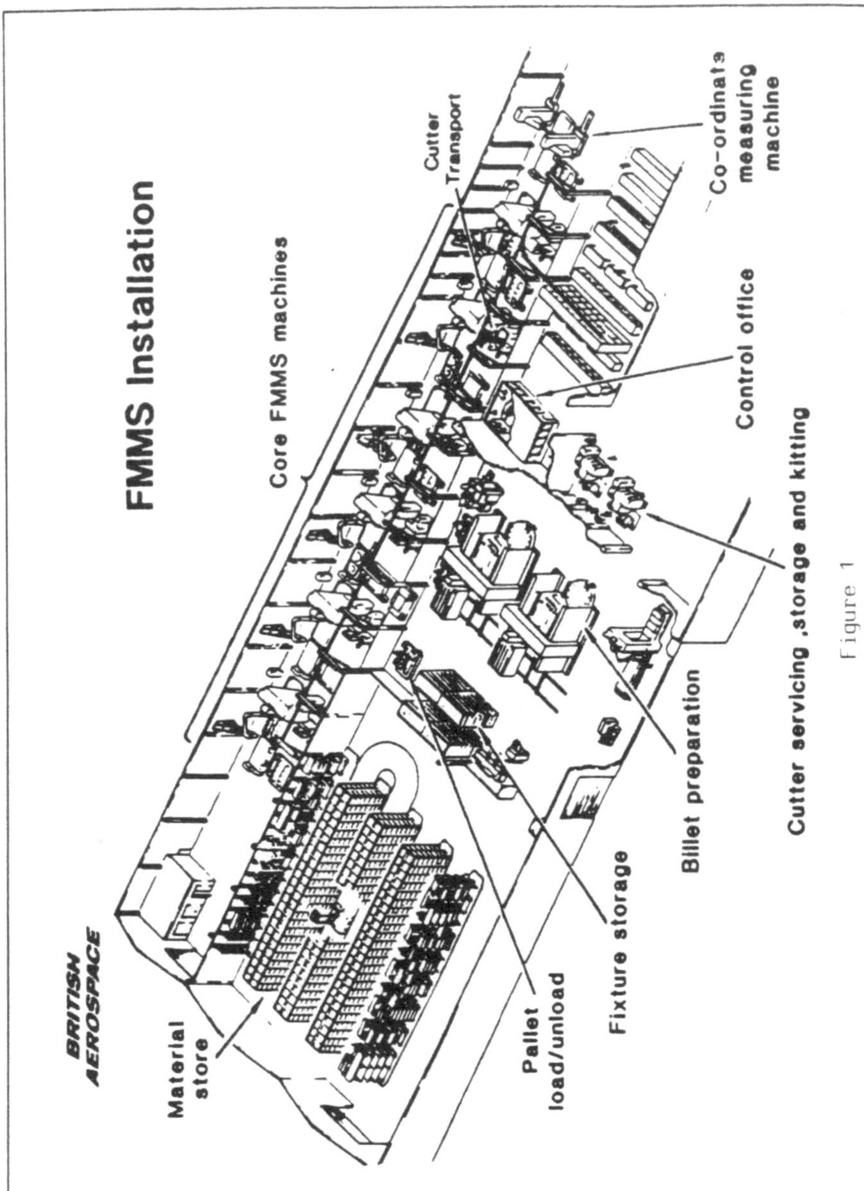

FMMS Installation
BRITISH AEROSPACE
Core FMMS machines
Cutter Transport
Co-ordinate measuring machine
Control office
Cutter servicing ,storage and kitting
Billet preparation
Fixture storage
Pallet load/unload
Material store

Figure 1

4. System Integration

Where the F.M.M.S. environment differs from all other areas which have previously been monitored is in the integration of all elements of production from stores and ordering through to inspection. However, the maintenance function is largely ignored by Production Departments, who blithely assume that the machine manufacturer's promises of reliability will be upheld. Under these conditions there is a need to integrate the maintenance planning and manufacturing systems in a dynamic manner. This will require up-to-date and accurate maintenance information, the object being to produce a fully integrated manufacturing environment.

In the absence of a fully integrated system, maintenance must develop a parallel system with as much interface as possible.

5. Summary

So the requirements for Condition Based Maintenance in the F.M.M.S. environment can be summarised:-

1. It should be based around anticipated failure modes and use a mix of preventative and condition based maintenance.

2. The system should be able to operate remotely for 24 hours a day.

3. The system must integrate with the N.C. controller and take account of the perceived conditions.

4. The system should automatically analyse the data and flag up oncoming faults with meaningful messages.

5. The system should integrate with Manufacturing systems to ensure proper planning of maintenance tasks.

APPROACHING THE PROBLEM

Having defined the requirements for monitoring the F.M.M.S., the problem was approached from several different angles.

1. Breakdown Analysis/Preventative Maintenance

A breakdown monitoring programme was set up to analyse the failure modes once the machines were installed. This would allow preventative and condition based maintenance to respond to changing breakdown rates.

This quickly showed that the 'core' mechanical components had acceptable reliability, i.e. below 10% failure rate, but that the peripherals added to achieve the F.M.M.S. concept were the most likely cause of failure. Tool changers, pallet loaders and tool probing caused, and still are causing, a constant succession of small stoppages, often with no fault to be found. Such failures are seen as development faults to be removed by re-design or a change in working practice. (See fig. 2). Of those elements of the machine which could be monitored, it turned out to be mainly elements on the core machine which offer the greatest savings. The peripheral items tended to have random or unknown faults, whereas the core items, heads, gearboxes and bearings, are monitorable and cause long breakdowns when they fail.

Arising out of the breakdown figures it was possible to establish a preventative maintenance programme. The programme started with such obvious things as changing filters and oil and moved using the breakdown analysis to include such things as:-

Changing the drawbars every 3 months.
Checking/changing pallet load chains every week.
Clearing the cable runways of swarf.

2. Specifying a Monitor

Based around the breakdown data available, it was then possible to design a monitor for an F.M.M.S. machine and ensure that from the second machine onwards the wiring and instrumentation for condition monitoring were included in the original build specification of each machine. Screened cables were run to all monitoring points and mounting holes were drilled at those monitoring points ready for the final implementation of monitoring.

The need to match the integration of the F.M.M.S. system in the maintenance system led to a specification for a monitoring computer for each machine. Looking at the proprietary products for machine health monitoring, it quickly became apparent that no method was entirely reliable and that the use of two methods of detection for each fault was highly desirable. Most methods were found to be based around vibration measurement, e.g. shock pulse, kurtosis or full-scale analysis. Each, however, required the purchase of a proprietary "black box" operating independently.

It was decided to look for a device capable of integrating all methods of monitoring, including temperature, pressure or any other measurable parameter. Added to this would be needed an onboard micro-computer to do acquisition analysis and prognosis in real time.

Originally, it was envisaged that B.Ae. would have to develop this, probably based around the S.T.E. Bus, but at this point, Stewart Hughes were approached and found to have developed an almost identical philosophy. They had also produced under Esprit 504 an intelligent data acquisition system of the type required, which had already been demonstrated on a test F.M.S. Cell in Spain. This cut short the development time needed for data acquisition, leaving the more complex problem of diagnosis and decision making to be addressed.

The Stewart Hughes Data Acquisition and Analysis System (D.A.A.S.) contains all the elements of a fully integrated condition monitoring system. It is based around the Motarola 68000 family of processors running on a G96 bus and contains the core elements of communications, signal acquisition, analysis and on-board diagnostics as a series of plug-in cards and software modules. Instead of buying a series of "black boxes" for each monitoring method, it is now possible simply to add software to the D.A.A.S. to simulate such things as resonance testing, band pass filtering, F.F.T. analysis, etc. (See fig. 3).

3. Preliminary Trials

Whilst waiting for Capital approval for a D.A.A.S., it was decided to pursue the diagnostic problems from several angles.

Portable equipment was hired at Brough to take baseline readings on the equipment in situ, so that when the D.A.A.S. arrived much of the initial data gathering would have been done. The hire of portable equipment had the added advantage of giving shop floor maintenance personnel an opportunity to gain an understanding of condition monitoring techniques before a full-scale system was installed. The portable monitor was so successful that one has since been purchased for use on non-F.M.M.S. machines around the factory and is now accepted by maintenance personnel as a diagnostic tool when a machine fault proves difficult to locate.

MEAN TIME BETWEEN FAILURES REPORT

FOR TR3 BETWEEN 01/07/86 AND 01/09/88

A typical Mean Time Between Failures Report used to determine preventative maintenance schedules. Faults are coded by type and the most common and regular faults are scheduled for repair on a regular basis. Other faults become the subject of condition Monitoring methods.

Page No 1
01/09/88

FAULT	TOT TIME	% of B/Down	no B/Downs	MTBF	Ave Repair Time
TOT	4268.49		757	15.69	5.64
TCT	975.79	22.86	241	49.29	4.05
HEA	580.91	13.61	70	169.70	8.30
MEC	459.92	10.77	31	383.20	14.84
RRR	395.10	9.24	15	791.94	26.34
GEO	371.08	8.69	33	359.97	11.24
PRO	347.77	8.15	86	138.13	4.04
CON	280.43	6.57	41	289.74	6.84
PAL	262.73	6.16	91	130.54	2.89
BAD	78.84	1.85	12	989.93	6.57
DNC	75.77	1.78	18	659.95	4.21
PTF	71.81	1.68	1	11879.10	71.81
DRA	53.86	1.26	26	456.89	2.07
BED	49.68	1.16	3	3959.71	16.56
ELE	38.40	0.90	14	848.51	2.74
CDS	37.00	0.87	6	1979.86	6.17
HYD	32.55	0.76	9	1319.90	3.62
SPI	28.45	0.67	3	3959.71	9.48
STA	27.20	0.64	4	2969.79	6.80
CVR	24.35	0.57	4	2969.79	6.09
COO	20.96	0.49	24	494.96	0.87
VAC	16.90	0.40	7	1697.02	2.41
BRG	8.22	0.19	1	11879.10	8.22
BOX	7.90	0.19	1	11879.10	7.90
TAB	6.80	0.16	3	3959.71	2.27
AIR	6.00	0.14	5	2375.83	1.20
	5.00	0.12	1	11879.10	5.00
DEC	2.50	0.06	2	5939.57	1.25
MSC	2.05	0.05	4	2969.79	0.51
OIL	0.50	0.01	1	11879.10	0.50

Figure 2

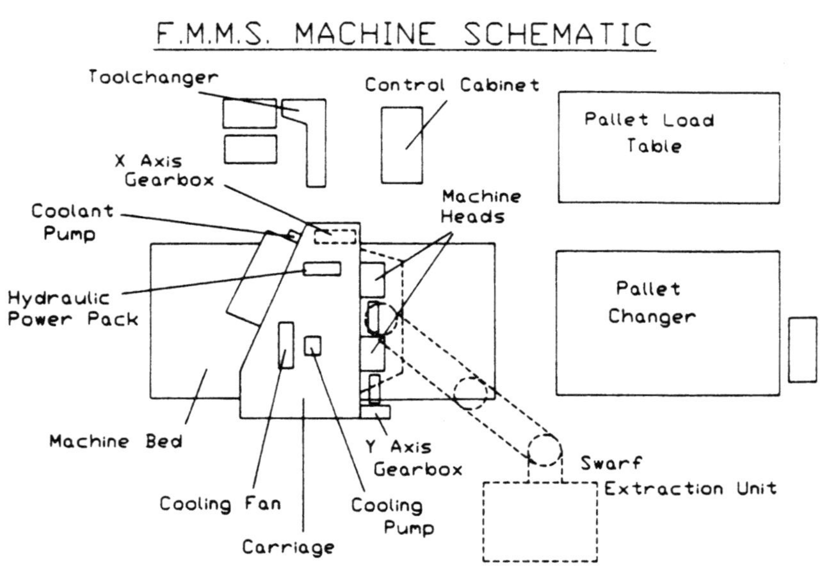

Figure 4

The original portable analyser was a Bruel and Kjaer BK2515 Analyser, chosen as a rugged, battery powered F.F.T. analyser which could be linked to an I.B.M. P.C. This was later found to be too heavy. The final selection for a portable analyser is the Endevco Microlog System with Entek software for analysis and route control. This analyser can monitor far more parameters and has a route for operators to follow programmed within itself for ease of use. It is also light enough to be carried up ladders in safety to reach inaccessable parts of the plant.

The portable equipment was used on dry cycle runs during shut-down periods with a certain degree of success. However, this period merely emphasised the need to take on-line readings with analysis at frequent intervals.

It was found that proprietary methods of wear detection with absolute indications of good and bad were unreliable. Often a component would be rebuilt as new and would read as 'bad' even though capable of adequate service. A more useful method is to take a reading when new and look for a rise above this level, even if the original level is high to start with.

At the end of this investigation, the major components which would be monitored were known and initial monitoring methods were determined.

The monitoring points and methods are as follows:- (See fig. 4).

Machining Heads	Full Scale F.F.T. Analysis. Order analysis. Torque measurement.
Gearboxes & Axis Drives	F.F.T.Analysis. Band pass. Response curves.
Carriage Way Bearings	Shock detection F.F.T. Analysis Torque/Power Monitoring.
Hydraulic Valves	Response Curves.

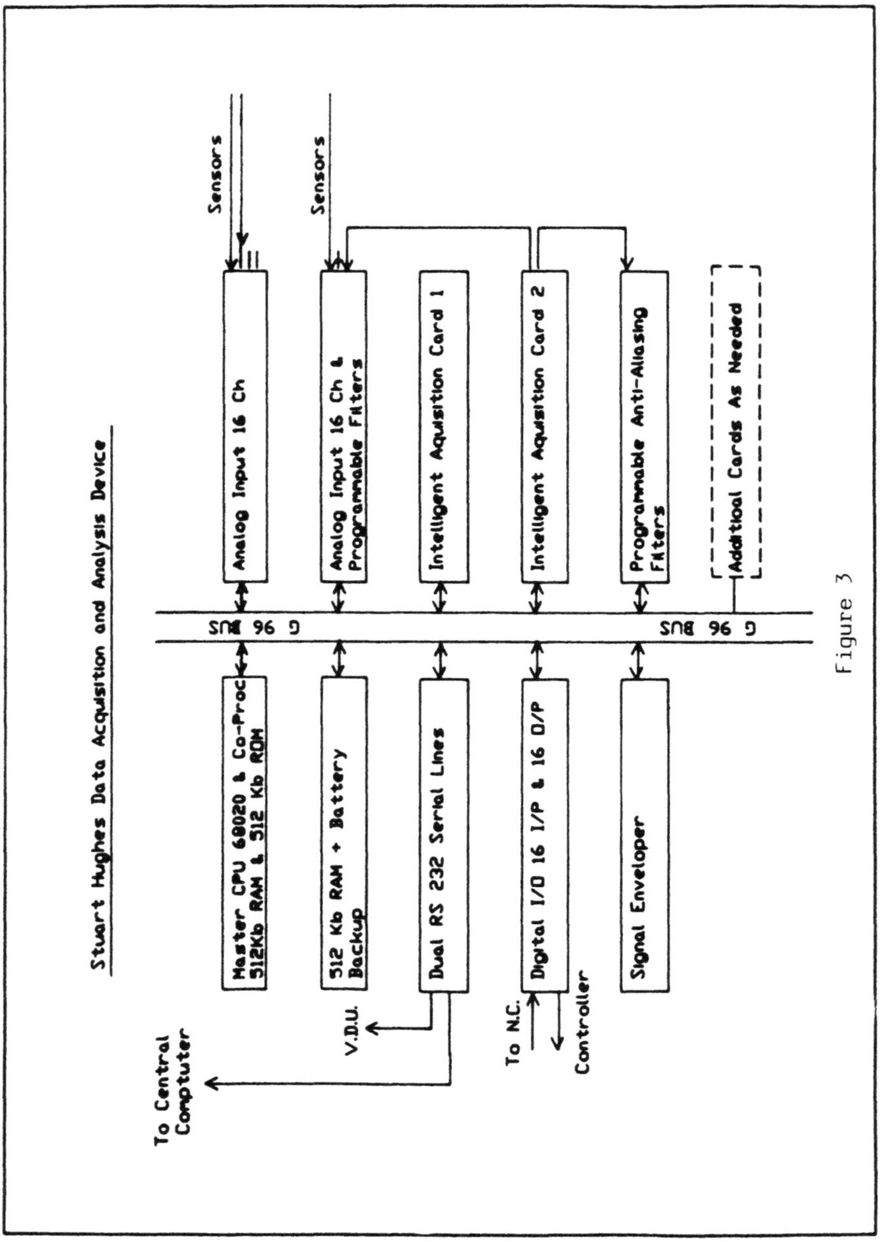
Stuart Hughes Data Acquisition and Analysis Device
Sensors
Sensors
Analog Input 16 Ch
Analog Input 16 Ch & Programmable Filters
Intelligent Aquisition Card 1
Intelligent Aquisition Card 2
Programmable Anti-Aliasing Filters
Additional Cards As Needed
G 96 BUS
G 96 BUS
Master CPU 68020 & Co-Proc 512Kb RAM & 512 Kb ROM
512 Kb RAM + Battery Backup
Dual RS 232 Serial Lines
Digital I/O 16 I/P & 16 O/P
Signal Enveloper
V.D.U.
To Central Comptuter
To N.C.
Controller

Figure 3

Hydraulic Pump & Motors	Debris analysis - On-line Vibration Analysis F.F.T. & Band pass. Period Lab. tests on oil samples. System pressure response.
Cooling & Coolant fans & pumps.	Vibration level. System Back Pressure.
Swarf Extraction Fans.	Vibration level & Band pass analysis.

MONITORING METHODS

The most useful method seems to be measuring vibration, which allows several different analyses on each point using the same equipment, e.g. kurtosis, shock pulse or F.F.T. analysis. In addition, the amount of data contained within the N.C. controller itself can give warnings of incipient problems. For instance, monitoring the control signals for an axis when subjected to a step change in velocity command, will give a measure of the axis damping ratio, response time and overshoot/backlash. There is a lot of good information locked away inside the N.C. controller which, if linked to a few sensors, can be used to determine the machine condition.

The hydraulic oil on the machine is kept at a very clean standard by constant filtering in the tank. By installing an on-line oil debris monitor which automatically samples various points on the hydraulic circuit, it is possible to detect debris production early and isolate the area of the circuit at fault. The original debris testers were based on the magnetic plug, eddy current method. This proved unsuccessful since the majority of debris is non-ferrous, consisting of either brass and chrome from motors and fittings or rubber particles from flexible pipe frettage. The Lindley Flowtech fluid condition monitor now installed after a 6 month trial uses a filter blockage method to measure debris and has been very successful. On-line debris monitoring is backed-up by a laboratory oil analysis using spectrometry to determine the elements comprising any debris and give a diagnosis.

The amount of data produced by an on-line monitoring system is immense and much local data processing and reduction must be undertaken. To facilitate this the system operates in three modes. (See fig. 5).

Mode 1. Protection.

Most of the time the system sits monitoring the N.C. controller, whilst intelligently polling through the monitoring points looking for gross errors which would indicate failure of a component or tool. The last few seconds of data are always retained so that if a problem occurs then the machine status at breakdown can be determined. This mode will be capable of implementing a machine shut-down if a gross error is detected, though this facility will not be connected until enough confidence has been established in the system.

In time it is hoped to use this mode to adapt the machine parameters to the conditions experienced, e.g. by slowing down a machine to cope with a faulty machining head or blunt tool. In this way, a fault tolerant machine as envisaged by Esprit Project 504 will be developed.

Mode 2. Prediction.

The system has programmed within itself those parameters which constitute suitable conditions for taking a "constant conditions" measurement for trending purposes. Whilst in 'protection' mode the machine is recognising cutting and non-cutting cycles which are suitable for such readings. The data taken during such cycles is analysed more deeply to look for smaller incipient faults and once compressed is transferred to a central computer which handles detailed analysis and trending. In this way, only measurements suitable for trend analysis are stored at the central computer and the data output is restored to manageable limits.

The system reacts intelligently to rising levels in that as alarms are approached, or even exceeded whilst waiting for a planned shutdown, so the machine will take readings more frequently, e.g. a perfectly good coolant pump may only produce one trend reading a day, but as failure is approached this may rise to a reading an hour or even more.

This mode provides the necessary information for the planning of maintenance tasks on a weekly basis by the central computer.

Mode 3. Diagnostic.

It is impossible to completely eliminate all breakdowns so this mode allows a maintenance operator to download the machine status immediately prior to breakdown and to programme a dry cycle test on a machine with detailed analysis for investigation purposes. Often a breakdown will not be due to a failure, but could be due to a degredation of surface finish. Such faults often require lengthy investigation with a "suck it and see" attitude being adopted. Diagnostic mode advises which component is most likely to be the cause of the problem, thus saving unnecessary investigation time.

This mode was invaluable in the early stages in delivering the base-line measurements necessary to set up the system.

WORK SCHEDULING

As a part of the whole maintenance function, the condition monitoring system injects jobs to be done directly into the preventative maintenance scheduler. This scheduler works in conjunction with Stores and Production Planning to ensure that maintenance tasks can be performed immediately a machine becomes available.

With the introduction of Direct Numerical Control (D.N.C.) to the F.M.M.S. it is now possible to determine the production status of each machine in real time and take advantage of unscheduled failures by fitting upcoming scheduled jobs into such extra time as comes available.

The impact of scheduled maintenance upon the Production Department's thinking should not be underestimated. Personnel who in the past have seen maintenance as a necessary evil must now be prepared to work alongside maintenance crew. They must also see maintenance as a necessary part of the production process. If large automated plant is to function efficently then all components must be working, for the failure of one component may well destroy the carefully worked out schedule and confuse the whole system.

A change of this magnitude to production thinking will, however, have to come from the very top, with Management commitment to breaking down the traditional rivalries between maintenance and production staff.

SUMMARY

In summary then the implementation of a successful condition based maintenance package for an F.M.M.S. system relies on the following criteria:-

The data acquisition must be remote, automatic and relatively intelligent.

Data acquisition must be linked to the N.C. controller of each machine so that measurement parameters are known.

Data must be automatically analysed to reduce the permanent data storage for trending.

Oncoming faults once detected must be subject to scheduled repair at the optimum time giving regard to available resources and production schedules.

Most importantly.

The commitment to implement an effective maintenance policy must reach right to the top in all areas of the organisation.

Predictive maintenance programme for GEN sets and reciprocating MUD pumps of oil fields

R. Pandian
Oil and Natural Gas Commission, Madras, India
Dr B.V.A. Rao
Indian Institute of Technology, Madras, India

Abstract:
Power and high pressure mud management system is a core function in any oil field operation. Consequential cost of any failure results in an exponential manner through out the chain of drilling activity. Under an open pay zone condition, the effect would compound leading to complications in oil recovery. Where such far flung effects are involved in terms of cost of failure, the demand of availability and reliability is not the final requirement of a maintenance manager. Monitoring the trend of all the achievements and failure also becomes an important activity to device a means for all the time injection of dependability. In this trend analysis the diverse and concurrent behavior of different group of equipments are to be monitored in a manageable manner for setting up the hypothesis structures to derive fairly repeatable and accurate predictions.

1. Introduction:

The drilling for hydrocarbons is one of the difficult operation both in terms of cost as well as technical. The drilling operation contains more uncertainties. For the drilling of hydrocarbons the mud plays major role.

The functions of mud in drilling are:

- effective & complete clearing of the bottom hole of the cuttings to the surface
- prevent the precipitation of the cutting at the bottom when circulation is discontinued
- help augment the formation pressures but avoid the fracture of the formation
- help augment the stabilities of the rocks of the hole wall
- cool & lubricate the friction surface particularly the bit
- not impair the oil trapping properties of formation
- etc., etc.,

From the above it is seen that the mud pump which is used for pumping the mud at the DG Sets which the power source for all the equipments are the critical ones.

2. The risk and catastrophe in drilling:

The drilling encounters uncertainties and the risk contributing factors if accumulates mainly to a blow out which will spoil the rig, the well and also the reservoirs some times apart from spoiling the environment. In drilling the risk is in-built right from the beginning of exploration of hydro carbon and stays until the well is successfully completed for production.

The wild cat wells costing may give some disturbing trend in case all the drilled wild cat wells in a region/formation turned to be dry. In order to avoid the possible risks of blow out the major equipments of the drilling rig should not fail at the required juncture.

3 Dependence of DG Set & Mud Pump in drilling activity.

Oil drilling is the most important and harsh activity next to geological field work. Drilling establishes the communication between the reservoir and surface. To carry out the topic the following systematic engineering activities are carried out.

Hole making.
Hole consolidation.
Establishing communication between the pay zone and surface.
Production activities.

Hole making is the drilling activity carried out in the following sequence:

Cutting (drilling) the soil.
Removal of the cuttings from the hole.
Hydrostatic balancing of the hole for establishing the hole physically to counter the subsurface pressure
Casing the hole.

The mud system is the central point for the above operations and mud pump is the most vital equipment equipment of the mud system. In case of the failure of the high pressure mud system during the open pay zone it will lead to blow out. Blow outs are catastrophe to environment. During the drilling operations also the high pressure mud should be available without any interruption. Failure of this will lead to may complications

Thus it is clear that the mud pump which is the vital part of the mud system is very important in drilling. The source of power for these mud pumps are the DG sets. In recent past the drilling rigs are of AC-SCR type because of its numerous advantages. The mud pumps are triplex single action type and DC motors for used for powering then

4 Various Methods of Monitoring:

The health and condition of the equipment are monitored by analysing the symptoms, to predict its likely failure/break down by any one or combination of the following condition monitoring methods.

Corrosion Monitoring.
Contamination Monitoring.
Pollution Monitoring.
Particle Testing & Monitoring.
Force Monitoring.
Temperature Monitoring.
Sound Monitoring.
Vibration Monitoring.
Performance & Performance Trend Monitoring etc.

Some of the companies had developed engine performance monitoring methods & instruments and patended them. These are based basically on the methods described above. Some of such monitoring available are

- Sulzer engine monitoring.
- Wartsila's Encom Engine Condition Monitoring.
- Cooper energy services - Engine Analyser etc.etc.

The main draw back of the above monitoring system are that it can not be used in all make/type of engine and also it required planning at the stage of ordering the engine itself.

5 Suitability and selection of condition monitoring methods.

Combination of minimum monitoring techniques should give a complete insight to the existing problems so that the diagnostics can be completed in minimum possible time &cost for an effective diagnosis of the problems.

5.1 For D-G sets

The best monitoring method for any engine is using vibration monitoring coupled with particle analysis. By judiciously combining Lub Oil monitoring with Vibration monitoring more than 98% of major problems can be diagnosed well in advance. Sound monitoring can be used to supplement the vibration monitoring to increase the prediction of probable faults/deteriorating health to about 93% reliability.

The vibration monitoring at the following locations of the DG Set will provide the clues leading the diagnosis of defects/faults to maximum extent.

1. Engine base fixing locations.
2. Crank Case Fixing Location.
3. Engine Crank Center Lines.
4. Cylinder body & Cylinder Heads.
5. Turbo Charger.
6. Pillow Block Bearings.
7. Cam and idler gear locations

8. Generator Bearings.
9. Generator fixing locations etc.,

It could be seen that the data collected from the above mentioned locations shall be quiet voluminous and therefore a proper interpretation is always complex. Hence computerisation of the data storage and interpretation for effectiveness is done.

While assessing the problems in engines the influence of inertia forces, mass acceleration, piston slap excitation, force generated from combustion etc.,are to be specifically considered.

Due to the effect of partial balancing of inertia forces, always a significant unbalance force shall be existing in engines. The frequency equal to the major component shall also be a strong excitation frequency in the engine. Many times the harmonic excitation frequency shall coincide with the major order component and therefore the assessment of inaccuracies resulting to the simple harmonics can be confusing unless other supplementary data are made available.

The wear particle content from the Lub Oil analysis can help to bridge this gap in many cases. If the vibration frequency resulting from increased bearing clearances is dominant the same can be correlated with the wear method from the bearing or from the engine oil pressure drop. The piston slap frequency if dominant the same can be reconfirmed by wear content from the oil of piston as well as the cylinder liner. The deterioration of gears in the system can normally be identified in the vibration spectra itself.

The influence of firing orders, misfiring, incomplete combustion will result in unsteady and non-synchronous vibration frequency. These influences cannot be supplemented by lub oil analysis whereas the engine performance will reveal such situation.

It has been found in practice that a combination of vibration analysis and engine performance analysis is the best suited package to monitor the engine driven generating sets for repeatable&accurate predictions.

5.2 For reciprogating Mud-pumps.

The reciprogating mud pumps falls in the category of low speed category equipment as the stroking speed is 140 per minute (max) but the drive and the intermediate chain transmission falls in the category of medium speed. Thus making these piston type reciprogating mud pump monitoring highly complex. Apart from the above the main bearing of the crank shaft system is not assessable from outside and therefore special provisions for the data collection are to be made either by incorporating permanent transducers or by opening the specified inspection doors for reaching the inaccessible locations.

The vibration monitoring is suited for the monitoring of the mud pumps but care is to be taken during interpretation or measurement stage to analyse the health from the gear chain transmissions, medium speed motors, low speed reciprogating & rotating piston and crank shaft. Special care is also to be taken to isolate the vibrations induced by the closed system high pressure mud and being handled by the pump.

6 Accuracy of diagnosing the fault by the selected method

As indicated in the previous Chapter the methods selected for monitoring reciprocating engines are:

1. Vibration Analysis.
2. Lub Oil Monitoring.
3. Engine Performance Monitoring.

The specific defects to be identified in a reciprocating engine are:

* Foundation stability.
* Adequate matching & fixing of engine bases and crank base.
* Criteria of Engine bearings.
* Piston slap and cylinder wear.
* Major firing disorders.
* Alignment
* Unbalance
* Electrical defects in generator
* Gear problems.

It is possible to accurately predict problems like Misalignment, Unbalance, Electrical defects in generators etc. by adopting the normal diagnostics procedure as in the case of rotating equipments. The effect of partially balanced forces are definitely to be considered while making effect of the unbalance in the rotor parts like shaft, fly wheels etc.

The foundation stability shall be revealed by the vibration distribution pattern and the directivity. Measurement at each foundation bolt locations should indicate a uniform distribution within +/- 2.5% variations. The normal vibration amplitudes for a healthy engine should be in the range of 15 to 20 microns of displacement. The velocity will vary with the speed of the equipment.

The accuracy of identification of instabilities from the foundation is in the range of 80 to 90% if the vibration measurements are carried out in all the three directions at the foundation as well as at the fixing locations.

Insufficient area contact between the mating surfaces is one common defect being noticed in high capacity engine. In most of the case this is a manufacturing defect, therefore during the inspection stage, the adequacy of area contact between the foundations & base frame and between the base frame & crank case should be critically reviewed. If the area contact is not adequate this will result in "Spring Back" and hence will induce looseness over a period of time. If the spring back is significant, this can result in the failure of foundation bolts. Vibration measurements above and below the mating surfaces alongwith the phase observations can reveal this defect.

A review of more than 200 case studies has revealed that the accuracy of identification of the defects falling in the category of improper matching & fixing is as high as 95%. One major complexity in this regard is the very high number of vibration data to be collected and interpreted.

The engine bearings are normally not accessible for analysis and hence measurements are to be taken at the engine body along the crank centreline. A fairly good reflection of the vibration characteristics is obtained for interpretations. The wear debris analysis carried and the lub oil pressure drop if any will confirm the bearing excessive clearances.

The piston slap excitation resulting from cylinder or piston wear will be indicated in the form of frequency equal to twice the number of strokes per minute. These symptoms can be obtained by carrying out measurements at the cylinder walls and at the cylinder heads. Supplementary data from Lub Oil Analysis to identify cylinder metal (or piston metal) content in the lubricant can be assessed from the performance analysis, the influence of cylinder and piston wear can be assessed by carefully monitoring the reduction in peak pressure.

The firing disorders shall be indicated in the form of transient conditions, Erratic frequencies and high frequency components.

The alignment and unbalance in the engine system can be identified from the normal vibration characteristic, but while making the assessment, the influence of partial balancing of inertia effects in the engine should be specifically considered.

The electrical defects in Generators can be analysed by collecting the vibration characteristics, attributable to individual electrical defects.

The accuracy of over all diagnostics by combining the three methods can be as high as 95% but one major practical problem being encountered is the inadequacy of effective Lub Oil Analysis procedures to keep pace with the simplicity and fastness of Vibration Analysis. Many times the Lub Oil data after the analysis shall be made available much later than the problem is first diagnosed from Vibration Analysis. Of course Engine Analysers are portable units which are being carried to various sites to have an on the spot analysis of the performance.

7. Remarks & Conclusion:

The condition based maintenance i.e. Predictive maintenance is to stay in the industry. This condition monitoring is the best tool available for the maintenance engineer/manager to assess the condition and the maintenance plan for the corrective action is taken which will eventually lead to zero break down baring sudden death cases. But a word of caution is that these methods are only tools for the maintenance & the non-machine relationship is not to be lost, for the best maintenance.

** ** **

A case study of the maintenance of Second World War military jeeps in Bhopal: a study in the resourcefulness of the local mechanics

Professor M.A. Qureshi
Department of Humanities and Management, Maulana Azad College of Technology, Bhopal, India
Mr Qamar-Us-Zaman Ansari
Ex-Sales Manager and Administrative Officer, Bhopal Motors Pvt. Ltd, Bhopal, India
Mr Ahmad Farhan Qureshi
A Jeep Enthusiast, Bhopal, India

Abstract
Bhopal, the city of lakes, is topographically a city of alternative steeps and downs and inhabited by a romantic and daring people. The older generation who still have affection for the late His Highness Hameed Ullah Khan of Bhopal have an interesting saying that President Roosevelt has presented the first military jeep manufactured in US, on the request of Sir Winston Churchill, to the Nawab to defeat the Axis Powers. And the Nawab did. It was in this background of simple people and their simple beliefs that this vehicle has its roots. Whats more important is that they and their later generations are still sentimentally attached to their II World War Jeeps. This paper is an humble effort to study the maintenance of these war model jeeps. Dispite the rising cost of spares and also in the face of their unavailability how they are able to maintain these vehicles? What are the special methods and techniques that the mechanics have adopted to keep them fit and fighting.
Keywords: Military jeeps, Mechanics, Maintenance, Second World War, Radiator, Gramflour, Needle Roller Bearing, Shikar.

1 Introduction

Not many people in Bhopal, a town of about 75,000 souls in 1940 had heard about a unique go-anywhere-vehicle which through a single engine and gearbox twin propeller shafts which drove both the rear and the front wheels. The makers were Willys Overland Corporation and the prototype was being tested by the U.S. Army. Bhopal was a princely state about size of wales. Much of the state was wooded plateau with primitive roads or cart tracks, few and for between. The ruling class and most of the large land-holders were warriors of yester years who had been persuaded with difficulty to exchange the swords with plough but the spirit of adventure had not entirely left them and their exurberance found an outlet in hunting or shikar as it is called in India. A jeep was the answer to their prayers if one could be had.

2 Bhopal saw its first jeep in 1943 when Bhopal aerodrome was converted into an RAF Figher base and the RAF moved in. A Jungle warfare training centre was also established in the deep-woods about 50 Km. from Bhopal under the famous colonel Jim Corbett of 'Man Eaters of Kumaon' fame. Those who could steal a ride in the jeep with their Army or RAF friends became instant jeep converts and enthusiasts. His late Highness Nawab Sir Hameed Ullah Khan, the ruler of Bhopal was an enthusiastic flyer and held the rank of Air Vice Marshal serving with distinction in Kenya and Eritrea and Ethiopia Campaigns during the 2nd World War and the belief is held among the simple folk of Bhopal that the first jeep was presented to him by President Roosevelt at the request of Sir Winston Churchill himself.

3 The end of II World War

During the War, the world saw no equals even a third rate of the American Jeeps, despite the best efforts of the Axis powers. The end of World War found a huge surplus of Army Vehicles and mountains of other equipment left behind at the bases by units of Armed Forces departing for home and the Army surplus vehicles became available to public by auction.

Thus the jeep stayed and continues to stay and its popularity has never diminished. People found the vehicle extremely sturdy capable of taking hard knocks and still continue in operation. The short wheelbase of 80 inches and a short turning circle with a high clearance of 8 inches made the vehicle extremely manourable off the track and the box shape utility body made the vehicle a really go-anywhere in the trackless jungles and hence immensely popular.

4 The Indian Version

So popular did the jeep become with the public and with an assured demand from armed forces for whom it was indispensable that a plant was set up in Bombay with the collaboration of Messrs. Willys Overland Corporation in India. The new vehicle became freely available to public but it was soon discovered that though the War time jeeps were supposed to be war grade and hence not upto American and European Standards for first class civilian vehicles, they proved sturdier and more reliable in service and the old models retained their popularity.

5 The Local Mechanics and American Jeeps

The maintenance and repairs of the jeeps called for much local ingenuity and improviosation in those early stages. Bhopal had its share of Bentleys, Duesenbergs, Pierce Arrows, De Sotos, Packards and every other popular make of cars so that the repair and maintenance know-how was not lacking. Replacement of parts was quite another thing however. So large was the demand that stocks began to give out and indigenous manufacturers both medium and small scale setup business usually with foreign collaboration and knowhow. The manufacture was concentrated in the area around Delhi in Punjab and Haryana and in due course Delhi became and has remained the capital of spare parts trade.

6 The Problem of Maintenance

6.1 Meanttime the old vehicles were kept running by the local talent with feats of improvisation. The local mechanics were unlettered almost to a man. None knew English in which most of the technical literature and instruction and operation manuals were available. Most could not read or write their own mother tongue which in Bhopal meant Urdu and Hindi but they were an ingenuous lot all the same.

6.2 They could rewind starter motor and dynamo armature and give an un conditional guarantee of six months but if asked why so many turns were given to the rewinding wire and why a turn or two more or less they had in reply except so I was taught.

6.3 The car batteries were also reconstructed in the same manner. As good workmen they liked jobs without any why, how and whats.

6.4 Radiator leaks were taken care of by the simple device of dumping a handful of gram flour in the radiator. It settled in the cracks and hardened when the water cooled down and made an effective seal even if not permanent as the resultant clogging required more careful and durable attention later.

6.5 The Fuel Pump in close proximity to engine block often got overheated in the hot weather and the diaphgram stopped functioning. This was remedied by lossely t ying two or three folds of towelling or any thick cloth filled with cowdung over the fuel pump and keeping the padding coal by frequent wetting.

6.6 The needle roller bearings in tie rod ends and wherever else were replaced by gramophone sound box needles and worked perfectly.

6.7 There were a few turners in Bhopal with ancient lathes and they could and usually turned out good grinding jobs on crankshafts and camshafts. Cutting and grinding valve seats and faces, presented no problems done manually.

6.8 The under size bearing pieces not been available pieces of bearings from Dodge Cars trimmed to size were made use of and did well.

6.9 So did the pistons and rings from the same source ground to size on lathes. Contact Breakers were freely available and were generally carried as spares on the vehicle as also a Rotor a Distributor cover, set of plugs and a spare fuel pump diaphgram washer and these generally were adequate for long and tough runs.

6.10 The jeeps offered no weather protection and every owner had his own ideas translated into action about the design and type of hood and seating arrangement.

7. A most gross and unsporting misuse of these jeeps. This was night shikar on the move which was an extremely rare occurrence before Shikar parties in over growing numbers invaded the jungles regularly every night. With a search light in the hands of a person composing the jungle raiders to shine over and dazzle the hapless animals. The slaughter was immense. Within a couple of decades of the arrival of the jeep on the scene game became scarce and many species had to be declared protected.

While giving due credit to the jeep as the most versatile, sturdy and dependable vehicle it must be squarely laid at its door (and it has no doors) that it was instrumental in the dscruction and wholesale slaughter of game animals to near vanishing points.

Acknowledgements

Thanks are due to Dr.B.L.Mehrotra, Principal and Professors Y.G. Bhave and Dr.S.Rawtani of M.A.C.T., Bhopal.

Maintenance strategy in the '90s

P. Upshall
Coopers and Lybrand Deoloitte, Birmingham, England

Abstract
This paper discusses the need for an integrated maintenance strategy. The method of developing this strategy is then explored, highlighting the matching of key success factors for maintenance to the overall business objectives. A vision for the future is then developed taking into account reliability requirements of key equipment and the trends and advances in condition monitoring and diagnostic engineering. An analysis of the differences between the current position and the vision becomes the basis for maintenance strategy, setting goals and timescales.
Keywords: Maintenance Strategy, Strategic Process, Diagnostic, Philosophy, Mandate, Policies, Objectives, Implementation.

1 INTRODUCTION

Despite the existence of condition monitoring techniques for decades, predictive methods of maintenance have not generally been included as part of an integrated maintenance strategy, with some notable exceptions.

However this will change in the 1990s. Catastrophic failures, such as Piper Alpha and Bhopal, or major commercial failures, such as recent contaminations of food products, where it appears ineffective maintenance has played a part, have meant that breakdown in many industries is no longer an acceptable circumstance. Business competitiveness, combined with advances in condition monitoring, make it a financial necessity for all significant operations to develop an overall maintenance strategy designed to eliminate unplanned equipment stoppage.

2 THE PRESSURES FOR IMPROVEMENT

Pressures to develop a maintenance strategy designed to improve effectiveness come from both inside and outside the organisation.

Internal pressures come from many sources. The extensive use of mechanization and automation reduces direct labour costs but increases maintenance costs. The availability of low cost IT brings out, previously hidden, information relating to downtime costs. Modern manufacturing techniques and philosophises like Just-in-Time, Total Quality or even MRPII cannot be successfully implemented without high standards of equipment reliability and availability.

The sources of external pressure are just as varied. The competitive threat goes without saying, but what of globalization and the liberalization of trade both in Western Europe in 1992 and in the rapidly changing Eastern bloc. Closer to home the pressures of government legislation and the environment lobby are ever increasing, and of course customers are expecting shorter lead times, increasing flexibility and improved service.

These pressures apply in varying degrees to the competition in the rest of the industrialized world and so should not be feared - provided that in all aspects of manufacturing we are better than most and striving to be best.

3 HOW GOOD ARE WE?

Intuitively we would perhaps expect that we are less effective than Japan and West Germany in our maintenance efforts but a recent survey by IMEDE ranked the UK 16th out of 22 industrialized nations.

The airlines, because of a combination of legislation and high operational costs need to keep aircraft flying, utilize some of the most advanced maintenance strategies with extensive predictive maintenance and advanced parts planning and scheduling. Some large industrial companies with high maintenance spends or high downtime costs have similar proficiency. In general however British industry operates a system of planned maintenance which is not geared to need and is not given the priority it requires.

There can be little doubt that a maintenance strategy supporting the overall business objectives is an essential element of a business plan to increase profitability.

4 THE STRATEGIC PLANNING PROCESS

4.1 The Current Position

An overview of the strategic planning process for physical resource management is shown in Fig.1. The first step in this process is the examination of the environment and context in which maintenance functions operate. This will provide a backdrop against which the ensuing work will be carried out.

To determine those factors which must be improved requires a comprehensive and objective review of the current activities. The effectiveness of maintenance can be viewed as a continuum, from a state of "innocence " through "understanding" to "excellence". The objective of the review is to determine where the maintenance function lies in the continuum and how far it can be taken towards excellence over a period of time.

Figure 1 : THE STRATEGIC PLANNING PROCESS MODEL

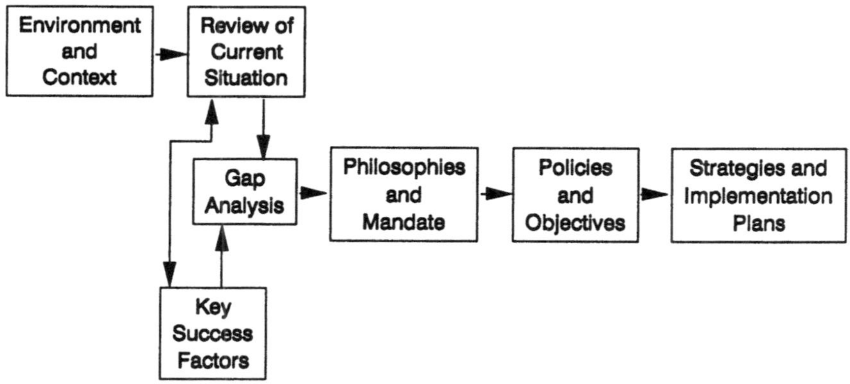

The areas analyzed include:

Maintenance standard and state of repair.
Organization and administration.
Human resource management.
Planning procedures.
Work order control.
Preventive and Predictive Maintenance.
Equipment Records Management.
Purchasing, stores and stock control.
Performance monitoring.
Computer system support.
Interdepartment Relationships.

Findings and conclusions are developed in each of these areas, in conjunction with an understanding of the environment, the key success factors are ascertained.

4.2 The vision for the future

The key success factors begin to develop the vision for the future. They are chosen, not to necessarily yield the best-engineered plant, but the most effective in overall business terms. It is impossible to generalize about the future of maintenance in all businesses but some trends can be extrapolated.

The general profile of maintenance will be raised with costs and effects becoming more visible. Equipment operators will take on more of the first line maintenance, but the highly skilled maintenance workforce will still be required in the long term. There will be a significant shift towards maintenance prevention through redesign, scheduled discard and failure prediction, condition monitoring, and periodic inspection. The use of Reliability Centred Maintenance will become far more widespread, further reducing the levels of unplanned maintenance. Instead of trying to become more responsive maintenance departments will exercise more control over their resources.

Once the key success factors have been selected the gap between these and the current situation can be determined and this gap forms the basis of the maintenance strategies.

4.3 Philosophy and Mandate

The maintenance philosophy to be applied must be chosen. Since the early 1970s much study has been carried out in this field but two philosophies are becoming more widespread; these are Total Productive Maintenance, which builds on total quality concepts, and Reliability Centred Maintenance which forces a structured evaluation of failure modes and consequences.

The mandate is a clear definition of the role that maintenance will play in the business identifying its responsibilities. Pragmatic questions must be answered. Is maintenance subordinate to production? Does maintenance manage spare parts stocks?

4.4 Policies and Objectives

The policies and objectives should be determined for each of the major areas covered in the diagnostic review phase.

The policies outline the way in which the key success factors will be achieved. The objectives are clear, precise and measurable targets against which the progress of the improvement process can be measured. They must describe what is to be achieved, when and where.

4.5 Strategy and Implementation Plans

The strategy is the statement of how the objective will be achieved. A variety of strategies may be proposed ranging from organizational changes to selection of an appropriate maintenance technique. Increasingly the techniques which will be used will be redesign or condition based maintenance.

With the maintenance strategies developed focusing on the key success factors, the maintenance issues clarified relating to mandate and policies and the objectives clearly defined, the maintenance improvement plan can now be documented.

The improvement plan should identify:

Major activity area eg. department.
Individual tasks for each area.
Those responsible for implementation.
Resource requirements.
Likely constraints or known problem areas
Start and finish dates (both planned and actual).

It is likely that this plan will be a major undertaking and so it is essential that it has board level commitment. It is advisable

Figure 2 : AN EXAMPLE OF THE STRATEGIC PLANNING PROCESS

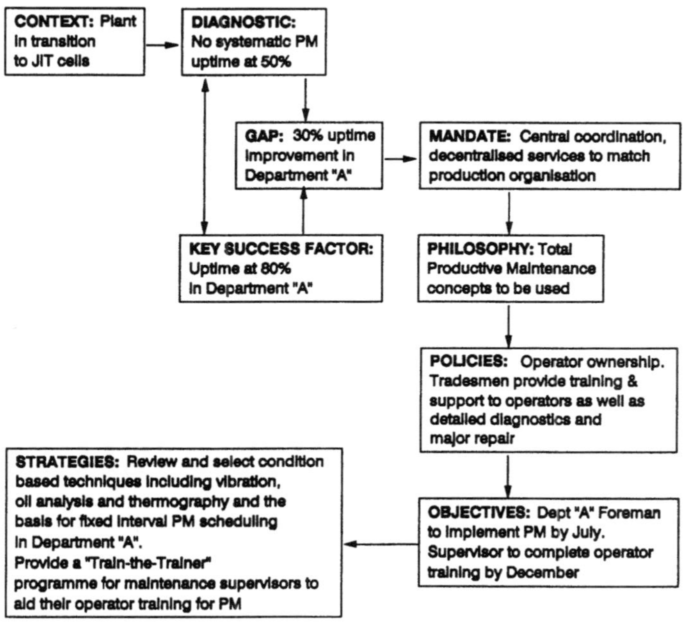

that a steering committee is set up to monitor implementation progress, involving representatives from all functions involved.

5 CONCLUSION

To illustrate the strategic planning process Fig.2. shows an example which relates to an electronics company.

The benefits of a comprehensive systematic maintenance improvement programme are substantially achieved through cost reduction, higher throughput and fewer quality defects. However making use of condition monitoring in an overall and structured strategy has the significant benefit of reducing painful surprises, reducing risk and allowing you to sleep better!

It is essential that maintenance strategies are built up from the total business key success factors so that, when successfully implemented, the results have a meaningful impact on the bottom line. As business development occur which change any of the premises on which the strategies have been built the process should be revisited to determine if the strategies need to be amended.

Risk and waste management in oil field

D.J. Barua and R. Pandian
Oil and Natural Gas Commission, Madras, India

Abstract

Risk and waste are never a visible element, assorted and identified and kept aside for arrest. They are logically interwoven in the systems as loading limits or an external efect countervetting safety limits. The mechanistic tools of calculation derived from the strength of material are not suffient tool for assessing all the risk situation. Any utilisation under nominally or overloaded condition is waste. Again this loading conditions are time intolerant factor. Since the input output relation cannot match unity and the conditions imposed by the nature also do not allow us any infinitely safe or wasteless system, we have to execute a high degree of caution to remain risk free situation.

1. Introduction

The environment of investment for Petrolium Industry was always changing. From "wild cats" to "scientific exploration" is a ocean of change. The development of the industry during the years is:

Prior to 1930:
a. erratic discoveries b. violent supply and demand trend
c. limited knowledge.

During 1930 to 1942
a. regular discoveries b. price established & demand steady c. begining of reservoir study and seismology.

During 1943 to 1947
a. years of war b. flog the field
c. shortage of manpower and technology

During 1947 to 1957
a. incresed demand b. gas identified as fuel
c. bad exploration practice d. technology waiting

During 1957 to 1966
a. surplus year b. taxes up
c. money value appriciated d. technology up

During 1966 to 1970
a. surplus ends b. oil companies diversify
c. technology steadily up
During 1970 to 1980
a. age regulation & oil shock b. scientific exploration & production c. uncertanity and risk linked profit estimates
d. high technology environment setout by offshore & artic finds.
During 1980 to till date
a. more regulations b. reduced finds c. deeper deapths
d. technology refined e. oil shock continues

2 Risk & risk analysis

Investment decisions also became more and more yardstick oriented like i. cash flow analysis ii. time element respected
iii. base accountancy concepts iii. related risks.

Risks may or maynot be used for calculating project ecconomics but i. risks are weighted for evalution
ii. risks are not enumurated, but covered and quantified
iii.risks are estimated from minimum criteria to offset unknowns.
iv. Montecarlo simulation and other mathematic tests are used.

A game is played to arrive at risk-weighed profitability. Here we are to scan through:

i.dryor wet or ecconomically dry ii.mostlikely size
iii. pay-off iv. multiple case v. multiple variables
vi. probability determinations vii. modelling.

TOTAL RISK OF PROJECT

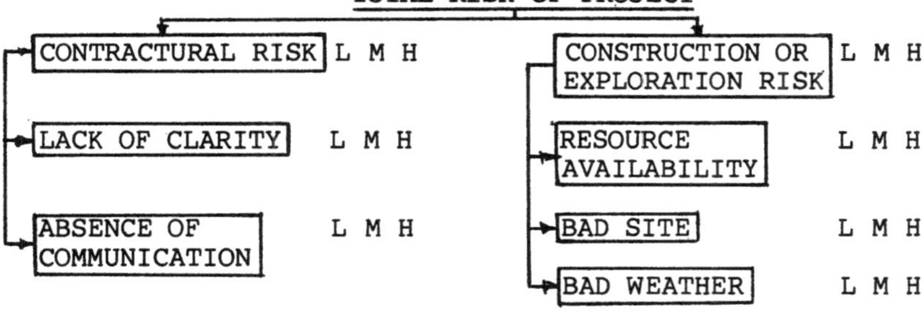

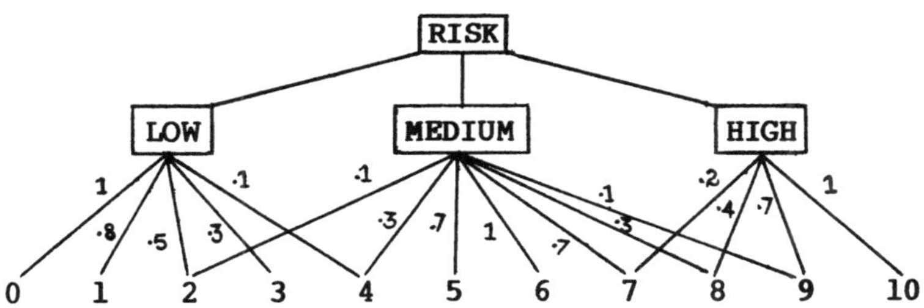

So Risk= an oppertunity for loss
uncertenity= outcome indeterminate but loss is not apparant.
the ecconomics of exploration and production decision = potential + economics of operation + risk

Where are these risks ? I think in every steps, samples are

a. geologic risks - structre , reservoir, environment
b. investment risks - wild cat or scientific, time tolerent probability positive
c. technology risk - adequqte design, adequate safety, adequate coverage, operation vs safety, maintanence vs safety, adequate upgradeability, cost vs risk conceptwell
d. human risk - skill and trained, strenth and tolerance of stress, diciplined and organised

All these factors have played their role so far like this (from accident analysis of mobile and fixed plteforms):

	mobile	fixed		mobile	fixed
blowouts	28%	39%	capsized	10%	9%
drift+grounding	9%	0%	fire+explosion	7%	25%
collision	6%	9%	others	4%	8%
equipment damage	10%	5%	structural damage	26%	5%
-do- due to war	0%	5%	casualities	310	333
oil spills(M.ton)	563	240	pollution	180*	mostly
-do- due to war	0	2%	not confirmed or reported		

Among the offshore rig types who & when is weak is as shown :

type	**% accident**	**type**	**%accident**
jackup	54.7	semi-submergible	22.0
drillship	12.6	submergible	10.7

activity	**mobile: all**	**jackup**	**fixed platform**
under change of site	22	35	1
wildcat	32	26	4
development	6	5	17
installation	11	19	5
production	1	0	52
standby (storm)	11	4	12
others	17	11	9

the above data came from

i. after 72 preliminary factors preceding the accident
ii. 74 different events making up a sequence
iii these preliminary factors leading to potencial dangereous situation, storm, towline breakagee, structural failure.

3 Waste & waste analysis.

All the time accidents leads to waste of (i) time (ii) money (iii) human resources (iv) reservoir (energy) (v) environment.

There is an interesting feature, we can find in jack-up storm damage, such as there were 13 accidents between 1980 &1985 but hurrican did not reach 5 catastrophe when the exposeure of rigtype as jackups 3500 rigyears, semi-submergibles 1150 rigyears, drillships 500 rigyears, fixed platforms 46500 rigyears.

The comparative risk of offshore installation is as under:

Canadian Standards serviceability	:	1.0×10^{-1}
Blowout : Semisubmersible	:	13.5×10^{-3}
Offshore worldwide blowouts(1955-1980)	:	8.3×10^{-3}
Blowout : jackup	:	8.0×10^{-3}
Tanker accident	:	2.3×10^{-3}
Air travel	:	1.5×10^{-3}
Canadian standards; safety class 2	:	1.0×10^{-3}
Jackup major damage due to storm (all precautions taken)	:	Less than 2.8×10^{-4}
Motor vehicle accidents	:	2.4×10^{-4}
Police killed in line of duty	:	2.2×10^{-4}
Fixed platform major damage due to storm	:	8.6×10^{-5}
Exploratory offshore blowout with oilspill	:	1.4×10^{-5}
Canadian standards: Safety class 1	:	1.0×10^{-5}

The consequence of failures are 10000 to 1000000 ($\$ \times 10^3$) There are reflections of easily accountable quest. The waste of resources already created are also accounted. What is not accounted under any other analysis are:

a) Cost of recreating resources again.
b) Cost of retrieval of debris.
c) Cost on global resource (These rates creates another demand)
d) Pollution of the environment due to activity, fall out, spillage of oil and chemicals, disturbing the bio-echology.

Any activity creates fall out, the effect of all these fall out are not accounted and extended to risk of damaging the environment. The classical example is the destruction of ozone layer from the industrial fall out. The second danger is the gree house effect due to rise in global temperature. Such non-realised effects snowball to create a challenge on survival itself.

4. The Thoughts:

1. Murphy's Law: If something go wrong it will
2. Risk can not be eliminated. Avoid or control it.
3. Risk is not easy to study.
4. So loss (waste)can not be eliminated.

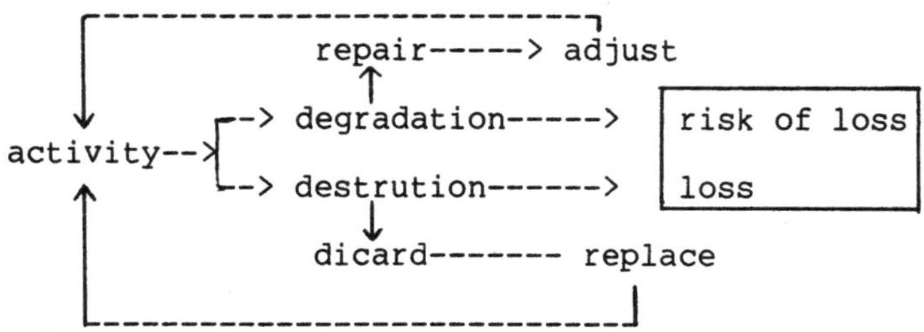

The sucess is
D = Probability of loss (total)
S = Probability of no loss (gradual)
Then D + S = 1
using binominal express $(D+S)^2 = (1)^2$

$$D^2 + 2DS + S^2 = 1$$

Then 25% chances of two DD
50% chances of one success (DS + SD)
25% chances of drilling 2S
75% chances of discovery (DS + SS + SD)

We shall move with hazard and risk analysis to reduce the waste and we shall do it for everything of exploration, drilling, production and maintenance.

It is not adequate to work with the standards. It is required to work the marginals and extreme cases also to find boarderline facts under the conditions like

a) Erection load
b) Static load
c) Environment loading.

This is a great help to find out limit of optimisation, maximisation, minimisation of all the types of variables.

For example:
Optimise = Trade of technique.
Minimisation = Cost of realisation project or down time
Maximisation = Output, production etc.

Then it becomes importance to try some co-relations between all the variables. The selection of technical system for the design of whole process is to be drawn up. Finally a data based control and acquisition system are injected as Tool for monitoring system by a management group. Then the whole state concept should be transportable to cost control.

Then the whole concept should be transferred to cost control.

From all the available data, we can build up probability to use in the decisions. There are various methods to fix up a model and calculate the impacts, but everything should be converted to cost data to bring the results to a common base for a decision help.

5 Conclusion:

Best available systems for 2000ft of water deapth are
i) subsea wells ii) tension leg platforms iii)semifloters iv) turret vssels
Best technology for cutting drilling and development cost are:
i) horizontal completion ii) extended reach wells iii)MWM iv) computer aided decisions v) top drive vi) drilling automation vii) steerable bottom hole systems
Best technology solutions for cutting production cost are:
i) remote control ii) subsea completion iii) multiple flow iv) enhanced recovery v) subsea separation vi) reservoir modelling

There is much scope to organise quality circles for resolving the risks and waste since all the different type of crews stay together. The modern technology improvement in risk analysis also helps to reduce the gap.

Biblography:

1. Practical drilling and production design by Dr.Douglas Bynum Jr.
2. Drillig Practices Manual by Preston L. Moore
3. Management of hazard and risk in petroleum industry by Howard Finlay Corporation U.K.
4. Offshore - December 1989
5. Offshore - September 1989
6. Engineering Management Vol.-36 Number 2 May - 1989
7. Ocean Industry - March, 1989

** ** **

Environmental pollution monitoring in oil exploration and exploitation

J. Chand
ONGC, Jorhat, Assam, India

Abstract
Oil & Natural Gas Commission has been exploring and exploiting for hydrocarbons in most diverse geographical locations on onshore & offshore in India. As a result of these activities, ONGC is interacting with the environment in various ways. In exploration and exploitation of oil & gas, the major environmental pollutants are i) effluent water contaminated with oily effluents e.g oil & grease, chemicals and solids from drilling fluid, ii) formation water produced along with crude oil and iii) gaseous emissions having CO, SO2, NOx, hydrocarbons, particulate from gasflare.

During drilling operations, the seepage/leakage of effluent water to the surrounding areas is a major source of pollution. To check the leakage/seepage of effluent water, a number of control measures are being taken viz.,

i) construction of ring bund on the periphery of drill site
ii) construction of peripheral trench along with boundary wall
iii) compartmentalisation of drill site
iv) construction of cutting pit, waste pit, oil pit etc.

The water produced along the oil is either evaporated in evaporation pit or treated in effluent treatment plant. The flaring of associated gas can cause air pollution and heat & glare produced can adversely effect the nearby fauna & flora. To check the air pollution, box flares are being used and to minimise the effect of heat and glare, multipoint flares of low height covered from all around with asbestose/cement sheet along with brick wall are being used.

The left over drill sites are being developed and trees all around the boundary of drill site area being planted to improve the ecological balance, which has been greatly disturbed.

At the offshore drilling rigs & production platforms, facilities exist to treat the sewage, burn the drill waste, treat the effluent water and collection of spilled oil from the unit etc.

1 Introduction

On the energy scenario oil & gas forms a dominent source of energy. India has made spectacular progress in this field during the last decade. Oil & Natural Gas Commission has contributed significantly in this direction and is set to beyond its frontiers to take the challenge of entering into 21st century.

The hydrocarbons are the basic inputs required for fertiliser and petrochemical plants in addition to being the major source of energy. But where oil is encountered, it is a potential threat to the environment, it always was and always will be.

In the long history of oil industry, which involves exploration, drilling, production, transportation and refining operations, no one will deny that industry has been a significant pollutant of land, water ways, oceans and atmoshere.

Pollution in offshore operations can result from drilling activity which involves discharge of cuttings and associated drilling fluid, blow out and accidents caused by human errors. When such incidents occur oil may be released in large quantities and float on the sea surface and thus driven away by the wind to areas, where it may create an amenity problems, for exmaple, a bathing beach. Sea birds appear to be most affected as they are attracted to the floating oil and heavy mortalities have been caused by oil spills, contaminated salt water interferes with the respiration of fishes, coats and destroys algae and plankton, thus removing a food source, interferes with spawning areas by coating the bottoms and destroying benthal organisims, interferes with the natural process of reaction and photo synthesis.

Oil & Natural Gas Commission has been exploring and exploiting for hydrocarbons in most diverse geographical locations on onshore and offshore in India. As a result of these activities, ONGC is committed to high standard of environment protection.

The major environmental pollutants in exploration and exploitation of oil and gas are i) effluent water contaminated with oily effluents e.g oil and grease, chemicals and solids from drilling fluids ii) foramtion water produced along with crude oil and iii) gaseous emissions having CO, SO2, NOx, hydrocarbons, particulate matter and heat and glare from gasflares.

2 Environmental control measures

2.1 Onshore operations

2.1.1 During drilling

At drill site the utility and/or rain water may get contaminated with oil, grease and chemicals etc., which may have some effect on the surroundings in case of leakage/seepage. To eliminate this effect on the surroundings, the following measures are taken :

(i) **Ring bund :** Cement plastered brick wall of 3-4 feet height with 1 foot deep plastered foundation on the periphery of drill sites is constructed so as to avoid seepage/leakage to the adjoining area. (Fig. 1)
(ii) **Peripheral trench :** Peripheral trench along with boundary wall is constructed to collect the effluent, if any leaked/seeped, from ring bund.
(iii) **Compartmentalisation of drill site :** Drill site is compartmentalised into three segments so as to avoid intermixing of drill fluid/chemicals, burnt oil and water etc. (Fig. 1)
(iv) **Cutting pit :** A cutting pit is constructed to collect the well cuttings separated from drilling fluid at surface alongwith washings of solid removal equipments. (Fig. 1)

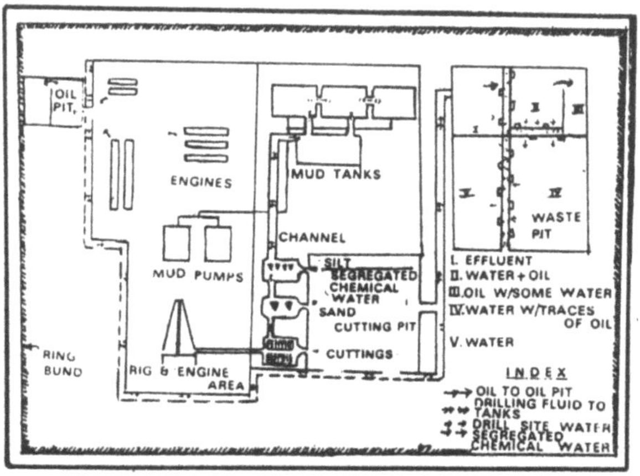

Fig.1. Block diagram of compartmentalisation.
(not to scale)

(v) **Waste pit :** The various streams of effluent are taken into a waste pit of capacity 4000 cubic metres and 3000 cubic metres in case of cluster and normal drilling respectively (Fig. 1). To check seepage/leakage from waste pit, (a) a distance of 1.5-2.0 metres is kept between waste pit and ring bund and (b) waste pit is made impervious with appropriate lining/cement pastering.

The waste pit is divided into five compartments. Compartments I & II, II & IV, IV & V are separated by 4" paddy hay filters placed between two honey comb brick walls as shown in Fig. 1. The filtered water obtained in compartment V is recycled for use in preparation of drilling fluid, washing of drilling equipments, channels and derick floor etc.

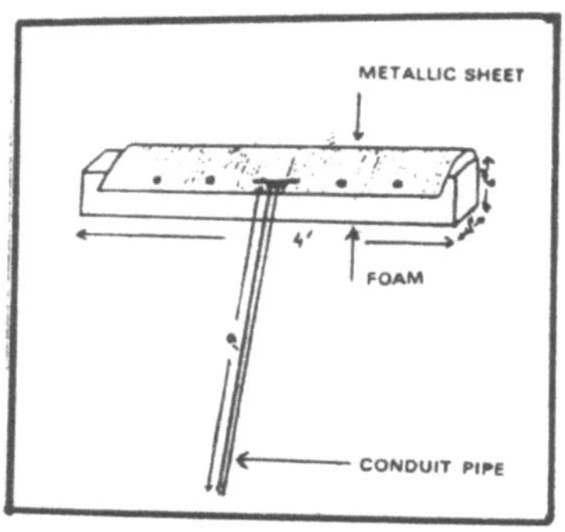

Fig.2. Oil skimmer

(vi) **Skimming of oil :** An oil skimmer as sketched in Fig. 2 is used for manual skimming of oil floating on waste water.
(vii) **Oil pit :** All the oily effluents from engine and other areas are collected separately in oil pit.
(viii) **Effluent treatment :** The effluent water is treated in effluent treatment plant and where effluent treatment plant is not available, the effluent water is treated with 600-700 ppm of alum and lime to make it fit for operational re-use.

2.1.2 During **production**

(i) **Effluent water :** The effluent/waste water separated from oil at GGS (Group Gathering Station) is treated in Effluent Treatment Plant (ETP) to render it suitable for surface disposal conforming to **IS-2490** specications. Alternatively, it is evaporated in a evaporation pit where ETP is not available. Flow diagram of an ETP is shown in Fig. 3. Data of untreated and treated effluent water is given in Table 1.

Table 1. Analysis data on effluent water at Lakwa ETP

S.N	Name of pollutant or characteristics mg/lt	Tolerance limits as per **IS-2490** mg/lt	Parameters of effluent mg/lt	
			Before treatment	After treatment
1	Suspended solids	100	-	-
2	Dissolved solids	2100	1700	1418
3	Oil and grease	10	2740	8
4	COD	250	2000	40
5	PH	5.5-9.0	8	7.5
6	Chloride (as Cl)	1000	596.2	497
7	Sulphate (as SO_4)	1000	67.2	76.8
8	Total alkanity	-	300	187.5
9	Carbonate	-	32	-
10	Bicarbonate	-	296	228.75
11	Total dissolved iron as Fe^{+++}	-	4	0.1

(ii) **Oil pit :** Oil produced during production testing of an exploratory well is recieved in a cemented oil pit and from development well it is dispatched to GGS through pipeline.
(iii) **Flaring of gas :** The gas associated with oil is flared where flaring is inevitable. The flaring of gas can cause air pollution and heat & glare produced during flaring can also adversely effect the nearby flora & fauna. To check air pollution, the flare is rendered smokeless by steam injection and box flaring is also done. The gas of South Bassein field of Bombay High which contains traces of hydrogen sulphide (H_2S) is treated in a sweetening & sulphur recovery plant.

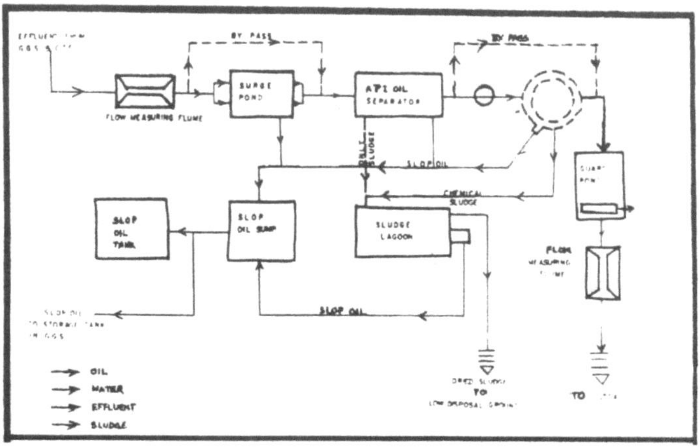

Fig.3. Lakwa effluent treatment plant.

To minimise the effect of heat and glare (a) a multipoint flare of 6-8 points and of low height of 1.0-1.5 metres are used instead of elevated single point flare and (b) flare area is enclosed completely by asbestos or cement sheet in addition to brick wall. (Fig. 4)

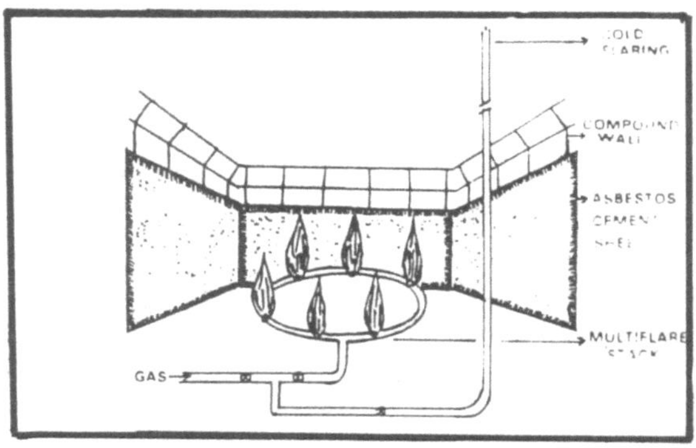

Fig.4. Gas flaring and its control

2.2 Offshore operations

2.2.1 During drilling

To avoid pollution, following facilities are existing on each offshore drilling rig; (i) Seawage treatment plant, (ii) A bilge pump to collect all possible spilled oil on the drilling unit, (iii) Incinerators for instant burning of the drill site waste, (iv) Formation cutting washing before discharge in to the sea, (v) Detection of H_2S and combustible gases and (vi) Left over drilling fluid transportation to the next location alongwith drilling unit.

2.2.2 During production

To check pollution the production platforms have been provided with ; (i) Surface controlled sub-surface safety valves, which in the event of uncontrolled oil flow are automatically closed limiting the oil spill in the area, (ii) Effluent treatment plant, (iii) Treatment of seawage of the living accommodation, (iv) Close circuit system to prevent leakage and (v) Automatic shut of valves in underwater pipelines.

3 Development of drill sites after completion of drilling

In the past, as soon as drilling rig was released, the rig was transported to next drilling location and the old drill site was left over as it was, leaving oily and muddy effluent in waste pit which in due course of time used to spread over other parts of the drill site. Sometimes, due to heavy rains, these effluents also resulted in leakage/seepage or overflowing to the nearby area. Even cattles used to make easy entry to this left over sites.

Presently following preventive measures to avoid such type of pollution in left over drill sites and to improve ecological balance around drill site area are being taken ;

Total oily effluents (mixed with burnt oil and diesel etc.,) are transported to the nearest GGS through tankers. Serviceable mud left over after drilling operation is not thrown at site but it is transported to the next drill site. All unwanted materials like scrap, wire ropes etc., are removed and sent to scrap yard. All the pits in the drill site area are filled in first with mud cuttings and then with the earth work in such a way that level of earth work becomes upto normal ground level, alternatively all the water is purified using mobile ETP. Leaving the area earmarked for X-max tree and oil pipeline connected to GGS, all the other area left over drill site is cleaned/levelled properly and made easily approchable.

Fig.5. Artist's impression of the site after tree plantation

All aound the boundary of the drill sites as well as with in a distance of 8.10 metres from boundary, two rows of trees with a distance of 2.0-2.5 metres between consecutive trees, are being planted leaving the area of main approch. Trees are also planted on the strip of roads leading to the well site. An artist's view of the left over drill site is shown in Fig. 5.

During 1985-86, 1986-87, 1987-88 and 1988-89 approximately 353182, 145834, 186000 and 387666 (Total 1072682) respectively trees were planted by ONGC at its various installations, residential colonies and office accommodations to improve the ecological balance which has been greatly disturbed over increasing pollution of water and atmosphere.

4 References

Bindu N. Lohani, Environmental quality management.
Robert R. Wheeler Maurine (Edited), From prospect to pipeline.
Purdy, G.A, Petroleum
Joshi, S.C. and Battacharya (Edited), Mining & Envoronment in India.
Diwakar Rao, P.L. (Edited) Pollution control hand book 1986.
Willium J. Cairns and Patrick M. Rogus. Onshore impacts of offshore oil.
Munn, R.E. Environmental assessment scope-5.

Comparison of the digraph and FAULTFINDER methods of fault tree synthesis for nested control systems

J.D. Andrews
Department of Mathematical Sciences, Loughborough University of Technology, Loughborough, Leics LE11 3TU, England
A.R. Khan
Department of Chemical Engineering, Loughborough University of Technology, Loughborough, Leics LE11 3TU, England

Abstract
This paper describes the development of a fault tree for an ethene–oxide reactor cooling system. Two methods have been used for the fault tree synthesis these are the digraph method and the FAULTFINDER computer code. The resulting fault trees are qualitatively analysed and the results compared.
Keywords: Fault tree analysis, Fault tree synthesis, Digraphs.

1 Introduction

Fault tree analysis is now a standard method for assessing the probability and frequency of occurrence of selected system failures. It has been extensively applied to determine the causes of unsafe or hazardous events associated with systems utilised in many industries. Whilst the advantages of using such techniques are well known, especially if applied at the system design stage, there are also disadvantages. The major disadvantage is the time taken to perform this kind of analysis. Once constructed, a fault tree can be analysed using one of the many programs currently available, Willie (1978), Lambert (1977). However, for all but the simplest of systems the fault tree construction will be a very time consuming process. There is therefore a strong insentive to devise approaches which avoid the errors and tedium associated with manual fault tree construction.

Over the last two decades several algorithms have been developed for the construction of fault trees. This paper considers two of these approaches; the digraph method and the FAULTFINDER computer program . The application of both of these methods to an ethene–oxide reactor cooling system is described. Fault trees produced during this work were qualitatively evaluated and the logical structures compared.

2 Ethene–Oxide Reactor Cooling System

The ethene–oxide reaction process operates very close to the safety limits as described by Piccinini and Levy (1984) and the brief description given by Kletz (1990) about the recent explosion in the BASF ethene–oxide plant at Antwerp in March 1989. The reaction is the epoxidisation of the ethene to ethene–oxide using pure oxygen on a silver base catalyst. There are undesirable side reactions such as complete combustion of ethene where the heat evolved is eleven times greater than epoxidisation. Therefore, the temperature in the reactor is critical and over temperature could result in the possibility of a runaway reaction.

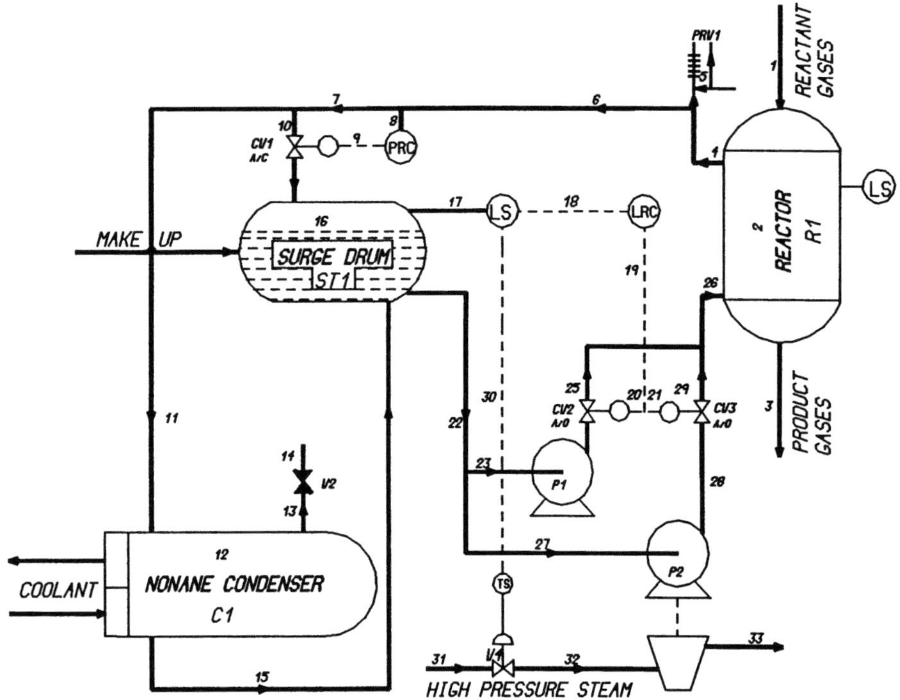

Figure 1. Ethene-oxide Reactor cooling system

The reactor cooling system (fig 1) forms a closed loop and uses liquid n–nonane to maintain a temperature gradient of less than 4° in the medium between the inlet and outlet of the reactor. The surge tank and the reactor are 6m above the ground–mounted condenser and pumps. This ensures satisfactory performance of the condenser. The reactor is fitted with a level recorder to make sure the reactor tubes are adequately submerged in the coolant. The temperature of the coolant is kept constant by controlling the evaporating pressure via the pressure controller. If the amount of vapour increases, the control valve closes, reducing the pressure in surge tank and hence the condenser. This exposes more of the condensing surface to condense the increased vapour. The surge tank is fitted with a level controller to maintain steady constant flow, thus keeping the coolant level in the reactor constant. There is another high level trip on the surge tank which, when activated starts the standby pump.

3 FAULTFINDER : Basic Method

The FAULTFINDER package, which has been developed in the Department of Chemical Engineering at Loughborough University, is written in Fortran 77. A detailed description of four aspects of this code have been given in previous papers; i.e. the modelling method, Kelly and Lees (1986a), the fault tree synthesis method, Kelly and Lees (1986b), the interactive facility created by implementing these, Kelly and Lees (1986c) and:– illustrative examples, Kelly and Lees (1986d), Mullhi et al (1988) and Khan and Hunt (1989). The modelling starts from the piping and instrument diagram of the plant, which is then decomposed into an equivalent block diagram configuration of units and connections. Typical units are pipe, sensor, vessel, heat exchanger and control valve. Each of these units is represented by a corresponding model, which gives the fault propagation characteristics of the unit. The model is based on simple functional equations (propagation equations, event statements and decision tables) which describe the way in which deviations of the process variables pass through the unit and the way in which such deviations are generated within the unit by faults, principally mechanical faults. The form of representation used is the mini–fault tree based on the concept of the input–output models of Fussel (1973), the digraphs of Lapp and Powers (1977), the decision tables of Salem (1977) and transition table of Taylor (1982). For each output variable there is a corresponding mini–fault tree, in which the top event is a particular deviation of that variable and the base events are deviations of the external input variables and faults occurring within the unit. An extensive range of models of various components of process plants have been written in a standard library which could be accessed by the user. In the case of a unique component a simple set of rules have been suggested to be used in the computer program (MODGEN) to generate the corresponding model.

In another program (MASTER) the configuration data, unit models and connections define the basic fault propagation structure of the plant. Some additional information on the relationship between certain groups of units is necessary for example control loops, trip loops, divider–header combinations, material failure, phase–change and sequential operations need to be identified. The synthesis of the tree is performed by selecting an undesirable event (Top event) and linking together the appropriate mini–fault trees from the individual unit models. There is a considerable body of rules which are applied to minimize duplication, to check consistency and rationalise in the tree structure for certain groups of units.

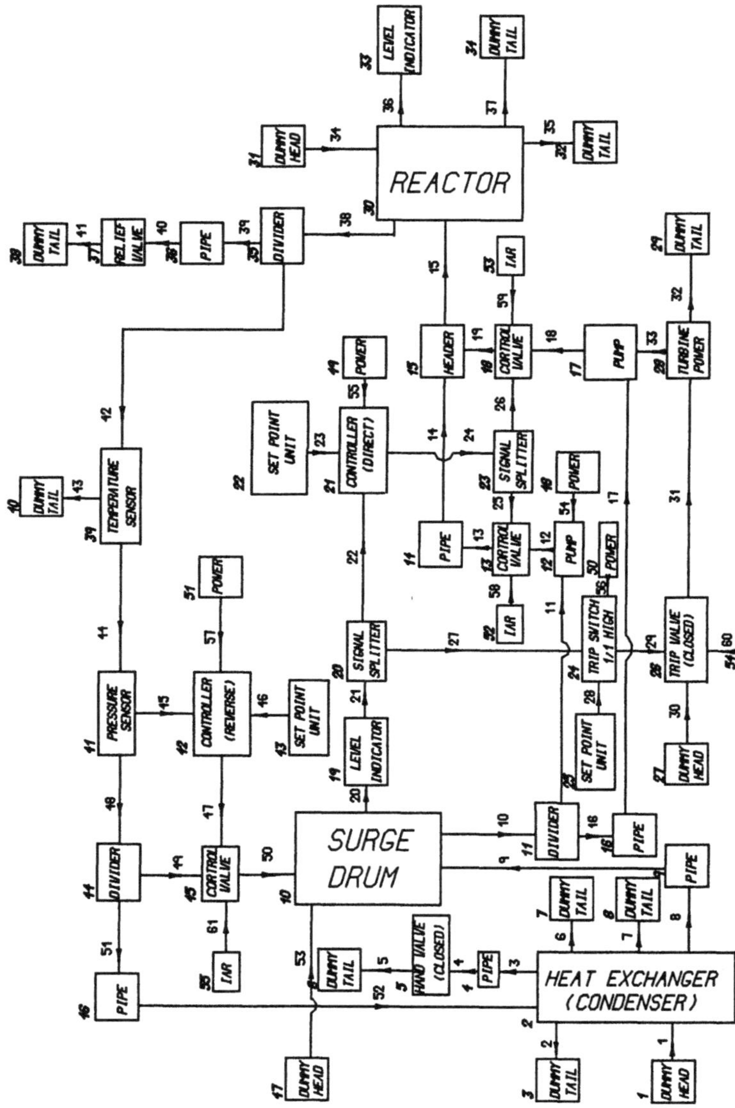

Figure 2. Configuration diagram of reactor cooling system

The block diagram of the ethene–oxide reactor cooling system used to represent the plant layout for FAULTFINDER is shown in figure 2.

Two runs of the FAULTFINDER code were made. The first time only pump 1 was considered working (LS, LRC, CV2 active) and the second time both pumps were working and all control and trip loops functioning. By this means the fault propagation in LS, LRC and CV3 control and trip systems could be observed.

4 The Digraph Method

Fault tree construction based on the digraph method is a two step process, (Lapp and Powers (1977) and Shaelwitz et al (1977), Lambert (1979), Andrews and Morgan (1986) and Andrews and Brennan (1989)). The digraph itself provides an intermediate step in formulating a fault tree for some particular undesired system event which is represented by the deviation in a process variable. Initially the digraph is constructed to represent both the normal function of the system and also the effect of component failures and deviations in the inputs to the system.

Fault tree construction is then carried out by tracing the causes of undesired deviations in the top event process variable back through the system.

A digraph is a set of nodes together with a set of edges or lines which are drawn from one node to another. The nodes are used to represent process variables in the system or events such as components failing. When one process variable has the ability to affect another, a directed edge is used to connect them. The direction of the edge is from the independent variable to the dependent variable. Each edge has an associated number, termed gain, used to represent the relationship between the nodes. Where numbers appear on their own the relationship is normally true. In some situations conditions which define when the relationship applies are also indicated.

The multi–valued logic system used for digraphs consists of the numbers –10, –1, 0, +1 and +10 and these are used in two distinct senses to represent both **deviations** in process variables and the **gain** associated with edges.

4.1 Deviations

These are the discrete states which are used to express disturbances in the process variables. When a variable achieves its normal expected value then the disturbance is represented by 0. A disturbance of magnitude 1 indicates a range of values that is considered moderate, that is an expected deviation which the system has the ability to control. Large disturbances are indicated by a magnitude of 10 and these are defined as being beyond the capacity of the system to rectify. The sign of the disturbance indicates whether it is above (+) or below (–) the normal value.

Where component failure states are represented by the nodes on the digraph the logic values 1 and 0 are used to indicate that the failure state exists or does not exist respectively.

4.2 Gains

Gains are the numbers placed on the directed edges to show the strength of relationships between two variables. Values of 1 and 10 indicate moderate and strong relationships respectively. If an edge links variable X to variable Y then the gain can be interpreted as $\partial Y/\partial X$. Therefore the sign is used to indicate the direction in which the dependent variable changes when the independent variable is increased. A zero gain is used to indicate a nullification of the relationship between two variables.

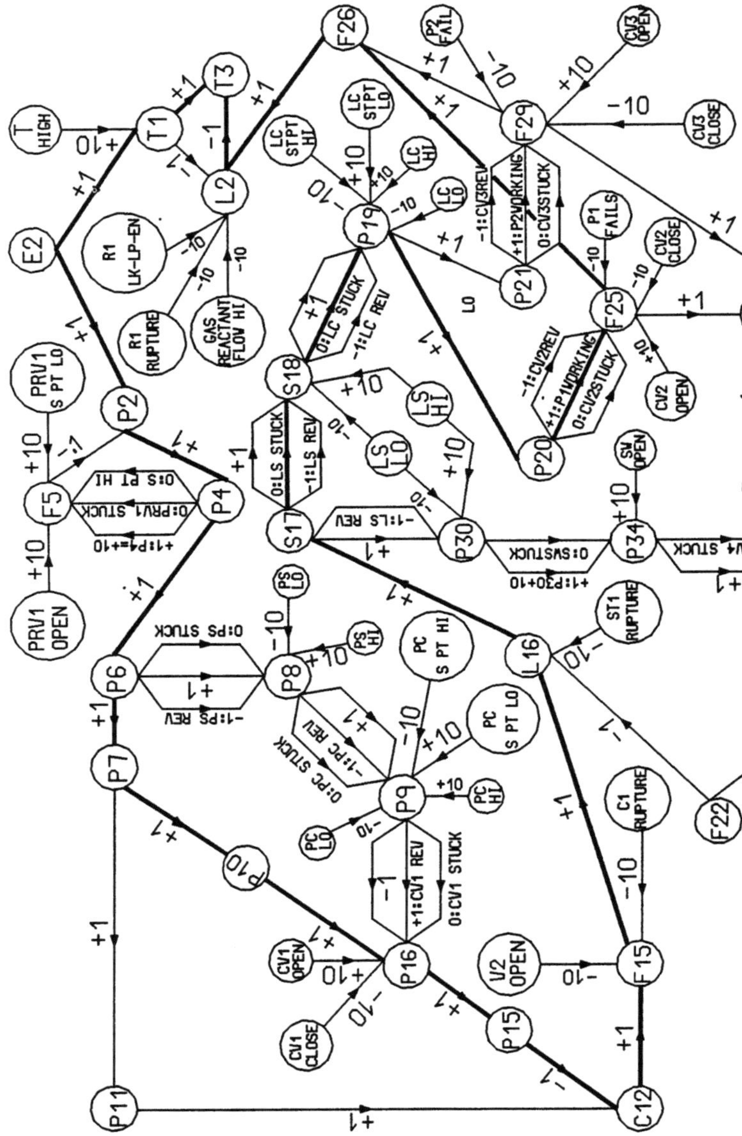

Figure 3. Digraph of reactor cooling system

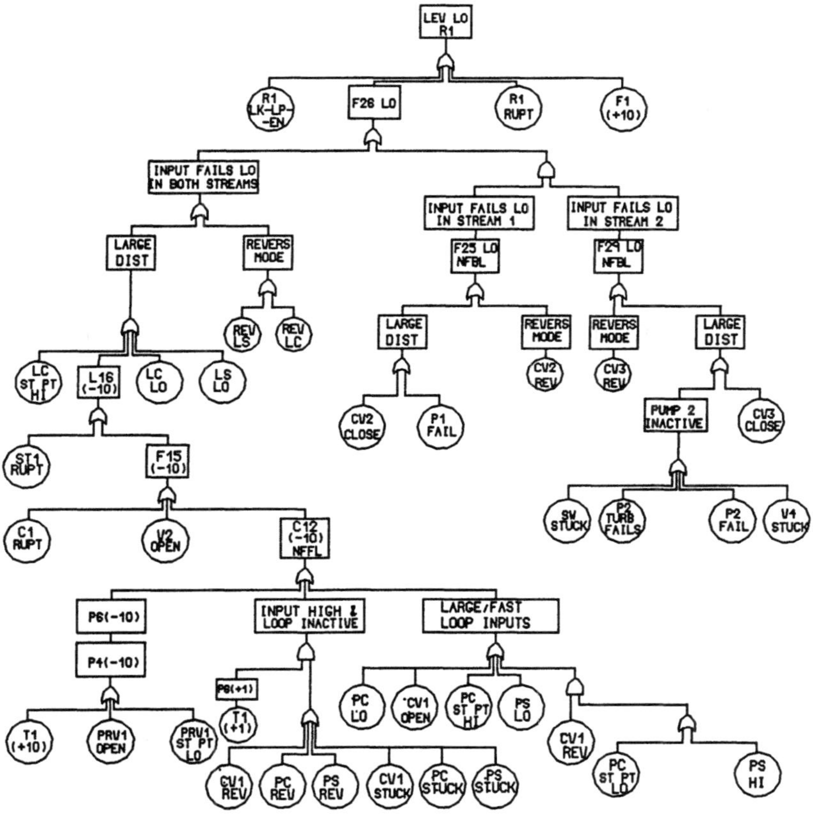

Figure 4. Fault tree for top event LEVEL-LOW in reactor R1.

To obtain the deviation in the dependent process variable caused by a deviation in the independent variable the general rule is to multiply the value of the independent variable disturbance by the gain. For the cooling system considered the digraph representation is shown in figure 3.

4.3 Fault Tree Construction

For simple systems which do not feature any control loops, fault trees can be developed very quickly by tracing the potential causes of a deviation in the top event variable back through the system digraph. In this situation the fault tree structure will be very simple with process deviations being combined by OR gates unless conditional edges are encountered. For a conditional edge the output variable deviation only occurs when the correct input deviation occurs AND the stated condition exists.

However the main advantage of the digraph method is to analyse systems which do contain control loops. For this type of system "operators" are applied to develop causes of control loops passing or generating disturbances. "Operators" were used to develop a fault tree for the top event "low coolant level in the reactor". The fault tree constructed using the digraph method is shown in figure 4. The fault tree constructed using FAULTFINDER is not included since the structures of the two trees were shown to be logically equivalent by comparing their minimal cut sets.

5 Conclusion

The fault trees produced by both the digraph method and the FAULTFINDER computer code for the top event "low coolant level in the reactor" were produced by different fault development algorithms. However the trees were logically equivalent in structure as was shown by examining their minimal cut sets. The example shows both methods are capable of constructing accurate fault trees for systems containing of control loops.

6 References

Andrews, J.D. and Morgan, J.M. (1986) Application of the Digraph Method of Fault Tree Construction to Process Plant. Reliability Engineering, 1 4, 85–106.

Andrews, J.D. and Brennan, E.G. (1989) Application of the Digraph Method of Fault Tree Construction to a Complex Control Configuration. Proceedings of Reliability 89, 4c / 4 / 1–28

Kelly, B.E. and Lees, F.P. (1986) The Propagation of Faults in Process Plants: 1. Modelling of Fault Propagation, Reliability Engineering, 16, 3–38.

Kelly, B.E. and Lees, F.P. (1986) The Propagation of Faults in Process Plants: 2. Fault Tree Synthesis, Reliability Engineering, 16, 39–62.

Kelly, B.E. and Lees, F.P. (1986) The Propagation of Faults in Process Plants: 3. An Interactive, Computer-Based Facility, Reliability Engineering, 16, 63–86.

Kelly, B.E. and Lees, F.P. (1986) The Propagation of Faults in Process Plants: 4. Fault Tree Synthesis of a Pump System Changeover Sequence, Reliability Engineering, 16, 87–108.

Khan, A.R. and Hunt A. (1989) The Propagation of Faults in Process Plants: Integration of Fault Propagtation Technology into Computer Aided Design. I Chem E symposium series No 114, 35–43.

Kletz, T.A. (1990) Lessons of another ethlene–oxide explosion. The Chem.

Engineer, 15, 469–470.
Fussell, J.B. (1973) Synthetic tree model – a formal methodology for fault–tree construction, Aerojet Nuclear report ANCR – 1098.
Lambert, H.E. and Gilman, F. (1977) The Importance Computer Code, Lawrence Livermore National Laboratory, Report No. UCRL – 79269.
Lambert, H.E. (1979) Comments on the Lapp–Powers Computer–Aided Synthesis of Fault Trees, IEEE Trans. Reliability, 28, 6–7.
Lapp, S.A. and Powers, G.J (1977) Computer–Aided Synthesis of Fault Trees, IEEE Trans. Reliability, 26, 2–13.
Mullhi, J.S., Ang, M.L., Lees, F.P. and Andrews, J.D. (1988) The Propagation of Faults in Process Plants: 5. Fault Tree Synthesis for a Butane Vaporiser System, Reliability Engineering and system Safety, 23, 31–49.
Piccinini, N. and Levy, G. (1984) Ethylene Oxide Reactor: Safety According to Operability Analysis.
Salem, S.L. Apostolakis, G.E. and Okrent , D.(1977) A New Methodology for the Computer Aided Construction of Fault Trees, Annuals of Nuclear Energy, 4, 417–433.
Shaelwitz, J., Lapp, S.A. and Powers, G.J (1977) Fault Tree Anaysis of Sequential Systems, Ind. Eng. Chem. Process Des. Dev., 16, 4.
Taylor, J.R. (1982) Automatic Fault Tree Construction with RIKKE – A Compendium of Examples, RISO M–2311, RISO National Laboratory, Denmark.
Willie, R. (1978) Fault Tree Analysis Program (FTAP), Lawrence Livermore National Laboratory, Report No. UCRL – 73981.

A study on the hierarchal quality control in CIMS*

Hu Changhua
Huazhong University of Science and Technology (HUST), Wuhan, China
Chen Zhixiang
HUST, Wuhan, China

Abstract

As an important part of CIMS, QAS is discussed in this paper. The hierarchal quality control(HQC) is the most effective control method in the modern manufacturing, so the dynamic quality control model is needed in HQC. The engineering feasibility of HQC is also analyzed in the paper. As an example, a 3–level quality control system is presented in this article, which is developed in author's qualitycontrol center. The main program methods are given by the authors.The result of tests is in agreement with design.
Keywords: Quality Assurance, Hierarchal Quality Control, Computer Aided Quality Control, Computer Communication.

1 Introduction

Two main demands are presented by the development of manufacturing automation: first, the manufacturing system must be suitable for complex and variable processing and products; second, the system must have higher reliability and stability. These demands are gradually becoming true thanks to the utility and popularity of the computer techniques, example for CAD, CAM, CAPP, CAQC, CAT,etc. CIMS(Computer Integrated Manufacturing System) is an application technique developed in recently years. Because the CIMS is suitable for the two demands, CIMS have been becoming ripper and commercial in the developed industrial countries. The machinery manufacturing technique, computer hard and soft technique, quality engineering, management technique are all integrated in CIMS. So it can be considered that the modernest techniques of engineering and management are all involved in the system.

However, the investment in CIMS often amounts to several millions U.S. dollars. Developing countries and small enterprises can hardly afford such gigantic sum of money. So a simple but useful hierarchal quality control system is researched. The principle of qualityassurance system of CIMS and the information communication method are discussed in this paper. The raised questions, such as processing control, data library and data processing, are also concerned.

* This project is supported by the National Natural Science Foundation of China.

2 Structure (Mode) of hierarchal quality control

Quality assurance system (QAS) can not be distinguished from other parts clearly. This is just the character of CIMS. In the modern manufacturing, there are neither simple processing without QAS nor simple CAQC parting with processing ground. Although the relative integrated manufacturing unit such as processing centre appears in the modern manufacturing, the processing, in fact, also can be considered as partial. As the assurance and supervision system of manufacturing, QAS must distribute to every working process; but the quality of product is classed according to whole finished product, the information of quality control must be relatively integrated. So the study on hierarchical quality control is very important.

Hierarchical quality control means: each machining process of components and each work process are under the control of computer communication net work or local area net so as to assure the whole product quality. The control can be classed into many levels according to the producing. The main method is: down–level station collects quality data and controls the process directly; up–level completes arbitration and diagnosis of the part's quality problems, calculate the quality cost, forecast the quality of products in the future, etc.

Based on the above points, 3–level hierarchical quality control system is designed in the tree mode. The system structure is shown in Fig.1.

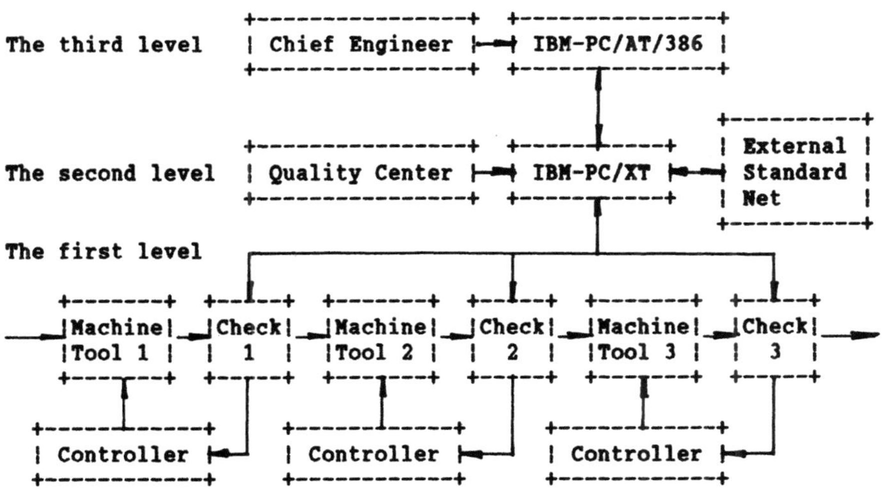

Fig.1 The system of 3–level quality control

The function of each level will be discussed later in detail.

The system communicates with operator at the first level, so does with quality manager at the second level, and communication with the chief engineer at the top level. According to this, the authority of each level are reduced from the top to the first level.

3 Realization of the hierarchal quality control

Based on the success of development of ZK–3, ZK–4, GC–3, and GC–4 and the utility of the STD–BUS, We can use them as the first level stations. The communication net is also designed by our CAQC group.

3.1 Communication net

There are many communication nets sold in computer companies, such as ETHER net, PC net, etc. But they are too expensive. The cost will be reduced if we developed communication net by ourself.

The communication net is illustrated in Fig.2.

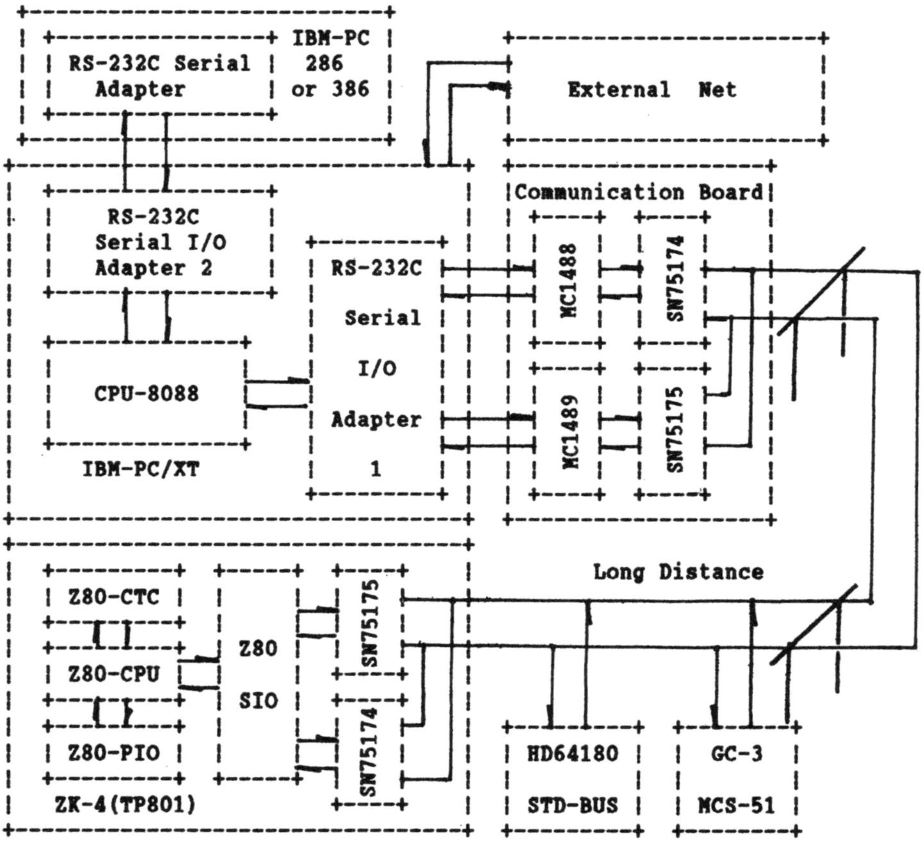

Fig.2 The communication net

The first level works in the manufacturing ground and the second level may be in the quality centre or workshop, so the communication between the two levels belongs to local area communication of different computers.

The protocol of communication is:

Serial asynchronous communication.
1 stop bit and 8 bits data.
Odd–even check (selected by user).

The CPU types of ZK–4, GC–2 and STD–BUS are the Z80 and INTEL MCS–51(96) series, so Z80–SIO and single chip computer's UART–POST are selected as the communication controllers. Their electric levels are compatible with TTL. The side of the second level is IBM–PC / XT, its communication port is serial adapter. But its electric level is RS–232c. According to the IEEE RS–232c the electric levels must be exchanged. 1488 / 1489 are used to satisfy this request.

The distance of communication is designed as 1 to 1.5KM, and the modem will increase the cost, so SN75174, SN75175 are selected as the line driver and receiver. They are quad differential line driver and receiver. The result of test, such as code error rate, is satisfied. Zk–4, GC–2, and STD–BUS can be called by IBM–PC / XT as NO.1, NO.2, and NO.3. To listen to the data bus, we can use the port B of SIO. Before asking the IBM–PC, SIO must inspect the data bus through the port B. If there are data on the bus, this computer must wait some time till there are not data. The waiting time are different from each other. This can be decided according to the position of this computer in the system. So the bus competition of computers is solved by this method.

The communication between the second and the third is the near or long distance communication of the same computers.

3.2 Duty distribution of each level

These are specified in TAB.1.

Table 1. The duties of each level

The first level	The second level	The third level
Check dimension	Histogram	Data library of crux
of internal	X bar R chart	quality target of
hole	Set AR(n) and forecast	Key processing and
Histogram	Cp and its of distribution	important workpiece
Cp coefficient	Linear regression	Middle and long period
Spoiled product	Un–linear regression	forecast of the
rate and	Diagnosis and arbitration	above targets
ordinal number	of processing breakdown	The quality cost of
X bar R chart	Cause select control chart	products
Check ellipticity	Reporting table of season,	Domestic and foreign
Surface roughness	month, week and day	quality data and the
(Ra,Rz)	Recording tolerance	contrasting with self
Check external	of form and position	factory
hole dimension	Record of equipment work	Weekly,monthly,season
	ability and maintenance	and annual reporting
	Maintenance of	Analyze and diagnose
	metrological instrument	quality breakdown

3.3 Dynamic quality control model

The current QAS studied in many countries is set up with statistical process control(SPC) method. But the control model of hierarchal quality control system must be set up based on the dynamic quality control method.

In fact, the SPC can be considered as two classes: statistics after processing and statistics in processing. The latter can be used to set up the dynamic control model. In processing, the workpiece is processed one by one, the quality state is indicated on the workpiece dimension or other inspects. The quality data are collected in real time and serially, then according to those data we can set up the AR or AMAR control model using the time serial analysis method. If the quality state changes, the quality data must change with it. So the control model can be modified by using the changing quality data. Furthermore, this model can forecast and compensate processing ahead of one step.

4 Design of the system software

The system software can be classified into two groups: communication program and QAS program.

4.1 Communication program

The interfaces between each level software are communication programs. The information flows in these programs.

In this part, the second level is connected with two levels, so its communication program is the most complex and important, its rationality is the key to the system. As an example, the detail of the second level's communication program is discussed. In this level two problems must be solved. The link between interrupt service program and communication program is the first one, the second one is the link between communication program and QAS program.

For the first problem, there are two methods: data files and memory data areas. Data files are set by interrupt service program, they can be used by communication and quality control program, but the setting time of data files is too long to control processing on line. Lots of time can be saved by using the second method, but the price is the computer memory casting. It also increases the difficulty of the second problem, because the memory map of IBM–PC / XT is not clear to advanced language, and the interrupt service program is designed in assemble language which the memory map is clear to. So the interrupt service program must not use the memory area used by advanced language. The last method is used in this paper.

The second problem can be solved according to the user's requests. So there are two methods to this problem too. One is hard interrupt, the other is soft interrupt. But the better is the soft interrupt. The operator can talk with the computer in this method. In our system the better method is adopted.

To check the effect of the system, many test have been performed in our quality center. The result is satisfied with the design. The communication net can transfer single data, pair data, group data, group of pair data, and group of group data, etc. The second level IBM–PC / XT can call and distinguish the first level computers. If the first level computers visit each other, they can talk through the second level computer. But in normal situation, this kind of communication is not necessary, and the possibility is also very small.

4.2 QAS program

This kind programs are different from each level and each station at the same level. Generally, these program can be classed into two groups: one is quality control program, the other is quality management program. The first programs are used to control machine tools or other manufacturing equipments. The second group is used to complete quality statistic work, such as reporting table or chart, etc.

At the first level most QAS programs are control and forecast program. In ZK–4, for example, this part is programed in Z80 assemble language. It realizes control on line and forecast using AR(n) model. At the second level most QAS programs are quality management programs, including: printing quality table, plotting histogram, middle level quality diagnosis, middle period quality forecasting, often used quality standards of factory, etc. The third level's QAS programs include: long period quality forecasting, high level quality diagnosis, expert system, quality standards, etc.

This part is very complex, and the contents in it are extensive, but the space of this paper is not sufficient, so it can not be discussed in detail.

5 References

Chen, Z.X. Hu, C.H. and Dai, Z.X. (1987) An Instrument for measuring Surface Roughness with Optical Fibre Probe and Its Intelligence System, 1988.1, **Journalof** Huazhong University of Science and Technology, 127–132.

Chen, Z.X. Zhang, H.H. and Chen, Z.C. (1989) Quality Control and Forecast in Cylindrical Centerless Gringing Process, **Proceedings of the 2nd International Metal Cutting Conference**, 711–729.

Cheng, J.B. Chen, Z.X. and Li, Z. (1989) ZK–3 Type Statistical Quality Control of Manufacturing Operation, **Proceedings of the 1st International Symposium on Measurement Technology and Intelligent Instrument**(ISMTII), 69–71.

Harrington, J.Jr. (1973) Computer Integrated Manufacturing Process, **Industrial Press**, New York.

HITOMI, H. (1979) Manufacturing System Engineering, **Taylor & Francis Ltd**, London.

Jone, A. and Mclean, C. (1986) A Proposed Hierarchal Control Model for Automated Manufacturing Systems, **Manufacturing System**, Vol.5, 1986, NO.1.

Warnecke, H.J. Melchior, K. and Kring, J. Integration of Quality Control in the Information Flow in Production, **CIRP**, 1986.

An experiment in automation of statistical process control using fuzzy logic

M.H. Lim and T.H. Ooi
School of EEE, Nanyang Technological Institute, Singapore

Abstract
This paper describes the development of a system to aid in maintaining a successful and reliable quality control program in a manufacturing environment. Such a system besides acquiring process related data will have the basic functions for the handling of massive statistical data which are necessary in order to monitor the performance of a production line. Starting with raw production data, the system outlined besides flagging for out-of-control case, has the capability to analyze unnatural statistical trends to determine the possible assignable cause(s) for the out-of-control situation. Based on knowledge acquired from a local expert, the system may suggest preventive or corrective actions. To achieve this, the technique used in the analysis of statistical trends or patterns is based on a fuzzy logic approach which has proven to be more suitable compared to conventional expert system.
Keywords: Statistical Process Control, Fuzzy Logic, Chart Analysis, Expert System, Quality Control.

1 Introduction

Many commercial spreadsheet packages such as dBASE and Lotus 1-2 3 are available for statistical computations and chart plotting. Such tools are powerful when used by a person with sufficient statistical background. In statistical process control (SPC), making useful deductions out of numbers and charts requires knowledge and experience in the processes and production environment. For example, a process engineer confronted with a control chart that shows a cyclic pattern may be able to reason from experience that worker fatique may be the likely cause for the unnaturalness. What is apparent is that the presence of a "local expert" is essential in maintaining a successful SPC program. To reduce the reliance on the local expert, a system that is able to deduce automatically the occurrence of unnatural patterns in control charts may be useful.

Automation is seen as a means of increasing throughput by industries. Accordingly, it would be beneficial to incorporate SPC in a production line with as little human intervention as possible. It should be realized that automation does not imply that the service of a local expert is no longer needed. It is only appropriate that any software system for SPC be viewed as an aid to the overall operation. The obvious outcome is perhaps an improvement in the overall productivity. Ideally, it would be appropriate that the presence of a local expert will serve to "educate" the software since a production line is essentially a dynamic environment. Hence, the knowledge base used in reasoning should be specific to the plant or ambient environment in order to be reliable.

The main issue here is not merely determining in-control or out-of-control situation from a given sample of raw production data. Under such circumstances, it is sufficient to use the basic statistical analysis tools and charts that most people are familiar with. Within the SPC circle, it is common to analyze the integrity of a process based on a set of guidelines established by Western Electric Company (WECO). Even though the rules are very specific, it cannot be denied that the nature of the problem is vague. Such uncertainty may perhaps be reflected in terms such as "very much in -control", "somewhat out-of-control", "going to be out-of-control" and so forth. In other words, there is a gray area involved in the continuous range from being in-control to out-of-control. Such vagueness is inherent in human reasoning and can be useful if it is incorporated into the inference mechanism of a computer-aided analysis system.

Ideally speaking, if such a production environment as described above can be realized, a production line will not go into an out-of-control state. This is achieved based on the premise that before a process becomes statistically out-of-control, the points on the control chart will show an unnatural pattern.

Because of the inherent vagueness involved, the problem becomes more accommodating when handled with a fuzzy logic approach. Before describing the analysis of control charts for unnatural patterns, a brief background on SPC will first be presented. The implementation of the fuzzy SPC system involves the use of a FSIM (Fuzzy shell/SIMulator) for the prototyping of fuzzy rule-based system. Hence a brief introduction on FSIM is in order followed by description on the overall system and the technique in analysing statistical control charts.

Table 1: Traditional versus SPC approach.

Statistical Process Control	Classical Process Control
• Control Charts • Uses Statistically derived Control Limits • Believes in Process Capability • Variable Data Preferred • Drives Controls Upstream	• Lot Acceptance • Based on Sampling Plans Specification Limits • Arbitrarily Defined Limits • Almost Exclusively Attribute Data • Checks End-Product for Conformance

2 SPC Review

There are a few differences that are apparent between traditional quality control and SPC. It can be briefly summarized in Table 1 [Tan(1989)].

2.1 WECO Rules

To determine if a set of sampled data is in statistical control, WECO rules can be employed. The application of these rules is straightforward and requires at least 8 points. The four basic rules are stated as follows and a typical control chart (violating all rules except the first) is shown in Figure 1 for illustration.

- one point outside the 3-sigma (s) control limits
- 2 out of 3 consecutive points lie beyond the 2-sigma warning limits (Zone A of Figure 1)
- 4 out of 5 consecutive points appear beyond the one-sigma limits (Zone B of Figure 1)
- 8 consecutive points appear on one side of the center line (Zone C).

Although not explicitly stated, the implication is that any distribution which appears to be non-random is sufficient doubt that the process is not in statistical control. In other words, SPC emphasize on the random nature of a process rather than strict adherence to control threshold limits.

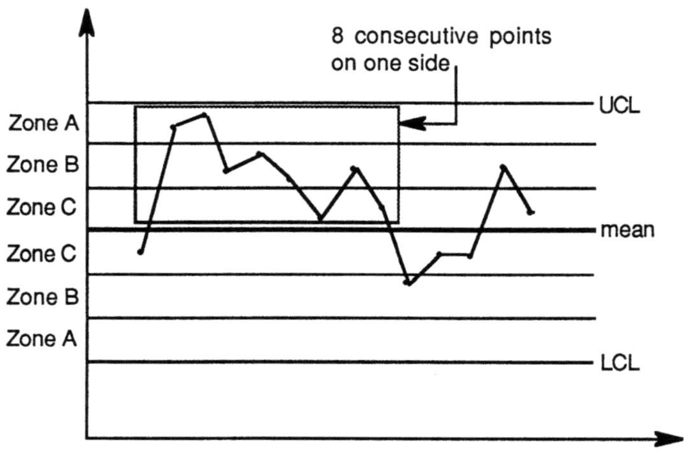

Figure 1: Zone test based on WECO rules.

2.2 Procss Capability Index

Another criteria that is of importance in SPC philosophy is an indication of process capability. By realizing the capability of a process, engineers may learn to make expectations that are achievable based on the resources available and the current state of affairs. One good indicator of the capability of a process in a production line is the Process Capability Index (C_{pk}). The capability index of a process can be determined as folows:

$$C_{pk} = \min(C_{pL}, C_{pU})$$

where

$$C_{pL} = | u - L | / 3\sigma$$

$$C_{pU} = | u - U | / 3\sigma.$$

Both *U* and *L* are upper and lower specification limits respectively which are set by management. The value *u* is the process mean.

3 Introduction to FSIM

FSIM is a software simulator for developing prototype fuzzy production systems. It is a tool that bears similarity to most expert system shells or tools such as EMYCIN, EXSYS, PC-Plus, and OPS5 just to name a few. An obvious difference is that FSIM is targeted for knowledge-based systems in the fuzzy domain. The inference mechanisms used in FSIM is characterized as approximate rather than exact, hence the term approximate reasoning. Another significant difference is the way elements are represented as tokens of fuzzy sets each with an associated grade of membership distribution. Therefore, instead of a straightforward matching of strings as employed in most non-fuzzy tools, the strategies for matching and manipulation of elements in the working memory are more complex in FSIM. Hence to use FSIM effectively, some basic understanding on the theory of fuzzy sets which has been well covered in literatures is required.

The main feature of FSIM is its generality. It consists of various built-in inference mechanisms to enable a user to rapidly develop a prototype knowledge-based system. The original version of FSIM was developed on UNIX-based machines. In using FSIM, commands are invoked interactively to control and monitor the progress of simulation. Typically, the steps involved in the development can be broadly categorized into 5 stages:

- creating a rule base
- creating a dictionary
- parsing the rules
- specify input
- simulate.

For a detail exposition to FSIM, readers are referred to [Lim and Takefuji (1990)].

4 Integrated Production Environment

Nowadays, the capability of personal computers (PC) have achieved a status that is comparable to what used to be only within the realm of expensive workstations and minicomputers. A wide range of application softwares are available for the PC users. In addition to that, it is possible or rather common to find PCs being connected together by means of Local Area Network (LAN). Plug-in cards for networking PCs in a LAN environment are widely available. It is therefore not surprising to find the wide usage of PCs for data logging and process control chores in a production line. Hence a PC-based system seems to be appropriate.

The eventual target of our work is to be part of an integrated EPROM production environment. At present, we are also currently prototyping a software package using on the expert system shell PC-Plus. Similar work on expert system for SPC has also been reported by [Evans and Lindsay (1988)] who used EXSYS in their implementation. With the PC-Plus approach, certain modules for data consolidation have to be realized by dBASE and a lower level programming language such as C. Note that complete automation is not achieved since the main limitation is that the trend or pattern analysis is left to the engineer. This is due partly to the fact that pattern analysis of statistical control chart is not easily realized in conventional expert system shells such as PC-Plus.

These limitations, are added advantage for resorting to a fuzzy logic approach which enables all the modules to be realized in C programming language. The application of WECO rules are fairly straightforward and can be easily realized. As for the pattern analysis part, it was developed using FSIM. After all the fuzzy rules and definitions have been tuned to function according to requirements, it can be easily ported to a PC. Not all the C codes of FSIM need to be incorporated. Once an application has been realized, the relevant modules of FSIM can be extracted with a reduction of at least 50% in C codes for the final system. From the standpoint of performance, this is significant when the program is to run on a PC.

5 The Fuzzy Rule-Based System

To distinguish between the 5 basic patterns of the control charts, 5 rules are required. The rules to realize the reasoning with the corresponding pattern for each rule is included in the Appendix. Three input variables are used in the matching of the lefthand side of the each rule; $CROSS, $Deff and $Dabs. $CROSS represents the number of times the distribution crosses the central line of the control chart. $Deff is calculated by summing the distances of each point from the centre line. Distance above the line is taken to be positive and below the line to be negative. The third variable, $Dabs is the sum of all the absolute distance of each point from the centre line.

All the fuzzy labels used in the rules such as MANY, SMALL and SHORT must be defined in the dictionary file. The labels are defined as fuzzy sets as shown in the Appendix. For example, the label MANY is defined such that any chart having more than 8 crossover points are considered to be many. Of course these definitions are subjective, in line with the vague and approximate nature of our approach. These definitions will have to be tuned or modified whenever conditions change. This means that suppose the number of points monitored is increased, than all the fuzzy set distributions may have to be modified.

6 Conclusions

In this paper, we have discussed the development of a software system to automate the SPC program in a production line. If sufficient past control points are monitored, it is possible to automatically analyze the statistical controllability of the process by means of fuzzy logic inferencing technique. This involve pattern analysis of the control chart which is not so practical if we rely on conventional expert system shells, such as PC-Plus. Such shell -based systems are cumbersome when run on a PC. To this end, we have demonstrated that automation can be achieved if we resort to a fuzzy approach.

Acknowledgements

The authors wish to acknowledge the contributions of two individuals, Mr Chan Wei You and Mr Chew Khiang Kee. Their dedication in their work made this paper possible. The help from Mr David Chin is also acknowledged.

References

Evans, J.R. and Lindsay, W.M. (1988) Expert Systems for Statistical Quality Control, in **Expert Systems**, (eds N.A. Botten and T. Raz), Industrial Engineering & Management Press, American Inst. of Industrial Engineers, 131-136.

Lim, M.H. and Takefuji, Y. (1990) Implementing Fuzzy Rule-Based System on Silicon Chips. **IEEE Expert**, February.

Montgomery, D.C. (1985) **Introduction to Statistical Quality Control**, John Wiley & Sons, Inc., New York.

Tan, C.H. (1989) Statistical Process Control: From Theory to Practice, **Proceedings, 3rd International Symposium on IC Design and Manufacture**, Nanyang Tech. Inst., S'pore, September.

APPENDIX

```
# FILE TITLE : SPC.KB
#Rule 1: Pattern A
#    (stratification)
IF <$CROSS IS MANY>
AND <$OUT IS VERY NOT FEW>
AND <$Dabs IS SHORT>
THEN <$PATTERN IS A>
#Rule 2: Pattern B (mixture)
ELSE IF <$CROSS IS MANY>
AND <$OUT IS VERY NOT FEW>
AND <$Dabs IS LONG>
THEN <$PATTERN IS B>
#Rule 3: Pattern AB (mixstrac)
ELSE IF <$CROSS IS MANY>
AND <$OUT IS VERY NOT FEW>
AND <$Dabs IS NOT OR
            (SHORT, LONG)>
THEN <$PATTERN IS AB>
#Rule 4: Pattern C (trends)
ELSE IF <$CROSS IS ONEP>
AND <$OUT IS NOT FEW>
AND <$Dabs IS NOT VERY LONG>
THEN <$PATTERN IS C>
# Rule 5: Pattern D (cycles)
ELSE IF <$CROSS IS NOT OR
            (ONEP, MANY)>
AND <$OUT IS NOT FEW>
AND <Dabs IS NOT VERY LONG>
THEN <$PATTERN IS D>
#Rule 6: Pattern E (freaks)
ELSE IF <$CROSS IS NOT ONEP>
AND <$OUT IS FEW>
AND <$Dabs IS NOT VERY LONG>
THEN <$PATTERN IS E>
```

```
# FILE TITLE : SPC.LABEL
# Memberships for number of crossover points.
# 1   2   4   6   8   10   12   14   16   18   10
~ONEP 11
  1.0 0   0   0   0    0    0    0    0    0    0
~MANY 11
   0   0   0   0  0.2  0.4  0.8  0.9  1.0  1.0  1.0
# Memberships for number of out of control points.
# 0   1   2   3   4   5   6   7   8   9   10
~FEW 11
   0  1.0 1.0 0.8 0.6 0.4 0.2 0.1  0   0   0
# Memberships for absolute distance.
# 0   4   8   12  16  20  24  28  32  36  40  44  48  52  56  60
~SHORT 16
 1.0 1.0 1.0 0.9 0.8 0.6 0.4 0.2  0   0   0   0   0   0   0   0
~LONG 16
  0   0   0   0   0   0   0   0   0   0  0.2 0.5 0.9 1.0 1.0 1.0

# Memberships for patterns.
# A   B   C   D   E
~A 5
  1.0 0.4  0  0.3  0.7
~B 5
  0.4 1.0  0  0.3  0.7
~C 5
   0   0  1.0 0.7  0.2
~D 5
 0.3  0.3 0.7 1.0  0.6
~E 5
 0.7  0.7 0.2 0.7  1.0
~AB 5
 0.8  0.8  0  0.3  0.7
```

Application of syntactic pattern recognition techniques to condition monitoring of machines

M.M. Ahmed and R.D. Pringle
Napier Polytechnic of Edinburgh

Abstract
A syntactic pattern recognition system for recognition of normal/abnormal vibration spectral patterns of machines is described. Vibrational spectral data is preprocessed and reduced to sentences of terminal strings. A general purpose software 'shell' infers a regular grammar from a set of sample patterns and converts the inferred production rules into a parser in the form of a finite state automaton. A search algorithm based on depth-first method with backtracking extracts those patterns from noisy input data which conform to the inferred grammar rules. Large amounts of periodic vibration data can be analysed efficiently for the occurrence or nonoccurrence of abnormal conditions.
Keywords: Vibration, Spectral Pattern, Syntactic Pattern Recognition, Software.

1 INTRODUCTION

This paper describes an implementation of the technique of syntactic pattern recognition for the automatic analysis and extraction of characteristic patterns from vibration spectra. The primary emphasis in this research is to demonstrate the utility of the syntactic method through the design and implementation of general purpose software tools which can be customised for a particular task. Syntactic methods have received increased attention recently because of their ability to describe patterns in picture processing, scene analysis, fingerprints identification etc and to model systems for processing natural languages[3].

The benefit of vibration based condition monitoring

programme using computers has been reported in literature[5,6,7]. The techniques employed use automatic spectrum comparison and statistical calculations based on trend analysis of different parameters. Digital Fast Fourier analysis of the time waveform has become the most popular method of deriving the frequency domain signal. The signature spectrum so obtained can provide valuable information with regard to machine condition[4].

Syntactic pattern recognition techniques are extremely useful when input data is noisy and contains a great deal of information relating to normal running of the plant in addition to information relating to damage and wear. Syntactic approach would be useful in cases where the characteristic spectrum for a machine is not available and in these circumstances the examination of relative amplitudes of group of component changes expected due to atypical conditions is carried out[8]. The inferred and hence the parser constructed is dependent on the completeness and accuracy of the sample patterns. If the sample patterns represent abnormal conditions, then the pattern extracted would be classified as a potential faulty condition and if present at an early stage, it could warn of the danger at very early stage.

2 PATTERN PRIMITIVES AND SAMPLE PATTERNS

The first step in formulating a syntactic model for pattern description is the determination of a set of primitives which will describe the patterns of interest. This determination is largely influenced by the nature of the data, the specific application in question and the technology available. For vibration spectral patterns amplitude levels at particular frequency component are important because they indicate machines' rotating elements condition. From physical conditions we know that not all the measurements points are of equal importance in determining which conditions exist. However, any one fault will effect some other parts of the spectrum significantly, others only slightly and many parts not at all. If this is generally true for all the faults of interest, then substantial data compression can be achieved by defining each fault by those set of pattern primitives which are useful in discriminating that fault from others. A vibration pattern is obtained by selecting only a subset of the various signal descriptors that might be derived from the measured vibration data. Engineering knowledge concerning to individual equipment, critical fault events etc are employed in selecting pattern samples. Sample patterns in terms of above primitives, of various operational conditions of the machine, should also be included since it has been demonstrated that variation in speed and load can introduce vibration level exceeding those associated with anomalies. The success of the syntactic technique in this case is dependent on the completeness of sample patterns. Nevertheless, the number of pattern primitives should be restricted so that the parsing process is efficient and computer memory is not exhausted.

Example of pattern primitive

X_1Y_1 ==> 'a'
X_2Y_2 ==> 'b'
X_2Y_3 ==> 'c'

where X_1 is the frequency component and Y_1 is the amplitude level.

3 EXPERIMENTAL DETAIL

A software routine extracts peaks from raw vibration spectral data and produces strings of sample patterns. In our case data came from different positions of a pump and a compressor running at steady speed collected over a period of 6 months. Grammatical inference routine utilise these sampled strings to infer regular grammar.

Example

R^+ = { abcd, abd, abce, abcde}
A regular grammar G = {V_T,V_N,P,S} where
V_T = {S,A,B,C,D};
V_N = {a,b,c,d,e} and inferred productions are P:

```
S -> aA
A -> bB
B -> d
B -> cC
C -> d
C -> dD
D -> e
```

The above sample set will produce the state transition diagram shown in figure 3.1.

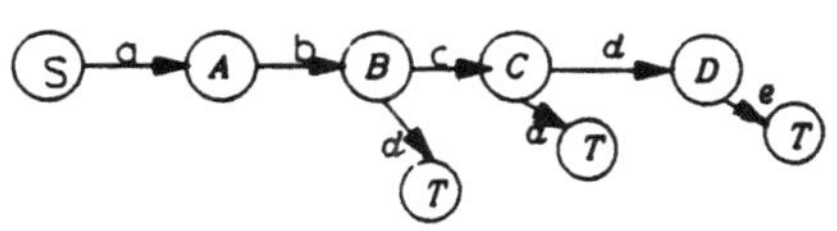

Figure 3.1 Transition diagram.

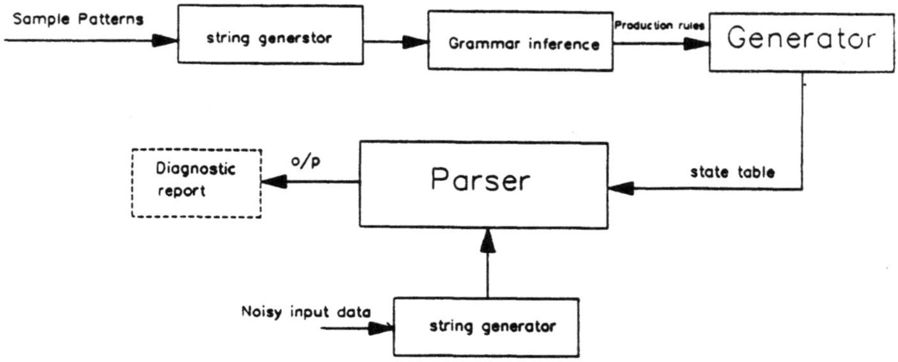

Figure 3.2 Shell structure.

The finite state automaton based on the minimised grammar rules is automatically prepared by the shell. The parser parses input noisy data and extracts patterns which conform to the inferred grammar using depth-first search with backtracking. Figure 3.2 shows a block diagram of the 'shell' and figure 3.3 shows an example of a pattern extracted from a noisy input. A tolerance factor has been introduced to extract patterns which are close to sample patterns. Stochastic pattern recognition techniques have also been employed by assigning probabilities to production rules which has improved the search speed. Error recovery routines will also be included in the shell to enhance the capability of recognising noisy patterns.

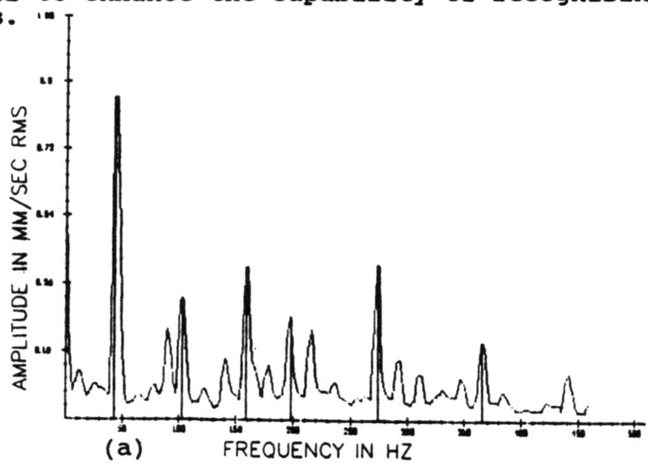

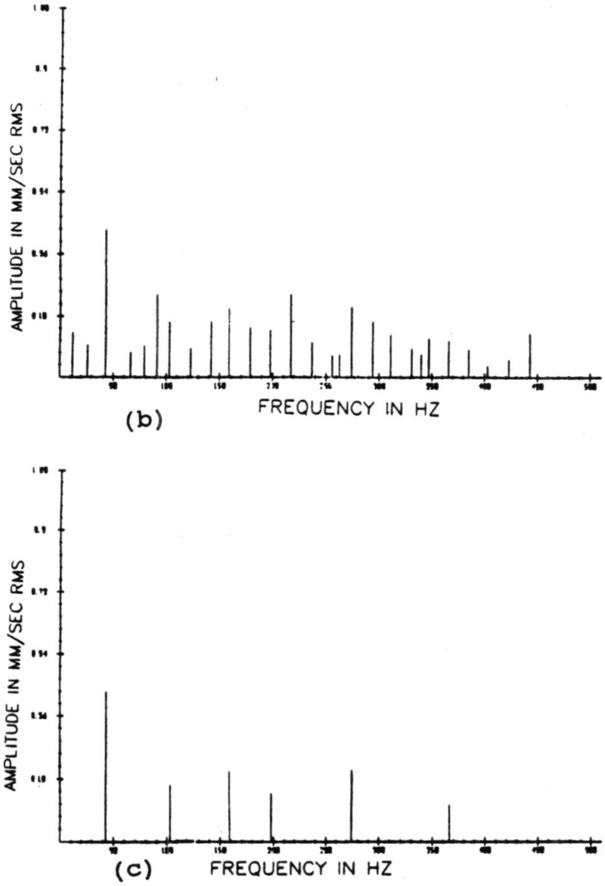

Figure 3.3 (a) Sample pattern (b) digitised input noisy data (c) Extracted pattern.

The software shell, based on syntactic techniques, is extremely useful for analysing a vast amount of data as once loaded with appropriate sample pattern files and data files it will explore for all the possible fault patterns. If a hierarchical database structure could be established as suggested[1] syntactic method could model the whole machine condition monitoring system if satisfactory pattern primitive solution to other monitoring parameters such as current, temperature, pressure etc could be obtained. A machine's current condition could be monitored without recourse to 'healthy vibration signature' of the machine. Furthermore, when the task of identifying harmonic patterns and structural resonance frequency pattern from each other is difficult[2] syntactic technique could be useful.

4. CONCLUSION

As software plays fundamental role in modern predictive maintenance, techniques of syntactic pattern recognition are applicable and valuable provided sample patterns of normal or abnormal conditions of machine's vibration spectra are available. A software shell is able to automatically analyse masses of noisy date without expert help for occurrence or non occurrence of faulty patterns of vibration and extract those pattern from noisy vibration data.

5. ACKNOWLEDGEMENT

We are grateful to YARD Ltd for providing the data.

6. REFERENCES

1. Carey, J. (1987) The role of software in predictive maintenance. Maintenance conference, 1987.
2. Dyer,D; Stewart, R.M.(1978) Detection of rolling element bearing damage by statistical vibration analysis. Journal of Mechanical Design, April, pp 229-235.
3. Fu, K.S. (1982) Application of pattern recognition. CRC Press, 1982.
4. Mathew, J. (1987) Machine condition monitoring using vibration analysis. Accoustic Australia, Vol 15, part 1, April, pp 7-3.
5. Nichollis, C. (1986) Condition monitoring using data collectors. CME Chart Mech, Vol 33, Part 5.
6. O'Dea, D.M. (1975) User experience with computerised machinery vibration analysis. Hydrocarbon Processing, Dec, pp 81-84.
7. Stewart, R.M. (1984) Machinery management based on the application of modern diagnostics, control theory and artificial intelligence. Condition monitoring Conf Proc, April 10-13, Swansea, pp 33-55.
8. Tait, A.J. et al (1986) Condition monitoring of electrical drives. IEE Proc, Vol:13, Part-B, No:3, May, pp 142-148.

An engineers's assistant for balancing of flexible rotors

R.J. Allwood
Department of Civil Engineering, University of Technology, Loughborough, England
C.N. Cooper
Formerly Department of Civil Engineering, University of Technology, Loughborough, England
G.D. Wale and M.H. Walton
GEC ALSTHOM Engineering Research Centre, Lichfield Road, Stafford, England

Abstract
Flexible rotors usually require to be balanced for multiple speeds using several balance planes. Since the sensitivity to unbalance varies with speed the selection of the locations for the balance masses are usually based on the deflected shapes of the rotor as it passes through its critical and service running speeds. When trim-balancing a machine on site the choice of locations is frequently restricted to non-enclosed parts of the rotor and a compromise must be sought to achieve satisfactory running. As part of a larger Alvey project on the use of expert systems in engineering environments, the application of expert system techniques to flexible rotor balancing has been investigated. Areas where these techniques can be beneficial include the initial assessment of the quality of vibration signals and a check whether unbalance is a likely cause of high levels, identification of critical speeds and mode shapes, and selection of the most effective balance planes from those available. A simple system has been developed to demonstrate some of these areas. In order to gain acceptance with practising balancing engineers emphasis was placed on providing information in a familiar and easily assimilated form. The nature of the task resulted in a hybrid system with procedural and rule based sections as appropriate to control acquisition of vibration data from the transducers, assess the results and make recommendations on positioning of balance weights. Due to memory size and processor speed limitations the functionality of the demonstrator was limited to simple rotor configurations but offers potential for enhancement as improved processors and software become available.
Keywords: Expert System, Balancing, Flexible Rotors

1 Introduction

Vibration is detrimental to the integrity of rotating machines, possibly causing loosening, rubbing, fretting and fatigue of components, eventually leading to failure. Analysis of the manner in which the vibration level of the machine has changed over a period of time is often used to assess fitness for further service and use of expert systems in the diagnosis of problems in high value machines has been of increasing interest. In a recent collaborative exercise involving CEGB, GEC (as domain/knowledge experts) and Carnegie Group using a development of their software package Knowledge Craft, a demonstrator expert system for diagnosis of difficulties in large turbine generator sets was successfully implemented. One of the many possible diagnoses arising from an investigation of high vibrations could be need to adjust the current state of balance of the machine.

Whether a machine is being rebalanced to correct for benign changes in its condition or to allow its continued use with a known or suspected fault until an outage is possible, there will be intense pressure to complete the rebalancing in as short a time as possible to allow revenue earning service to continue.

This paper describes a system to assist the balancing engineer to meet this challenge. The work was part of a larger ALVEY project on the use of expert systems in an engineering environment and applied to the balancing of flexible rotors, typified by the high revenue earning machines used by electricity utilities and the petrochemical industries. The system is able to function at site where little prior information on critical speeds and mode shapes may be available but may equally be used for balancing within an overspeed laboratory. It was not intended that the system should replace the balancing engineer, but to relieve him of routine tasks such as data collection and filing, provide presentation of data in a familiar and meaningful form and assist in interpretation of the results. In this way the engineer could devote more time to assessment of results and consideration of possible actions, so leading to improved decision making and reduced outage time.

2 Balancing of a Flexible Rotor

Rotors may be classified as rigid or flexible. A rigid rotor does not deflect under service forces and its first natural frequency of flexure is above the normal running speed. The balancing of such a rotor is straightforward and quasi static methods suffice. The rotors of large machines are normally an extreme case of a flexible rotor system where the system flexibility is so low that several natural frequencies (critical speeds) exist below the normal running speed. The machine designer will normally have arranged that running speed is not too near a critical speed as this results in a machine very sensitive to load induced variations. For these flexible rotors, the balance must not only be suitable for normal running, but also allow the machine to be run up and run down without excessive vibration. The vibration level at any speed depends on the distribution of unbalance and the way this interacts with the mode shape. Not unexpectedly, unbalance has more effect where the mode shape deflections are higher. As the mode shapes encountered on run up and run down differ at each critical speed, the balancing problem is to provide an array of masses distributed along the machine in a way which achieves adequate vibration control over all the speed range.

In theory, perfect balance can only be achieved if correction masses are fitted all along the rotor to counteract the existing unbalance. In practice this is unnecessary and only a limited number of locations for fitting balance masses will be provided by the machine designer. The balancing engineer thus has to select masses which, when fitted in the limited number of positions available, will give acceptable vibration levels at normal running speed over the normal load range, during run up and during run down. In order to reduce the outage time, the smallest number of investigative runs possible should be used.

At the start of a balance exercise, the only information normally available will be vibration levels under various conditions, as monitored by the machine user. The balancing engineer will then have to decide the most appropriate method of measuring the machine vibrations for his purposes. This will normally entail the fitting of additional transducers and/or analysis equipment.

The first step in balancing is then normally a run of the machine in the as found condition, with a view to obtaining as much information as reasonably possible, encompassing if possible a cold run up, normal running under various load conditions and a hot run down. The variable speed runs may be of value in assessing the reliability of the data obtained and enable potentially corrupt data ranges to be eliminated. The next step is selection of a trial balance change in the machine, for which there may now be sufficient evidence as to the most appropriate plane to make a start. Associated with this is the choice of trial balance change for which the balancing engineer attempts to make significant changes to the vibration pattern but without exceeding vibration levels acceptable for the tests.

A further monitored run enables the changes in vibration vectors at the various conditions brought about by the mass change to be identified and their reliability assessed.

A new balance plane and/or mass is now chosen and the process repeated until the balancing engineer is confident that sufficient control over the machine's vibrations can be obtained.

At this stage the engineer starts to fit arrays of masses as calculated from the results so far and another test run commissioned. Further information resulting from this run, combined with previous data, may indicate the need to refine the balance either by a new array of masses using existing planes or moving on to any additional planes which may be available. When an acceptable balance is obtained, securely fitted and documented, the task is complete.

It is thus seen that balancing is a highly interactive enterprise, with decisions needing to be taken each time new information arises. Any aid to the engineer in presentation of facts that need to be considered will lead to a better quality of decision, and the expert system to be described fulfils this function.

3 Aspects of Balancing Where Expert System Techniques may Potentially be Useful

The balancing of flexible rotors has been regarded by many people as something of a "black art", however it is basically a procedural process. The expertise lies in the ability to recognise patterns in the behaviour of the rotor from measurements at a restricted number of, not necessarily ideally located, points along the rotor, and hence select an appropriate balancing strategy.

The calculations for optimising the final correction masses assume that the measured running speed vibrations arise purely as a result of unbalance.

If there is any corruption of the measurements, either from other vibration sources or electrical interference, incorrect results will be obtained. The system must therefore help the engineer to recognise when corruption is likely and either make corrections or avoid using such data.

4 Balancing Strategy Adopted for the Demonstrator System

The demonstrator system is based around an existing influence coefficient program developed by the GEC ALSTHOM Engineering Research Centre. The aim is to assist the engineer acquire meaningful data for the balance program so as to achieve a satisfactory result with a minimum of trial runs. Its main features are as follows:-

4.1 Database

On starting the system the user is prompted to enter details of the rotor, such as the lengths of the spans, the bearing positions and the locations of the available balance planes. He is then asked to specify the locations, orientations and calibration factors of each vibration transducer. As the balancing trials proceed the system automatically updates the database with the positions of the trial masses and the corresponding measurements from each transducer for inspection or passing to the balancing program as required.

4.2 Data acquisition

The system acquires vibration data at uniform speed increments over the speed range during a run up or run down. Two modes of data analysis are provided, spectral analysis and harmonic analysis. For spectral analysis a frequency spectrum of the vibration signals is acquired at each speed increment to show their overall content. For harmonic analysis the data acquisition is synchronised to a once per revolution shaft marker so that the amplitude and phase of the running speed component of vibration may be extracted. If proximity transducers are used the data may optionally be corrected for the slow roll runout of the rotor.

4.3 Datum run analysis

The datum run is usually carried out with the machine in the as-found condition and provides the reference against which the changes produced by the trial masses are compared.

In this system a detailed analysis of the datum run is carried out to provide an initial assessment of the quality of the signals. It is also used for choosing the order in which the available balance planes will be used for trial runs and selecting the speeds at which data will be extracted for passing to the balance program.

4.3.1 Signal assessment

Data is acquired in spectral analysis mode and plotted as a map showing the frequency content of the signals at each speed increment. This may be inspected to check for a significant running speed component, if there are none on any channel this is not a balance problem. The maps also aid identification of any speeds where data is likely to be corrupted by interference.

4.3.2 Balance speed selection

Ideally the rotor should be perfectly balanced at all speeds. However, unless the unbalance coincides exactly with available balance planes this is impractical. Balancing is therefore carried out for a restricted, usually small, number of speeds at which the rotor is most sensitive, in the expectation that the vibration will also be reduced at the other speeds. In this system the running speed data from harmonic analysis of the datum run is plotted to show amplitude and phase versus speed. The chosen speeds are then selected using the graphics cursor. Speeds where data corruption was suspected during inspection of the spectral maps are highlighted to warn the user. When speed selection is complete the corresponding amplitude and phase data is entered into the system database. Modal analysis techniques may be used to help extract this data but were not implemented in the demonstrator.

4.3.3 Balance plane selection

Ideally balance planes should be chosen which provide independent control of each of the major modes of vibration. While this may sometimes be possible on an exposed rotor in the overspeed laboratory, it is rarely so during site trim balancing. The system examines the data from each transducer at each of the selected speeds and attempts to estimate the mode shapes, using also the rotor lengths, bearing positions and transducer locations from the database. Where two proximity probes are mounted close together either side of a bearing, rotor slopes as well as deflections are taken into account. For each mode the likely amplitude at each balance plane is estimated and they are ranked accordingly for order of use in the trial runs. The estimated shapes are displayed schematically for inspection by the user.

4.4 Trial balance runs

The system advises the user to calibrate each balance plane, in turn, in the order of their ranking. However an experienced user may decide to use combinations of planes rather than single ones if he wishes. For each calibration run trial masses should be fitted in the chosen planes, the machine run under the same conditions as for the datum run, and a harmonic analysis of the vibration signals performed. The system extracts the results for the selected speeds and adds them to the database. The vibration vector changes may be plotted for inspection.

After each trial the user may elect to calibrate another plane or to run the balance program. For the latter option the system uses the available information to calculate an array of balance masses to minimise the residual vibrations. If the residuals are within the specified limits the recommended masses may be fitted and a test carried out to confirm satisfactory running, otherwise more planes may be necessary.

4.5 The influence coefficient balance program

This program allows both least squares and minimax optimisation. For rotors whose unbalance response is temperature variable, data may be entered for cold and hot running and the optimisation carried out for cold, hot or compromise conditions. The program also permits weighting factors to be applied to indicate the relative importance of the vibration at each transducer position and speed; however this facility is not used in the demonstrator.

5 System Implementation

The system was developed as a hybrid of procedural and expert system techniques. An expert system consists of two parts referred to respectively as the knowledge base and the shell. The knowledge base contains an explicit statement of the knowledge relevant to a specific application. The shell is a program which can be used in many different applications. It carries out reasoning with the contents of the knowledge base and generates questions and instructions to a user. Following a review of commercially available shells the one chosen for this project was Goldworks. This is written in Lisp and executes on a personal computer with 8 megabytes of extended memory. A facility is provided to call external Lisp functions. This was extended to further call programs written in C and Fortran to execute the procedural tasks.

A separate computer based data acquisition system was used for performing the spectral and harmonic analyses under the control of the expert system on the PC. The Fortran influence coefficient balancing program was also executed on this computer. Serial line communications between the two computers were implemented using the Kermit protocol and special routines were developed to ensure correct synchronisation. The data acquisition computer provided only two channels of simultaneous data sampling so a multi-track tape recorder was required as an intermediate signal store.

The final system was menu driven so that the user could select which operations he required to perform and only the appropriate rules would be activated. All user interaction with the system was via the PC keyboard and screen. A mouse could also be used for menu selection and as a graphics pointer.

A model flexible rotor rig was constructed as a means of testing the system. This consisted of a long thin rotor with disc masses supported by three hydrodynamic bearings. The number of masses, their positions and those of the bearings could be altered to give a variety of configurations. Vibration measurement was by means of eddy current proximity probes. A shaft marker was provided for speed sensing and as a phase reference. The rig was used during the development of the system for testing the data acquisition routines and various balancing strategies. It also proved invaluable for helping to demonstrate and explain the balancing process to the expert system programmers so that they could translate the engineer's reasoning into rules.

6 Results and Conclusions

The project has shown that a combination of procedural and rule based languages can successfully be used to develop an expert system to assist the balancing of flexible rotors. The system had sufficient expertise to guide the balancing of a simple rotor in a logical way. By relieving the engineer of many of the routine tasks, and providing facilities to review the progress of tests it is easier for him to assess the results when exceptions to the rules do occur.

The Goldworks shell provided the necessary flexibility to build a hybrid system but was found to be very slow in execution and required very large amounts of memory to store information. This made the system rather cumbersome to use and restricted the complexity of a balancing exercise to that of the model rotor. A more powerful computer with a multi-channel data acquisition capability and more efficient shell software would be required for a practical system.

During the development of the system it was found that in specifying the processes in sufficient detail to allow rules to be defined for the expert system, their procedural nature became more apparent.

In this type of application perhaps the appropriate role of a conventional expert system shell is during the development stage when rules can be defined and the explanation facilities used to verify correct reasoning. Then, when the system is complete and everything has been defined, the production version could be re-coded in C or Fortran which are more efficient in storage requirements and faster in execution. This approach merits further development.

On-line decision making for diagnostic and monitoring purposes

N. Ardjmand and Z.M. Ranic
Department of Electronic Engineering, Coventry Polytechnic, Coventry CV1 5FB

Abstract
Monitoring of systems can be achieved using a hierarchical formation of monitoring functions.
This paper describes a method for monitoring a system for diagnostic purposes based on a hierarchical structure.Whenever a failure is detected in a system,a library of diagnostic programmes will be consulted to determine the nature of the failure.
The process of inference is done on two levels.The LOW LEVEL is mainly procedural knowledge and the HIGH LEVEL is declarative knowledge.Petri nets are used at both levels to model the problem.
Keywords:Diagnostic,Monitoring,Petri net,Inference,Propositional ,Predicate,Prolog.

1 Introduction

Systems controlling and coordinating nuclear power plants and jet aircrafts are among the most complex ever created by man.The task of diagnosing these systems is complex [6].To carry out diagnosis on-line requires fast and efficient methods.

The software required for this purpose needs to satisfy certain fundamental specifications.Formal specification employs mathematical notation in order to achieve precision and conciseness.The description of intended behaviour requires that the meaning of its operations be specified.Some notation and conventions beyond standard mathematics are required in writing formal specifications.VDM and Z [4],[14] use such conventions.

The method which is used here is Application Specific Advanced Programming(ASAP)[3].In ASAP the problem is decomposed into subprocesses.The combination of these procedures form a library of routines.

The inference is split into two levels.The combination of these two levels diagnose the system and inform us about any malfunction. Petri net [5],[7],[8],[10] has a dynamic nature and therefore it is very useful for modelling the diagnostic domain.The complete structure is shown in fig1.

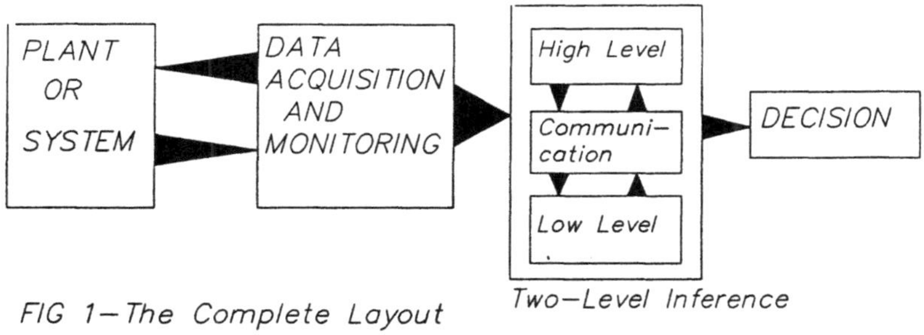

FIG 1-The Complete Layout

2 Data Acquisition And On-Line Algorithm

2.1 Implementation

Obtaining information from the system while it in operation requires an on-line algorithm [2]. This would extract information from the system and store it for processing by inference engine.

Since the algorithm performs an on-line monitoring function, interrupt technique has been used to implement the diagnostic function. The sequential procedures of the algorithm has been decomposed in different small programme called 'states'. Thus,the on-line algorithm can be seen as a state machine which progresses from state to state.

3 Hierarchical Reasoning

3.1 Low Level

Propositional logic [11],[12] is used in this layer for reasoning process. Propostional logic is appealing because it is simple to deal with .

Consider the formula:

P IMPLIES [Q AND[[P OR Q] IMPLIES [R EQUIVALENT P]]]

The truth assignment,P:=T;Q:=F;R:=T then the above formula can be replaced by;

T IMPLIES [F AND[[T OR F] IMPLIES [T EQUIVALENT T]]]

and finally it reduces to;

T IMPLIES F

which by the truth table (for IMPLICATION),F is the outcome.

A proposition can only be true or false.This property and petri net principles are used in an integrated form to represent the knowledge and guide the reasoning process.

The algorithm is split into two components [9].These are LOGIC and ACTION.The ACTION component determines what is to be done.The LOGIC component specifies how is to be done.These are represented as a set of procedures.A procedure has a Head and a Body.The Head identifies the form of the problem which the procedure can solve.The

Body is a set of procedure calls and other related arguments.

Petri net is used here to model the above.It therefore has a Head and a Body.The Head is a 'place' in the petri net and the Body is a set of arguments.The first argument is always a Logic procedure .The second argument is the Action routine.The execution of this results in an outcome.The third argument contains the name of the next place if the outcome is true and the fourth one has the name of the next place if the outcome is false.

The first part of the implementation is the definition of petri net which involves creation of a petri net table or matrix. A row matrix contains a number of bytes to suit the requirement.The pointers in the row are allocated as defined below:

a-A pointer to a position in a list of places.
b-A pointer to Logic procedure.
c-A pointer to Action procedure.
d-A pointer to a place if the outcome is true.
e-A pointer to a place if the outcome is false.

In addition to petri matrix,a name list is also formed.This list contains the name of places and procedures in the petri net. Names are necessary for man-machine communication during programme development and for on-line intervention.
The definition of a petri net model according to the above algorithm is as follows:

```
NODE(LOGIC,ACTION0,OUTCOME1,OUTCOME2)
OUTCOME1(LOGIC,ACTION1,NODE,OUTCOME3)
OUTCOME2(LOGIC,ACTION2,OUTCOME2,OUTCOME3)
OUTCOME3(LOGIC,ACTION3,OUTCOME3,OUTCOME2)
```

The corresponding petri net is shown in fig 2.

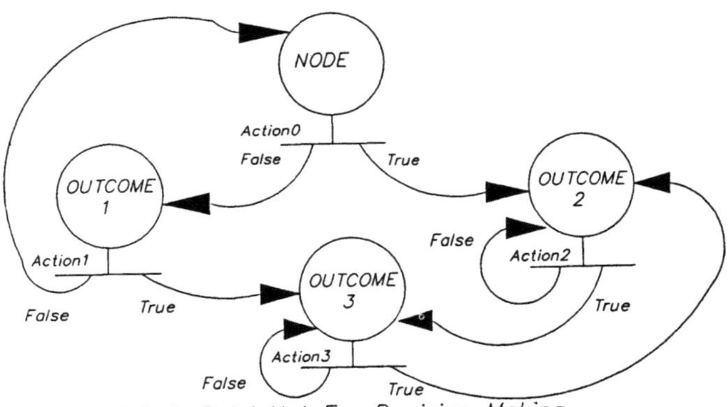

FIG 2–Petri Net For Decision Making

3.3 High Level

At this level,predicate logic [11],[12],[13] is used to represent knowledge and carry out reasoning.Predicate logic permits representations of things that cannot reasonably be expressed in

propositional logic.The propositional logic lacks ability to represent many statements (such as "All lights are in order") in useful manner.
Predicate calculus overcomes this by having extra facilities such as variables,universal and existential quantifiers.For example the sentence "All lights are in order" can be represented by:

$(\forall X)[light(X) \rightarrow in\text{-}order(X)]$

or the sentence "There is a light out of order" can be expressed as:

$(\exists X)[light(X) \rightarrow out\text{-}of\text{-}order(X)]$

where X is the quantified variable.

Prolog models the problem in a declarative way.Prolog's power is in its ability to infer (derive by reasoning) new facts [1],[2],[13].

At this level,we define the problem in terms of facts and rules. The satisfaction of certain facts or rules at this layer would result in an outcome for the misbehaviour.

4 Communication And Translation

Both levels have a medium for communication.This medium creates a link between levels.The communication programme is itself a subprocedure and therefore it has a place in the petri net [2]. When this place is executed,it first activates the relevant prolog programme.Once this is achieved it expects the high level to ask for data.Prolog does this by sending a relevant pointer.This is the address of a particular place in the global petri net.The execution of this place results in some specified action(e.g. reading a port or a data location).The information is then sent to prolog for processing.This process continues until high level can deduct an outcome.

Information in both levels is structured in different forms as well as different codes.In order to have understandable communication between both levels,there is a translation facility. This enables the data to be converted to a form which is useful for a particular level.

5 Example

This example shows how the method explained above can be used to diagnose a car.

5.1 Low Level Knowledge

This level tests the validity of the following observations:
Engine_Misfires(P01),Engine_Cranks(P02),Strong_Gasoline_Smell(P03) Engine_Stalls(P04),Engine_Stalls_When_Cold(P05).

It can conclude the following outcomes:
Grade_Of_Gasoline_too_Low(Fault_01),Out_Of_Gas(Fault_02), Carburetor_Flooded(Fault_03),Choke_Stuck(Fault_04),Carburetor_Icing (Fault_05),Consult High-Level.

The corresponding petri net is shown in fig 3.

5.2 High Level Knowledge

At this level the following obsevations are checked: Engine_Turns_over(P1),Electrical_Power(P2),Spark_Delivery_System (P3),Battery_Connections(P4),Relay(P5),Starter_Motor(P6), Battery_Water(P7),Battery_Charge(P8).

The following causes can be deduced: Faulty_Ignition(Fault_11),Faulty_Starter(Fault_12),Needs_Tuning (Fault_13).Tighten_Battery_Connection(Fault_14),Add_Battery_Water (Fault_15),Charge_Battery(Fault_16),Replace_Battery(Fault_17), Check_Spark_Plugs(Fault_18),Check_Gas_Gauge(Fault_19).

The corresponding petri net is shown in fig 4.

If P01 in low level is true then Fault_01 is the conclusion.If P01,P02,P03,P04 and P05 are false then high level will be consulted. If P1 and P3 are true in high level then the outcome is Fault_19.

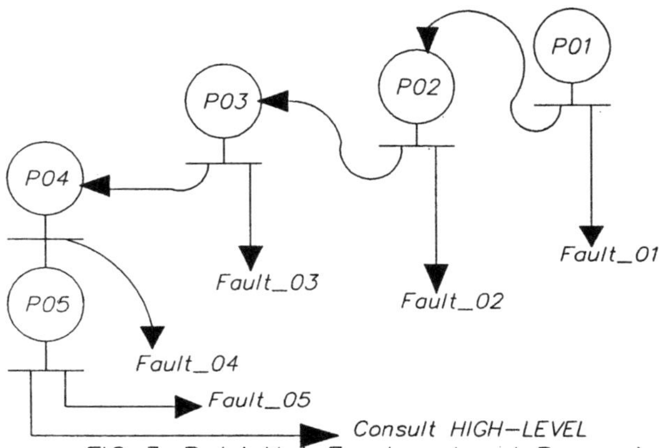

FIG 3-Petri Net For Low Level Reasoning

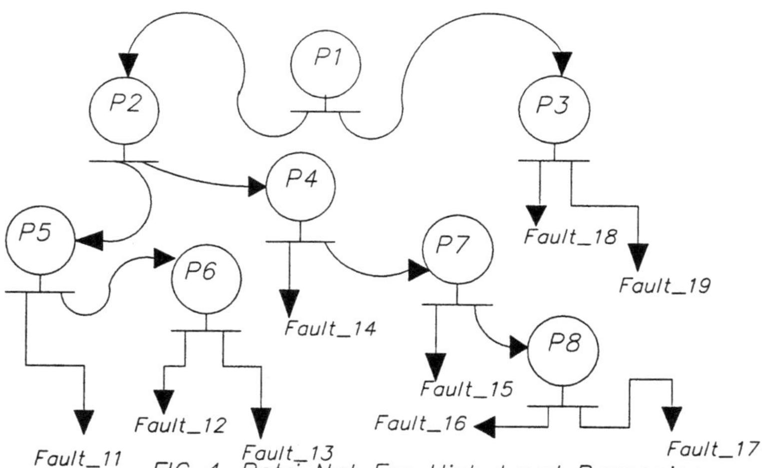

FIG 4-Petri Net For High Level Reasoning

Conclusion

In this paper,we approached the tasks of diagnosis and monitoring by splitting them into a set of subtasks.This resulted in smaller tasks which were easier to and deal with.It also improved the clarity of the problem.The splitting of the algorithm at Low Level gave us greater control over its speed and efficiency.It also enabled us to represent the knowledge with greater flexibility. Logic was used due to its decision making capabilities.Two level reasoning can deal with problems with different degree of complexity in terms of knowledge representation.Extending this to multilevel reasoning can have a great degree of flexibility for problem solving.

REFERENCES

[1] Ardjmand, N.(1987) Expert System for Diagnostics.Msc Thesis.Coventry Polytechnic.

[2] Ardjmand, N.(1990) Seminar Notes.Coventry Polytechnic.

[3] Brajovic, V.(1988) ASAP for declarative And procedural reasoning.International Conference On Systems Diagnostics,Prague.

[4] Dijkstra,E.W.(1976) A Discipline of Programming.Prentice-Hall, Inc., Englewood,Cliffs,N.J.

[5] Fidelak, M.(1986) Petri net for logical representation. German Research Institute For Mathematics And Data Processing,Scholb, Birlinghovev.

[6] Genesereth, M.R.(1984) The use of design description in automated diagnosis.Artificial Intelligence 24,pp. 411-436.

[7] Looney, C.G.(1981) and Abdulrah, A.F.(1981) Logical control via boolean rule matrix transformation.IEEE Transactions on Systems, Man and Cybernetics,vol 17,no 6 , pp. 1077-1081.

[8] Sandhu, S.(1988) Petri Net For FMS.Msc Thesis,Coventry Polytechnic.

[9] Kowlaski, R.(1979) Algorithm=Logic+Control.Communications of ACM,vol 22,no 7,pp. 424-435.

[10] Peterson, J.L.(1981) Petri Net Theory And Modelling Of The Systems.Prentice-Hall,Inc.,Englewood,Cliffs, N.J.

[11] Rich, E.(1983) Artificial Intelligence.McGrawhill International.

[12] Thayse, A.(1988) From Standard Logic To Logic Programming. John Wiley & Sons Ltd,Chichester.

[13] Townsend, C.(1987) Mastering Expert Systems With Turbo Prolog.Howard W.Sams&Company,Indianapolis,IN 46268 USA.

[14] Jones, C.(1976) Systematic Software Development using VDM.Prentice- Hall,Inc.,Englewood,Cliffs,N.J.

Expert systems for quality management in glass container production

M. Gianotti, P. Molinari and A. Pozzetti
Dipartimento di Meccanica, Politecnico di Milano, Italy

Abstract
This paper describes three expert systems, developed in GoldWorks, for the detection of defects in glass containers and for the identification of their causes.
Results of tests carried out on real cases are also presented.
Keywords: Expert Systems, Quality Management.

1 Introduction

In glass container production, the defects can be classified into four categories (see for example Budd et al. (1961), Clark-Monks and Parker (1980), Jasinski (1969)):
- stones, consisting of minerals of a different composition, trapped inside the mass of glass proper;
- geometrical defects, such as a bent neck of the container, non-uniform thickness of the glass, cracks, etc., which normally arise during the process of forming;
- bubbles, i.e. inclusions of gas inside the glass;
- cords, i.e. zones of glass which have a different chemical composition from the normal one in the container.

As a percentage of total defects, bubbles and cords are much less frequent than stones and geometrical defects. The identification of their causes is also normally quite simple.
As for stones and geometrical defects, the problem presents three different aspects:
- identification of the defect (nature of the mineral constituting the stone or nature of the geometrical defect);
- search for possible causes;
- application of corrective measures.

The identification of the mineral constituting the stone is possible only after laboratory examination, which makes it possible to identify certain characteristics of the stone itself (refractive index, crystalline system, presence of a viscous sac, etc.).
Even on the basis of the characteristics found, the identification of the stone and of its causes is not automatic, but requires considerable experience because, with about fifty different stones described in the literature on the subject, there are about sixty possible causes in total.
The detection of geometrical defects is, on the other hand, carried out on the production line using automatic machinery. There is, however, no unique link between defect and cause. With about forty different defects described in the literature on the subject, there are about two hundred possible causes in total. Corrective action must also be chosen and applied with great accuracy, so as not to give rise to further defects.

In practice, both for stones and geometrical defects, each cause generates numerous defects and each defect can originate from a number of causes.
Up till now, the management of the above problems has been entrusted exclusively to experienced human personnel, as an "automatic" solution was not judged possible.
Today, the employment of expert systems offers new possibilities in this respect.
In what follows, two expert systems used for the identification of stones and of their causes are described.
Mention is also made of a third expert system, for the identification of causes of forming defects.
The expert systems have been developed in GoldWorks, Gold Hill Computers (1987), on an IBM PS/2-80, at the Department of Mechanics of the Politecnico di Milano.

2 Expert system for the identification of stones

The input to this expert system consists of results of laboratory analyses, which make it possible to identify certain optical characteristics of the stone. It is not necessary to input all the data requested. Clearly, the more data fed in, the more dependable is the response of the system. Conversely, when faced with uncertainty, it is possible to insert several values for the same characteristic parameter.
Fig. 1 shows a summary of the expert system operating logic.
After the introduction of the data, the expert system, using a suitable set of rules, creates a model of the stone having the characteristics fed in by the user.
The model thus created may not correspond to any existing mineral. In other words, there may not exist a mineral which exhibits the same characteristics as the model of the stone. This is easily explained, bearing in mind that some of the data introduced may have been derived from subjective observations, or even from faulty analyses. In addition, as already stated, some characteristics may be defined in a non-unique manner, or even not defined at all.
The problem of identification does not present itself as a deterministic one.
In a deterministic problem, when starting from certain data, one arrives at a solution which is both unique and certain. In the case of the expert system, one can talk of many possible solutions with different relative probabilities. For this reason, the logic of the system for the identification of stones is not based on the successive reduction of the domain of possible solutions, but on the allocation of a probability to each mineral contained in the knowledge base of the system.
Nevertheless, there exist also some rules which reduce the domain of the possible solutions and identify their subset.
These rules, however, are specific and apply in particular cases only. This means that only in specific cases it is possible to identify minerals which definitely do not constitute the stone, without however being able to say anything about the others (for example, if the stone is not magnetic, it certainly doesn't contain iron).
In order to recognize the stone, a scoring-type logic has been chosen.
A suitable set of rules compares the model of the stone with all the types of stone known by the system. After this operation, a numerical rating is given to each possible stone. The value of this rating is linked to the probability that the given stone represents the solution to the problem.
The value assigned to each possible stone is obtained by summing the number of points which it has been given when compared with the stone model.
If a given characteristic of the possible stone agrees with that in the stone model, a number of points is given to the possible stone, depending on the nature of the particular characteristic.

The value assigned to any possible stone has no absolute significance, as the number of data inserted by the user is not fixed. It becomes significant only when compared with the values assigned to other possible stones.

The scoring-type method has allowed, amongst other features, to regulate accurately the number of points attributed to single characteristics. It is clear that a numerical quantity obtained from an instrument reading is more dependable than a subjective judgement by the operator. Because of this, a higher number of points is assigned in the first case.

The basic distinction made is between:

a) numerical or qualitative characteristics;

b) objective or subjective characteristics.

First of all, each characteristic has been classified on the basis of four possible combinations existing in a) and b) above.

The difficulty in determining the characteristics has also been considered. Birefringence, for example, whilst being a numerical and objective quantity, is difficult to measure and depends also on the thickness of the stone, which cannot be measured under a microscope.

The expert system also takes into account the fact that some characteristics have a particular importance if they assume specific values.

For example, the crystalline forms DENDRITIC 90° and DENDRITIC 60° are particularly significant in the identification of the mineral constituting the stone, when compared with other possible formations.

The system is also provided with particular judgements whose aim is to minimize the disturbances caused by a faulty operator evaluation. For example, when observing zirconia under the microscope, the normal form expected is DENDRITIC 90°. However, zirconia can also present itself in the form of grains or needles. Because of this, the knowledge base of the system contains the DENDRITIC 90° form, the GRAIN form, the NEEDLE form and also the generic DENDRITIC form, in case the operator does not know, or cannot find out, that the ramifications of this crystal are 90° instead of the 60° typical of other minerals. While this judgement reduces slightly the certainty of identification, as the data remain more generic, on the other hand the operational usefulness of the expert system is increased. This is because it provides a plausible opinion even at the moment of maximum utility, i.e. in the cases in which difficulties of observation or operator inexperience do not permit him to recognize certain characteristics immediately.

At the end of the phase in which points are assigned and added up, minerals with the highest score are identified. These are then displayed to the user.

Whenever two minerals score the same number of points, the system displays some laboratory tests that may indicate which of the two minerals constitutes the stone. This operation is carried out by comparing the characteristics which make it possible to distinguish the two minerals from one another. For example, if the two minerals are mullite and beta-alumina, the system suggests the determination of form and birefringence, the characteristics which make it possible to distinguish the two minerals from one another.

In order to describe the knowledge base, two types of knowledge representation technique have been employed: frames and rules (see for example Waterman (1986)). For example, a FRAME called <STONE> (fig. 2) has been defined. This contains, as SLOTS, all the characteristics of a generic stone. <STONE> contains the INSTANCES pertaining to each possible stone. The general characteristics of the SLOTS are defined for the FRAME and are automatically inherited by the INSTANCES, while the peculiar characteristics are defined inside each INSTANCE. Many SLOTS are MULTIVALUED; for example, the SLOT <FORM> is MULTIVALUED because the

mineral constituting the stone may often crystallize in different ways.
In addition to the INSTANCES which form part of the knowledge base right from the start (static knowledge), others are created by the rules during the inference procedure (dynamic knowledge, which evolves right up to the point of finding a solution, i.e. to the identification of the stone).

3 Expert system for the identification of causes of stones

The input to this expert system consists of stones (type, quantity) detected in a sample consisting of a suitable number of glass containers and identified by means of the expert system described above.
The logic of this expert system (fig. 3) is based on the assumption that the causes which should be investigated most urgently are those responsible for the greatest number of stones.
Because of this, in a first phase, the system links the importance of a cause to the percentage quantity of the stones, as reported by the user, which can trace their origin to this cause.
For example, if stones consisting of molybdenum are present in 70% of the containers constituting the sample, 70 points are assigned to the possible causes of molybdenum stones, i.e. impurities in the raw material and the crumbling of the electrodes.
If stones of different natures found in the sample have the same cause, the number of points assigned to this cause is equal to the sum of the percentage quantities of the stones.
If a stone has several possible causes, the same number of points is assigned to each of them.
From this point onwards, the system considers only those causes which have a non-zero score, i.e. a non-zero sum of all the points.
Thus, the domain, within which the solution to the problem is sought, is reduced accordingly.
The score thus assigned to the causes is rather primitive, in that it is derived only from the evaluation of stones. It is hence necessary to "refine" the scores by means of introducing further information which takes into account the real operating conditions in the glass-melting oven.
This information, which is normally available or easily obtainable, consists of the following:
- age of the oven;
- volume flow rate;
- firing method (methane or fuel oil);
- presence of electric heating;
- recent repairs to the oven, to the ceiling or to the regeneration chambers;
- recent removing of tin tetrachloride from aspiration hoods;
- variations in glass density.

It should be noted that requests for additional information are not made as a matter of routine, but are activated only when necessary. If, for example, none of the stones contains tin, it is useless to know whether tin tetrachloride have been removed from aspiration hoods. Not only would this piece of information add nothing useful to the search for a solution to the problem, but it would unnecessarily prolong the calculation.
The score assigned to certain causes is suitably increased when it is possible to establish, on the basis of additional information, that in their case there exists a higher probability of being the cause of the observed defects. On the other hand, the introduction of additional data can lead to the exclusion of other causes from the set of

possible solutions.
For example, methane firing of the oven excludes the possibility that sulphur-containing stones might be due to the fuel. Conversely, the use of fuel oil increases the probability that the stones originated from the sulphur present in the fuel itself (the score assigned to this cause is, accordingly, suitably increased).
At the end of this second computation, the user is presented with a list of causes, compiled by choosing those with the highest score.
In order to avoid the display of too many causes, thus confusing the user with too long a list, a threshold value has been introduced. If the total score for any given cause falls below the threshold level, the cause is not displayed.
This method suffers from the drawback of not presenting all the possible causes of stones which exist in any given case.
However, as the expert system has been prepared for applications in industry, it has seemed more important to concentrate on the display of the most important causes, i.e. those with the highest overall score. In practice, the efforts of the operators to apply corrective action will be concentrated on those causes.
The threshold is the higher of two values, one of which is fixed and the other adjustable.
The fixed value arises from the fact that, if the total score assigned to a given cause is too low, it can be maintained that the simultaneous presence of stones which are potentially generated by this cause is merely accidental and enters into the realm of normal errors connected with the technological process. However, above this value, it is probable that there exists a cause of the stones which has to be kept under control.
The adjustable threshold, on the other hand, is a fraction of the highest score obtained. This judgement makes it possible to have a relative comparison between the possible causes. A relative comparison is necessary because the highest score is subject to considerable variation.
In fact, according to the case, it is necessary to consider differently a cause which has obtained a rather low score. This is because such a score has no absolute meaning but has to be compared with the highest score to determine the importance of the cause being considered.
In addition, in order to facilitate the interpretation of the output, the displayed causes are sub-divided into three classes, on the basis of the relative scores obtained.
In this way, the user obtains an immediate insight into the importance of the causes.
At this point, there exists the possibility of knowing which causes have been responsible for each single defect, thanks to an option called <EXPLANATION>. In the explanation phase, the display shows only causes already displayed previously, with the name of the stone generated appearing next to each of them.

4 Tests on the expert systems developed

The expert systems developed have been tested by comparing, in several hundred cases, the indications provided by them with the identification of the stones and of their respective causes carried out by industrial experts. In some cases, the latter has been supplemented by non-routine analyses.

4.1 Tests on expert system for the identification of stones

As regards the expert system for the identification of stones, the test results are as follows.
A) In 73% of the cases, the expert system has identified the real mineral constituting

the stone as the most probable one. This class includes the cases in which the reply displayed several minerals, but in which the difference between the most probable mineral and the others was evident.

B) In 21% of the cases, the mineral constituting the stone was shown as the most probable one, but was not the only one to show a high probability. In other words, there were two minerals with a high probability of being responsible for the stone, but the system was not capable of making a clear distinction between them on the basis of available data.

C) In 5% of the cases, the mineral constituting the stone was displayed together with other minerals, but it was not clearly distinguished from them.

D) In 1% of the cases, the mineral constituting the stone was not displayed.

The replies of type A are clearly correct.

Replies of type B should also be considered correct. This statement is based on the analysis of the tests carried out. It has been observed in practice that, in certain cases, the two minerals are due to the same cause. Therefore, the detection of the presence of one or the other mineral leads to the diagnosis of the same cause.

For example, in some cases, sillimanite and beta-alumina have been shown as the two minerals with the highest probability of constituting the stone.

As both of these minerals originate from the crumbling of refractory materials in the oven, the distinction between sillimanite and beta-alumina would not add to useful knowledge in this case. With a more careful analysis, it is also possible to observe that, in general, these two minerals are distinguishable by their crystalline form. Both of them can, however, appear in needle form, in which case there exists a high degree of resemblance between the two types of crystal. Hence the impossibility of distinguishing between them by the system.

Answers of type C make it possible to at least reduce the set within which to search for the mineral responsible for the stone. Answers of this type can be normally traced to the introduction of uncertain data (for example, several values for the same characteristic). The expert system functions in this case too, but the uncertainty in the input data is accompanied by an uncertainty in the system response.

Replies of type D are clearly unacceptable.

The percentage of correct replies has been the same for unusual stones too.

4.2 Test on the expert system for the identification of causes of stones

The results furnished by this expert system must be considered very good, bearing in mind that in 90% of the cases examined the causes displayed as most probable by the system coincided with those put forward by industrial experts.

In the remaining 10% of the cases, the system has produced a partial answer, i.e. it has only indicated some of the causes of stones.

The percentage of correct replies has been the same for unusual causes too.

The achievement of outstanding results has been made possible by the fact that the system takes into account the real operating conditions of the plant.

5 Constitution of the knowledge base

A long effort was necessary to construct the knowledge base. During this time, various prototypes of expert systems have been built in order to check the degree of completeness of the knowledge inserted.

Detailed knowledge has been derived from the following sources:

A) discussions with experts in this field;

B) examination of specialist texts;
C) statistical analysis of past data concerning the defects encountered and their causes; this analysis has made possible the linking of the presence of certain stones to particular conditions of the glass-melting oven and to the quantity of scrap of outside origin used as the raw material;
D) direct analysis of particular stones using appropriate instrumentation.

This approach has required a considerable effort in ordering, sifting and aggregating the information. It has also been necessary to separate significant from irrelevant information and to eliminate potentially contradictory items.

Using this method, however, it has been possible to develop a knowledge base (which in any case is capable of further enrichment) which represents not just a copy of the experience of experts consulted, but may add further competence to that already existing.

6 Brief account of the expert system for the diagnosis of forming defects

This expert system identifies the possible causes of forming defects. It should be remembered that the connection between defects and their causes is complicated not just by the large number of defects and their possible causes, but also by the multiplicity of the connections, that is to say each cause generates numerous defects and each defect can originate from a number of causes.

The system logic is summarily described in fig. 4. After the description of defects present in the sample studied, the system singles out a set of possible causes. Within this set, the operation of eliminating contradictory causes is carried out, as clearly, it is not possible that a defect is generated simultaneously by opposite causes. In this way, the number of causes, among which those constituting the final solution is sought, is further reduced.

On the basis of data supplied, the system assigns to each remaining cause a score related to the total number of defects deriving from it.

After the display of the most probable causes, the system provides information on the consequences of "overcorrections" which might be carried out in order to eliminate the causes of the defects.

The expert system can also operate in the opposite direction, indicating the defects which could be expected to follow from the causes presented to the system.

7 Conclusions

The expert systems developed have a structure which permits an easy extension of the knowledge base. In detail, it is very simple to:
- expand the knowledge of the possible types of defects and of their characteristics;
- increase the number of rules which correlate the defects with a particular cause;
- modify the points assigned, so as to change the influence of single characteristics on the final judgement;
- modify the acceptable tolerances on the numerical data fed in;
- modify the output information.

In addition, the expert systems, in their user application, do not require any knowledge of GoldWorks, which is their operating environment.

Acknowledgments

The authors wish to thank the firm A.V.I.R. (Aziende Vetrarie Italiane Ricciardi) for their invaluable collaboration.
Particular thanks are due to Prof. M. Garetti of the Politecnico di Milano for his helpful suggestions.

References

Budd, S. M., Exelby, V. H. and Kirwan, J. J., (1961) **The Formation of Gas Bubbles in Glass at High Temperature**, Group Research Department, United Glass Limited, London.

Clark-Monks, C. and Parker, J. M., (1980) **Stones and Cord in Glass**, Society of Glass Technology, Sheffield.

Gold Hill Computers, Inc., (1987) **GoldWorks. Expert System User's Guide**.

Jasinski, J., (1969) **Stone Identification by Visual Methods**, Owens-Illinois Inc., Glass Container Division, Research Engineering, Toledo, Ohio.

Waterman, D. A., (1986) **A Guide to Expert Systems**, Addison-Wesley.

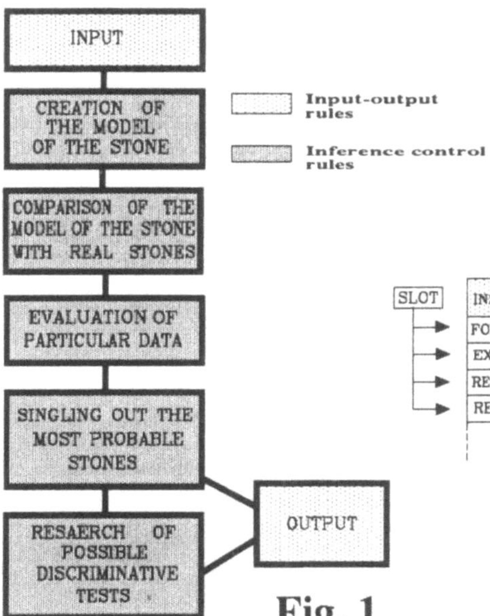

Fig. 1

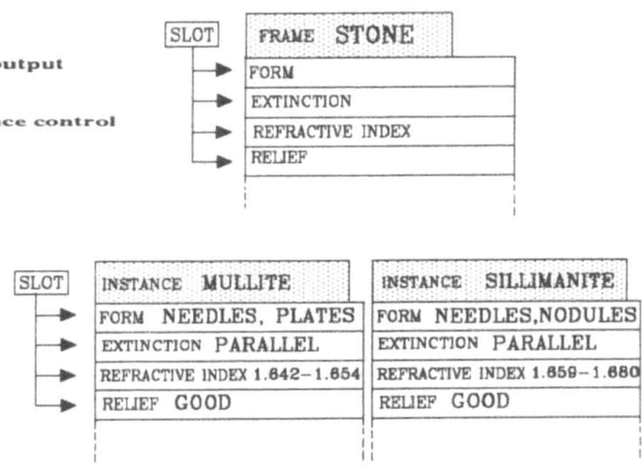

Fig. 2

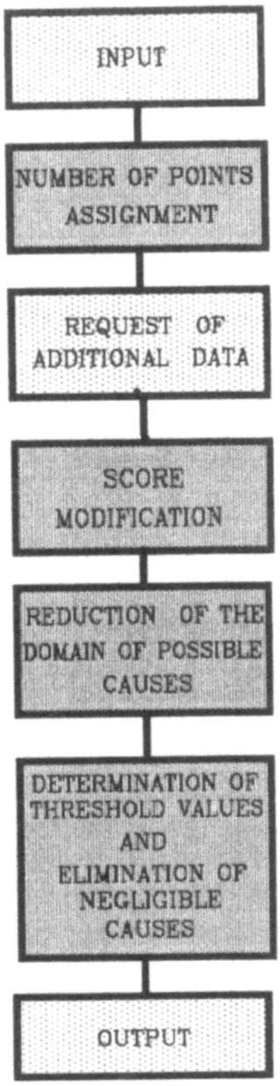
INPUT
NUMBER OF POINTS ASSIGNMENT
REQUEST OF ADDITIONAL DATA
SCORE MODIFICATION
REDUCTION OF THE DOMAIN OF POSSIBLE CAUSES
DETERMINATION OF THRESHOLD VALUES AND ELIMINATION OF NEGLIGIBLE CAUSES
OUTPUT

Fig. 3

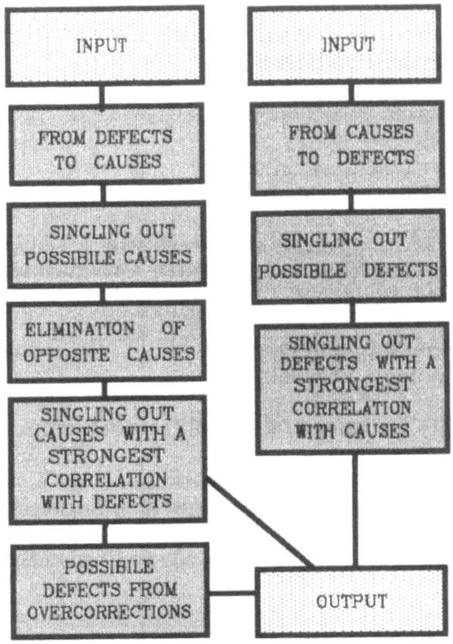
INPUT
FROM DEFECTS TO CAUSES
SINGLING OUT POSSIBILE CAUSES
ELIMINATION OF OPPOSITE CAUSES
SINGLING OUT CAUSES WITH A STRONGEST CORRELATION WITH DEFECTS
POSSIBILE DEFECTS FROM OVERCORRECTIONS
INPUT
FROM CAUSES TO DEFECTS
SINGLING OUT POSSIBILE DEFECTS
SINGLING OUT DEFECTS WITH A STRONGEST CORRELATION WITH CAUSES
OUTPUT

Fig. 4

System-independent approaches to diagnostic knowledge reasoning and representation for mechanical and thermal systems

H.K. Ho, S.S.G. Lee, M.K. Lim and Y.W. Wong
School of Mechanical and Production Engineering, Nanyang Technological Institute, Singapore

Abstract
This paper presents and reviews methods and approaches for generic and system-independent representation of diagnostic reasoning and knowledge, particularly as applied to mechanical and thermal engineering systems. Generic diagnostic reasoning involving generic sub-tasks and the use of diagnostic meta-knowledge are presented. Methods for the generic modelling of the structure, function and behaviour of devices and components at various levels are described. Several generic attributes to describe observations and tests are proposed. Finally, areas requiring further research and study are outlined.
Keywords: Knowledge-based systems, Diagnosis, System-independent representation, Generic modelling, Mechanical engineering

1 Introduction

The use of knowledge-based or expert systems in diagnosis has become increasingly common [1]. Early knowledge-based diagnostic systems typically used a collection of *system-specific*, shallow, "if-then" rules in the knowledge base. The MYCIN program developed at Stanford [2] can be considered the archetype of such systems. Even in those early systems, there was already some evidence of generalisation which enabled, for example, the derivation of the generic shell EMYCIN from MYCIN. The generic features of EMYCIN and other similar rule-based system shells have been used successfully to develop a large number of knowledge-based diagnostic systems, albeit, with some amount of (largely ad-hoc) tuning and improvement of the basic control/inference mechanisms.

In recent years, there has been increased interest in the *generalisation* of the methods and knowledge used in specific knowledge-based diagnostic applications. At the same time,there is a trend towards *specialisation* of general-purpose development tools/methods into diagnostic domain-specific tools are taking place [3].

The purpose of this paper is to present and review methods and approaches for generic and system-independent representation of diagnostic reasoning and knowledge, particularly as applied to mechanical and thermal engineering systems. The study of these methods and approaches promotes the understanding of the diagnostic problem in these domains. Applying them to the *development* of knowledge-based diagnostic systems facilitates the creation of consistent, explicit, accessible, and easily updated diagnostic knowledge bases.

Interest in system-independent approaches has increased with the desire to make knowledge-based diagnostic systems more powerful and flexible than can be achieved with the earlier, shallow, experience-based (heuristic) approach. *Causal modelling*, "deep" reasoning, and the accompanying emphasis on reasoning based on structure and function has made generic diagnostic reasoning and knowledge representation even more important and relevant.

2 Characteristics of diagnosis

Fault diagnosis can be defined as the determination of the root cause of observed and detected abnormalities. It basically requires the mapping of a set of observations with a basic fault (or set of basic faults) that can explain these observations.

If the diagnostic reasoning uses knowledge in a *compiled* and *explicit* form (e.g. in the form of a fault tree, fault table, symptom-fault diagnostic rules etc.), it is called "shallow" or experiential. Typically, this form of diagnosis is more efficient, but will not be effective if the compiled knowledge base is incomplete. On the other hand, if, during diagnosis, the associations between observations and basic faults are *ascertained* based on the structure, function and behaviour of the faulty device/system, then the diagnostic reasoning is termed "deep" or "causal". This form of diagnosis is more powerful, since even faults which have not occurred or were not forseeable can still be diagnosed. However, it is typically less efficient, and more information and knowledge-intensive.

It should be noted that a basically shallow system (from the diagnostic reasoning point of view) can make use of knowledge compiled from a deep model of the device/system. It is also possible for a system based on deep reasoning to make use of "heuristics" (for example, the well-known "half-split" diagnosis heuristic). This distinction between diagnostic *reasoning* and *knowledge* will be used in our discussion on the generic nature of diagnosis.

3 Generic diagnostic reasoning processes

3.1 Decomposition of diagnostic reasoning into generic sub-tasks

Generic cognitive aspects of diagnostic reasoning have been studied by researchers of various background and specialisation. Clancey [4] proposes that *heuristic classification* (of which diagnosis is an example) can be decomposed into the tasks of data abstraction, heuristic matching, and refinement. In a similar vein, Chandrasekaran [5] proposes a *generic task* (GT) architecture for diagnosis comprising four components: hierarchical classification, hypothesis matching, abductive assembly, and data abstraction/inference. The work of Johnson and Keravnou [6] and Rasmussen [7] are also very relevant.

A common thread in these approaches is the attempt to define a set of basic cognitive processes which is common across a spectrum of diagnostic problems. It is therefore necessary to provide means for incorporating, integrating and applying these various cognitive tasks in a wide range of knowledge-based diagnostic systems. Figure 1 presents an example of how the sub-tasks in Clancey's heuristic classification approach may be used to represent diagnostic reasoning associated with a faulty engine component.

3.2 Generic meta-knowledge for control of diagnostic reasoning

Although some aspects of the *control* of the diagnostic reasoning process are already implicit in the approaches mentioned above, there exists a separate set of commonly-used system-independent principles for controlling diagnostic reasoning (sometimes called "meta-knowledge"). These are not part of the generic sub-tasks mentioned above but are often utilised to make diagnosis more efficient and effective. Figure 2 lists several of these which will be useful in many knowledge-based diagnostic systems.

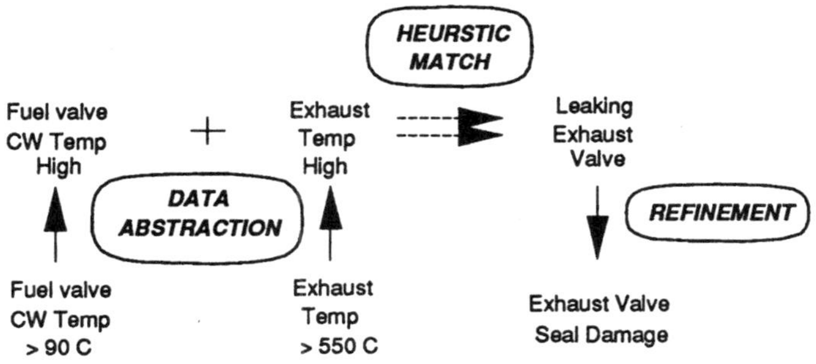

Fig. 1. Diagnostic Reasoning Using Heuristic Classification

"If the number of observations available on the system is ***large****, initially monitor only the most pertinent among them, and* ***focus*** *on relevant sets of other observations only as necessary"*

"If there exists a body of ***basic*** *and easily accessible observations which are* ***particularly useful*** *for diagnosis, obtain and analyse them before proceeding further in the diagnosis"*

"If there are several (abnormal) observations requiring diagnostic attention, process the observations in the order of the ***criticality*** *of the possible basic faults associated with those observations."*

"If there are several competing hypotheses to consider, then consider them in order according to the ***ease*** *with which confirmatory observations can be obtained for each possible hypotheses"*

"If it is not possible to discriminate between possible hypotheses with additional observations or tests on its ***present*** *state, make use of* ***historical*** *information on the system's failures to rank the possible hypotheses".*

"If the output from a system (comprising a series of smaller components) is abnormal, but the output from a component mid-way along the system is normal, attention should then be focussed on components downstream of that mid-way component"

"If the input to a component is normal, but the output is abnormal, then focus on the component itself and perform a deeper level of diagnosis on it."

Fig. 2. Typical Generic Diagnostic "Meta-Knowledge"

4 Generic device/system modelling and representation

4.1 Objective of generic device modelling

Commonalities in the behaviour of engineering components such as valves and switches; pumps and compressors; electrical, hydraulic and pneumatic motors; storage tanks and capacitors, are but a few examples pointing to the usefulness of generic device modelling.

Thus, one would hope that by applying *causal reasoning* on instantiations of generic models, it would be possible to develop a diagnostic knowledge base for any device/system simply by providing information on its *specific structure* (in terms of connected instances of generic components/ sub-systems), *actual performance* (if required), and *available observations* (sensors, measurements or tests). Figure 3 illustrates such an idealised development environment.

Even if shallow reasoning were used, generic device modelling will be useful in ensuring the completeness and consistency of the compiled diagnostic knowledge base.

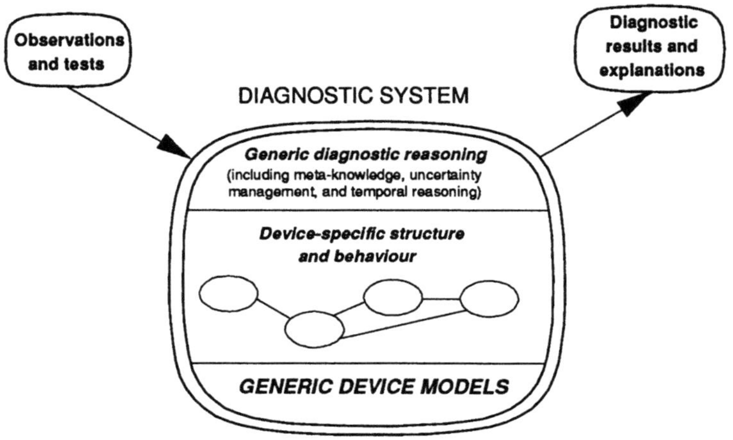

Fig. 3. Generic Models in an Ideal Development Environment

4.2 Generic representation of device/system structure

The *structure* (topology) of a device/system describes the *physical or functional connectivity* between components or sub-systems. It defines the interactions that can take place within the device, for example, with respect to mass and energy transfers.

The structure of a particular device is, in general, different from other devices. However, the method of representing structure can be made generic to some extent. For example, for a flow system comprising a series of components, UPSTREAM and DOWNSTREAM connectivity attributes (indicating type, direction, maximum throughput, etc.) for components will facilitate reasoning about the system's behaviour.

4.3 Representation of function and behaviour in terms of mass/energy transfers

At a relatively low physical level, the *function, behaviour and failure* of a device/system can ultimately be stated in terms of its ability or inability to *transfer or store energy, mass or momentum at the required rate*. Different forms of energy (mechanical, electrical, thermal, light, electromagnetic etc.) are known to be applicable to different devices/systems.

For example, the failure of a pumping system to deliver the expected amount of liquid can be diagnosed by an analysis of the mechanical/ thermal energy, and mass relationships in the system. This is because basic faults such as a faulty pump, flow blockage, pipe leakage etc. will affect the energy/mass interactions along the system differently. On the other hand, the failure of a bearing supporting a rotating system can be detected by the (dynamic, extraneous) mechanical energy created by such a failure, as reflected in the vibration and noise patterns.

4.4 Modelling behaviour at the engineering level

Engineers often make use of a higher level of abstraction than at the basic mass/energy level. Thus, mechanical and thermal engineers think about systems in terms of components such as valves, pumps, tanks, filters, gears, levers, etc., whose general behaviour and performance characteristics are quite well understood.

For example, specifying the performance of a specific pump often only requires specification of its *Flow Quantity versus Pressure Head (Q-H) characteristics, its Flow versus Power Consumption characteristics,* and perhaps the *Minimum Net Positive Suction Head (NPSH)* required to operate the pump.

4.5 Qualitative modelling

The practical level of modelling described above is obviously important and useful. However, it may *not* be generic enough, and may be inadequate for sufficiently *detailed* diagnosis. Furthermore, they typically require *device-specific quantitative* data to complete the description of a particular device. To overcome these limitations, modelling at a lower level, and using only *qualitative* reasoning has been attempted [8].

Several researchers have made use of qualitative models to describe the behaviour of devices/systems for diagnostic purposes [9,10]. For example, Fink [9] makes use of a set of *"functional primitives"* (including the *transformer, regulator, reservoir,* and *conduit*) to describe how a specific device functions. Each of these "primitives" incorporates its own set of behavioural characteristics, which can be used to build up a qualitative simulation of more complicated devices.

5 Representation of observations and tests

Observations and tests form the basis for the detection of faults and the subsequent confirmation or rejection of possible basic causes. They include *human observations, sensor readings, condition monitoring equipment/systems, and special tests*. During diagnosis, the type, relevance, accuracy, availability, cost, validation status, and time-related characteristics (period of validity, maximum rate of change etc.), of observations and tests have to be considered. Including these attributes in a generic representation of observations and tests, together with "meta-knowledge" on how to process these attributes would contribute substantially to the efficiency and effectiveness of diagnosis.

Some examples of system-independent meta-knowledge related to observations and tests have already been presented in Figure 2.

6 Research needs and future developments

The analysis and understanding of generic cognitive activities involved in diagnostic reasoning requires more attention and research. This is particularly so for diagnosis using *causal reasoning*, involving *multiple basic faults*; multi-level *control loops* and complex sub-system interactions; and in the *integration* of shallow and deep reasoning approaches. This would be a logical extension to what has already been learnt from the research on diagnostic reasoning described in Section 3.

Generic device behaviour modelling requires a *multi-level* approach as different levels of granularity and accuracy would be needed for different fault situations. Studies have to be made on the models which should be included and used under different circumstances, and how generic models can be integrated with system-specific behavioural models.

Understanding and modelling the *generic steady state* behaviour of systems will have to be supplemented with modelling of their *dynamic* behaviour in order for diagnosis to be effective. Temporal reasoning and real-time performance issues are areas which require further investigations.

The increased interest of generic approaches will promote *automated knowledge acquisition and "learning"* of diagnostic rules. Possible sources of knowledge/ information for such purposes include CAD databases and design calculations.

Generic methods for *managing uncertainty* can be used in conjunction with relatively imprecise (but computationally economical) models for efficient diagnosis in the presence of incomplete data and knowledge. Further work on appropriate representations of uncertainty in diagnosis is required.

6 References

[1] Scherer WT and White CC, A survey of expert systems for equipment maintenance and diagnosis, in *Fault detection and reliability*, Singh MG et al (eds), Pergamon Press, Oxford, 1987.

[2] Buchanan BG and Shortliffe EH (eds), *Rule-based expert systems*, Addison-Wesley, Reading, Massachusetts, 1984.

[3] Ho HK, Development tools for knowledge-based diagnostic systems in engineering applications, to be presented at *COMADEM 90*, Brunel University, UK, July 1990.

[4] Clancey WJ, Heuristic classification, *Artificial Intelligence*, vol 27:3, 1985.

[5] Chandrasekaran B, Generic tasks as building blocks for knowledge-based systems: the diagnosis and routine design examples, *The Knowledge Engineering Review*, 1988

[6] Johnson L and Keravnou ET, *Expert systems architectures*, Kogan Page, London, 1988.

[7] Rasmussen J, Strategies for state identification and diagnosis in supervisory control tasks, and design of computer-based support systems, in *Advances in man-machine systems research, Volume 1*, Rouse W (ed), JAI Press, 1984.

[8] Cohn AG, Approaches to qualitative reasoning, *Artificial Intelligence Review*, vol 3, pp 177-232, 1989.

[9] Fink PK and Lusth JC, Expert systems and diagnostic expertise in the mechanical and electrical domains, *IEEE Trans. on Systems, Man and Cybernetics*, Vol SMC-17, no. 3, May/June 1987.

[10] Iwasaki Y, Qualitative, causal reasoning about device behaviour, *IEEE Intl. Workshop on AI for Industrial Applications*, Hitachi City, Japan, May 1988.

[11] Umeda Y et al, Model-based diagnosis using qualitative reasoning, in *Computer Applications in Production and Engineering*, Kimura F and Rolstadas A (eds), Elsevier Science/IFIP, 1989.

Intellectualization of faults diagnosis of rotating machinery – expert system

Jiang Xingwei and Wang Shujian
Harbin Institute of Technology, Harbin, China

Abstract
This paper describes the application of an important branch of artificial intelligence —— expert system in faults diagnosis of rotating machinery. By using the fuzzy sets theory and the method of confidence, a model of inference of non-precise is first introduced. Then, an expert system (MMMD) for faults diagnosis of turbogenerator set is described in detail. MMMD has used some idea about static state variable and dynamic variable and semantic chain and so on. MMMD has consulting and explaining subsystem. Finally, some practical examples of faults diagnosis are given.
Keyworks: Expert system, Faults diagnosis, Rotating machinery.

1 Introduction

It very often needs special knowledge of many fields to diagnose faults of rotating machinery. Because of the limitation of filed conditions and the complexity of faults, it is very difficult to diagnose fault correctly. In order to make the best use of the special knowledge, there is a natural tendency to build a popular knowledge bank for faults diagnosis of rotating machinery by organizing the special knowledge that is from lots of experts.

2 Application of fuzzy sets theory

2.1 Fuzzy sets and membership degree

There are some facts whose boundaries are fuzzy in faults diagnosis, for example the big or small about amplitude and the heavy degree of faults etc. Therefore, it has become a new manner to apply fuzzy sets theory in faults diagnosis of rotating machinery.

Generally, the relation of some element X or other and a set A in classical sets theory is: X belongs to A or doesn't. This is a simple two value logic {0,1}. Fuzzy set makes two value logic {0,1} in classical sets theory into a continuing value logic in closed interval [0,1]. The relation of some element X or other and a fuzzy sets B is not simply two value logic. It is described by the membership function $\mu_{(x)}$.

Due to the membership function depicts a gradual process of some element from belonging to the fuzzy sets to those not belonging to, it is necessary for faults diagnosis of rotating machinery to select a proper membership function. For instance, if the variable X describes the vibration amplitude of rotating machinery, the bigger X is,the more serious the rotating machinery may be. For this reason, we select the formula (1) as membership function. We can change the fuzzy variable into quantitative variable by selecting coefficient K. If we select coefficient K to equal 1/2500 , $\mu_{(x)}$ equals 1/2 when X equals 50 μm. Since the value of $\mu_{(x)}$ describes the conditions of amplitude, we let

$$\mu_{(x)}=\begin{cases} 0 & x\leqslant 0 \\ \dfrac{kx^2}{1+kx^2} & x>0 \end{cases} \tag{1}$$

the confidence of amplitude equal the membership degree $\mu_{(x)}$. By using membership fuction $\mu_{(x)}$,we can proceed the fuzzy facts in quantitative analysis.

2.2 Inference of non-precise

The symptoms of machinery's faults often do not correspond simply with the faults. In addition to this, the symptoms may be frequently fuzzy yet. In the face of those fuzzy questions,we adopt a model of inference of non-precise by combining fuzzy sets theory with confidence theory in MMMD. For example, the tenth rule in MMMD is:

IF: the amplitude in daily operation rise suddenly(CF),
THEN: the fault may be lost parts rotor(0.6).

This rule means: if the amplitude of turbogenerator set rise suddenly in daily operation, the set may have the fault of lost parts rotor. The coefficient CF is the confidence about the amplitude rise suddenly. CF equals membership degree $\mu_{(x)}$ calculated by the formula (1). The coefficient 0.6 is experience of expert.The confidence of fault of lost parts rotor equals 0.6×CF. When there are more evidences,the synthetic coefficient of fault is calculated depending on probability sum law.

3 Design of MMMD expert system

MMMD expert system is composed of a main program and eight modules. The main program is linked with every module by the interfaces of modules. The frame of MMMD is shown in fig.1.

The main program 'Main' manages every module. Others are subprogram. 'IntroMMMD' introduces the function and operation of MMMD. 'InitMMMD' organizes the system into initial state.'Spectrum' processes the static state variables about freguency spectrum of vibration. 'Controller' controls inference procedure of the system. 'Proparevars' selects dynamic variables depending on certain rules. 'Doscreen' operates dynamic variables in the question tables of the system by the form of man-computer dialogue. 'Propvars' matches the facts with diagnostic rules. 'Finalp' shows the diagnostic results.

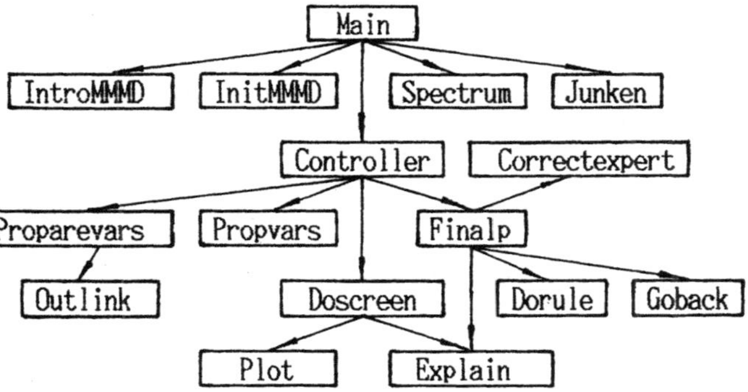

Fig.1. Structure of MMMD expert system

4 The brief introduction of operating principle of MMMD

4.1 The variable

The symptoms of turbogenerator set faults are called variable in MMMD. Every variable has seven recording fields. These recording fields are applied for recording the conditions of every variable. 'Number' field records, for instance, the number of variable, 'Valcon' field records the value of variable, 'Varlink' and 'Link' and 'Semantic' are fields of pointer.The 'Varlink' pointer is used to form a chain between every variable. Because the 'Varlink' chain does not change in the working process of MMMD,it is called static state chain.Comparatively speaking, the chain formed by 'Link' is called dynamic chain. The 'Semantic' pointer is used for forming semantic chain that links a series of variables which have some relations to each other.

The variables which record conditions of vibration spectrum are called static state virables. Because those variables can obtain their values from data bank which have been provided by its user, the static state variables are first managed by MMMD. The variables except static state variables are called dynamic variables. When static state variables have been managed, MMMD will select dynamic variables depending on 'Occupying coefficient'. The 'Occupying coefficient' is the amount occurring variable in current candidate faults set. Because the 'Occupying coefficient' of every variable decides the contribution of variable, the bigger 'Occpuying coefficient' shoud be treated with priority.

4.2 Inference of MMMD expert system

After disposing of a set of variables provided by 'Proparevars', MMMD will match diagnostic rules by calling 'Propvars'. The fuzzy confident set of faults is defined by the fuzzy confident set of expert knowledge and the fuzzy confident set of variables(facts). A reduced faults set will be obtained since the faults which value of confidence is less than the threshold value have been rejected.

5 Consulting & explaining subsystem
Transparency, the ability to understand each other between system and user, is an important characteristic. MMMD is provided with consulting & explaining subsystem.

5.1 Consulting & explaining in the process of inference
MMMD often needs to put forward questions to a user for replenishing essential evidences. Sometimes, the user may be surprised at the questions and hope to know why these questions are raised. According to the guide in question tables of MMMD, a user can start the function of consulting & explaining. At this moment, the system will show the user how many faults have been removed, and how many candidate faults are concerned with this question.

5.2 Explanation of result
When MMMD has completed diagnosis and shows the diagnostic result, the user may have doubts in the result and expect MMMD to explain it. When the user press the function key to explain the result in diagnostic end menu, MMMD will show the results per reference that is stored in a 'dynamic state blackboard'.Having observed varying process of candidate faults per reference, the user would naturally understand where the diagnostic result is from.

6 Practical examples of faults

6.1 Misaligement
This fault signal is from the machine set of a certain factory. The waveform of vibration in time domain is shown in Fig.2. The direct current component is 1.17 volt, the alternating-current component is only 96 millivolt. The spectrum graph of this signal is shown in Fig.3. The diagnostic result shows that the fault is misaligement and the confidence equals 0.87(Fig.4).

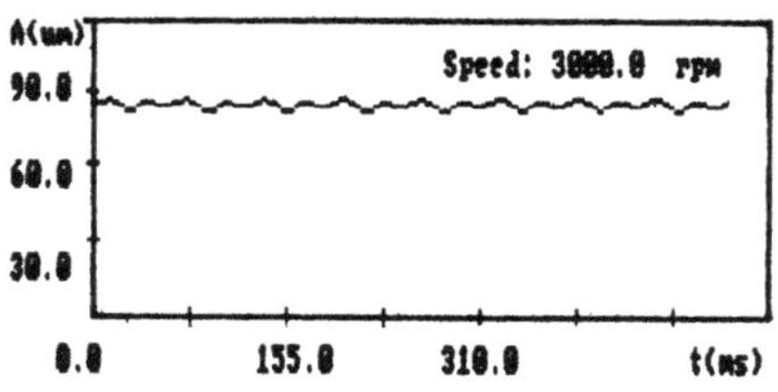

Fig.2 Waveform of virbation

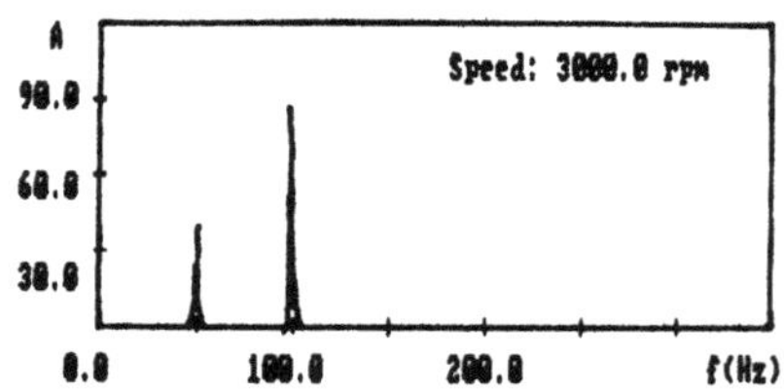

Fig.3. Graph of frequency spectrum

6.2 Oil whirl
This fault signal is from a compact type high-speed steam turbine. The value of amplitude is bigger than 50 μm at that time(Fig.5), yet the permitted value of amplitude is only 30 μm. Its graph of frequency spectrum are shown in Fig.6. The diagnostic result shows the confidence of oil whirl equals 0.92(Fig.7).

MMMD 诊断结果	
故 障	可信度
不对中	0.87_

Fig.4. Diagnostic result

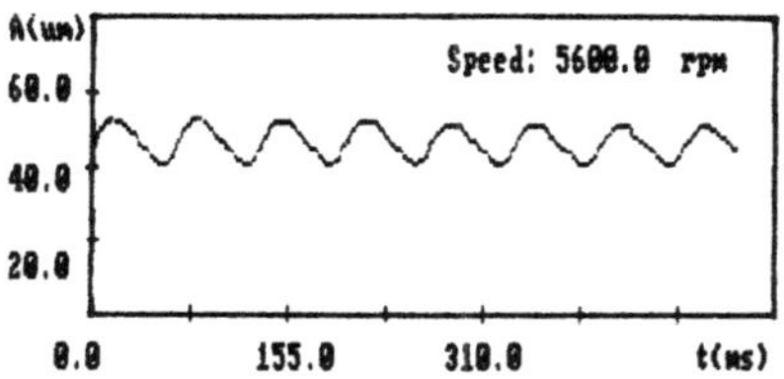

Fig.5. Waveform of virbation

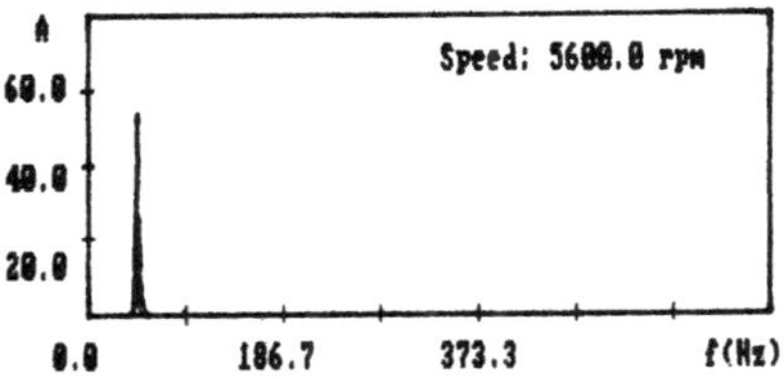

Fig.6. Graph of frequency spectrum

MMMD 诊断结果	
故 障	可信度
油膜振荡	0.92_

Fig.7. Diagnostic result

6.3 Conclusion

The examples mentioned above show the diagnostic results of MMMD tally with the facts on the whole.MMMD has a fairly good level.Because faults diagnosis of rotating machinery deals with knowledge of many subjects, the experience accumulated year after year over can't be known well in a short time. With the development of computer science and artificial intelligence technology, it is possible to solve this problem.

References

[1] Shao Chengxun, The application of fuzzy set theory in the fault

diagnosis for vibration of trubomachine, CSMDT'86, June,1986
[2] S.M.Weiss & C.A.Kulikowski, A practical guide for designing expert system, JiLin university press, August,1986.
[3] Fu Jinshun, Artifical intelligence and application, Qinhua university press, September,1987.
[4] Lin Xiaorui, The principle and practice of expert system, Qinhua university press, April,1988.

Amethyst: an expert system for the diagnosis of rotating machinery

Dr Robert Milne
Intelligent Applications Ltd, Livingston Village, West Lothian, Scotland

Abstract
Amethyst automates the interpretation of rotating machinery vibration data in order to facilitate condition based maintenance. This results in considerable time savings, greatly reduces the needed skill level and provides considerable benefits to the end user organisation. **Amethyst** has been delivered to a large number of companies in the UK and the USA, resulting in considerable cost savings for industry. Because it is an easy to use product, it has the potential to address a large number of companies throughout industry worldwide.
Keywords: Manufacturing, Diagnosis, PC Based, Rulebased

1 The Problem Being Addressed

Every item of rotating machinery such as a pump, compressor, fan, motor and shaft all generate vibrations as they rotate. Although these items are carefully constructed and balanced, very small vibrations occur due to imperfections in the bearings, the mounting of the equipment, the shape of the shaft, build up on the blades of fans, or within the pump, or other problems due to the materials. The key part of condition based monitoring is to determine two items. The first is that the vibration indicates that there is a problem with the machine and so some action should be taken. The current data collection packages and software do this very well. The second, and more critical activity, is to determine the exact problem, this impacts the repair that is necessary. Some problems are very simple to resolve, others require a shut-down of the plant and major overhauls. It is very important to understand what repair is needed.

The problems addressed by **Amethyst** include bearing problems, possibly requiring a replacement of a bearing; gear problems, reflecting gear teeth damage and gear misalignment, possibly requiring an overhaul of the gearbox; blade problems to do with pumps and fans, possibly requiring a simple cleaning of the blades or a more expensive replacement of the blades; problems of balance of main rotor shaft requiring a particular balancing action to be conducted; problems of misalignment of the machine, generally requiring the machine to be bolted down more properly; problems of misalignment between the driving motor and the prime unit such as a fan, requiring a basic misalignment sequence to be performed;

problems of cavitation in pumps; problems from hydraulic or aerodynamic difficulties; problems from resonances; and finally; and perhaps most importantly, problems from the bad use of the data collector and bad data collection. These problems indicate that there might not be a problem with the machine at all, but that the operator did not do the data collection properly and accurately enough.

The primary value of the condition monitoring system is derived when the end user knows whether to tighten a bolt, re-balance the main shaft, replace a bearing or overhaul the gearbox.

In order to determine which fault it is, it is necessary to examine the vibration spectrum. Each physical part of the machine will vibrate at a different frequency. The main shaft rotates at what is called the fundamental frequency, if there are 6 blades the vibration will occur at 6 times the fundamental frequency. Bearings are generally around 20 times the fundamental frequency and gear mesh problems at about 50 to 75 times the fundamental frequency.

Traditionally, the operator must look at the FFT graph to determine for each peak what physical part of the machine it represents. He must also then have knowledge of the overall vibration level and the amplitude of each frequency. There are standard guidelines for when the amplitude is considered to be severe and he must make the calculations to determine whether a high level of vibration represents a problem or not. In diagnosing many problems it is necessary for the operator to also look for side band vibrations, they are vibrations at a frequency such as the gear mesh frequency plus or minus the RPM frequency. This again requires some careful calculations.

The only tool provided by existing software products is an **X Y** cursor capability with the graphs. Once the end user has identified the peaks and their frequencies he must then apply fundamental knowledge about rotating machinery and how faults manifest themselves.

One of the major limitations for most companies attempting to use condition monitoring is having adequate knowledge of how to interpret the data, in order to use the system properly [1]. For large corporations, experienced and trained mechanical engineers perform this diagnosis. The basic training course will cover simple problems, but is not adequate to cover common, but more complex deviations. As a result, there is a classic problem of not having enough skilled personnel with enough experience to actually interpret the data. Yet this is the most critical and valuable part of the entire condition monitoring system.

The purpose of **Amethyst** is to automate this diagnosis and interpretation of the spectral data. This results in allowing lower skilled people to use vibration based condition monitoring, providing cost savings to existing customers and eventually widening the base of companies which can use condition monitoring. As can be imagined, it is a very slow process to actually diagnose a fault. On a typical 8 hour route collecting data, a dozen measurement points will often be in alarm and require diagnosis. It can easily take an experienced person 4-8 hours to manually go through the steps needed to develop the diagnosis. **Amethyst** does the same diagnosis fully automatically, the end user must spend approximately 30 seconds starting the analysis and then a report is produced with the results (see figure 1). This leads to a direct times saving of 4 hours potentially for every route collected. This also represents a reduction factor of 500 in the time needed for the task. For large corporations in active use, this represents a half man years savings immediately.

	Alarm: IN/S Alarm: g/SE
Route STEEL	Total numer of faults: 3
Machine SCREW COMPRESSR	
Pos 5 Dir H IN/S	Bad bearing with MAJOR faults
Mach Type pump-horz cent	BSF and/or BTF Sidebands Exist
	Hydraulic or Aerodynamic
Rotating Speed RPM 1525	Misalignment
Other Shaft RPM 1775	
Overall Ampl 0.179 IN/S	
Alarm Limit 0.314 IN/S	

Fig 1: An Example Diagnostic Report

2 The Solution Being Adopted

Amethyst involves two software packages integrated together to an interface of a third software package. The IRD Mechanalysis condition monitoring software, known as 7090 is used as the main database. The expert system rules are implemented in the Crystal expert system shell from Intelligent Environments. The ***Violet*** product from Intelligent Applications is used to provide the interface between Crystal and the IRD database [5]. It is also used to provide the extra functionality needed such as the manipulation and extraction of the vibration spectrum. For those of you who are not geologists, an **Amethyst** is a ***Violet*** coloured Crystal.

Intelligent Applications worked with IRD Mechanalysis to perform two major software tasks. An interface was developed between the ***Violet*** software package and the IRD Mechanalysis database. ***Violet*** is able to access virtually all the information in the database. It had to know how to integrate with the database, to scan all points on a particular route to identify whether the point was in alarm or not, and to be able to pick up different types of measurements. This portion of ***Violet*** is written in C and is standard software for accessing a standard database. ***Violet*** then provides a number of access functions which are callable from the expert systems shell to access the information which has been extracted from the database.

The knowledgebase development was focussed by discussing the types of machines for which the knowledgebase should be applied. For the group of machines, a list was also made of the common faults for which the knowledgebase should deal with. For each fault, we then discussed how it would manifest itself and be identified. After only one day of joint knowledgebase development, the expert would typically spend several hours writing more rules to cover larger classes of faults. The author would then edit and restructure the rules so that they were cleaned up. As the resulting rulebase was run on more and more examples, inconsistencies, or gaps in the rulebase were discovered, and rules were developed to cover these faults.

It is very important for this application that the one rulebase can assist so many potential end users. Because a rotating shaft is fundamentally what is being monitored and diagnosed, and the vibration of the rotating shaft is actually independent of whether it is moving air, water or driving a motor.

The system uses the Crystal expert system shell, which is a pure backward chaining rulebased system [7]. It was necessary to implement various mechanisms for causing rules to fail explicitly in order that it had the correct behaviour for dealing with exclusive sets of problems or combinations of problems. In general, we would always make the system find all possible rules that match. To accomplish this, the behaviour of the backward chaining system was altered, such that it found all rules rather than stop on the first rule.

The tasks which were better performed by the C code were implemented as standard software. The resulting expert system only manipulates the expert knowledge level and not any of the fancier database accesses. The resulting system therefore is a good integration of expert systems technology with standard programming and database technology.

The total expert system rulebase is composed of 780 rules according to the Crystal expert system shell. Of these 780 rules, 269 rules are for diagnostic interpretation. The difference illustrates the number of rules needed to control and organise such an expert system. There are also many rules needed for initialisation and editing of the user options.

The machinery monitoring system was tested on a large number of machinery databases. Data for every fault was identified and the system was tested to be certain that every fault did occur. For several large databases, the system was also checked to make sure that the diagnosis it developed was the same as a human expert would have developed. As many combinations of user options and data examples were used as possible to make sure that the system was robust.

3 The Annual Savings Or Benefit

Currently, **Amethyst** reduces a task which should take from 4-8 hours to only a few minutes. Most large companies, particularly those using **Amethyst**, currently conduct this type of analysis every day. If we take the lower end of that estimate; 4 hours per day, this results in one half man year of savings directly. Based on a man year of £60,000 that is a saving of £30,000 per company. In actual fact, the savings can be much higher by allowing better use of personnel and preventing other problems as discussed under indirect costs below.

Because condition monitoring is fundamentally orientated towards preventive maintenance, it is extremely difficult to accurately assess the cost savings. Because **Amethyst** increases the effectiveness of condition monitoring, it helps to magnify already established condition monitoring savings.

4 Lessons Learned

With regard to technology, standard PC based rulebased expert systems were more than adequate to provide the functionality needed for this application area. This project was successful because the developers were not allowed to explore new technology. In a climate of advancing AI technology, this was actually very difficult to achieve. It was quite embarrassing for the senior members of IA to show such a simple application to a large number of their friends, many of whom are considered some of the worlds leading experts in AI; but by focussing on a simple technology, one was then able to focus more completely on the man machine interface, product marketing, packaging and functionality issues. Simple applications do work, and in fact are perhaps the only ones that work well.

A final lesson learned underscores the differences between condition monitoring and traditional process industry applications. In condition monitoring all end users have fundamentally the same problem, determining the condition of a rotating shaft. Intelligent Applications has also worked extensively in the process industries. In these areas every process is different and as a result every expert system product requires a new knowledgebase to be developed, this leads to very expensive systems and no opportunity to have one system circulated very widely. The opportunity to have a significant impact on industry is much greater where the industry is standardised. The time spent in looking for such an opportunity and developing it is well worth the search.

5 Applicability Across Industry

A major impact of **Amethyst** is that it has the potential to affect the many thousands of companies conducting condition monitoring. It is not a large application at a high price requiring extensive development time. It is a simple, but effective, application at a low cost that is very easy to use. Although the initial customers have tended to be very large corporations, the price and usability of **Amethyst** is well within reach of every manufacturing facility in the United Kingdom that uses condition monitoring currently. The product clearly has the potential to impact a very large number of small manufacturing operations.

Currently, one of the major reasons preventing companies from using condition monitoring who could benefit from condition monitoring, is that they have to train the expertise in collecting and analysing the data [6]. It is estimated that only a very small percentage of the potential condition monitoring market has been penetrated to date. There have been estimates that suggest that only 10% of the companies that could benefit from condition monitoring currently do so.

One of the major restrictions on those companies is having the skilled expertise available. **Amethyst** does not help in providing the expertise to collect the data, but does provide a simple means of interpreting that data. As a result, **Amethyst** has the potential to help the condition monitoring area mature another step. This could result in a considerable increase in the number of companies able to benefit from condition monitoring with the resulting consequent savings to UK industry.

If all companies were using condition monitoring properly, not only would there be tremendous direct savings in engineering maintenance cost, but there could be considerable knock on improvements in manufacturing efficiency, throughput, energy cost and reduction in product cost.

6 Conclusion

In this paper we have looked at **Amethyst,** an expert system for the diagnosis of faults in rotating machinery. **Amethyst** was constructed with straight forward rulebase technology and classic expert system development procedures. By being carefully focussed on a well defined application, it is now having a widespread impact. Not only are many engineers able to use sophisticated expert system technology, but a major revolution is taking place with regard to how they perform their day to day task. The technology is already prompting further development and the lessons learned from this application will certainly lead to many new and better capabilities.

7 REFERENCES

[1] Hill, J.W., and Baines, N.C. (September, 1988) Applications of an Expert System to Rotating Machinery Health Monitoring, Institution of Mechanical Engineer Conference, Heriot-Watt University.

[2] Hills, P.W., (1988) Vibration-based Predictive Maintenance Systems for Rotating Machinery. The Factory Efficiency and Maintenance Conference, PEMEC 88, Birmingham

[3] Green, L., (1988) Condition Monitoring - Its Impact on Plant Performance at BSC's Scunthorpe Works. The Factory Efficiency and Maintenance Conference, PEMEC 88, Birmingham.

[4] Milne, R.W., (1987) Artificial Intelligence for On-Line Diagnosis, IEE Proceedings, Vol. 134, Pt. D, No. 4..

[5] Milne, R.W., (1989) Artificial Intelligence for Vibration Based Health Monitoring. Journal of Condition Monitoring. Published by BHRA, The Fluid Engineering Centre. Editor, Ruth Higginbotham, Vol 2, No 3, P213-218.

[6] Milne, R.W., (1987) Artificial Intelligence for Vibration Based Health Monitoring, 4th European Conference on Non-Destructive Testing", London, England.

[7] Winston, P. (1984) Artificial Intelligence, Addision-Wesley Press.

Integration of a PC based expert system shell and production line tracking system for failure diagnosis and rework definition

B.G. Oak
IBM UK Ltd/Brunel University, Uxbridge, England

Abstract
The key to the success of data driven manufacturing systems lies in the collection, storage and manipulation of production process data. This paper outlines a project undertaken to integrate a commercial PC expert system shell with a production line tracking and data collection system, to perform failure diagnosis and rework definition. The volume of potentially useful data in this, as in many manufacturing operations, is too great for most production personnel to assimilate. Expert systems offer a tool to improve failure diagnosis, and thus rework efficiency by applying expert engineering knowledge and judgement to available data. However, such systems must facilitate change and maintenance if they are to be successfully deployed in the manufacturing environment. An assessment is given of the types of knowledge required to perform the task and the applicability of expert systems. Maintenance considerations are highlighted, namely knowledge base structure and domain vocabulary. Finally, these points are summarised with reference to DISPEX, an application developed using the TI Personal Consultant Plus shell.
<u>Keywords:</u> Expert Systems, Manufacturing, Knowledge Engineering.

1 Introduction

In the manufacture of high value products, specifically fixed disk drives for computer systems, there is a need to ensure that the product conforms to stated specifications. During the functional testing of product, large amounts of parametric and logistic data is collected and stored on a production database. If a unit fails during testing, all this information has potential uses to diagnose the cause of failure and the most appropriate rework path; the disposition process. The decisions arrived at need to satisfy the criteria of: high probability of fixing the problem, be cost effective and nonrepetitive, the product should not be directed along the same path on successive attempts.

The capability of line operators to evaluate all potential information is limited, because of the volume of data and technical content, and often leads to inefficient decisions, cycling units and sub-assemblies and operator bias. A decision was made to integrate an expert system into the production line environment to perform the dispositioning operation. So that maximum benefit can be obtained from the system, it has been introduced early in the product life cycle. This has meant developing the knowledge base and data interfaces in parallel to process change and increasing product understanding and expertise, placing more emphasis on flexibility and maintenance.

2 Classification of knowledge

The dispositioning process (failure diagnosis and rework definition) is a multistage task requiring various types of knowledge and levels of expertise at each stage. These are illustrated in Table 1.

Table 1. Dispositioning and knowledge classification

Stage No.	Description	Knowledge
1	Failure -> Probable Cause(s)	Static Domain Knowledge
2	Cause(s) -> Possible Rework(s)	Product Expertise
3	Rework(s) -> "Best" Rework	Heuristic

2.1 Stage 1 - Failure diagnosis

Parametric data is collected at each stage of testing, examples being return codes from file electronics, various data head measurements and mechanical resonances. The meaning of most of this data can be determined by consulting a variety of documentation; performance and mechanical specifications, on which the testing is based and listings to interpret product return codes. In the layered, conceptual models of knowledge and expertise, in particular the Knowledge Acquisition and Document Structuring (KADS) four layer model, Neale (1988) and Weilinga (1989), this documentation forms the static domain knowledge.

2.2 Stage 2 - Cause and rework identification

To proceed to the next stage and establish probable causes, and alternative corrective actions, requires significant expertise. Highly skilled, and therefore expensive, engineers are needed to apply knowledge and understanding of the product, its function and constituent technologies, servo control systems, magnetic data storage and microelectronics. Only with this expertise can the cause of failure be established and hence corrective rework be defined. This stage of the task embodies the KADS inference layer: defining the reasoning process undertaken to understand the significance of different parameters; how parameters interrelate and identification of the mechanism of failure - which together define possible rework actions.

2.3 Stage 3 - Rework selection

The selection of best rework path is primarily an heuristic process. Rules of thumb that state general criteria for rework selection based on previous rework attempts; probability of success, based on statistical analysis of historic data, and unit history. Historical evidence on current and previous products has shown that the volume of useful data, available to carry out this stage, is large and often awkward to obtain. Human nature is such that we attempt to simplify processes to make them quicker and easier to complete. The consequences of this are poor decisions based on personal bias, incomplete analysis and a strong element of trial and error.

3 Application of Expert System Technology

3.1 Applicability

The dispositioning task, essentially, falls into the category of a diagnostic process, an area in which many expert systems have been successfully deployed, Waterman (1985). If this process were to be characterised, with reference to the application of expert systems, the key features would be:

Data intensive	- analysis of a large volume of data is required to be effective
Limited goal set	- finite number of possible rework paths
Low heuristic content	- rules of thumb are constrained to the final section of the "best" rework strategy

3.2 Expected Benefits

From the analysis of the various substages of the dispositioning process and the knowledge used in each, it is possible to derive the benefits expected from the application of expert systems.

Labour requirements. A reduction in skilled labour requirements by allowing an automatic system to replace operators.

Consistent operation. Human operators can give unpredictable results, use different problem resolution strategies and exhibit bias in decision making.
Permanent, central record of knowledge. In any organisation resource is constantly being moved, people are assigned to different tasks and with them the knowledge and experience moves too. The pressure to move highly skilled engineers onto new products is increasing in the competitive market environment. Expert systems provide the opportunity to centralise failure analysis knowledge in a permanent form.
Complete information assessment. All possible data can be considered by the rule base, without the concern that key information might be missed or even ignored. Important information might be awkward or tedious to obtain manually, the tendency to ignore data is not exhibited by a computer program.
Improved rework selection. The application of expert knowledge and judgement should lead to improved rework selection and associated reductions in scrap and less impact on capacity.

This list is by no means unique to the dispositioning task and a more general summary is presented by Waterman (1985).

Whilst considering the advantages offered by this technology, sight must not be lost of its limitations, and the benefits of maintaining human involvement in the process. The human operator, even with limited experience, is able to respond to a wide variety of sensory inputs that may change their perception of the environment. Experience and historical information will lead to modifications in the actions and, perhaps, the strategy employed to arrive at them. Alternatively, changes in the process or data collected could generate new rules and cause others to become obsolete. The underlying point is that humans are adaptive and flexible to changes, and the environment is dynamic; an expert system is not. New rules and modifications to existing rules must be explicitly defined; proficiency and effectiveness is entirely a function of the knowledge base. In the manufacturing environment the successful deployment of expert system technology will be dependent on ease of maintenance and modification.

4 Maintenance Considerations

4.1 Knowledge base structure

In any software system, maintenance is assisted by imposing structure on to it. Conventional programs are written and executed in a strictly sequential order predetermined at the time the software is written. In the expert system domain, structure can be viewed from two perspectives:

organization of knowledge objects, rules and parameters; and secondly, the inferencing strategy, the methods employed that lead to goal resolution, level 4 in the KADS model.

The task, itself can be viewed as separate stages; broken down to subgoals each of which require different resolution strategies and knowledge types. Maintenance is made easier if the knowledge base can be organised to reflect these subgoals.

Structure can be influenced in two ways, both dependent upon the tool; organisation and strategy. The organisation of knowledge objects in some logical manner can be achieved, for example, by using rule and focus groups, groups that hold rules and parameters relevant to specific components of the problem. Secondly, the inferencing strategy can be influenced by specific ordering of groups of rules, setting priority levels and invocation of metarules, rules that can modify the inferencing process based on information received or derived during a consultation.

4.2 Domain vocabulary

Every domain of expertise has its own terminology, or domain vocabulary; words or phrases that have specific meanings when associated with that domain. In expert systems the correct and logical use of such terminology can assist maintenance by providing easily read rules and ease understanding of inferencing processes through the explanation facilities provided by most commercial tools.

In some conventional programming languages parameter or variable names are limited to a specific number of characters, causing designers to develop cryptic acronyms instead of providing the system maintainer with self explanatory, English like names. Expert system shells, generally, allow more flexibility in naming conventions. For example, in the dispositioning project a parameter is stored at the host as DEARESx, the label is limited to 10 characters by the programming at the host, in the expert system this parameter can be referenced with its exact meaning, i.e. HEADx_AMPLITUDE_NOISE_RESOLUTION. When the rule is viewed, it is immediately obvious what the rule is about and therefore, simplifies the process of identifying where modification is required.

Internal to the knowledge base, the system designer can provide further assistance for future modification by detailing the intention of rules, or specific details of parameters using description properties. A group of rules instead of appearing as a bland list of, can be given added meaning by a simple description of the intent of each rule. This descriptive content can also be used as the basis of explanation facilities available during consultation.

5 DISPEX dispositioning expert system, a summary

The selection of an appropriate tool was determined by the considerations highlighted above and automatic, unattended operation; external language interface; integration with the existing production database and some constraints on acceptable products. Maintenance requirements and development time, emphasised the need to use a commercially available expert system shell, and processor system overheads restricted this choice to PC based packages. The tool selected was Texas Instruments Personal Consultant Plus, a rule based expert system shell, with the ability to structure the rule base through rule groups; limited control of inferencing strategy through metarules and procedural and data interfacing through Scheme, the TI LISP dialect.

The knowledge base currently consists of about 200 rules, which are grouped in two ways. Firstly, for stages 1 and 2, Table 1., each process operation has an associated focus group, keeping any rules and parameters together in one place. Stage 3 of the dispositioning is handled separately, through a different focal group, using metarules and subgoals to invoke these rules once a "cause" has been identified.

In the current manufacturing environment, modifications to the rule base are being made at least weekly and enhancements to available information has meant that the original expert system will soon require a major design iteration if it is to remain a useful and effective part of the process.

The system attempts to incorporate the thoughts highlighted above. Eventually, it will be handed to the manufacturing personnel to maintain without engineering involvement. Its success will depend not only upon the savings achieved, but primarily the ease with which it can be changed to reflect the process in which it operates.

6 References

Laufmann, S.C. (1987) A strategy for near-term success using knowledge based systems. **The Knowledge Engineering Review** , 3, 179-183.

Neale, I.M. (1988) First generation expert systems: a review of knowledge acquisitions methodologies. **The Knowledge Engineering Review** , 2, 105-145.

Waterman, D (1985) **A Guide to Expert Systems** , Addison-Wesley, Wokingham.

Weilinga, R. and Schreiber, G. (1989) Future directions in knowledge acquisition, in **Research and Development in Expert Systems VI** , (ed. P. Hammersley), Cambridge University Press, Cambridge.

An expert system with circulating notice for emergency in underground coal mine

Y. Tominaga, K. Ohga and Z. Yang
Hokkaido University, Sapporo, Japan

Abstract
An expert system capable of circulating notices to send/receive letters and pictures in facsimile has been developed in order to consult human experts in case of unexpected accidents in underground coal mines. It is shown that, as information with graphic transmits the features more exactly than linguistic expression, human experts can direct the countermeasures for the accident provided that the conditions of the disaster is not changed widely during the response of the expert.
Keyword: Expert System, Underground Coal Mine, Mine Safety, Rock and Gas Outburst, Circulating Notice, Information with Graphic.

1 Introduction

Expert systems are used in diagnosis and design mostly (cf. references). To make the system, there is an implicit condition that manual of solving the problem, knowledge concerning fields and input data are known.

In mine safety field, many countermeasures against the predictable accidents are prepared. But it is not avoidable that an unforeseen accident happens in underground. Therefore the expert system used at coal mines should be flexible, such as allowing for human processing when required.

When an unforeseen accident occurs, information about it are collected by any means at first. Then the experts of each field are consulted for the countermeasures to be taken. In order to run the above procedures by a computer, an expert system with circulating notice has been developed. Both documents and pictures are transmitted through terminal sets in the university, the institute and the coal company by using the system in file transmission or facsimile.

When it is recognized in the system that the information to take countermeasures is lacking, the information about the accident are transmitted to the specialists registered on each field using the circulating notice. The answers from the specialists are displayed on CRT and the manager directs the system to either take account of the answer or not.

In this paper, an introduction to the system and its application to take countermeasures for rock and gas outburst at Horonai coal mine, Japan are described.

2 Hard ware of the system

This system consists of three personal computers with Star Fax board which communicates with each other by telephone cable. Operation system is MS-DOS Ver.3.3 for a 16 bits computer. The program for communication is consisted of some batch files. The speed of transmission is 9600 bps. In order to protect the system, host computer checks the users number and pass ward. This system cannot be used without Star Fax board made by Mega Soft Co., Japan. Therefore the countermeasure to prevent hacker, etc. is well considered.

3 Expert System

In this paper, the exert system for rock and gas out burst are explained as an example. The system was developed by using Guru made by Micro Data Base Systems, Inc., U.S.A..

Goal variable is what to do as countermeasure for a problem. The factors and the procedures from data input to the determination of countermeasure (goal variable) are described below:

3.1 Factors and input data

In this coal mine, rock and gas outbursts occurred at one specific sand seam. From these experiences, the factors affecting the rock and gas outbursts are taken as followings.

1) The change of ratio of aromatic hydrocarbon gas flow from the face.
2) The distance from the face to the specific sand seam.
3) Observation value of acoustic emission from the surrounding faces during the drivage of roadways.
4) The smell of the emission gas from the blasting holes.
5) The counts rate of acoustic emission after blasting and before the blasting.

3.2 Rules

If-then rules are used in the study. An example of applied rules are as follows.

RULE 1: If gas pressure and flow rate in an advanced boring hole is low, gas in the rock mass is exhausted thoroughly or gas is isolated in the rock.

RULE 2: If core discing happens in the advanced boring hole, stresses in the rock are large.

RULE 3: If the advancing face approaches to the specific sand seam, or count rate of AE changes with every blasting, the advancing face is located in the danger zone of gas and rock outbursts.

RULE 4: If the aromatic gas smells from the blasting hole in the danger zone is high, then the rock and gas outbursts may happen with high probability.

RULE 5: If atmospheric pressure on the surface is low, or the methane concentration at the upcast shaft is high, gas condition in underground is considered to be bad.

RULE 6: If the probability of occurrence of rock and gas outburst is high, a countermeasure with highest certainty factor is selected in a group of procedures.

RULE 7: If certainty factor of the countermeasure is not high and a manager asks human experts for advice, circulating notices are delivered with the request of time limit for the answer.

RULE 8: If advice from the human experts is in time, the system displays the advice and queries for the manager who may take account of the advice or not.

3.3 Dependency diagram

The relation between measured items and goal factors is called dependency diagram, which is shown in Fig.1. In this figure the items which are written at both sides of the triangles show the rules as knowledge base. The items at the left side represent a part of the prerequisite in the rules. The items at the right side show a part of conclusion in the rules.

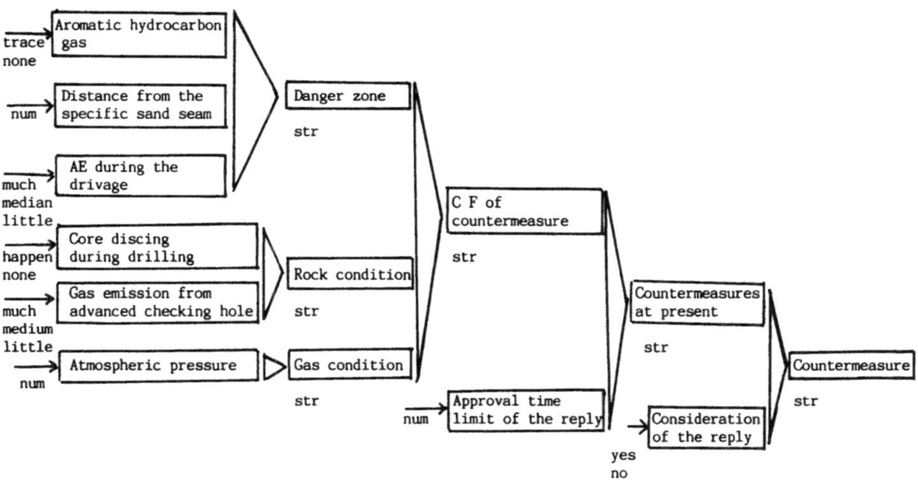

Fig. 1 A dependency diagram

3.4 Flow chart of consultation

The flow chart on the consultation is shown in Fig.2.

4 Application

The main road ways in the Horonai coal mine are shown in Fig.3. The mining depth is about 1200m from the surface. The annual production of coal had been 1 million tons. But this coal mine was closed for the economic reasons by the government, because the cost of production was three times as much as in other countries.

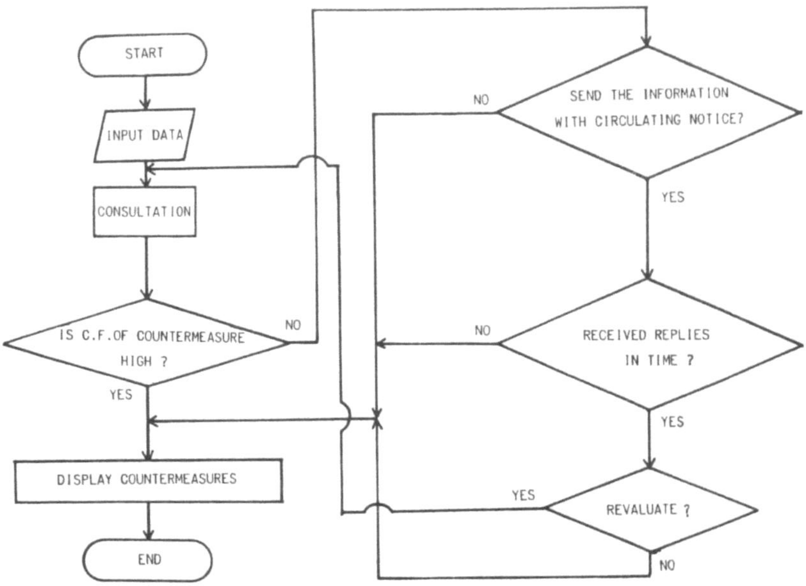

Fig. 2 A flow chart on the consultation

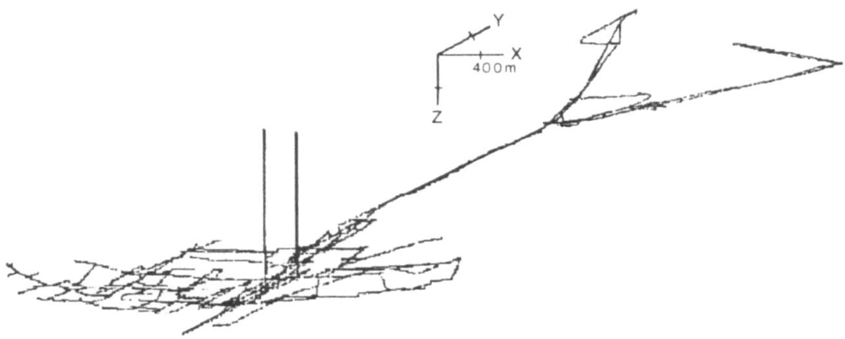

Fig. 3 Display of the network

In this coal mine, while driving the roadway in the specific sand seam, the rock and gas outbursts have occurred several times, but during drivage of the roadways in other seams, it has never occurred.

About 10 years ago, mine fire occurred due to the rock and gas outbursts and two-thirds of the roadways were filled by water to extinguish the fire. Therefore, when the roadways in the specific sand seam are driven, some countermeasures to prevent, predict and take refuge are to be taken.

The values used in this simulation were assumed as followings.

1) Trace of hydrocarbon gas was detected in the emission gas from the advanced boring. Certification factor was estimated as 90 (CF=90).
2) The specific sand seam appeared in the face and drilling operator smelled aromatic hydrocarbon gas, during drilling the blasting holes. (CF=100)
3) Counts rate of AE during the blasting increased. (CF=80)
4) Gas emission from the advanced check borings was little. (CF=50)
5) The disking phenomenon occurred during drilling the advanced borings. (CF=100)
6) The methane concentration in the upcast shaft was 0.7% and the atmospheric pressure was 920 mm bar. (CF=70)
7) Approval time limit of response from human experts was 15 minutes.

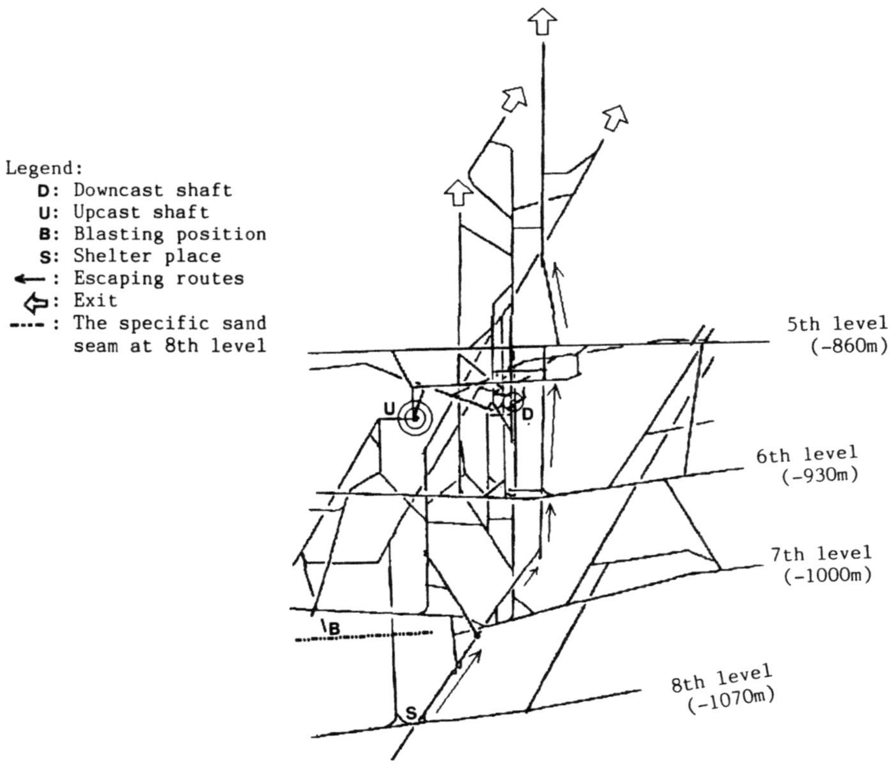

Fig. 4 Plane view of the network in the vicinity of the shafts and escaping routes

The results of consultation as shown below were displayed on CRT.

1) Provide the shelter place during blasting, far away from the face. (CF=60)
2) Secure and confirm the routes for escape. (CF=50)

And these information with figure were transmitted to all the specialists. From these replies the manager could get the escaping routes providing that the condition of accident is not changed.
The routes to take refuge are shown in Fig. 4.

5 Conclusion

The information with pictures is an effective measure to consult and direct the countermeasures. The expert system with a circulating notice capability is an effective one in the case that flexibility is required as a supporting measure on mine safety.
But using this system is limited to the cases where the conditions of the accident do not change rapidly.

Acknowledgements

The authors wish to thank to Professor Higuchi, Hokkaido University, for the helpful advice, to Mr. Kazem Najm for his useful correction of the text and to Horonai Coal Mine for giving us the data.
This work was supported by Arai Science Technology Foundation, Japan.

References

Bodkin, K.E. (1988) Expert system for colliery problems, Colliery Guardian, Vol.236, No.7, pp.216-219.
Denby, B. and Atkinson T. (1988) Expert system applications in the mining industry, The Mining Engineer, Vol.147, No.320, pp.505-509
Fries, E.F. and Welsh J.H. (1986) Expert systems and real-time mine monitoring, 19th APCOM Symposium, Penn. St. Univ., pp.802-808.
King R.L. and Kissell (1988) The pro series of expert systems from the U.S. bureau of mines, Proc. of 8th ICCR, Tokyo, pp.82-91.
Tominaga Y., Umeki Y., Arakawa M. and Takahashi K. (1989) The application of an expert system to control mine ventilation at the Taiheiyo coal mine under the open sea, Proc. of 4th U.S. mine ventilation Symposium, Berkeley, pp.317-321.
Tominaga Y., Kon N., Arakawa M. and Yamaguchi S. (1989) Development of an expert system for climate control underground, MMIJ/IMM Joint Symposium, Kyoto, pp.467-474.

Detection of periods of significant activity in noisy signals

A.J. Charles and A.S. Sehmi
Department of Engineering, University of Leicester, UK

Abstract
A method of detecting periods of significant activity and classifying information in long sequences of noisy data is described. These segments of activity are found despite the possibility of fluctuating baselines and unwanted trends. A second signal, called the *transformed signal* is created from the incoming raw data; this is then used for detecting significant active segments of the original input. The transformation algorithm has the property of automatically aligning characteristic features in similar segments, and this greatly enhances classification and extraction of those features using simple methods.

An application in the extraction and classification of features in bioelectric signals is described, however it will be evident that the technique is widely applicable in medical signal processing and monitoring industrial processes.
Keywords: Signal Processing, Pattern Recognition, On-line Feature Extraction.

1 Introduction

The type of signal to be considered here is one which contains repeated, randomly-occuring, noise-like, structured components, such as those recorded from animal tissues. When recording from a single site in such tissue, the signal voltage potentials may contain components which vary greatly according to the proximity of their source to the recording site, the strength of the signal produced at each transmission site and the number of active sources. The location of the recording site in relation to the sources is not known. The objective is to automatically recognise as many of those portions of that noise-like signal which contain consistent and nearly time-invariant structure as possible. Fixed bandwidth filtering cannot always be applied to this task since features in the original signal may vary greatly in both amplitude and frequency according to their distance from the recording site, recording methods, and effects of electrophysiological reactions. Using an amplitude threshold test to collect all of those segments is, in most circumstances, fruitless. It is possible, with the method we describe, to retrieve and classify much of the required raw signal components whilst preserving their raw quality, and to classify them using simple and fast techniques.

2 The Algorithm

The algorithm has been based on a method proposed for the detection of QRS segments in the electrocardiogram. (Hamilton and Thompkins, 1986). As the raw data is collected, a second signal, which we have called the *transformed signal* is constructed from it. The most important property of the transformed signal is that it must preserve the temporal information contained in the raw data, since this is the key to its ability to temporally align any segments in the raw data that will be detected. A schematic representation of all the steps taken in the algorithm is shown in Fig.[1].

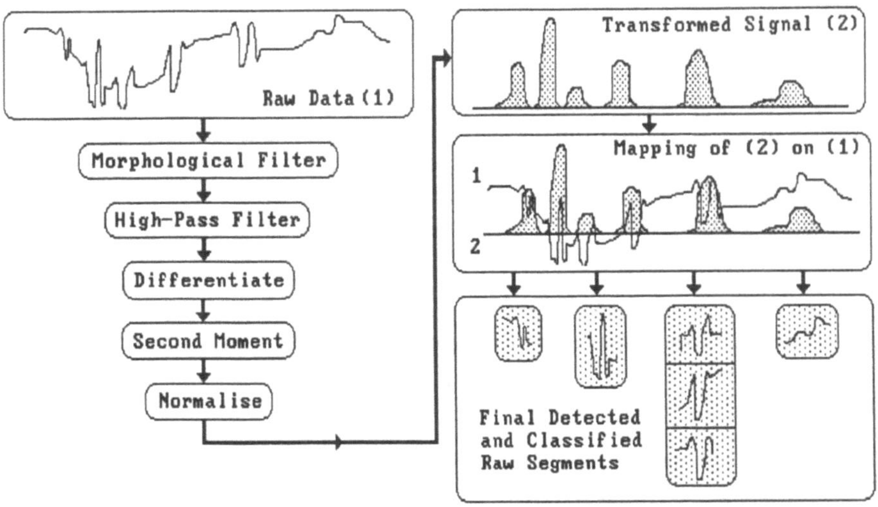

Fig.[1] Schematic diagram showing the steps taken to extract periods of significant activity in noisy signals.

The transformed signal is constructed using a combination of linear and non-linear filtering of a *copy* of the original signal. The parameters chosen for each stage of processing depend on the nature of the signals under consideration, the sampling rate used etc. and are discussed in section three. The initial stage of the process is a morphological filter, (Chu and Delp,1989), which has a structuring element with a constant value of 1. The length of the structuring element is discussed in section three. This morphological filter is used to remove any very high frequency noise. The resulting signal is then high-pass filtered using a linear-phase recursive filter (Lynn, 1971, 1977). This is followed by quadratic fit differentiation, squaring, a centred moving average and normalisation.

The choice of width of the moving average is discussed in section three. The transformation results in a signal which contains smooth, square-wave like pulses, and has a stable baseline. These square-wave like pulses indicate the presence of a feature in the original signal. The normalisation mentioned above is performed on the transformed signal so that a preset amplitude threshold level can be used to select as many features as possible. The amplitude threshold level is discussed in section three.

If the processing is to be performed on-line, then normalisation is performed over a window containing at least three pulses. Those segments of the original raw signal, (which was copied earlier), corresponding to the length of time when a pulse is above the threshold are collected for analysis, together with a number of points on either side. The times when the transformed signal crosses the threshold level, the location of the pulse's peak and the width of each pulse at half its height are noted for each segment. The transformed signal and the remainder of the original signal can then be discarded unless required for some other purpose.

The segments are classified according to the following strategy. The first segment enters an initial bin, subsequent segments are then compared first on amplitude, then on timing constraints, (in certain applications there must be a minimum delay between emissions from the same source), and finally by calculating the approximate absolute area difference between the segment in the bin and subsequent segments. In practice this is done by summing the absolute differences in amplitude between the two segments being compared. The smaller the area, the closer the match. Segments not satisfying a matching criterion indicate a new bin which will be used in the remaining segment classifications. Subsequent segments are compared with all bins and where a match with more than one bin is indicated, enter the bin where the match is the closest.

3 Selection of parameters

The length of the structuring element for the morphological filter should be selected according to the sampling rate used and the highest expected frequency components in the features of interest. For example, at a sampling rate of 20kHz, to select features with a highest expected frequency component of 5kHz, we used a structuring element length of 5 data points. This was found empirically bearing in mind that there is a speed penalty associated with long window lengths.

The moving average width should also be chosen according to the sampling rate and the amount of smearing of feature durations that can be tolerated. This was again found empirically. We used, for example, a width of 16 data points at 10kHz and 32 data points at 20kHz to collect features with normal durations of between 3 and 20msec. The preset threshold level can be set very low. It is used only to ignore small fluctuations in an otherwise stable baseline.

4 Experimental results

The top trace in Fig.[2] shows a section of muscle activity recorded from the biceps of a normal human subject. Several stages in transforming the signal are shown in the traces beneath it. The classified segments are shown in Fig.[3]. These segments were

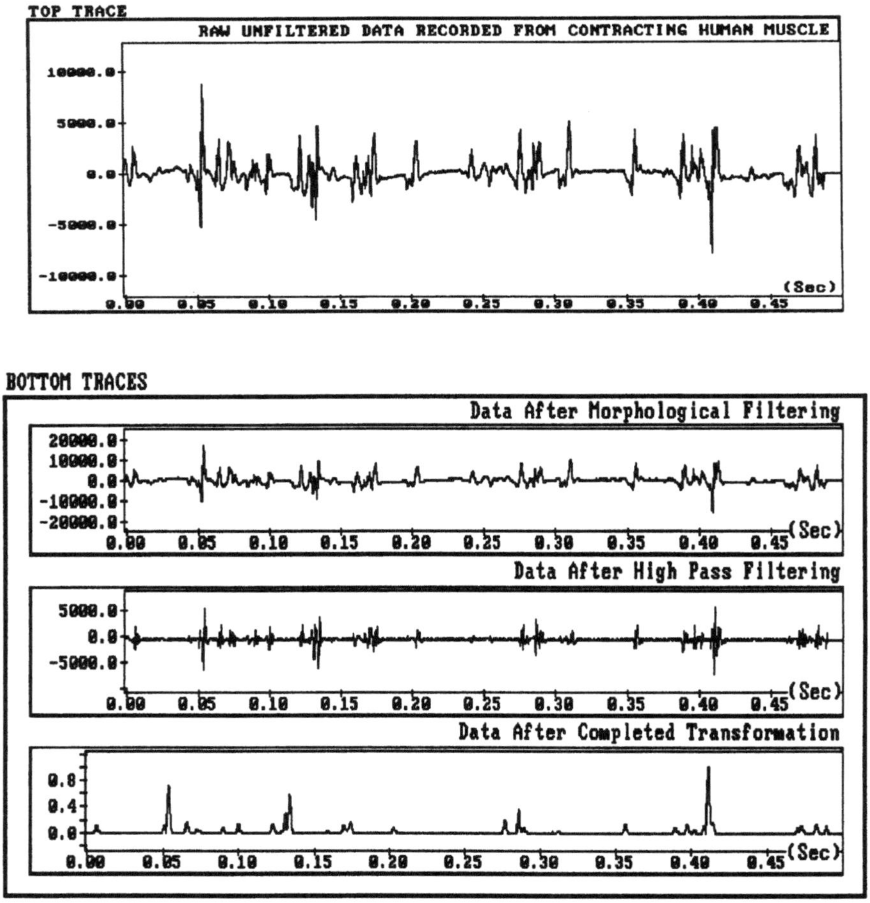

Fig.[2] Top trace: Muscle activity recorded from the biceps of a normal human subject.
Bottom traces: Results of some stages in signal transformation.

classified as described in section two. We have also investigated classification strategies based on quantitative measurements of the transformed pulses. These are still under investigation, and so a more detailed qualitative exposition of the current classification strategy is presented here.

The pulses correspond to those segments of the original signal where significant activity occurs. The width of each pulse at half its amplitude is consistently close to the moving average width, and therefore it is this portion of the pulse that contains the most information. Introducing a second threshold level, this time at half the amplitude of each pulse, (i.e. the second threshold level varies according to the pulse amplitude), we have found that the time between the leading edge of the pulse crossing the first

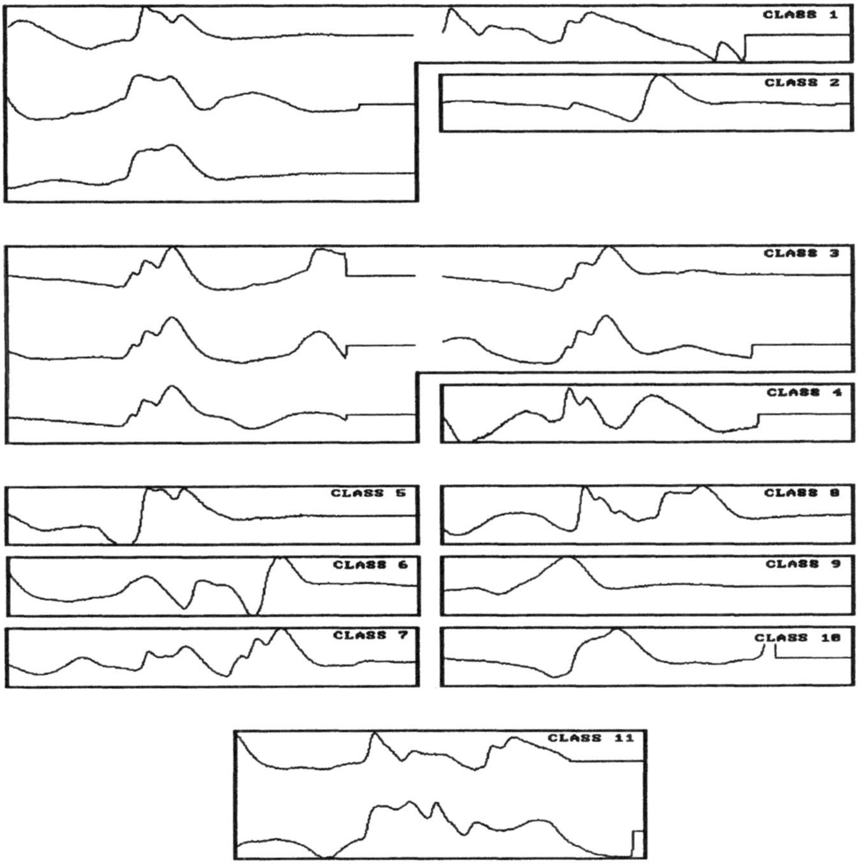

Fig.[3] Eleven classified segments from the raw data of Fig.[2].

threshold and the trailing edge crossing the second threshold is as good, if not a better method for classifying the features as is the area difference. The half-height width can also be used in some circumstances to pinpoint segments of the original signal that contain features arising from overlapping emissions at the recording site. If the half-height width is much greater than the moving average width, then the underlying segment is composed of overlapping features, (a complex). Also, if the half-height width is greater than the total duration of the pulse, then the half-height point lies below the first threshold level. We have found that this is usually indicative of a complex. The complexes, once identified, can be decomposed into their constituent parts, (e.g. Loudon et al., 1989). Because the pulses in the transformed signal are temporally aligned with the original signal, the occurrence times of features can easily be extracted and point processes generated for further analysis (Lago and Jones, 1983). In addition the latency variability associated with specific features of segments in a class can be computed readily (see Sehmi, 1989).

5 Conclusions

It has been shown that this algorithm can be used to extract and classify features from a signal recorded at a single site but containing random features originating from several transmission sites. Classification of the segments according to their morphology can be used to identify repeated and overlapping emissions contained in the data arising from the different source sites. It should be of use in applications where it is only possible to make a recording of several sources at one recording site. The algorithm has been developed to allow real-time segmentation and classification of the signal and may therefore find applications in areas such as real-time monitoring and intermittent fault analysis in the process industry and foetal heart rate monitoring under noisy conditions in the medical industry.

6 References

- Chu, C.H.H. and Delp, E.J., (1989), Impulsive Noise Suppression and Background Normalization of ECG signals using morphological operators, IEEE Transactions in Biomedical Engineering, Vol BME-36, No.2.
- Hamilton, P.S.and Tomkins, W.J., (1986), Quantitative Investigation of QRS detection rules using the MIT/BIH Arrhythmia Database., IEEE Transactions in Biomedical Engineering, Vol BME-33, No 12.
- Loudon, G.H., Jones, N.B.and Sehmi, A.S., (1989), Knowledge based decomposition of myoelectric signals, IEE colloquium, Biomedical Applications in Digital Signal Processing, IEE Savoy place, London.
- Lago, P.J.A.and Jones, N.B., (1983), Turning Point Spectral Analysis of the Interference Myoelectric Signal, Med.Biol. Eng.Comput., Vol 21, pp 333-342.
- Sehmi, A.S., (1989), New Environments for Neurophysiological Investigations, PhD Thesis, University of Leicester, Leicester, U.K.

Application of an expert system on the monitoring of human lungs

A.K. Jawad and A.I. Khalil
Engineering Division, Humberside College of Higher Education, Hull, England

Abstract
This paper reports on progressive research work to develop an expert system for the condition monitoring of patients' lungs. Conventionally, monitoring lung operational condition is carried out by obtaining related data, while the patient is on mechanical ventilatory assistance. Interpretation of the acquired data is then made by human medical knowledge and experience to reach an initial or final conclusion as to the lung condition and the type of action needed to maintain or otherwise improve that condition. The intended expert system produces interpretations of physiological clinical knowledge about the diagnostic implication of the data. The system to be presented in the paper is designed to assist rather than substitute human medical expertise in the management and control of lungs treatment. The work being carried out in co-operation with the Intensive Care Unit staff of Hull Royal Infirmary.
Keywords: Medical Expert System

1 Introduction

Expert systems, also known as intelligent knowledge based systems (IKBS), are one of the main applications of artificial intelligence (AI).

An expert system may be regarded as the embodiment within a computer of a knowledge based component from an expert skill in such a form that the machine can offer intelligent advice and/or take an intelligent decision about a processing function.

Many expert systems are currently being applied to a number of medical domains, most notably diagnosis, treatment planning and to monitor physiological variables. Also to manage information storage and facilitate the interpretation of data. Medicine has also been one of the preferred fields by knowledge engineers. Their role is to assist the medical practitioner by giving ready access to the level of skill shown by experts in a particular field.

Extensive research work has already been done, yet only a few systems (such as PUFF [1], ONCOCIN [2] and MYCIN [3]) have reached routine medical use.

The aim of the present work is to design an expert system which can be used to interpret physiological data related to the lungs of a patient receiving post-operative mechanical ventilation in the Intensive Care Unit.

2 The basic structure of an expert system

An expert system typically has the structure shown in Fig.1

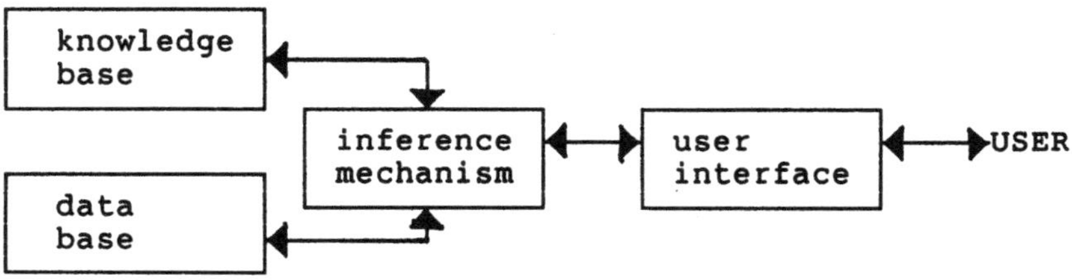

Fig.1. Structure of a typical expert system.

The knowledge base stores specific knowledge related to a particular field and comprises a series of facts and rules from which the system draws its expertise (the knowledge is about problem domain). An expert system which is designed to aid medical practitioners in the diagnosis of asthma, for example, would have a knowledge base which contained facts and rules about the likely causes of such disease and symptoms related to each of these causes. The data base stores facts about current problems and for later consultation. The more data there were to be processed, the more obvious the power of the computer becomes. The inference mechanism has the ability to look through the knowledge base and apply the rules to the solution of a specific problem in co-ordination with the data base, whilst the user interface provides means of communication between the user and the system.

3. Element of the expert system

3.1 The knowledge base sub-system The knowledge required to carry out the treatment process is transferred from the medical experts into the knowledge sub-system. The relevant expertise is represented by a set of rules in the form of IF...THEN... The presence of ventilatory failure can be detected by the rise in the arterial pC02 which can be written according to these rules. The failure, however, may occur as a result of a reduction in total ventilation or a failure to increase ventilation to compensate for the impaired C02 elimination resulting from ventilation/perfusion inequalities. The sub-system also contains the relationship between various parameters and objectives. It provides rules, which are represented in the form IF...THEN...

IF alveolar ventilation increases
THEN pC02 falls

3.2 The data bank sub-system

The approach to the management of a patient under mechanical ventilation takes into account the underlying clinical condition and the various functional abnormalities which lead to inadequate tissue oxygen delivery and carbon dioxide removal. Initially, the treatment starts by gathering data required to set up the ventilator. These include tidal volume (VT), respiratory rate (RR) and inflation pressure (IP), which are guessed on the basis of the patient's condition and personal details; namely age, sex, weight and height. It is the data bank which handles this information, shown in the table below, after being entered via the user interface.

Patient's condition:	respiratory and cardiac arrest
Age:	68
Weight:	17 stones
Sex:	female
Height:	5' 3"
Date of admission:	11.1.1990
Time:	1800 hours
Tidal volume:	800 ml
Respiratory rate:	10 bpm
Inflation pressure:	38 cmH20
PEEP:	5
pH:	7.4 kPa
pC02:	7.2 kPa
p02:	12.2 kPa
HC03:	35.4 kPa
Bx:	9.5 kPa
Temperature:	$36.5^{o}C$
I:E ratio:	1:2

3.3 The inference mechanism

The inference mechanism is the part of an expert system that puts the expertise contained in the knowledge base to work in solving the problem preesented by the user. An important part of this task is to draw inferences by combining what is known about the domain. The inference techniques usually adopted to build the mechanism are known as 'forward chaining' and 'backward chaining'. The forward chaining starts with the set of known facts and tests all the hypotheses in which these facts play a part, whilst the backward chaining attempts to find data to prove or reject a suggesteed explanation for a particular fact. The system first identifies the appropriate condition that would be enough to meet a need of the specified goal, then it will carry on to try these conditions by examining the data base to check whether to be true or false. Thus the goal is resolved into a number of sub-goals and so on until the proposition is reached. The latter technique has been chosen for the present work.

4. Implementation of the expert system

Fig.2 shows the patient's treatment path in the presence of the expert system.

The treatment of a respiratory failure starts with the admission of the patient to the Intensive Care Unit by gathering information related to the present and past medical record of that patient. This information, which is usually conveyed to the medical experts, is entered instead into the data base of the expert system via the user's interface. The task of guessing the initial values of the parameters required to set the ventilator connected to the patient, is performed by the expert system instead of being directly obtained from the human expertise. It is noteworthy that the accuracy of this guesswork depends conclusively on the quality of information transferred to the knowledge base sub-system and the power of infereence technique used in the process. Having set the ventilator, the user will then feed the new set of measurements acquired from it back into the data base sub-system, following the same procedure used previously. At this stage, the blood gases results obtained from the blood analyser are ready to be entered into the sub-system. A revised set of data is then issued by the expert system and the ventilator is re-adjusted accordingly. This process continues until stable conditions are reached and displayed on the monitor.

It is anticipated that the initial implementation of the expert system should be carried out and carefully monitored by the medical experts, who should also act as users of the system at this stage.

It is anticipated that the initial implementation of the expert system should be carried out and carefully monitored by the medical experts, who should also act as users of the system at this stage.

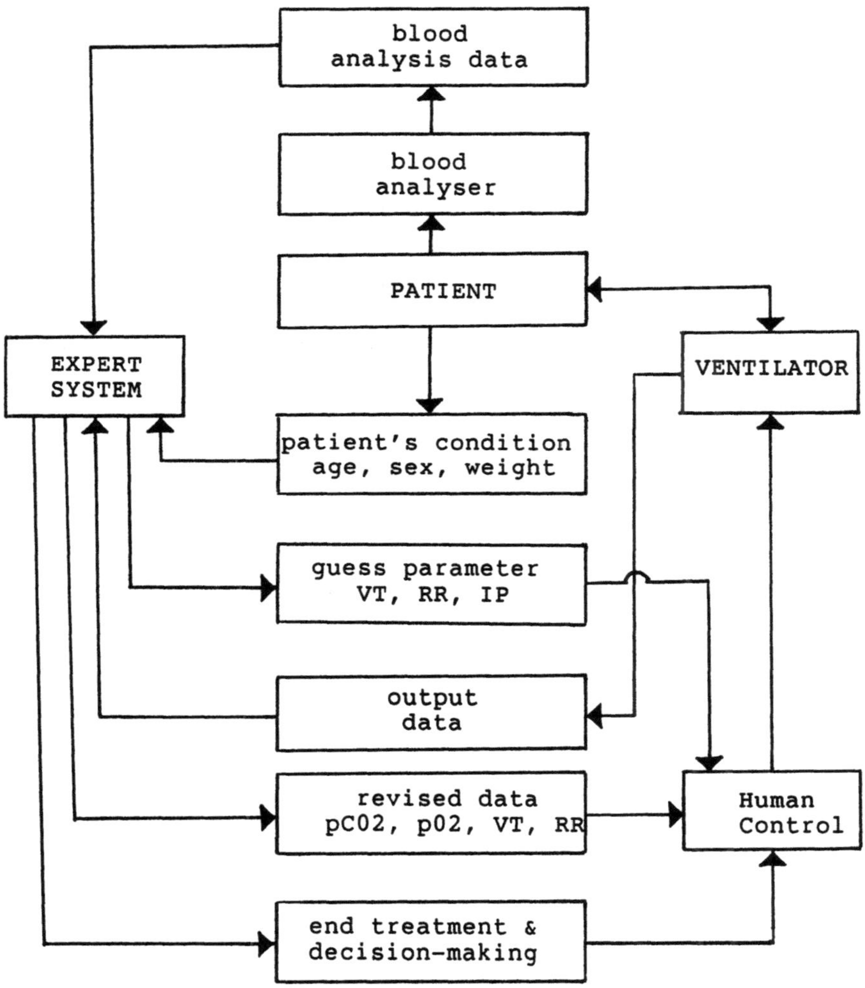

Fig.2. Implementation of expert system.

5 Conclusions

It is concluded that the expert system can play a significant role in improving the treatment of respiratory failure. The advantages of the new approach are: (a) to speed up the treatment process, (b) to increase the accuracy of the initial setting of the ventilator, and (c) to produce better prediction parameters such as C02 production and dead space.

6 References

Aikins, J.S. et al (1983) **PUFF: An Expert System for Interpretation of Pulmonary Function Data.** ComputBiomedres, vol.16 pp. 199-208.

Shortliffe, E.H. (1976) **Computer-based Medical Consultation: MYCIN** American-Elsevier, New York.

Shortliffe, E.H. et al (1983) **ONCOCIN: An Expert System for Oncology Protocol Management** in proc. I.J.CAI 81, pp. 876-881

Intelligent classification in EMG decomposition

G.H. Loudon, A.S. Sehmi and N.B. Jones
Department of Engineering, University of Leicester, Leicester, UK

Abstract
This paper presents research relating to the use of computers for the intelligent decomposition of myoelectric signals (EMG). A knowledge based expert system is described which decomposes superimposed waveforms formed from overlapping motor unit action potentials (MUAPs) in a myoelectric signal using symbolic information provided by numerical recognition analysis. The system, written in Prolog, consists of some 30 rules in the knowledge base that are driven by an interpreter that incorporates uncertain reasoning based on fuzzy set theory. The expert system contains both procedural and declarative knowledge representations of the problem domain. The declarative rules contain a description of the relationships between the raw motor unit (MU) information collected by the numerical analysis and the superimposed waveforms being decomposed. The procedural rules interact with the declarative rules through rule attachments that activate demon procedures. The demon procedure computes fuzzy certainty factors for all the possible combinations of MUAPs that form a superimposed waveform.
Keywords: Knowledge based Signal Processing, Electromyography, Decomposition.

1 Introduction

The manual assessment of myoelectric data by human experts is based upon complex processes of data reduction and feature extraction, some of which are apparently subjective. The complete task of data evaluation and interpretation falls largely on the human expert who uses an holistic view of the data collected, in conjunction with some simple quantitative measures which can be calculated by computer based algorithms. Since our aim is to fully decompose and interpret myoelectric signals automatically in a computer, the holistic reasoning that a human expert uses must be simulated. The simulation of this reasoning is performed using knowledge based expert system techniques.

2 The Application Domain

The electrical activity recorded from a muscle when under contraction is known as the myoelectric or emg signal. The myoelectric signal results from the activation of groups of muscle fibres by impulses sent down motor nerves from the spinal cord. Each nerve impulse innervates a group of muscle fibres known as a motor unit (MU). The resultant waveform recorded from a MU is known as the motor unit action potential (MUAP) (Basmajian and DeLuca,1985). During a constant force contraction, trains of impulses are sent down the motor nerves at fairly regular intervals resulting in a train of MUAPs. The firing period distribution of nerve impulses is almost gaussian with a very small variance (Andreassen and Rosenfalck,1980). The resultant myoelectric signal recorded with needle or surface electrodes is then a summation of the individual MUAP trains from different MUs in the muscle. The signal is called the interference pattern EMG (IPEMG).

The shapes of the MUAP waveforms in a myoelectric signal are an important source of information used in the diagnosis of neuromuscular disorders.The objective of our research is to intelligently decompose the IPEMG into its individual MUAP trains under fairly high force level conditions (say 30% maximum voluntary contraction). This requires the classification of both non-overlapping MUAPs and the decomposed superimposed waveforms formed from overlapping MUAPs in the signal. Non-overlapping MUAPs are classified using a statistical pattern recognition method. The method first describes the MUAPs by a set of features and then uses diagonal factor analysis to form uncorrelated factors from these features (Gorsuch,1974). An adaptive clustering technique groups together MUAPs from the same MU using the uncorrelated factors (Tryon,1970).

The decomposition of superimposed waveforms is divided into two sections. The first section is a procedural method that finds a reduced set of all possible combinations of MUAPs which are capable of forming each superimposed waveform by using a template matching procedure. A more detailed description of this procedural analysis outlined can be found in Loudon et al. (1989). The second section is the knowledge based analysis of the candidate MUAP combinations forming each superimposed waveform. This latter analysis decides which combination is the most probable (see Fig[1]).

3 Intelligent Classification of Superimposed Waveforms

It is possible for a machine to decompose a superimposed waveform into its constituent MUAPs. However, with current methods of decomposition either errors will result, the time taken for decomposition is very long (LeFever and DeLuca, 1982) or much training is required to use the decomposition scheme. Errors could be due to noise in the signal or because of the similarity between results after a pattern matching analysis. Human experts are able to identify which of the possible MUAP combinations comprise a superimposed waveform by studying the firing times of the MUAP trains (Basmajian and DeLuca,1985). Typically they would arrive at a decision based on uncertain and incomplete evidence available from the MU firings already classified. This observation has led to the specification of an expert system which uses a fuzzy reasoning model to describe the decomposition protocol of superimposed waveforms.

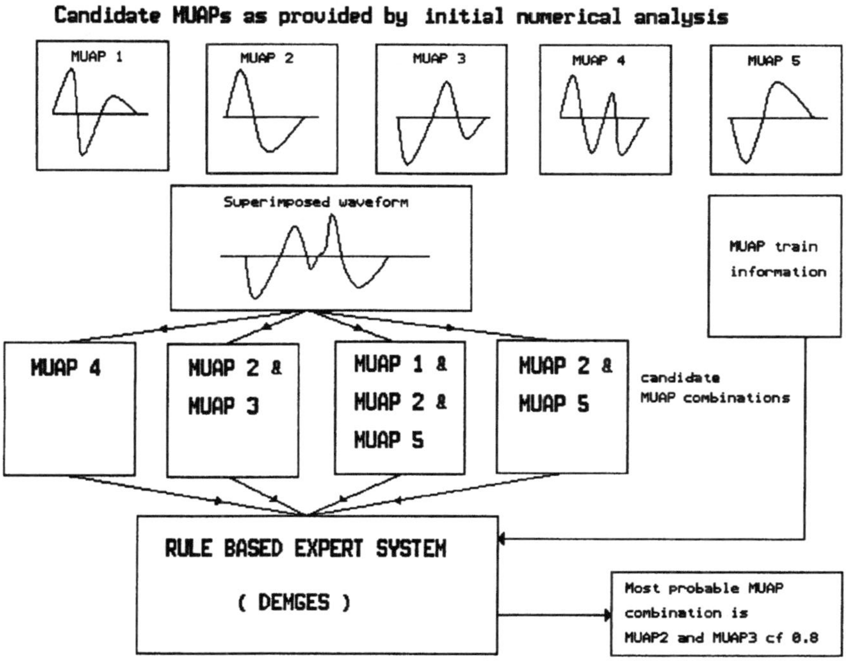

Fig.[1] Functional diagram of the DEMGES expert system.

3.1 Fuzzy Certainty Factor Calculations in DEMGES

DEMGES is an acronym for Decomposition of EMG Expert System. It is a suite of programs used to automatically decompose myoelectric signals and it includes the preliminary numerical analysis programs. DEMGES uses fuzzy certainty factors to model its uncertain reasoning mechanism so that intelligent classification of superimposed waveforms can be performed. The fuzzy membership functions are described by the firing period statistics of the partially classified MUs.

The rules that DEMGES contains are modelled on the judgemental processes that an expert uses for superimposed waveform decomposition. The information provided by the numerical analysis is examined by DEMGES rules in the early stages of a goal-directed reasoning process. MUAP candidates in a combination selected from the numerical analysis are given fuzzy values which are propagated through the search space towards the final goal of finding the certainty of a MUAP combination forming a superimposed waveform. Hence, the numerical analysis results are easily assigned qualitative descriptions for the expert system to reason with.

The method by which fuzzy values are assigned to the MUAP combinations relies on the definition of a fuzzy model describing the myoelectric signal. The model is described next, and is comprised of two parts, namely the procedural and the declarative model components.

3.2 The Fuzzy Procedural Model (Database)

The fundamental primitive for information modelling is propositional statements of the form : an attribute of an object has a particular value. This is represented in the Prolog language as the symbolic structure :-

Object Attribute Value

Again in Prolog, we may express that a MUAP candidate in a combination definitely occurs at exactly the position of a superimposed waveform X by writing :-

MUAP position_is X

This will not be the case in reality, because the firing times of a MU are not exactly regular. The MUAP candidates are given fuzzy values related to the *possibility* of a MUAP occuring at the position of a superimposed waveform. The fuzzy value is calculated using the fuzzy membership function shown in Fig[2]. This function can be described in terms of a fuzzy set (Zadeh,1965). Very briefly, fuzzy set theory states that a fuzzy set is a class which admits the possibility of partial membership in it. A fuzzy value of 1.0 represents full membership and a fuzzy value of 0.0 represents non-membership. Intermediate fuzzy values represent partial membership.

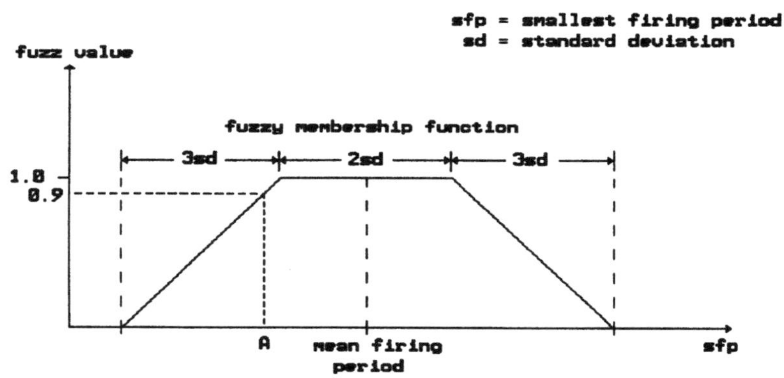

Fig.[2] The fuzzy membership function.

The fuzzy value of a MUAP candidate is calculated by mapping the smallest firing period (SFP) between the nearest neighbouring MUAP in the MUAP train and the position of the superimposed waveform onto the fuzzy membership function which describes the MU firing period distribution. In Fig[2] for example, a MUAP candidate 'A' maps to a fuzzy value of 0.9 indicating partial membership of the fuzzy set of all possible firing periods. The duration of the fuzzy function varies for each MUAP train (and hence each MU) in the myoelectric signal, depending on the mean and standard deviation of the firing period of the MU duration.

3.3 The Fuzzy Declarative Model (Knowledge base)

This model attempts to capture the expert decision making process used to decompose superimposed waveforms. The fuzzy procedural model above makes it possible to formulate propositions of the form :-

Object Attribute Value cf Fuzz

where the certainty with which a proposition holds is expressed with a propositional attachment called the certainty factor (**cf**) or fuzzy value. The fuzzy value **Fuzz** relates to the *possibility* of the proposition being true. Futhermore we can formulate the consequence of fuzzy propositions in Prolog by using fuzzy rules of the form:

```
Rule ::
    if Object Attribute Value cf Fuzz
    then ObjectX AttributeX ValueX cf CF.
```

The confidence in the rule being true can be expressed through the rule attachments **CF**. So the value **ValueX** of the object **ObjectX** is concluded with the combined fuzzy value computed from the fuzzy value **Fuzz** of the object **Object** and the rule confidence attachment **CF**. **Fuzz** and **CF** are combined by simply multiplying their values together.

The premise of a rule can contain both conjunctions and disjunctions of propositional clauses. The combined fuzzy value of a conjunction or disjunction of clauses is determined using fuzzy set theory. In a conjunction the minimum fuzzy value is taken from the computed fuzzy values in the set of clauses in the premise. In a disjunction the maximum fuzzy value is taken. For example, in the rule below, the fuzzy value of the conclusion will be computed by taking the minimum of **Fuzz1** and **Fuzz2** and then multiplying this by 0.95.

```
rule19 ::
    if
        Mean < (SFP / 1.5) cf Fuzz1 and
        Firing_period is (SFP / 2.0) cf Fuzz2
    then
        muap( MuapNo, Mean, SD, FiringPeriod )
        is_compared_with SFP cf 0.95.
```

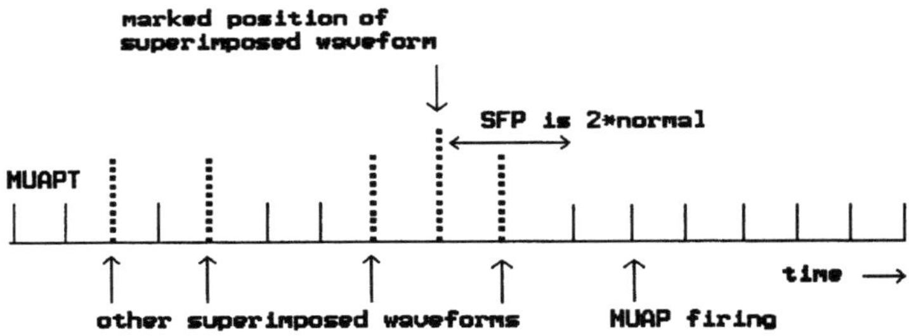

Fig.[3] Firing times of a motor unit (MU) before full decomposition of superimposed waveforms.

Some rules in the knowledge base (e.g. rule19 above) cater for problems that arise when the SFP is much greater than the mean firing period of a MU (due to unclassified MUAPs) as shown in Fig[3]. This is done by repeating the fuzzy membership function at multiples of the mean firing period. These rules are given certainty factors that reduce in value as the SFP increases with respect to the mean firing period of a MU. Fig[4] shows the effect on confidence in the rules at increasing multiples of the mean firing period of a MU.

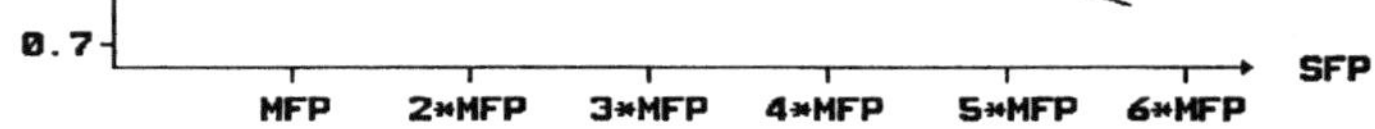

Fig.[4] The reduction in rule confidence as the SFP increases from the mean.

Rules are provided to calculate the overall fuzzy value of each candidate MUAP combination (Z) in a superimposed waveform. The formula used to find the overall fuzzy value is:

F(Z) = F(X) * NOT(F(Y))

where F(Z) is the overall fuzzy value of a MUAP combination, F(X) is the combined fuzzy value of the intersecting set of individual MUAPs in a combination and F(Y) is the combined fuzzy value of the exclusive set of individual MUAPs not in a combination.

3.4 The DEMGES Interpreter

This interpreter shell is backward chaining and provides uncertain inference and explanation capabilities on the declarative model described in the previous section. It also provides the interfacing with the procedural model through invocation of user-defined demon-procedures. The interpreter manipulates and executes Prolog expressions making up the declarative model (ie. knowledge base) and hence, it is possible to pass goals such as the demon procedures to Prolog for execution. This facility is necessary for the evaluation of mathematical constructs and for interfacing to the 'C' language which performs the initial numerical analysis. An important specification for this implementation has been separation of all DEMGES modules. In this respect the same interpreter can be used to execute knowledge bases that will be defined in the future for further interpretation of the results produced by DEMGES for diagnostic purposes. The structure of the interpreter shell is derived from the work by Sehmi (1988) and Sehmi and Jones (1989).

```
%Top call
solve(Goal) :-
    solve(Goal cf Fuzz, Fuzz, []).

%Is goal known
solve(Goal cf Fuzz, Fuzz, _ ) :-
    fact :: Goal cf Fuzz.

%Is goal solvable using a procedural call
solve( Goal cf Fuzz, Fuzz, _ ) :-
    demon(Goal,Demon_procedure),
    call(Demon_procedure).

%Is goal solvable using a rule
solve(Goal cf Fuzz, Fuzz, Stack) :-
    Rule :: if Premise then Goal,
    satisfy(Premise, Fuzz, [Goal+Rule|Stack]),
    conclude(Goal,Fuzz,Rule).

%Ask user for solution
solve(Goal cf Fuzz, Fuzz, _ ) :-
    askable(Goal),
    assert_solution(Goal).
```

Satisfy/3 attempts to solve the clauses in Premise by recursively invoking solve/3 with each clause in turn. A successful goal will cause satisfy/3 to calculate its certainty (or fuzzy) value and eventually a combined certainty value for all the clauses in Premise. Conclude/3 will then assert the Goal into the database with the combined certainty value of Premise.

3.5 The User Interface

The user interface provides the facility to query the results of the myoelectric signal decomposition. Fig[5] shows a simulated myoelectric signal containing four motor unit trains. The suggestion that a superimposed waveform contains MUAPs two and three has been queried by selecting the waveform in the myoelectric signal using a mouse pointing device. A pop-up window shows the fuzzy certainty values of the result and all the other possible combinations of MUAPs that could have formed the selected superimposed waveform.

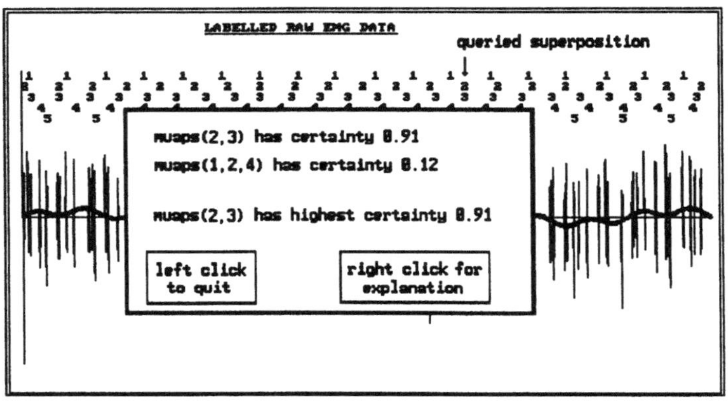

Fig.[5] The explanation user interface with DEMGES.

The user is given the option to study *how* the conclusions were formed. The *how* explanation displays the trace of reasoning taken by the expert system by interpreting the proof-tree built up in reaching that conclusion. The user is also able to study any rule invocations to seek a more specific explanation to a query.

4 Discussion

Tests are being carried out on both real and simulated myoelectric data. The effectiveness of the fuzzy reasoning mechanism of the expert system will only be determined through an extensive validation study. *Deep knowledge* in the form of pathophysiological models would have to be included to extend the usefulness of this knowledge based decomposition scheme in the actual diagnosis of neuromuscular disorders. The expert system would then be able to suggest a probable disease hypothesis and to suggest other tests that should be performed to confirm or deny that hypothesis.

5 References

- Andreassen,S. and Rosenfalck,A. (1980) A regulation of the firing patterns of single motor units - J Neurol Neurosurg and Psychiatry Vol 43, p 897 - 906.
- Basmajian,J.V. and DeLuca,C.J.(1985) Muscles Alive - The Williams and Wilkins company.
- Gorsuch,R.L. (1974) Factor analysis - W.B.Saunders company.
- LeFever,R.S. and DeLuca,C.J. (1982) Procedure for decomposing the myoelectric signal into its constituent action potentials - part 1: technique, theory and implementation, IEEE TRANS. BME., Vol 29 p 149 - 157.
- Loudon,G.H., Jones,N.B. and Sehmi,A.S. (1989) Knowledge based decomposition of myoelectric signals - IEE Colloquium, Biomedical Applications of Digital Signal Processing, IEE Savoy place, London.
- Sehmi, A.S., (1988), Epaxis: An expert system for automatic component labelling in evoked potentials, Turing Institute research memoranda, TIRM-88-32, Turing Institute Press, Glasgow, Scotland.
- Sehmi,A.S. and Jones,N.B. (1989) Intelligent Interpretation of Evoked Potentials - IFAC-BME 89 workshop on "Decision Support for Patient Management: Measurement, Modelling and Control".
- Tryon,R.C. and Bailey,D.E. (1970) Cluster analysis - McGraw Hill.
- Zadeh,L.A. (1965), Fuzzy Sets, Information and Control, Vol 8, p 338-353.

Octave band revisited – machine condition monitoring using octave band data collectors

J.H. Burrows
Dawson College, Montreal, Canada

ABSTRACT
In the field of machine condition monitoring, the range of instrumentation available to the millwrights and maintenance personnel in the primary resource and process industries, has become increasingly more complex and techniques more sophisticated. This is not only due to the spectacular developments that have occured in the electronics and computer industries during the last 25 years, but also to the fact that there is an increasing awareness that an improved maintenance philosophy manifests itself in the form of increased productivity and profitability.

This paper describes an approach, pioneered initially by the Canadian Navy but modified to suit the needs of industry, that has proved effective in a number of primary resource and process companies located in Quebec, Canada. The evolution of the instrumentation is described from an historical perspective and the technical merits of two types of measurement, Constant Percentage Bandwidth (CPB) and Fast Fourier Transform (FFT) spectrum analysis, are compared and contrasted in order to develop the rationale for the approach that is finding increasing acceptance in the maintenance departments of Quebec industry. A number of case histories and examples are described in order to support the techniques proposed in the paper.

1 INTRODUCTION

As the title of the paper implies, what follows is not a description of an exciting new technology that is going to sweep away past practices and introduce revolutionary new procedures, but rather a re-affirmation of some of the techniques used during the past 20 years that have been streamlined and made more effective by using the latest electronics and computer technologies.

Two papers by C.A.W Glew et al. describe the results obtained by the Canadian Navy using a portable octave band analyser for predictive maintenance in the 1970's. Vibration in general and octave band analysis in particular still ranks as one of the best techniques for fault detection and it is fundamental to the machine-condition monitoring approach described in this paper.

A large part of the Canadian economy is based upon the primary resource and process industries (pulp & paper, mining, chemical & petrochemical). These industries provide an ideal opportunity for companies who design instruments for the machine-condition monitoring market to test new types of instrumentation and new approaches to predictive and preventive maintenance.

An example of how one particular Canadian mining company has benefited from a machine-condition monitoring programme using instruments and techniques discussed in this paper is described in a Brüel & Kjær application note, "Machine Condition Monitoring using Vibration Analysis - A Case Study from an Iron-Ore Mine". The Quebec Cartier Mine at Mount Wright lies on the Quebec-Labrador Trough in Northern Canada and produces some 16 million tons of iron ore per year. The Mount Wright mine is an open-pit development 6400 m long by 1220 m wide and will ultimately sink 300 m below ground level. On the shovels used for ore excavation, the condition of the motor-generator set, the hoist swing transmission and the hoist's Magnetorque drive are monitored. On the haulage trucks used for transporting the ore to the crushing plant, the monitoring programme covers the diesel engine and generator. In the mine's concentrator, the autogeneous mills, pumps and conveyors are covered. Due to the success of the first part of the maintenance programme a decision has been made to include the monitoring of blasthole rotary-drilling units and other ancillary equipment.

Very often by implementing a systematic maintenance program, faults can be detected at an early stage, monitored, and repaired at the next scheduled shutdown, with the consequent large saving in production. Today many Canadian companies, anxious to enhance their machine-condition programmes in order to improve their competitiveness in the international market, have re-discovered one of the techniques pioneered by the Canadian Navy, i.e. Octave Band Velocity Spectrum Comparison.

2 USE OF VIBRATION MEASUREMENTS

There are a number of parameters that can be monitored, e.g oil debris analysis, temperature but without a doubt, one of the most powerful parameters which can be used to monitor rotating machinery is vibration. There is a large amount of information contained in the vibration signals that are obtained at the various key points of a machine but these signals can be very complex and even with today's state-of-the art measurement techniques, there is still much to learn in order to be able to measure, display, and utilize the vibration data to its fullest potential for predictive maintenance purposes. Often, the unsuccessful use of vibration measurements to assess machine condition in a given situation does not come from the fact that it does not carry the proper information, but rather from limitations of our analysis or data presentation techniques. Despite these limitations, generally due to a lack of understanding of machine dynamics and signal processing techniques, the measurement of vibration is still a very effective tool to determine machine condition, especially as it can detect abnormal operating conditions long before there is any permanent damage to the machine, which is not often the case with the use of other parameters such as temperature measurements, oil analysis, etc.. The emergence of more powerful instrumentation has greatly contributed to on-condition maintenance programmes and the continuing development and availability of more effective instrumentation is intimately linked to the development of better maintenance philosophies and predictive maintenance programmes.

3 EVOLUTION OF INSTRUMENTATION

The earliest machine-condition monitoring techniques used by maintenance personnel were based on using their eyes, ears, and experience. Observing certain parameters with their eyes, for example oil leaks, temperature, pressure, speed etc, is still an integral part of a machine inspector's maintenance procedure. However, the technique of listening to a key measurement point via the handle of a screwdriver pressed against the ear of an experienced millwright or maintenance engineer in order to determine the machine's condition is being replaced by the use of modern instrumentation that enables the vibration signal at the same key point to be accurately measured and analysed. This means that less experienced maintenance personnel can now be used to collect information and more objective judgments of machine condition can be performed.

It is not surprising that some of the first instruments used to measure the vibration signal were sound level meters and instruments based on sound level meters technology, since the vibration signals are generally located in the audio frequency range and using the ear was already a well-known technique for monitoring the condition of a machine. The Canadian Navy and many other navies around the world used instruments that were based on the sound level meters. One of the advantages of that type of instrumentation was that instruments had already been developed to obtain octave band and one-third octave frequency spectra for sound measurements, and the instruments were readily adaptable for vibration signal analysis by using an accelerometer and a charge amplifier instead of a microphone transducer.

As vibration measurements became more widespread in industry, more dedicated instruments were developed, such as simple portable vibration meters measuring only the overall level as well as manually swept filter vibration analysers for finer resolution of the various frequency components contained in the vibration signal. Dedicated vibration monitors using proximity probes and velocity transducers were also developed for permanent monitoring of critical machines.

In the late 1970's and early 1980's, the development of small low cost micro-computers became a reality due to Large Scale Integration (LSI) and other technological advances in the electronics and computer industry. These advances enabled the large data bases necessary for trending and analysis, to be handled with relative ease.

The emergence of more powerful diagnostic instruments, vibration analysers, incorporating recursive digital filters and Fast Fourier Transform (FFT) techniques, with advanced features like cepstrum, envelope analysis and constant percentage bandwidth (CPB) spectrum comparison using log-log scales, enabled diagnosis to be made more scientifically and reliably. However these instruments were usually fairly bulky, making it difficult to carry them in the field for extensive periods of time. They were also relatively complex to operate.

In the 1980's, the advent of portable vibration measuring instruments with built-in memory, commonly called data-loggers or collectors, was the latest major development in vibration measuring instrumentation to have a significant impact on machine-condition monitoring. The reasons why data collectors were popular are obvious. They are light, can be carried around the plant all day, and remove the burden of writing down the results in adverse environmental conditions. Thus the task of collecting data by a machine inspector during his daily inspection routine is simplified. As data-loggers are very easy and more practical to use, the routine inspections can be carried out more regularly and systematically by less specialized personnel. By using the software that normally accompany such instruments, reports can be generated quickly for distribution to the concerned authorities who plan the maintenance schedules. Only a few seconds per measurement point are required to acquire the data so several hundred points can be covered in a single day.

The following is an example of a report from a short inspection route carried out in the calander section of a newsprint paper machine at a Quebec mill. Over one hundred bearings are monitored using overall acceleration and RMS-PEAK trending is performed when in doubt. The trend can also be displayed on a severity chart (see figure 1 and 2).

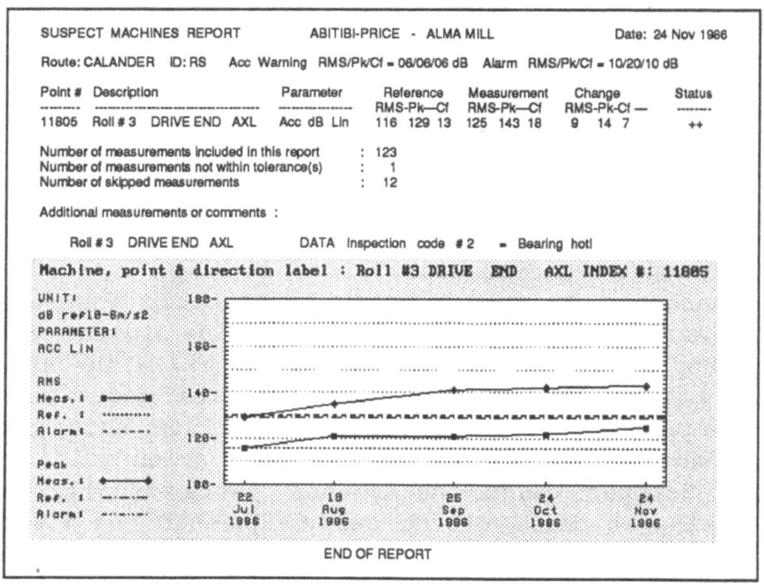
SUSPECT MACHINES REPORT ABITIBI-PRICE - ALMA MILL Date: 24 Nov 1986

Route: CALANDER ID: RS Acc Warning RMS/Pk/Cf = 06/06/06 dB Alarm RMS/Pk/Cf = 10/20/10 dB

Point #	Description	Parameter	Reference RMS-Pk—Cf	Measurement RMS-Pk—Cf	Change RMS-Pk-Cf —	Status
11805	Roll # 3 DRIVE END AXL	Acc dB Lin	116 129 13	125 143 18	9 14 7	++

Number of measurements included in this report : 123
Number of measurements not within tolerance(s) : 1
Number of skipped measurements : 12

Additional measurements or comments :

Roll # 3 DRIVE END AXL DATA Inspection code # 2 = Bearing hot!

END OF REPORT

Figure 1. *Report provided by a small data collector system at the end of a half-day inspection run.*

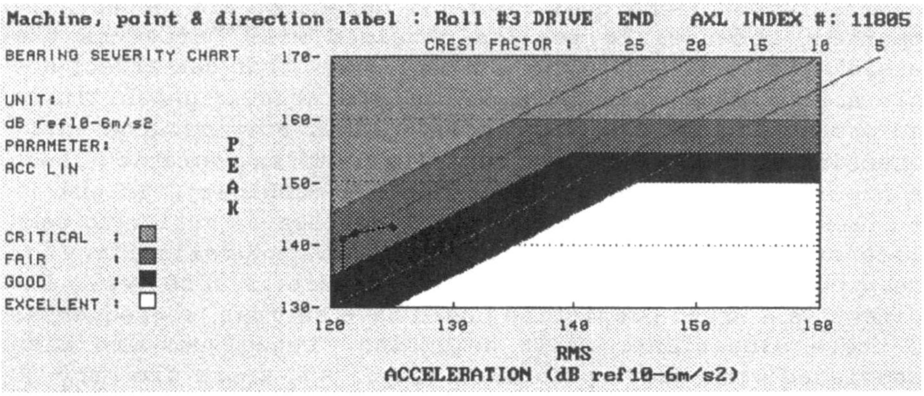

Figure 2. *Rolling-element bearing severity chart based on crest factor recently developed in Quebec.*

It would then seem that there is not much more that one would require from a data-logging system and that these instruments have achieved near perfection to detect machine faults. However, this is not the case. Even today, most data-loggers still suffer serious drawbacks, especially in the area of fault detection. Before explaining why, let us review the two main purposes of vibration measurements: fault detection and diagnosis. The first one is mainly concerned with detecting abnormal conditions in machines, a sort of screening process which will seek out of a population of machines only those exhibiting abnormal behavior. On the other hand, the diagnosis is the process of analysing the data in order to determine precisely what is wrong with a particular machine, once a fault has been detected.

This distinction is very important as out of the many analysis techniques available to the vibration analyst today, some are well adapted for diagnosis but not for fault detection and vice-versa. Inadvertant use of an inadequate measurement technique could result in unreliable results, precious time being wasted and sometimes complete failure to detect an important fault. This partly explains why many vibration analysis programmes in industry have failed or fallen short of meeting the expectations of those who implemented them, and has considerably slowed down the acceptance of vibration measurement in industry as one of the most useful maintenance tool.

Measurement techniques for fault detection are especially important because the results obtained will often be the determining factor in important decisions that can affect not only maintenance costs but also the operational capabilities and production savings of a given plant. For instance early detection of faults can be crucial on complex operations such as paper machines, continuous process machinery, if one is to allow enough time to plan for the corrective action to take place during a regular scheduled shutdown. The reliability of the prediction of the lead time to failure is also very important as a wrong prediction could have serious consequences in these situations. Fault detection techniques are also associated with quality control (QC), and there are many QC applications worth mentioning such as the verification of overhauled machinery, acceptance of new machinery, assesment of the running quality of a large population of machines, all of which play an important role in the mining industry as well as many other types of industries.

Although there has been a trend in the last few years to increase the diagnostic capabilities of data-loggers, principally with the use of FFT analysis, the use of more powerful detection techniques in commercially available instrumentation has been somewhat over-looked. For instance many data-loggers detect faults based solely on a change of the overall level. Since some machine faults do not necessarily result in an increase of the overall level, a strong vibration component from another source can mask the change for instance, the machine could very well fail long before the instrument can detect the fault. Therefore it is highly desirable to compare and trend the vibration data to a baseline spectrum.

An example is cited below. A spherical roller bearing supporting one end of a large roll in the dryer section of a paper machine was heavily damaged and close to the point of rupture. This could have caused an emergency shutdown of the machine, which would have resulted in very large production losses. The trend of the RMS-PEAK overall acceleration revealed no change in the bearing condition which was replaced on September 23rd 1987 (see figure 3).

Fortunately, systematic CPB spectrum comparison was carried out periodically until one day several multiples of the Ball Pass Frequency Outer race appeared in the spectrum (see figure 4). A FFT spectrum analysis of the signal measured on the defective bearing was also performed (see figure 5).

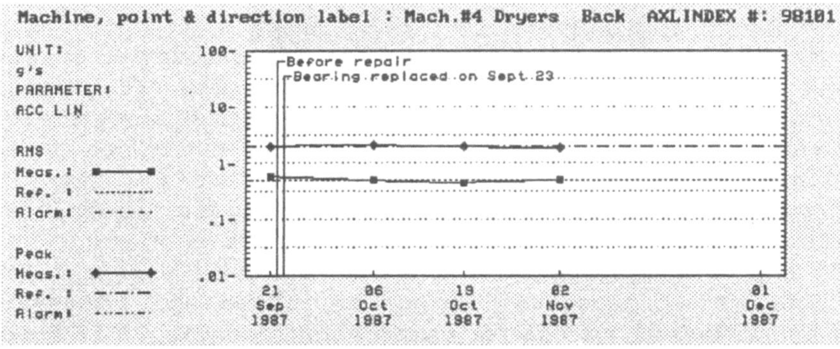

Figure 3. *Trend of the overall acceleration measured on the bearing housing of a large dryer roll.*

The two larger peaks in the spectrum are the gearmeshing components, they are the first and second harmonics, of the speed reducer from a drive located nearby the roll. These load-dependent components dominate the spectrum entirely and this explains why the overall acceleration did not change when the bearing went from a good to a bad condition. This example, which is not an isolated case since there are several such rolls on the machine, demonstrates clearly the need for spectrum comparison in fault detection. Spectrum comparison is especially important in the case of gearboxes, as we have witnessed several gearbox failures in which there was no change in the overall vibration level.

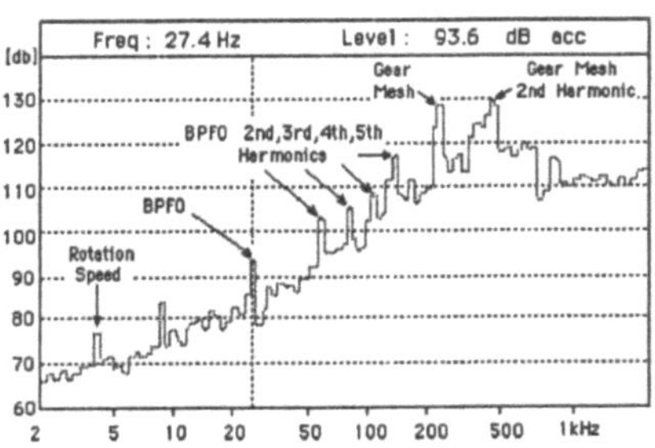

Figure 4. *CPB spectrum of the damaged bearing.*

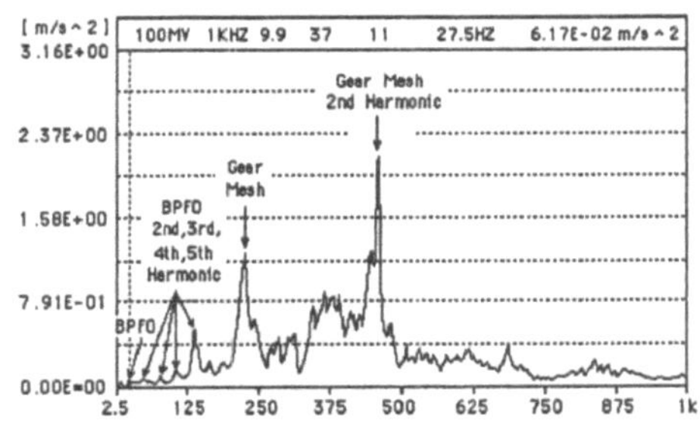

Figure 5. *FFT spectrum analysis of the same data.*

4 FFT VERSUS CPB

There are two main techniques available today to obtain a frequency spectrum: FFT and CPB analysis. The FFT method gives a constant bandwidth based on a linear frequency scale. The CPB method gives a constant percentage bandwidth based on a logarithmic frequency scale (see figure 6). The FFT spectrum is more suited to analysis and diagnosis, as it shows more clearly the harmonics and the sidebands pattern in the signal, and the CPB spectrum is more suited to trending and detection, as it covers a larger amplitude and frequency range, and is easier to use for comparison, especially if speed compensation is required.

In order to detect most machine faults, a broad frequency range must be used to include low frequency components such as a sub-harmonic of shaft speed, oil whirl, etc. and high frequency components, i.e. harmonics of tooth mesh, structural resonances excited by rolling element defects. Experience has shown that at least 60 dB of amplitude range and three decades of frequency information are required to display all the essential information in a vibration spectrum obtained with a good quality accelerometer. As a result, a baseline comparison is not very easily accomplished using FFT spectra. An FFT spectrum is computed with a linear frequency scale and unless a very large number of lines are used, only one decade of frequency can be adequately displayed on the baseband spectrum. A quick calculation of how many lines would be required on an FFT analyser to maintain a 3 % resolution at 10 Hertz and still maintain a full scale frequency of 10kHz would yield over 30,000 lines, which is beyond the capability of current analysers.

As can be seen by comparing figure 4 and 5, it is possible to see much more information on the CPB spectrum than on the FFT spectrum, even though the FFT spectrum appears to have a finer resolution. Because it covers a broader range of amplitude and frequency, the CPB representation should also be prefered over the FFT for spectrum comparison (see figure 8).

A special case of a CPB spectrum analysis is the Octave Band analysis. One of the main advantages of that approach is that the filters have been standardised in the ANSI SI.11-1966 class II specifications. Octave Band Velocity Spectrum Comparison is recommended for the following reasons: a) since the bands are quite wide, the amount of data is kept to a

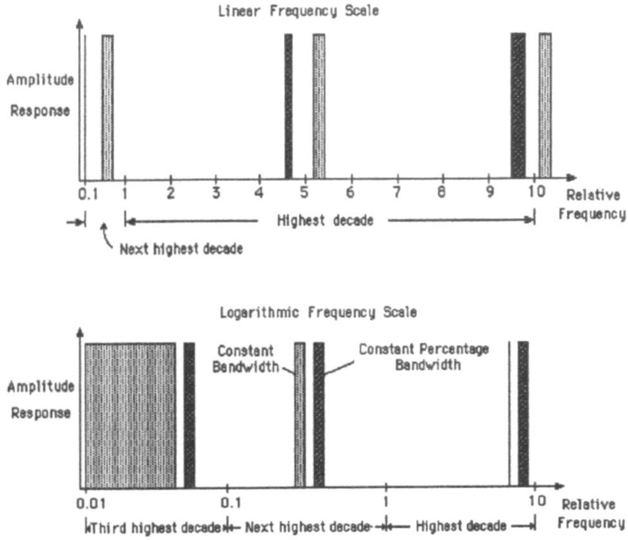

Figure 6. *A comparison of Constant Bandwith (FFT) and Constant Percentage Bandwidth analysis.*

minimum although there is still enough resolution to identify unbalance, misalignment or gear and bearing faults (see figure 7), b) time averaged Octave Band Spectrum Comparison can be trended much more easily than FFT data and c) more reliable spectral estimates will be obtained in the presence of random, impulsive and non-stationary signals. Also, the use of standardised frequency bands generates more universal statistics and simplifies the preliminary diagnosis of machine faults by less skilled personnel.

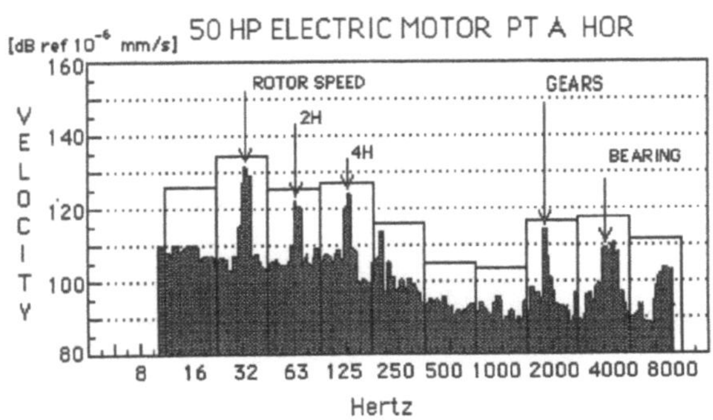

Figure 7. *4% and Octave band velocity spectrum of a small electric motor.*

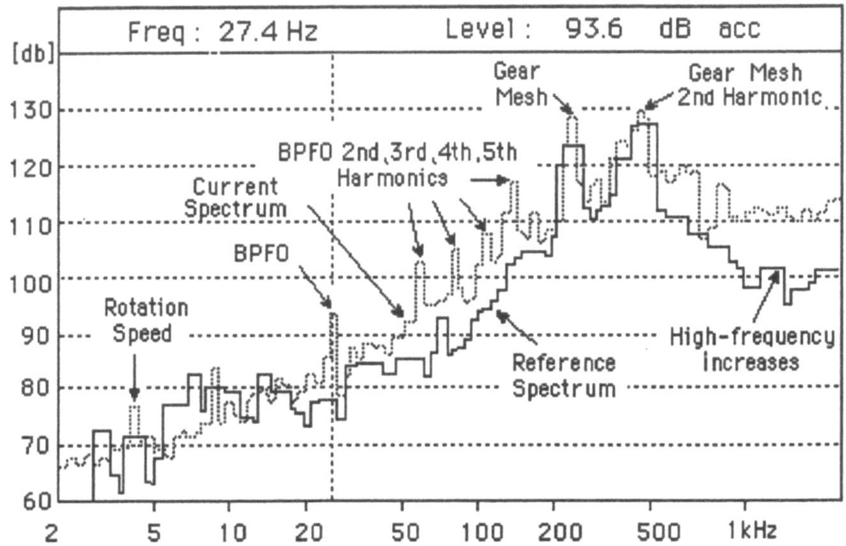

Figure 8. *CPB Spectrum Comparison showing changes in the spectrum for the same case as figure 4.*

Figure 9, shows that in the example of the dryer roll described above, Octave Band Spectrum Comparison is quite an effective fault detection technique, allowing to separate rotational speed component and its first few harmonics, bearing frequencies, gearmeshing components and high frequency signal change caused by the damaged bearing.

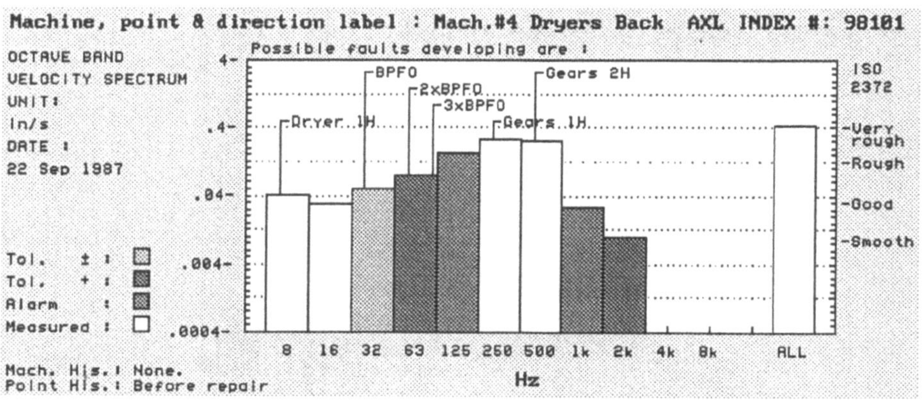

Figure 9. *Octave Band Velocity Spectrum Comparison for fault detection (same case as figure 8).*

5 TRENDING AND MACHINE PROFILES USING THE OCTAVE BAND VELOCITY SPECTRUM

The reasonable amount of data provided by Octave Band analysis allows more meaningful data to be represented on the same graph, thereby simplifying the detection, diagnosis and trending of faults. When weekly overall level measurements were replaced by monthly measurements of the Octave Band Velocity spectrum in a small chemical plant, the reliability of fault detection improved significantly, as well as much earlier warnings were obtained. Over 500 motors and pumps are monitored approximately once a month in this plant. On figure 10, is shown the evolution of a bearing fault on a small electric motor driving a pump. The use of a logarithmic frequency scale allows the visualisation of the component at the RPM of the motor (32 Hz band) as well as the high frequencies (8 kHz band) on the same spectrum. Just before the summer, large increases were found in the 1k, 2k and 4k bands and a bearing fault was diagnosed. However, no action was taken until after the summer holidays, when a sudden increase in the 8k Hz band on August 20th indicated that the fault had progressed and that the bearing was getting close to failure. It was changed at the subsequent shutdown, on September 27th, 1987, with the corresponding decrease in the 500, 1k, 2k, 4k and 8k Hz bands. This example illustrates the additional information provided by trending octave band data, when trying to estimate precisely the lead time to failure.

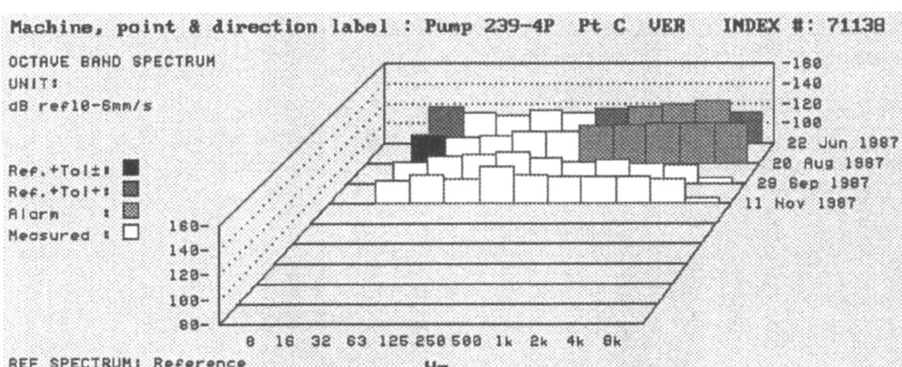

Figure 10. *3-D diagram of Octave Band Velocity Spectra versus time on an electric motor.*

Octave band data also lends itself very well to generate machine profiles such as the one shown below. This representation enables the machine inspector to get a better overall view of the vibration severity of a machine, as each band represents different categories of faults (figure 11). Changes in the profile are also very useful representations to the vibration analyst.

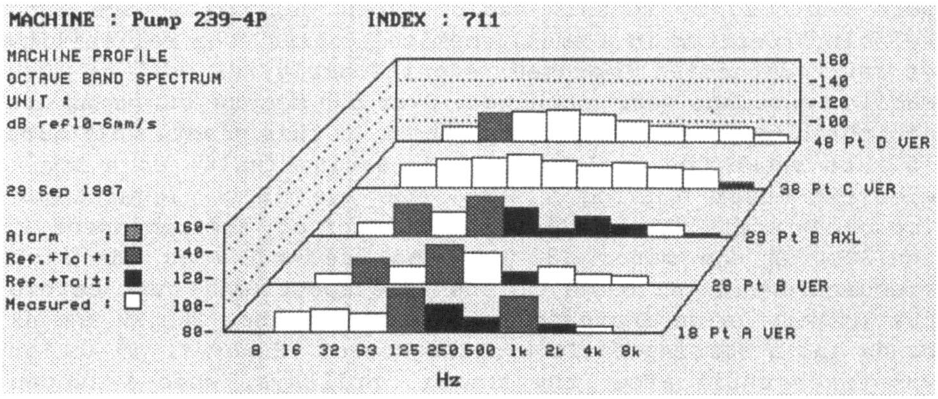

Figure 11. *3-Dimensional profile of Octave Band Velocity Spectra on an electric motor.*

Because its frequency bands are standardized, Octave band velocity spectra can be used as the basis to compute meaningful statistics on various categories of faults, such as the one shown below (figure 12), where the 31 Hz band corresponds to the rotational speed of a population of similar machines. From the distribution, it is possible to establish which machines deviate the most from the statistical profile.

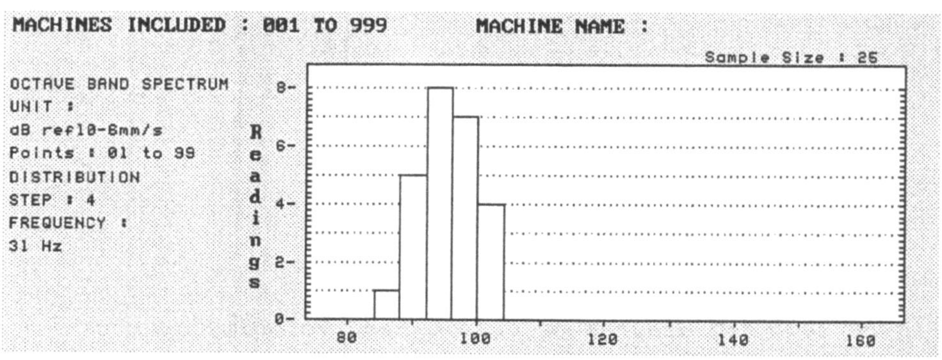

Figure 12. *Statistical profile for the rotational speed of a group of vertical pumps.*

6 SYNTHESIS OF CPB SPECTRA FROM FFT DATA

The synthesis of a CPB spectrum from FFT data does not always yield reliable results. Many factors render the operation difficult. Vibration data from machinery consists of deterministic, random, impulsive and non-stationary signals. The FFT handles the first type of signal very well, but for the other types, extra care must be taken by using special time windows, long averaging time, and special integrating cursors. Measurements of non-stationary signals requires extreme precautions, otherwise results could easily be off by more than 10 dB. One can be led into a false sense of security when measuring what appears to be a stable machine, as the introduction of certain types of faults will often create impulsive signals which, at a high sampling rate of the FFT analyser, will appear as non-steady signals. The limitations of FFT techniques when dealing with these non-steady signals comes from the fact that by its very nature, it operates on blocks of samples, instead of continuously processing the signal like analog or recursive digital filters, and one must insure by judicious choice of windows and sampling rates that there are no gaps in the data between the various blocks, otherwise important information could be lost. At high sampling rates, however, FFT analysers included in current data-loggers introduce very large gaps, as the speed of the processor is not able to cope with the high flow of incoming data.

Because CPB data covers a wide frequency range, and therefore requires high sampling rates, its measurement is not easily achieved with FFT techniques. Attempts to simulate octave band CPB data from FFT data can introduce significant errors due to windowing, asymmetry, non-standard slopes and ripple in the pass and stop bands. In order to obtain sufficient resolution at low frequency while still maintaining high frequency information, a very large number of lines would be required, as previously stated, or several analyses must be performed, with the consequent introduction of gaps in the data and potential for errors. In either case, if sufficient averaging is performed to obtain repeatable data (minimum BT product = 10 with stationary data and a much larger BT product if the data is non-stationary), the synthesis of a CPB spectrum using FFT techniques will be much slower than standard filtering techniques, with the potential risk of large errors.

7 RECOMMENDED APPROACH

Since fault detection and diagnosis impose such conflicting requirements on the instrumentation, the approach suggested in this paper is to divide the maintenance procedure into two parts, namely:-

a) DETECTION
&
b) DIAGNOSIS

A small portable, overall and octave band data collector with a built-in route capability is used to detect the faults (see figure 13). If a more elaborate diagnosis is required, then a real-time vibration analyzer incorporating zoom FFT, envelope analysis and cepstrum analysis is brought in to diagnose the fault. Especially with complex and critical machinery, the use of more advanced diagnosis techniques such as cepstrum and envelope analysis, gating techniques, intensity and operational deflection shapes is often warranted since regular FFT analysis does not always provide enough information to diagnose a fault. For a description of these powerful analysis techniques, please refer to the references. With this approach, equipment and personnel can be used more effectively and less skilled personnel can be brought into the process.

Many data-loggers available today perform both fault detection and diagnosis using FFT analysis. Unfortunately, some engineering compromises have to be made, usually to the detriment of the fault detection capabilities of the instrument.

Accurate transducers and proper mounting techniques are also essential in order to obtain repeatable data. Unreliable measuring techniques could invalidate the measurements, cause serious errors which could result in machine loss and may even discredit the vibration programme altogether. A scheduled shutdown of a machine that is not faulty will reduce the cost efficiency of a plant just as readily as unscheduled shutdown due to an undetected fault. Since vibration impedance paths may vary for similar machine by as much as 60 dB (1000 to 1), comparative rather than absolute measurement is recommended, the decibel unit being ideal for this purpose.

In an iron-ore mine in Quebec, the Mount Wright Mine, 4689 measurement points are monitored on various types of machinery e.g electric motors, gear trains, bearings, conveyors etc. Since 1987 the annual cost of repairs on electric motors alone has decreased from $181,800.00 in 1987 to $8,326.00. Depending on the type of machine, a different combination of acceleration, velocity and octave band velocity readings are collected and every point is trended in order to determine the condition of the machines in the mine and plan the maintenance schedules. From the experience gained in this mine and other plants throughout eastern Canada, it was found that the approach advocated in this paper is reliable and cost effective, and has resulted in substantial savings to the companies employing it. The basic approach is not new but the instruments and the manner in which the techniques are being implemented are.

A powerful fault detection capability is the back-bone of a good on-condition maintenance programme. Increasing the fault detection capabilities of their vibration measuring instrumentation by techniques such as Octave Band Velocity Spectrum Comparison has proven very profitable to many Canadian companies, even when advanced diagnosis capabilities were already at hand. In the 90's, techniques such as Octave Band Velocity Spectrum Comparison will probably spread throughout Canadian industry, enabling many companies to implement new maintenance philosophies based on the concept of monitoring and improving the running quality of rotating machines.

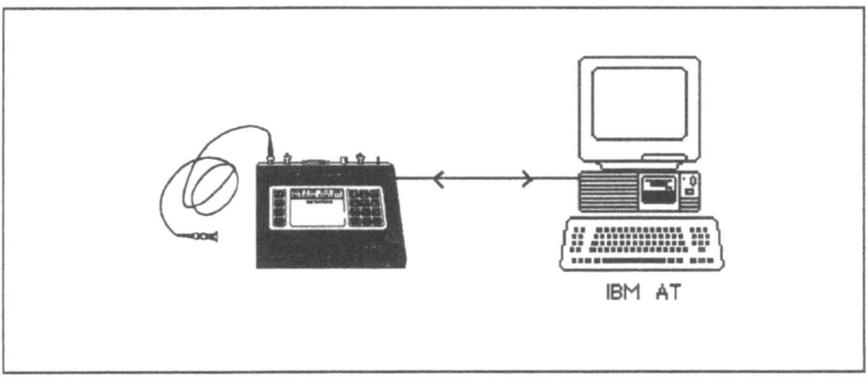

Figure 13. *Small system based on an Octave Band data collector and an IBM PC compatible computer.*

7 REFERENCES

Brown D.N. & Jensen T, The use of Spectrum Comparison for Bearing Fault Detection - A Case Study from Alma Paper Mill, Quebec, Canada, Bruel & Kjaer Application Note,1989.

Courrech Joelle and Gaudet Mark, Envelope analysis - the key to rolling-element bearing diagnosis, Bruel & Kjaer, 1987.

Glew C.A.W & Watson D.C, Vibration Analysis a Maintenance Tool at Sea, Institute of Marine Engineers,1970.

Glew C.A.W, The Effectiveness of Vibration as a Maintenance Tool, Institute of Marine Engineers, 1974.

Randall R.B., Cepstrum Analysis and Gearbox Fault Diagnosis, Edition 2, Bruel & Kjaer, 1986.

Serridge, M. Fault Detection Techniques for Reliable Monitoring, Application Note, Bruel & Kjaer.

8 ACKNOWLEDGEMENTS

The author is indebted to Bruel & Kjaer (Canada) Ltd & also to Mr. R. Archambault of Bruel & Kjaer (Canada) Ltd for his advice during the preparation of this paper.

Tool wear monitoring sensors

K.H. Hale and B.E. Jones
The Brunel Centre for Manufacturing Metrology, Brunel University, Uxbridge, England

Abstract
Tool breakage and uncontrolled wear are the main reasons for poor quality in machining of discrete components and expensive secondary costs. On-line tool wear monitoring is thus desirable in an unattended manufacturing system. It is now considered necessary to employ several types of sensors measuring different parameters associated with tool wear and to pass the information to an expert system which can make a judgement based on many factors. The environment of production line machinery is electrically very noisy and for this reason it may be desirable for many of the electrical type sensors to be replaced by optical equivalents. This paper reviews sensors suitable for tool wear monitoring.
Keywords: Tool wear, Monitoring, Sensors, Acoustic Emission, Expert System.

1 Introduction

Detection of the end of useful cutting tool life is necessary to initiate tool replacement to ensure maximum efficiency of automated advanced manufacturing systems. Also prediction of tool failure during machining is required because unexpected failure can cause damage to both machine tool and workpiece. Many different monitoring techniques have been tried and they may be classified into direct and indirect measurements.

Direct measurement

1. Optical and fibre optics with TV cameras and CCD arrays
2. Electrical resistance sensors
3. Radioactive sensors
4. Pneumatic sensors
5. Touch probes

Indirect measurement

6. Cutting force, thrust, torque, and power sensors
7. Vibration and sonic analysis (noise)
8. Acoustic emission.

All of these types of sensors have existed in electrical form since conception and are available commercially, but more recently optical fibre equivalents have been developed in prototype form. Optical fibre sensors can operate in hostile environments that cause problems for electrical sensors.

2 Optical

The simplest optical monitoring technique is to view the moving cutting edge with a synchronized stroboflash through a low power microscope. Progressive wear can be seen quite clearly because of the higher reflectivity of the wear land compared with the uneven surface.

Imaging is made easier by the use of a TV camera and displaying the image on a TV monitor screen. To overcome the problem of having to use bulky optical components between the cutting edge and the TV camera, a bundle of optical fibres (endoscope) can be used instead. Finally, to simplify the task of quantifying the tool wear, charged coupled device (CCD) arrays or cameras can be used to divide the image into individual pixels with areas of about fifty square microns. Thus the reflectivity or the profile of the worn tool can be digitized and a judgement can be made on the extent of tool wear based on numerical information.

Work on these systems is progressing in many laboratories but their widespread use has been delayed because the price of CCD systems has only recently fallen to an acceptable amount.

3 Electrical resistance

A change in electrical resistance due to wear of a cutting tool can be used to generate a change in an electrical output signal. The change in electrical resistance occurs because the contact area between the tool and the work piece increases with tool wear. There have been problems with resistance changes being caused by variations in cutting speed, feed and cutting forces.

Another system is to deposit a thin film resistor on the cutting tool either by printing, for example graphite ink, or evaporating chromium and using a thin insulating film of heat resisting paint. The additional cost and availability of tools with surface resistors together with additional setting up procedures and costs are obstacles to their use.

4 Radioactive sensors

An alternative technique to monitoring electrical resistance changes with wear is to monitor changes in radioactivity levels of a thin film of radioisotope deposited on the tool. A decrease in detected radioactivity means that the tool has worn and so a correlation can be obtained between radioactive particle count and tool wear. Although a good idea in principle, radioactivity in normally nonradioactive environments is not advisable practice.

5 Pneumatic sensor

Pneumatic gauging is an established technique for dimensional measurement. A suitable jet of low pressure air is directed at the surface to be measured and the change in back pressure with changing nozzle to surface separation allows dimensional changes to be measured with accuracies of the order of $\pm 1\ \mu m$. A refinement is to transform a pressure change into an electrical signal which operates a stepping motor to keep the nozzle to surface separation constant. The displacement of the stepping motor/nozzle is a direct indication of tool wear.

6 Touch probes

Commercially available touch probes have now become very sophisticated but at the same time are easy to use with short setting up times. Essentially the probe is a simple switch, although a highly accurate one. Ruby ball stylii are suitable for most applications, being highly visible against metallic surfaces. A highly spherical (better than 0.15 μm) industrial ruby ball is drilled and bonded on to a nonmagnetic stainless steel stem. When the stylus contacts a surface, a trigger signal is transmitted to the machine control. A kinematic location ensures high repeatability, relocating the probe stylus to within 3–5 μm (2 sigma). Unidirectional repeatability (2 sigma) is 1 μm or better, (0.25 μm at best). The probes are ideal for detecting tool wear or tool breakage when the tool holder is stationary. Electrical data transmission from the probe can be replaced by an optical, cable free system.

7 Cutting force, thrust, torque, and power sensors

Studies of the correlation between cutting forces and tool wear have been in progress for about twenty–five years. Although the actual details of tool wear are very complicated as are their contributions to increasing cutting forces, it has been possible to develop several commercially available systems which monitor tool wear. One such system uses a feed force sensor which converts the mechanical forces generated at the tool cutting edge during machining into electrical signals by means of strain gauges. The nominal force range is from 1 kN to 100 kN, the linearity is ± 1% and the sensitivity is about 1 mV/V. It has been found that the axial force component increases significantly as an initially sharp tool wears.

Another type of sensor is the plate sensor available for turning, drilling and special metal cutting machines. The plate sensor comprises several sensitive zones each containing complex strain gauge configurations. The design and positioning of these zones provide optimum feed force measurement. The plate sensor is installed close to the cutting edge, directly in the path of the transmitted feed force. For example it is installed between the turret and cross–slide in a CNC turning machine to provide monitoring of all cutting tools in all axes. The operational load range and sensitivity is similar to that of the cylindrical force sensor

A popular strain gauge sensor for use on spindle driven cutting machines is the torque sensor. However, in tests carried out to compare increases in axial feed force and torque as a function of tool wear, the increase in axial feed force using the same tool is more clearly defined.

Another sensor is the electrical current sensor. When the cutting edge of a tool becomes worn or broken, a change in power consumed by the feed motor normally occurs. A single conductor from the power cable supplying the feed motor is fed through the current sensing ring (one or more turns). The current carried by the conductor passing through the sensor ring is transformed into a proportional voltage signal. This voltage results in a sensor output signal which is relative to the feed force. For drilling and turning applications, monitoring the feed forces in this way will normally provide a good indication of the tool condition. It is also possible to monitor the torque exerted on the tool by the current sensor but this may not be so successful for monitoring tool condition.

The sensors with their signal processing instrumentation continuously monitor several alarm levels:

(a) Level for tool wear. When the signal exceeds a set level this alarm can be used to initiate a tool change at a convenient time after completion of an operation.
(b) Higher levels for tool breakage and crash protection. This alarm should be used to stop the machine immediately as soon as the signal exceeds appropriate set levels.
(c) Lower set level for in–cut detection. This alarm is used to monitor tool in–cut or not in–cut. This may be a tool or a component missing. Either a warning can be given to an operator or the machine can be shut down.

The tool monitoring system "learns" during machining of the first work piece when the tool, or in the case of a multi–channel system when the tools are sharp. In the learn mode, the system will automatically store the peak value of the sensor signal for each tool or operation. These values are stored in the memory until required for a subsequent identical machining operation in monitoring mode. New values can also be manually entered into the tool memory or the previous values changed from the front panel controls. During monitoring, these stored peak values form the fixed references from which the tool monitor sets the various alarm levels. The tool wear alarm signal can be given a relatively long reaction time (normally 0.1 to 1 s) while for tool breakage a short reaction time can be set (normally 1 to 10 ms). Electrical systems are available from several suppliers.

Optical fibre versions are under development in several laboratories..

8 Vibration

The measurement of vibration is widely used for the condition monitoring of the bearings of rotating machinery. Vibration is produced by worn bearings because of lack of balance. Most commercial vibration meters are equipped to measure acceleration, velocity and displacement. The piezoelectric accelerometer is the universal transducer for vibration measurement. It has very wide frequency and dynamic ranges with good linearity, is robust and remains stable over long periods. Acceleration measurements are used for high frequency and small amplitude vibrations; displacement measurements for low frequency and large amplitude vibrations and velocity measurements are best for intermediate frequencies between 10 and 1000 Hz.

There have been numerous investigations on the correlation between vibrations of a machine tool and tool wear. The piezoelectric sensor can be fastened to a suitable part of the machine by a magnet or adhesive. The natural frequencies of the transducer (49 kHz) and the fastening system (20 kHz) are high enough not to interfere with the useful monitoring frequency (2.5 kHz).

There are also commercial non-contacting vibration sensors. They are based on Mach-Zehnder or Michelson interferometers. Light from a laser is directed at the vibrating surface and the reflected light is then received by instrumentation which detects the Doppler frequency change.

9 Acoustic emission

Acoustic emission (AE) can be defined as the transient elastic energy spontaneously released from materials undergoing deformation, fracture or both. Such deformation or fracture occurs in materials under stress hence the phenomenon is also referred to as stress wave emission.

The energy contained in the AE signal is strongly dependent on parameters such as the rate of deformation (strain rate), the applied stress and the volume of participating material. A process can therefore be monitored using AE if process characteristics can be directly related to one or more of the parameters.

Five potential mechanisms of AE in metal cutting operations have been proposed.

1 plastic deformation in the shear zone;
2 chip sliding in the rake face;
3 chip fracture for discontinuous chip formation
4 chip entanglement with the tool;
5 tool work rubbing on the flank face.

It is evident that tool wear has two main effects on the above mechanisms. First, wear changes the tool geometry through flank, crater wear and chipping or deformation of the cutting edges, thus affecting the mechanics of the metal cutting process. Secondly, the changes in tool geometry will be accompanied by an increase in chip-tool work rubbing.

Signals recorded during cutting operations can be analysed in several ways:

(a) time evolution of a characteristic usually representing an evaluation of energy typically RMS, total energy, cumulated number of counts, cumulated number of bursts;
(b) statistical distribution of bursts according to a given parameter, e.g., amplitude distribution, rise time distribution, duration distribution, energy distribution.

The majority of commercial AE sensors are piezoelectric devices, employing lead-zirconate-titanate (PZT) as the piezoelectric (active) element. Piezoelectric sensors may be broad-band or resonant devices. Recent developments include piezoelectric film (polyvinylidene fluoride) to replace the PZT element. PZT sensors are extremely sensitive devices with quoted resolutions of 10^{-12} m for broad band (50 kHz to 1MHz) and 10^{-13} m for resonant devices. Piezoelectric film with suitable brass backing is capable of a flat frequency response between 1 and 20 MHz and has considerable potential for tool wear monitoring applications.

Optical transducers for AE monitoring are still very much at the development stage. They are attractive because of the possibility of using both contacting and non-contacting measurement probes and because of their potential for flat frequency response over a very large bandwidth. In addition certain optical configurations offer direct displacement calibration in terms of wavelength of light. They are also insensitive to electromagnetic interference. Bulk optical devices are not sufficiently robust for machine shop applications but optical fibre sensor equivalents would be.

A non-contacting optical fibre Michelson interferometer is undergoing trials. Measurements have been made of amplitudes of 10^{-9} m, the resolution is 10^{-11} m with a flat response spectrum to 10MHz. The interferometer has already been applied to the detection of AE generated by face milling.

10 Design of the expert system

An expert system is a computer program using integral domain knowledge and reasoning strategies to solve problems normally requiring expertise.

An expert system usually consists of the following components:

(a) the data base containing data structures from which conclusions can be drawn - these data structures may be static, that is unchanging, or dynamic, that is capable of being updated;

(b) the knowledge applied to the data;

(c) the inference engine consisting of routines that are used to follow a certain rule-selection strategy in order to arrive at the desired conclusion or solution.

The expert system used for tool wear monitoring must be able to absorb data from many different types of wear sensors, to assess probabilities of true wear from each sensor and to make a decision as to whether a cutting tool has reached the end of its useful life time and should therefore be replaced by a new tool. There is a choice of several expert systems on the market.

11 Conclusions

There are many different types of cutting tool wear sensors but none of them in isolation is able to give an unambiguous decision as to whether a tool has reached the end of its useful life. A number of useful sensors both available now and still in prototype form have been briefly described which used together with an expert system, could form a reliable tool wear monitor.

12 Acknowledgements

This work is part of a BRITE project. The authors wish to thank the European Commission, project partners and industrial collaborators for support.

System detection instrumentation

I.A. Henderson and J. McGhee
Industrial Control Unit, University of Strathclyde, Glasgow, Scotland

Abstract
Binary signals have advantages over other test signals in their ease of microcomputer generation and application to a component, a control system or a process. **M**ulti-frequency **B**inary **S**equence, or MBS test signals offer the advantages of excellent signal-to-noise power ratios for their dominant harmonics, zero offset and a wide variety of power spectral distribution. Compact MBS add powerful new interrogation signals for monitoring a wide range of industrial systems, or their constituent parts, using frequency estimates. An analogy of optical zooming has led to the development of compact zoom MBS which may be applied when narrowband detail with a high spectral resolution is required. The accurate estimation of the open-loop gain and phase crossover may be crucial in the assessment of the condition of a process. In this paper, a prototype system detection instrument is used to test a component, a closed-loop control system and an open-loop process using both wideband and narrowband frequency estimates.
Keywords: System Identification, Condition Monitoring, Binary Test Signals, Multi-frequency Binary Sequences, Automatic Frequency Response, Control Systems.

1 Introduction

The concept of an identification channel and the application of information theory to system identification has resulted in two identification coding theorems. These theorems have been responsible for the discovery of new compact **M**ulti-frequency **B**inary **S**equence test signals (Henderson et al., 1987). Baseband versions of these signals concentrate the test signal energy in a known relatively small number of dominant harmonics which include the fundamental. In contrast to **P**seudo **R**andom **B**inary **S**equences, or PRBS, the higher energy per MBS dominant harmonic allows a lower magnitude of binary interrogation of the process under the same noise conditions. When compact MBS are used by a monitoring instrument, they will ensure minimum interference in the normal operation of the process. They are also ideal computer generated monitoring signals where the calculated frequency estimates may be used to establish satisfactory working of the process.

Previous MBS microcomputer test signals have been successfully applied to reactors (Buckner and Kerlin, 1972; Chen et al., 1972), an electrical resistance furnace (Plaskowski and Sankowski, 1984) and a plastics extruder (Bezanson and Harris, 1986). The channel identification coding theorems have presented clear design guidelines in terms of the compactness of the generating code and maximisation of the power per component of specified dominant harmonics. This ensures fast, economical and efficient digital generation (Henderson et al., 1987) and signal processing (McGhee et al., 1987).

Signal engineering techniques have allowed Strathclyde compact baseband MBS codes (Henderson et al., 1987) to be highlighted. Their dominant harmonics have either octave or decade frequency relationships. Recently, the design of maxent zoom phase shift keyed signals has been proposed by Henderson and McGhee (1988), and applied to control systems (Henderson and McGhee, 1990). These use phase shift keyed modulation with a finite squarewave carrier to concentrate the test signal power in a narrow bandwidth about a specified central frequency. A zoom factor determines the size of the bandwidth to be highlighted.

These novel compact MBS signals are the basis of the design of **SY**stem **D**etection test equipment for generating and signal processing MBS test signals. In this paper, a prototype **SYD** is used to evaluate both baseband and zoom frequency estimates for a dc servo motor, a closed-loop dc servo and an open-loop warm air process. Their suitability for condition monitoring of a process or an important component is discussed.

2 Generating codes and compact MBS

The binary version of the generating code of an MBS test signal has two symbols "1" and "0", which correspond to a chosen DAC output voltage level of + V or - V respectively. As a **max**imum **ent**ropy or maxent sequence must have the same number of "1" and of "0" symbols in the binary code, the signal will not alter the steady state condition of the process. In a time-domain format, the maxent code defines the zero crossover. Hence, the generating code may be also given in a decimal zero crossover form. As an example, the full code for the 16 bit compact short octave MBS is 1110001001000111 or 3,3,1,2,1,3,3, if it is written in its decimal form.

Although the possibility of both asymmetrical and symmetrical maxent codes exist, previous work has concentrated on the neater symmetrical coding. A code with an axis of even symmetry has $f(t) = f(-t)$ while one with an axis of odd symmetry has $f(t) = -f(-t)$. Each code change over approximately corresponds to the zero crossover of an equivalent composite test signal with a sum of fundamental and defined harmonics of either sine or cosine terms. The associated phase shift is 0 or 180 degrees and is easily accounted for by a plus and minus sign. Any rotate right or left of the bits of the binary code alters the phase angles of the spectral signature. Although the generating code is in one of its many asymmetrical forms, the power spectral density is not changed. Most symmetrical maxent codes have two axes of either even or odd symmetry. However, some codes have dual symmetry, which means they have four axes of symmetry.

Modulation by digital phase shift keying may be applied to a finite squarewave carrier. The modulated signal as well as the carrier and the modulating signal are all maxent binary codes. A psk binary sequence has a spectral signature which consists of upper and lower side bands about a central suppressed carrier frequency. In this way, the test signal energy is concentrated into twice the original number of dominant harmonics which are available with the chosen modulating signal. It is possible to detail a portion of a frequency response record in a manner similar to that of an optical zoom lens with the aid of a zoom factor. This is achieved by making the carrier have a 2^n zoom factor, Z, which is the number of bits per modulating code bit. Zoom MBS test signals allow narrowband frequency estimates with high spectral resolution to be made (Henderson and McGhee, 1988; 1990). These are important new signals for monitoring sensitive narrowband frequency information.

3 An instrument for system detection

The design of the test equipment for the software evaluation of frequency estimates depends on the universal nature of the identification coding theorems. A calibration procedure is included in the design of SYD. This allows the DFT algorithm (McGhee et al., 1987) to be used to obtain the spectral signature of the chosen test signal. It also acts as a check on the operation of the instrument.

Once the test signal is chosen only one fast calibration run, which is independent of the time required for the test run, is required. When the unknown system is tested, the spectral signature of the test signal is modulated by that of the identification channel. As well as the VDU display, a hard print copy of the test signal details and the magnitude and phase of output/input for the system under test is available. The frequency information may be taken as plotted points in either a Nyquist, inverse Nyquist, Bode or Nichols plot.

Waveform, eye pattern, generating code, percentage binary power contained by the dominant harmonics, average percentage power per component and other details of the compact MBS test signals are required. These details of the MBS test signals, which are used to obtain baseband and narrowband frequency estimates, are given by Henderson and McGhee (1988; 1990). As minimum signal levels are of special importance for diagnostic test signals, an examination of the dominant harmonic signal-to-noise power ratios is required. However, in order to make comparisons a standard is required. It must be realised, that the poorest signal-to-noise power ratio at all frequencies is obtained, when the signal is emulating white noise. PRBS test signals meet this requirement and they provide the necessary standard for a comparison to be made. The improvement in dBs, S_n, above the signal-to-noise ratio for a PRBS may be evaluated at MBS dominant harmonic number, n. A PRBS signal with the minimum number of bits, which will include the MBS dominant harmonics in the signal processing is used. Assuming the binary signals have the same amplitude and process noise, the values for S_n will allow comparisons to be made between the infinite variety of MBS test signals. Hence,

Table 1. Harmonic signal-to-noise improvement of compact MBS over PRBS

Compact MBS	Name	Dominant harmonic improvement on PRBS, S_n (dB) (where n = harmonic number) * **not used**
Baseband	Dpsk2-Z2	S_1 = 9.20, S_3 = 6.25, S_5 = 2.88, S_7 =-13.60
Component		
Zoom	Dpsk2-Z16	S_{29} = 5.17, S_{31} = 14.30, S_{33} = 13.87, S_{35}= 3.89
Baseband	Dpsk3-Z2	S_1=-10.81*, S_3 = 5.60, S_5 = 8.41, S_7 = 1.06 S_9= - 0.62, S_{11} = 3.06, S_{13} = -4.60, S_{15}=-30.7*
System		
Zoom	Dpsk3-Z8	S_{25} =-9.0*, S_{27} = 8.79, S_{29} = 12.82, S_{31}= 6.64 S_{33} = 6.22, S_{35} = 11.54, S_{37} = 6.65, S_{39}=-12.1*
Baseband	Octave Extended	S_1 = 5.00, S_2 = 4.77, S_4 = 4.34, S_8 = 4.42 S_{15} = 6.79, S_{16} = 2.90, S_{24} = 3.25, S_{30} = 4.29
Process		
Zoom	Dpsk2-Z16	S_{28} = 8.89, S_{30} = 8.48, S_{31} = 9.85, S_{33} = 9.43 S_{34} = 7.63, S_{36} = 7.18

$$S_n = 10 \log_{10} [P_{n\ MBS}/(P_{n\ PRBS}] = 20 \log_{10} [E_{n\ MBS}/(E_{n\ PRBS}]\ \text{dB} \qquad (1)$$

Table 1 gives the values of S_n for the compact MBS test signals used in this paper. This table clearly indicates the future potential of compact MBS and their frequency estimates in the condition monitoring field.

4 Condition monitoring using MBS frequency estimates

To illustrate the potential of these signals for condition monitoring

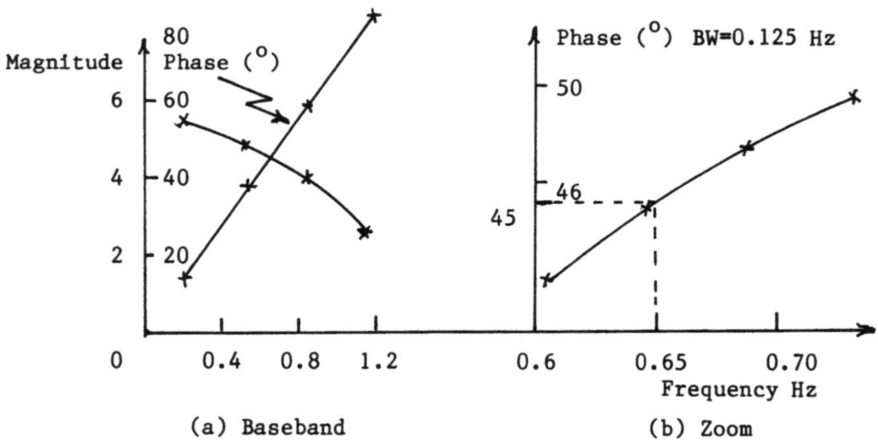

Fig. 1. Frequency responses of a dc servo motor

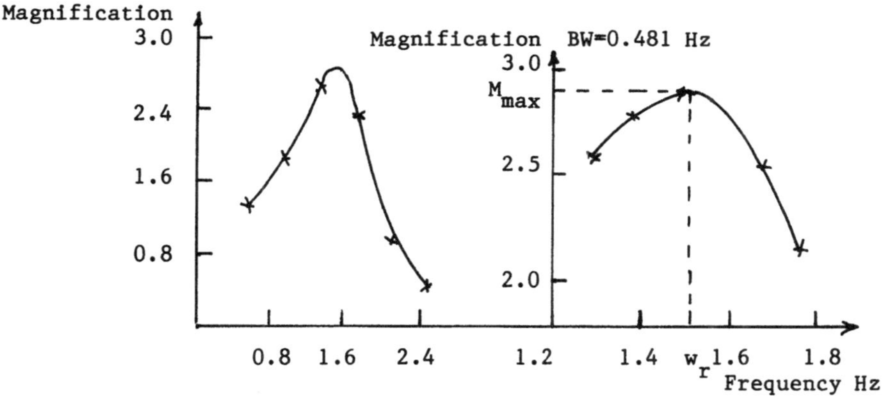

(a) Baseband (b) Zoom

Fig. 2. Frequency responses of a dc position control system

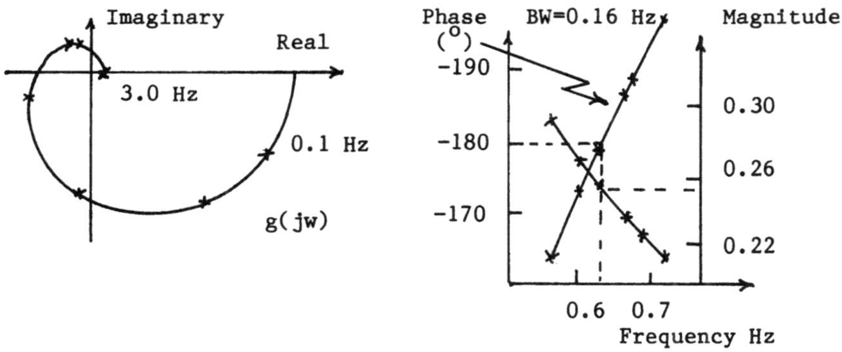

(a) Baseband (b) Zoom

Fig. 3. Frequency responses of an open-loop warm air process

by a digital computer, the experimental results for a component, a system and a process are shown in Fig. 1, 2 and 3. The initial diagrams give baseband responses. However, all systems have an important sensitive portion of their frequency response and three examples are illustrated in the second diagrams. The names of the MBS signals, which are used in these experiments, are given in Table 1 and full details may be found in Henderson and McGhee (1990).

In Fig. 1, the component investigated is a dc servo motor with the output signal taken from a tachogenerator which is attached to the motor shaft. The 45 degree phase shift which allows the single time constant to be monitored is emphasised in Fig. 1(b). Fig. 2 examines a very underdamped closed-loop dc servo. The detail of the peak of the closed-loop magnification is given in Fig. 2(b). In Fig. 3 an open-loop warm air system is examined and the phase crossover and gain margin may be estimated from Fig. 3(b). All the experimental results involve real systems. A simulation software package has also been developed. Table 1 illustrates the excellent signal-to-noise power ratios obtained by digital phase shift keyed zoom MBS. This is confirmed in the accurate detail which is obtained from the three zoomed frequency estimates.

5 Conclusions

The compact MBS test signals developed in the Industrial Control Unit at Strathclyde University offer new opportunities for testing and condition monitoring. In particular, compact zoom MBS have been shown to be excellent signals which highlight sensitive features of the frequency response. Their good signal-to-noise power ratios for the dominant harmonics is especially valuable. If a process has been shown to accept a step (not manual) or a PRBS input, MBS test signals should also be acceptable with no additional difficulties. They add even greater interrogation power for the microcomputer evaluation of frequency estimates. Although further applications are required, these novel signals must offer many new opportunities for condition monitoring of control systems and their constituent parts.

6 References

Bezanson L.W. and Harris S.L. (1986) "Identification and control of an extruder using multi-variable algorithms" IEE Proc. Pt. D, 133, PP. 145-152.

Buckner M.R. and Kerlin T.W. (1972) "Optimum binary signals for reactor frequency response measurements" Nuc. Sc. & Eng., 49, pp. 255-262.

Chen C.H., Kerlin T.W. and Fry D.N. (1972) "Experiences with binary periodic signals for dynamic testing at the HFIR" IEEE Trans. on Nuc. Sc., 19, No.1, pp. 828-835.

Henderson I.A., Ibrahim A.A., McGhee J. and Sankowski D. (1987) "Assembler generated binary test signals for process identification" in Microcomputer Application in Process Control, IFAC Proc. Series, 1987 No. 7, Pergamon Press, pp. 77-82.

Henderson I.A. and McGhee J. (1988) "PSK maxent MBS test signals for narrowband electro-thermal identification" Preprints IMACS 88, 12th World Congress, Paris, 3, pp. 351-353.

Henderson, I.A. and McGhee, J. (1990). "A digital phase-shift-keyed technique for narrowband system identification" to be published in Trans InstMC.

McGhee J., Fisher G. and Henderson I.A. (1987) "A fast DFT algorithm for on-line MBS process identification" Preprints of 7th Int. Conf. on Cont. Sys. and Comp. Sc. (CSCS), Bucharest, pp. 111-128.

Plaskowski A. and Sankowski D. (1984) "The use of multi-frequency binary sequences (M.B.S.) to on-line identification in electro-heat" IMACS European Simulation Meeting, Eger, Hungary, pp. 285-296.

Real-time model and kinematic control of machine tools

C.Y. Li and C.R.Zhang
Department of Mechanical Engineering, Shandong Polytechnic University, Peoples Republic of China

Abstract
A computer software compensating technique is developed to acquire higher accuracy of machine tools. An improved variable step least-mean-square (VSLMS) adative algorithm is proposed for estimating the parameters of an autoregressive AR(p) model, then, a MCS-51 single chip computer-based control system is projected to implement the above algorithm. The experimental results on model Y38 gear-hobbing machine indicate that about 65% of stochastic error can be compensated for.
Keywords: Machine tool accuracy, Error correction, Real-time control, Forecasting model, Adaptive estimation algorithm.

1 Introduction

Accuracy requirements in manufacturing industries are becoming very stringent due to machine and equipment's development towards high speed, big power and full automation. Consequently, higher accuracy requirements to the machine tools are needed. Machining accuracy of parts may be affected by many factors, of which the transmission error of machine tools is main one of them. With the extensive applications of microcomputer, computer aided error compensation and correction (CAEC) technology has been developed for improving the machining accuracy. A lot of work has been done by many researchers in the field of CAEC, especially, after the methods of modelling and forecasting are introduced into CAEC, the time lag of system between the compensating action and occurrence of error can be eliminated and the determinstic and stochastic errors can be compensated at the same time. So the good compensating effect has been achieved. But up to now modelling in most of research work is still off-line (H. Z. Bin et al. 1984, Li C. Y. et al. 1987, Q.X. Zhou et al. 1986) because of complexity of modelling algorithms, in other words, the model of error is first fitted off-line from the batch error data by microcomputer, and the fitted model is then transferred into another control-purposed computer to forecast and compensate error on-line from the measuering error data. In this way there are three main defects:

(1) Two computers are needed, which may be make the system be too expensive for the practical purposes.

(2) It is necessary to fit a new model when the machining conditions change, because error model (order and parameters) is changeable due to variety of cutting conditions. Thus, there is too much modelling work to be done and it is impractical.

(3) The model which is fitted from the error data without control action (off-line) will not well adapt the error data after given control. So reduction of error will be limited. In order to overcome the above shortcomings, a new real-time model and forecasting compensation technique is used in this paper. A VSLMS algorithm is proposed and implemented by mean of a MCS-51 SCC. The experiment conducted on Y38 gear-hobbing machine indicated that about 65% of short periodical error can be compensated for.

The paper will be described as follows: In next section we will introduce briefly a VSLMS algorithm; section 3 is devoted to the description of the new method and system; in section 4 we depict the experimental results on Y38 gear-hobbing machine.

2 An improved VSLMS algorithm

2.1 LMS algorithm

The LMS algorithm was developed by Widrow-Hoff in 1959. The updating formula of AR(p) parameters (weight vector) in the LMS algorithm is expressed by

$$\mathbf{W}_{j+1}= \mathbf{W}_j+2UE_{fj}\mathbf{X}_j \tag{1}$$

forward predictiod error $E_{fj}=x_{j+1}-\mathbf{W}_j^{\tau}\mathbf{X}_j$ (2)

where $\mathbf{W}_j^{\tau} = (w_{1j}, w_{2j}, \dots , w_{pj})$

$\mathbf{X}_j^{\tau} = (x_j, x_{j-1}, \dots , x_{j-p+1})$

The p different weight relaxation time constants τ_i and misadjustment M of LMS algorithm are given by [E.Walach et al. 1984]

$$\tau_i =1/2U\lambda_i \qquad (i=1, 2, \dots , p) \tag{3}$$

and

$$M = U \sum_{i=1}^{p} \lambda_i=\frac{1}{2} \sum_{i=1}^{p} 1/\tau_i \tag{4}$$

where λ_i is a eigenvalue of autocorrelation matrix R_x of stationary time series X.

2.2 An improved VSLMS algorithm

From the expression (3) and (4), we can see the constant step U in the LMS algorithm controls the convergence rate of weights but also determines the misadjustment. A large U will get fast convergence, however, will also result in increased misadjustment or error residual. We can also know that the effects of step U on τ_i is before convergence and on M after convergence. If we can find a variable step U_j which decreases corresponding to the convergence, the algorithm will obtain fast convergence and small misadjustment. Since the square of the forward error E_{fj}^2 decreases as the convergence, we let the variable step U_j changes according to E_{fj}^2, hence

$$U_j = U_{j-1}F + E_{fj}^2K = K\sum_{i=1}^{j} E_{fj}^2F^{j-1}+U_0.F^j \tag{5}$$

where F, step forgetting factor, $0 < F < 1$; U_0, primary step; K, step coefficient,which keeps U_j not too large, assures the convergence of the algorithm.

If the constant step U in the formula (1) is replaced by variable step U_j, we get an improved LMS algorithm, i.e. VSLMS algorithm.The VSLMS algorithm has almost no increase in computation and storage space compared to the constant step LMS algorithm. Theory analysis verifies that the VSLMS algorithm has a faster convergence rate and smaller misadjustment than the LMS algorithm with $0.8 < F < 1$, when the signal is disturbed by white noise of uniform or normal distribution.We will not give the proving process here because of the limited pages.

The performances of LMS, VSLMS algorithms are compared with by computer simulation of learning curves.The signal is generated by the below model:

$$X(t)=\sin(2\pi.0.1.t)+a_t$$

where a_t is a zero mean Gaussian white noise with signal-to-noise ratio (SNR) 3 db. The coefficients are chosen as: U=0.003, U_0=0.015, F=0.96, K=0.001 and the order of AR model p=8. Fig. 1 is the learning curves got from the average of 60 forward error curves. As a result of simulation, it has been found that VSLMS algorithm outperforms the constant step LMS algorithm. The former has much faster convergence rate and smaller misadjustment than the latter.

3 Description of the new method and experimental setup

A schematic diagram of the experiment setup is shown in Fig. 2. The error signal of machine tool from the seismometer on the table of the machine,through an amplifier, is converted into digital data by A/D converter of computer subsystem, in which an adequate time AR(p) model is fitted from the data and the parameters Wj of AR(p) are estimated by VSLMS algorithm adaptively. The model is then used to

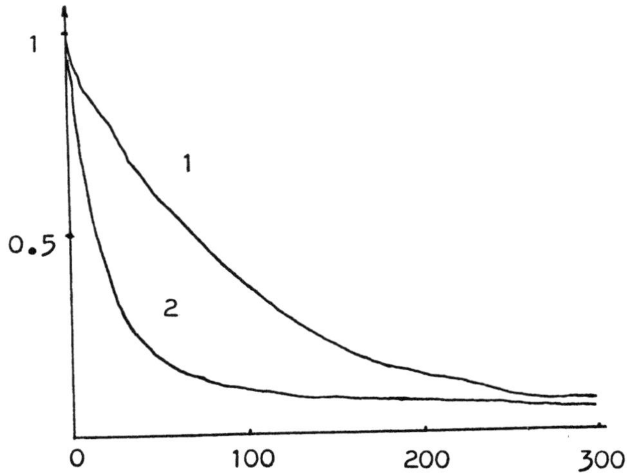

1-LMS 2-VSLMS

Fig. 1 Comparison of learning curves

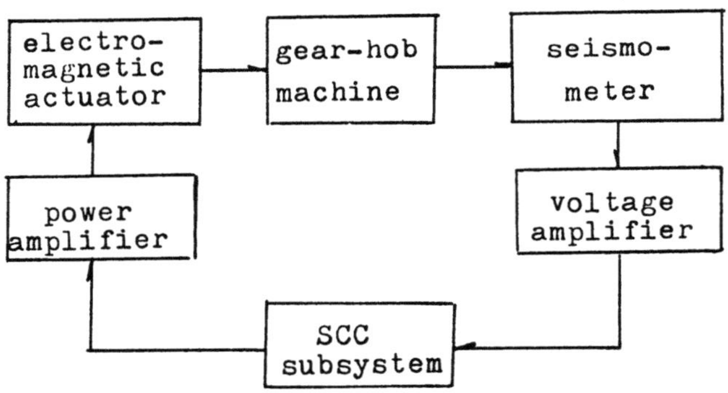

Fig. 2 Schematic diagram

forecast one step ahead error of the machine tool, i.e.

$$x_j(1)=w_{1j}x_j+w_{2j}x_{j-1}+\ldots+w_{pj}x_{j-p+1} \tag{6}$$

where $x_j(1)$ is the predicted value of X at (j+1)th moment. the predicted value is further converted into analogue voltage sending to the special electromagnetic actuator so that the error of next moment can be compensated and reduced. Equation (6) provided one step ahead predicted value $x_t(1)$ of x_{t+1} at jth moment. Prediction makes it possible to eliminate the lag of the system. The control action is added to the movement of the machine table through the differential shaft of the gear-hobbing machine. The procedure is as follows:

(1) obtain jthsampling value of error.

(2)estimate or modify the AR(p) parameters W_j adaptively based on the p sampling data before the jth moment.

(3) forecast one step ahead error at jth moment.

(4) compensate the error according to the prediction error to keep the correct movement between the table and the hob shaft.

4 Compensation experiment

The compensation experiment was conducted on Y38 (800mm) gear-hobbing machine under the race operation. The stochastic error of hobbing machine is composed of a lot of harmonic components and the frequencies of error are higher, because of complex of transmission chain in hobbing machine. So gear-hobbing machine is a classic experimental object for compensation. The experiment work was carried out under various operation conditions and different model parameters. To keep this paper within reasonable bounds, we report here only one example in Fig.3 with hob shaft speed n=97rpm; sampling frequency Fs=60Hz; ahead time of given control T=8ms and model parameters U0=0.1416, K=0.0139, F=0.9375, p=15. The term "original error" used in refers to error without control,which is relatively stable and "residual error" to error after given control. From the experiment result, about 65% of stochastic error can be compensated for.The good compensation result has been obtained on the gear-hobbing machine.We believe that the system and principle can be applied to the other machine tools.

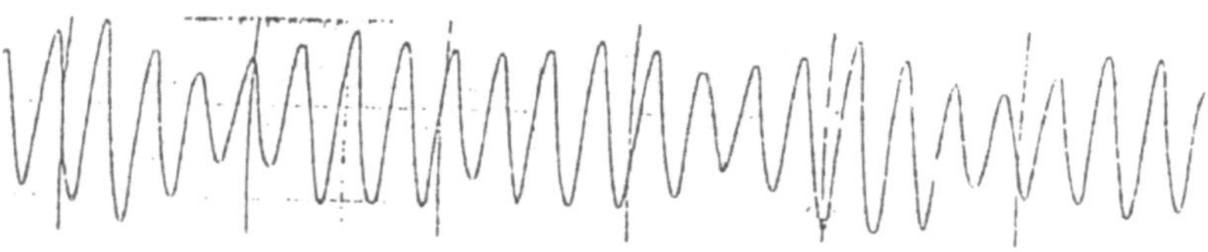

(a) Original error

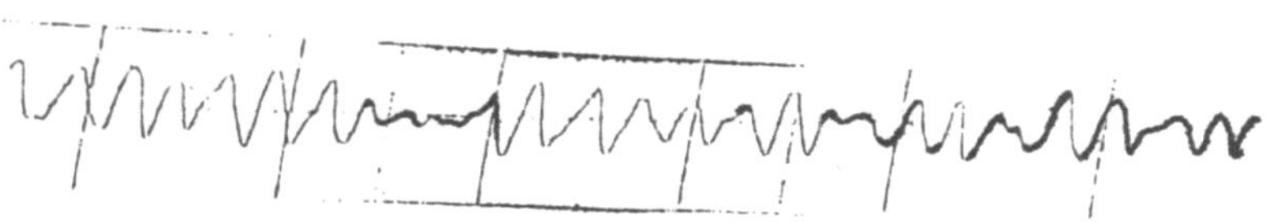

(b) Residual error

Fig. 3 Original error and residual error

5 Conclusion

Based on the present work, the following conclusion can be drawn out:

(1) The VSLMS algorithm has a fast convergence rate and small misadjustment. Moreover, its high computational efficiency and small storage requirement make it easily implemented by a single chip computer in real-time control.

(2) About 65% of the Stochastic error of gear-hobbing machine can be compensated for by one step ahead forecast compensation.

(3) The control system, based on MCS-51 SCC, is characterized by flexibility, low price, and high reliability. It can be applied to practical production.

(4) The principle and methodology used in this paper can be used for improving the accuracy of other machine tools.

6 References

E. Walach & B. Widrow, (1984) The least mean fourth (LMF) adaptive algorithm and its family. IEEE Trans. Vol. IT-30 No. 2 , pp275-283

H.Z. Bin & M. Dervies, (1983) 24th Int. Machine Tool Design and Research conf.

Li C.Y.& Yue M.J. (1987) Compensation and control of kinematic errors of gear-hobbing machine by a forecasting technique. Proceedings of CAPE2, Edinburgh, UK, April, pp321-324

Quingxian Zhou, Omer Anbagem and Kornal Eman, (1986) A new method for measuring and compensating pitch error in the manufacturing of lead screws. Machine Tool Design and Research, April, pp359-367

Digital capacitance displacement measuring instrument

Li Hongyan, Yang Lizhi and Li Jun
Changsha Institute of Technology, Changsha, Peoples Republic of China

Abstract
A new type of digital capacitance-based precision instrument is developed for measuring micro displacment and vibration. The measuring principles and the effects of the various errors on the accuracy of overall system are presented in this paper. The representative application of this instrument is used for the on-line or in-situ measurement, it can also be applied to the control system, and the dynamic error compensation system, etc.
Keywords, Digital, Capacitance Sensor, Micro Displacement, Measurement, Accuracy, Parameter.

1 Introduction

With the development of the science and technology, high accuracy of both size and form of the workpiece is required. Since the method, which reduces the machining errors based on the theory of error compensation, was first put forward in 1970's, there have been many studies of it. Moreover, increased attention is being paid to this method recently. In the error compensation system, displacement measuring instrument plays an important role, its errors may affect the machining accuracy directly. So, a high quality instrument for displacement measurement is very essential.

This paper introduces a new type of digital capacitance-based precision instrument for measuring micro displacement and vibration. The details of this instrument are described in the below sections.

2 Measuring principle

The measurement system hardware is shown in Fig. 1.

2.1 Basic principle

* Finance supported by The Chinese National Nature Science Foundation

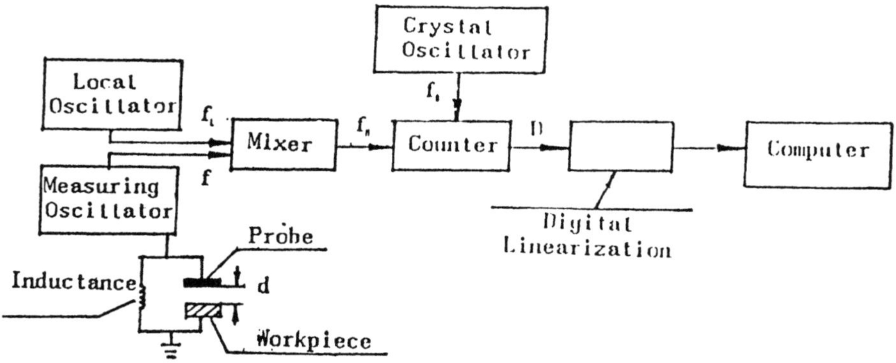

d, measured distance.
f, the frequency of the measuring oscillator.
f_l, the frequency of the local oscillator.
f_m, the frequency of the output from the mixer.
f_s, the frequency of the counting impulse.
D, the value of the counter.

Fig. 1. Block Diagram of the Measuring Principle.

The probe of capacitance sensor and a workpiece comprise the capacitor which plays a role in the resonant circuit. The variation of the distance between the probe and the workpiece may lead to the variation of capacitance, and lead to the variation of resonant frequency of the measuring oscillator. The relationship between the frequency of measuring oscillator f and the mixed frequency f_m is given by,

$$f_m = f - f_l \tag{1}$$

Let f_l, the frequency of the local oscillator, be a fixed value, the variation of mixed frequency f_m will be caused by that of measueing frequency f. After the counter, which output is related to the value of the mixed frequency, the measuring result is obtained with the digital hardware linearization. The output from the linearization is machine-readable, it is able to transmit to a computer directly without an A/D converter. Therefore, the quantization error can be avoided. As a matter of fact, after the counter, there are all digital circuits instead of the frequency detector and other analogue devices. So, the instrument can be with compact construction. On the other hand, the reduction of analogue devices means the reduction of error sources, and it is possible to get a high accuracy of measurement.

Here, the waveforms of the counter are shown in Fig. 2. During the two periods of the signal from the mixer, the counter records the number of the impulses which are produced by a high-frequency crystal oscillator. Let D is the output of the counter, it can be presented by,

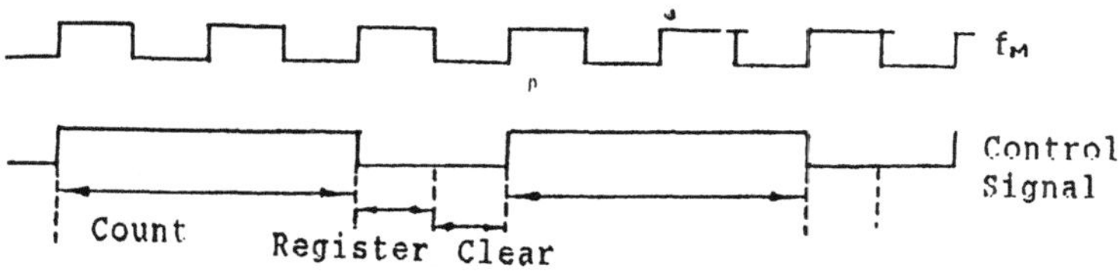

Fig. 2. the Sequence of Counter.

$$D = \frac{2f_s}{f_M} \qquad (2)$$

Within next period, the result of the counter is sent to the register, then the counter is clear so as to do an another cycle. Hence, a counting cycle needs three periods of mixed signal.

In addition, the linearization is in fact a procedure of table look-up based on the EPROM.

2.2 Mathematical derivation

Suppose measured area is a plane, we have,

$$C = \frac{\varepsilon S}{d} \qquad (3)$$

here ε = the dielectric constant of air.
C= the capacitance.
S= the dimension of the probe.

According to the principle of resonance, we have,

$$f = \frac{1}{2\pi\,[L(C_0 + C)]^{1/2}} \qquad (4)$$

here L = the inductance
C_0 = the stationary capacitance of the resonant circuit

If $d = \infty$ then

$$f_0 \triangleq f = \frac{1}{2\pi\,(LC_0)^{1/2}} \qquad (5)$$

and

$$K \triangleq \frac{\varepsilon S}{C_0} \qquad (6)$$

we get,

$$f = \frac{f_0}{\left(1 + \frac{K}{d}\right)^{1/2}} \qquad (7)$$

By using Eq. (1), (2), and (7), the measured distance d can be expressed as,

$$d = \frac{K}{\left(\frac{Df_0}{2f_s + Df_L}\right)^2 - 1} \qquad (8)$$

3 Error analysis

The errors of the measurement system involve static error and dynamic error.

3.1 Static error

Suppose d_0 is the initial clearance between the probe and the measured workpiece, the operation range is from $d_0-R/2$ to $d_0+R/2\mu m$ and R is the measuring range. In order to simplity calculation, the frequency of the measuring oscillator is supposed to be directly proportional to the distance d in measuring range. the static errors can be classified as follows:

(a) Frequency drifting error

In a certain time, if δf_M is the mixed frequency drift which is caused by the instability of the frequency of oscillators, the measurement error caused by δf_M is called $\triangle e_1$, and

$$\triangle e_1 = \frac{\delta f_M R}{f_{M2}-f_{M1}} \qquad (9)$$

here f_{M2} = the mixed frequency corresponding to the distance $d_0+0.5R$.

f_{M1} = the mixed frequency corresponding to the distance $d_0-0.5R$.

(b) Counting error

In general, the error of the counter is ±1 count word. The error due to this reason is related to the mixed frequency and is called $\triangle e_2$. We have:

$$\triangle e_2 = \frac{\triangle f_M R}{f_{M2}-f_{M1}} = \frac{R}{f_{M2}-f_{M1}} \frac{(f_M)^2}{f_s} \frac{1}{2} \qquad (10)$$

here $\triangle f_M$ = the deviation of mixed frequency equivalent to the error of ±1 count word.

(c) Mixer error

Since the mixer is never a ideal one, there are lots of interaction. They would lead to measuring errors, which now can be reduced less than $\pm0.005\mu m$.

3.2 Dynamic errors

In dynamic measurement, this kind of error is mainly owing to the time delay. Suppose that the measured displacement d is composed of many sine components, each of them has different amplitude and frequency. So d can be described as:

$$d = \sum^{N} d_i = \sum^{N} A_i \sin 2\pi F_i t \qquad (11)$$

here d_i = the sine component of measured displacement.

A_i = the amplitude of the component d_i.

F_i = the frequency of the component d_i.

Considering one of the components d_i, the derivative of d_i with respect to t is called M_i and

$$M_i\, dt = d(d_i) = (2\pi F_i A_i \cos 2\pi F_i t)\, dt \qquad (12)$$

So, the measuring error $\triangle e_i'$ can be given by:

$$\triangle e_4' = (2\pi F_1 A_1 \cos 2\pi F_1 t)\triangle t \tag{13}$$

From Fig. 2, we know that the max. time delay $\triangle t$ is equal to two periods of mixed signal, and

$$\triangle t = \frac{2}{f_s} \tag{14}$$

we obtian

$$\triangle e_4' = (2\pi F_1 A_1 \cos 2\pi F_1 t)\frac{2}{f_s}$$

Suppose Q is the derivative of $\triangle e_4'$ with respect to t. Let Q is equal to zero, we can find a point t_o. At this point, the error get its max. value, $\triangle e_4'$ can be expressed as follaw approximately,

$$\triangle e_4' \doteq \frac{8\pi A_1 F_1}{f_{s1}+f_{s2}} \tag{15}$$

If the component is independent of each other, the total dynamic error $\triangle e_4$ can be derived as,

$$\triangle e_4 = \frac{8\pi(\sum^{N} A_i^2 F_i^2)^{1/2}}{f_{s1}+f_{s2}} \tag{16}$$

4 The option of parameters

The errors caused by each of the four quarters is supposed to be equated. Using the formulas mentioned above, we can obtion the value of total measurement error $\triangle e$ and a group of reference parameters, they are given by,

$$\triangle e = (\triangle e_1^2+\triangle e_2^2+\triangle e_3^2+\triangle e_4^2)^{1/2} \tag{17}$$

$$f_{s2} = (\delta f_s \cdot 2f_o)^{1/2} \tag{18}$$

$$f_{s1} \doteq \frac{[4\pi(\sum^{N} A_i^2 F_i^2)^{1/2} - \delta f_s \cdot R] \cdot f_{s2}}{4\pi(\sum_{N} A_i^2 F_i^2)^{1/2} + \delta f_s \cdot R} \tag{19}$$

$$d_o \doteq \left(\frac{Kf_o \cdot R}{2(f_{s1}+f_{s2})}\right)^{1/2} \tag{20}$$

$$f_L = \frac{f_o}{\left(1+\frac{K}{d_o}\right)^{1/2}} - \frac{f_{s2}+f_{s1}}{2} \tag{21}$$

In practice, the drifting error $\triangle e_1$ can be decreased, but it is difficult to reduce the mixer errors. So, considering the specifical conditions, the distribution of errors may be adjusted in order to obtain the better option of parameters.

An instrument based on the principles above has finished after lots of experiments. Its specifications are given by,

(a)Static displacement measurement
Measuring range, 2μm, 10μm, 30μm, 100μm.
Resolution, better than 0.005μm.

Reliability, indicator drift $\leqslant$ 0.02 μm/hour.
Diam. of the probe, 3.6mm.
Measuring error $\leqslant$ 0.005μm (min. range).
(b) Dynamic displacement measurement.
Time for each measurement, 200μs.

5 Conclusions

The purpose of this paper is to introduce a newly capacitance-based precision instrument. Differing from a general frequency modulated or operation type capacitance-based displacement measuring instrument, this one based on the principles of frequency modulation, frequency mix, and digital technique The measuring results are machine-readable, which are able to be sent to a computer without an A/D converter. Since the sources of error are decreased, it is possible to get higher accuracy. So a general measurement system which composed by this instrument may have the benefits of low cost, light-weight, compact construction, and particularly, high accuracy

This instrument can be used for micro displacement and vibration measurement. Its representative application is used for the on-line or in-situ measurement, such as roundness and cylindricity measurement of a workpiece. It can also be applied to the dynamic error compensation and control system, etc.

6 References

Yang Lizhi and Li Jun ,(1988)
Digital capacitance-based instrument for measuring displacement and its application.
Thesis for M.SC. of Changsha Institute of Technology.

Yuan Hongxiang, Li Jun, Zhang Feng and Pan Peiyuan,(1989)
Roundness on-line measurement and error compensation technique in vltra precision machining.
Chinese Journal of Aeronautics, Vol.2, No.1, Feb. 1989.

Han Zhenyu, Li Jun and Yuang Hongxiang,(1989)
In-situ measurement of cylindricity and data processing.
Thesis for M. SC. of Changsha Institute of Technologe.

Optimisation and diagnosis for numerically controlled maching

M.A. Rahbary and G.N. Blount
Department of Combined Engineering, Coventry, England

Abstract
This paper outlines how productivity can be improved by reducing long idle time and increasing the efficiency of programming. Three important techniques are introduced which will result in reduced labour, lead time and programming cost. These techniques are : Control data verification and diagnosis, Three dimensional clash detection and avoidance, Dynamic optimisation . The paper describes an implementation of the techniques in a scaled time dynamic simulator with off-line cycle time monitoring capabilities.
Keywords: Dynamic Optimisation, Verification, Control lines, Simulation, Position control, Short-Path, Collision Free.

1 Introduction

When Numerically Controlled (N.C.) machines first appeared on the market, a great effort was needed and was made to show the potential economic advantages of using them in comparison with conventional machine tools. This was necessary because of the large price difference between N.C. and conventional machines. The price difference is likely to remain as manufacturers tend to be continually enhancing the specifications of their machines. The large cost difference between N.C. and conventional machines means that every purchase of an N.C. machine still requires a detailed financial justification in most companies. Kilmartin (1981), describes how the investment decisions being taken by companies are always well scrutinized and the larger the capital cost of any machine, the higher up the corporate management structure will the decision to purchase be made. Thus the situation involves optimisation of various factors that can effect the product manufacturing time, and cost, which in fact are numerous. Indeed, if the problem is to be optimised thoroughly, all the above factors should be incorporated in a single objective and optimization then carried out. The problem is therefore a complex one. Even if every stage is considered alone and a linking function used, the size of the dimension of variables in the problem will be very large.

The paper approaches the problem by considering the cases individually and finds the optimal solution(s). The optimal solution(s) mainly aimed at N.C. part-program optimisation by implementing two techniques defined as: N.C. part-program verification by simulation and dynamic optimisation.

2 Verification by simulation

Although machine tool technology and machine control data communication (MCD) techniques have advanced considerably, no appreciable changes in MCD verification methods have occurred since the introduction of cutter centre line plotting. N.C. programs can be easily and safely checked by dynamic simulation. After having input the coordinates of the unmachined part, the contour of the workpiece, chuck, turrets, and tailstock are displayed. Then the N.C. program is run step by step, and the tool is moved on the screen in scaled time . If the tool is in the cutting position, the contour of the workpiece changes accordingly. The actual state of the work in progress is updated continually, so that continuous snapshots are displayed. The system can also detect collisions between tools or various other parts using a swept volume technique, Collingwood, M.C. (1987).

3 Dynamic optimisation

Effective software optimisation techniques which could be implemented in numerically controlled machine tools are many and diverse. Optimisation techniques such as high speed machining, optimised process planning, and multi-axis operations are among the most recognised techniques considered for part-program optimisation. The need for numerical control part-program optimisation increases as the complexity of these machine tools, and component geometries, increase. The "Dynamic Optimisation" technique is introduced in this paper in addition to the "Verification by Simulation" technique, in order to reduce the gap for achieving the goal of optimum N.C. part-programming. For this purpose three techniques are implemented to achieve dynamic optimisation in N.C. They are as follows:

I. Collision free short-path generation.
II. Indexing position optimisation.
III. Tool tip position control.

3.1 Collision free short-path generation
The concept of collision free short-path generation initially was investigated in the area of robotics e.g., Brooks, R.A. (1983), Lazano-Perez, T. (1983), and Brooks, R.A. (1985). It is thought that for efficient use of industrial robots, minimum-time trajectory planning in an automated manufacturing system is a central issue. Minimum-time trajectories can lead to the reduction of cycle time of the task, and hence enhancement of productivity. The collision free short-path generation for numerically controlled machine tools must take a different format in comparison to robots, specifically it must cope with the cluttered environment of N.C. machine tools. The technique behind a "Collision Free Short-Path" algorithm for N.C. applications requires generation of optimised path(s) of cutting tool(s) in the presence of barrier(s) constraints. For this purpose an algorithm was developed, Rahbary, M.A. and Blount, G.N. (1990). this successfully generates a set of paths which are optimised based on accuracy, safety, and speed.
Figure (1) indicates how the system can be implemented for manufacturing purposes.

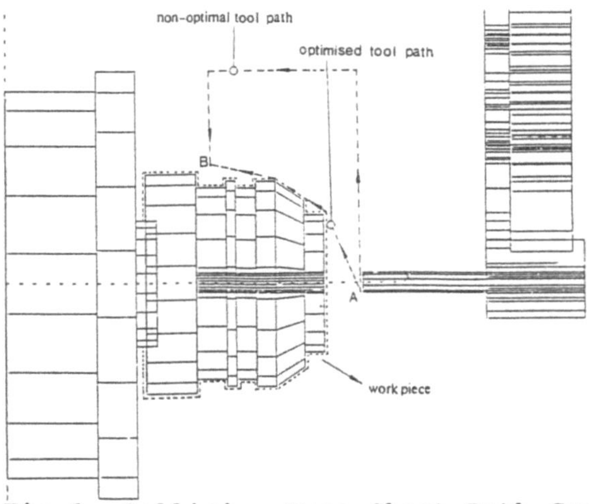

Fig.1. Collision Free Short-Path Generation

3.2 Indexing position optimisation

Machine tools which were operated in the early days often used single tooling turrets for cutting operations. When a particular cutting operation required a number of tools to be used, the operator had to stop the whole operation, change the tool and reset the machine tool for the next operation. The manual tool change meant a great deal of cycle time loss. Consequently with the arrival of numerically controlled machine tools, turrets were equipped with multi-tooling stations which provided them with the facilities of automatic tool changes during cutting operations. Most machine tool operations involve the use of more than one tool, and usually they vary in length and/or diameter. The tendency to return the machine slides to a set position before making a change, a practice that has merit only from a safety point of view, can add considerably to the total time taken to machine the part.

To eliminate this problem a technique can be developed to optimise the position of the indexing operation. The system could be developed by creating the following data and implementing them in an optimiser program:

1) Updated geometry of the component
2) Geometries of the turret(s) and tools
3) Direction and angle of rotation
4) Last and next position before and after indexing
5) Safety margin

3.3 Tool tip position control

This technique unlike the term more often used for a sensor based tool tip position error compensation system, is totally software based and aimed at optimising the tool movements in the part-program source code level. The concept of dynamic optimisation of an N.C. part-program in a simple form could be defined as a technique for reducing tool travel within working envelope.

A technique can be developed in order to achieve this, which with no doubt can reduce machining operation cycle time considerably. Such a technique is provided by creating an algorithm(s) which combines the short-path and visibility algorithms, Rahbary, M.A. (1990). The system takes a given safety margin and by getting the updated workpiece geometry, swept volume, and other relevant data, produces a set of points which are optimised based on safety, accuracy, and speed.

4 Design method

A system was developed, as part of a research program, Rahbary, M.A. (1990), which combines the capabilities of the dynamic optimiser and simulation systems, in order to generate optimised N.C. programs. The flow-chart in figure (2), illustrates how this system works. Some tests also carried-out in order to demonstrate the result of implementing the system. The programs used for this purpose were industry based, and they were carried-out on a five-axis Fanùc controlled machine tool. The result of the tests are illustrated in a graph representation format in figure (3).

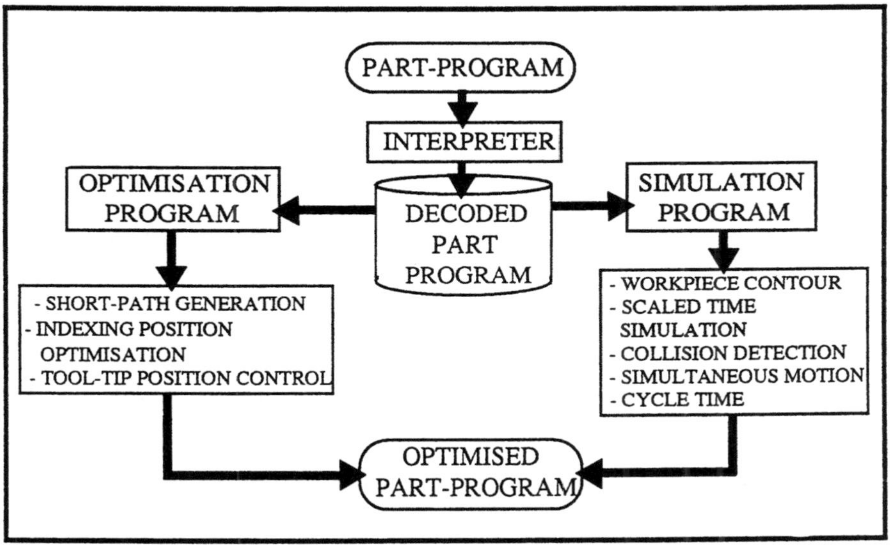

Fig.2. System Flow-chart

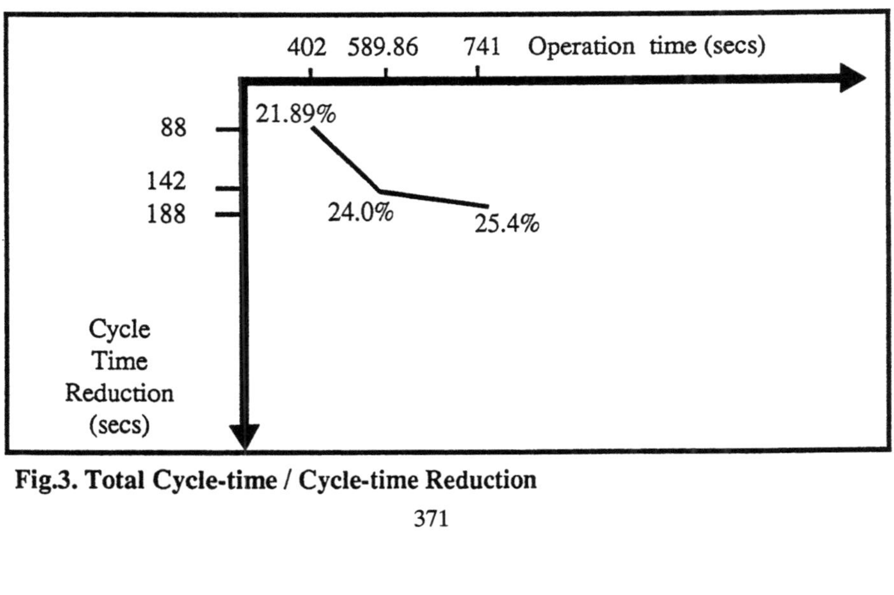

Fig.3. Total Cycle-time / Cycle-time Reduction

5 Conclusion

The implementation of the actual optimisation system developed according to the definition in this paper showed some interesting results.

The results of the tests demonstrated in figure (3), show that as the total cycle time of the part-programs increases, the potential cycle time reduction of these programs also increases. Therefore, it can be concluded that the implementation of this system can result in an average cycle time reduction of between 20 - 27%. It also can be said that the dynamic optimisation and N.C. part-program verification by simulation systems are useful aids for producing error-free optimised N.C. programs.

6 References

Collingwood, M.C., (1987) **A Study of Collision Detection in Multi-Axis N.C. Turning.** Ph.D. Thesis, Coventry Polytechnic.

Brooks, R.A., (1983) **Planning Collision-Free Motions for Pick and Place Operations.** The International Journal of Robotics Research, Vol. 2, no. 4, PP. 19-44.

Lozano-perez, T. (1983) **Spatial Planning: A Configuration Space Approach.** IEEE Trans. Comput., Vol. C32, PP. 108-120, Feb.

Brooks, R.A. and Lozano-Perez, T. (1985) **A Subdivision Algorithm in Configuration Space for Find Path with Rotation.** IEEE Trans. on Systems, Mani Cyberetics, Vol. SMC-15, no. 2, PP. 224-233, March/April.

Rahbary, M.A. and Blount, G.N. (1990) **A Minimum Trajectory for Machine Tool Dynamics.** International Journal of Production Research, "Submitted for Publication".

Rahbary, M.A. (1990) **Computer Assisted Machine Tool Part-Program Optimisation.** Ph.D. Thesis, C.E.S., Coventry Polytechnic, Coventry.

Kilmartin, B.R. and Hannam, R.G. (1981) **An in-Company Study of N.C. Machine Utilization and Its Improvements by a Systems Approach.** Int. J. of Prod. Res., Vol.19, no. 3, PP. 289-301.

In process measurements using an opto-electronic probe

I. Shams and C. Butler
Centre for Manufacturing Metrology, Brunel University, Uxbridge, England

Abstract
In recent years there has been an increasing interest in automation and automatic assembly. All manufacturing and assembly processes involve a considerable amount of handling and orientation of parts. Therefore, a fully automated manufacturing system requires inspection to be incorporated as part of the production operation.

This paper discusses the various aspects of in-process inspection and the tools available for this purpose. It also introduces a new device for dimensional gauging of parts.
Keywords: Measurement, Opto-electronic, Probe.

1 Introduction

Human operators use their sensory capabilities and judgement in various semi automatic manufacturing processes. It is this characteristic of the human operator which must be replaced when fully automatic production methods take over. In adding this capability to the automatic operation not only are the inconsistancies and errors in manufacturing reduced but also a feedback route is introduced. This permits corrective actions to be taken. Automatic inspection decreases the time spent on a process which adds no value to the product and hence decreases cost.

The purpose of the feedback is to allow compensating adjustments to be made in the processes to improve quality. This would act as a guide for adjustments for tool wear and other sources of variation. The automatic inspection procedure will help to determine not only whether the parts were within tolerance, but also to perform a trend analysis and to input this information into the machine tool control system so that compensating adjustments could be made to the tool path.

With an integrated inspection system, all the data from the operation is processed with statistical process control software, the analysis of which could be passed onto the control computer.

2 Inspection Procedure

There are three main choices for inspecting manufacturing procedures:

1. Off-line inspection.
2. On-line/in-process inspection.
3. On-line/postprocess inspection.

On-line/in-process inspection is the one this paper is mostly concerned with. The benefits of in-process inspection is that it may be possible to influence the operation producing the current part, thereby correcting a potential quality problem before the part is completed. The earlier that a defect is discovered in the automated manufacturing process, the less expensive the rework or scrap will be. It is possible that automatic part inspection may even eliminate certain breakdowns in the production equipment due to the jamming of the parts or similiar failures. The on-line inspection is typically done on a 100% basis using automated sensor methods. Using Statistical Process Control (SPC) the sampling rate could be decreased.

An additional feature which gets us closer to the factory of tomorrow is the distributed inspection. This is done by having automatic inspection stations located along the line of flow of work in a factory, placed strategically at critical points in the manufacturing sequence. This is especially relevant in assembled products where many components are combined into a single unit that cannot be easily taken apart. If one defective component would render the assembly defective, it is obviously better to catch the defect before it is assembled.

3 Sensors used in Automated Inspection

Automatic inspection procedures, employing sensors, are controlled by and communicate data to digital computers. There are two broad categories of sensors used in this respect:

1. Contact Inspection Method.
2. Non-Contact Inspection Method.

The contact method will be explored further in this discussion. Inspection systems used for this method are, in essence, some means of precise positioning within a three dimensional space and a probe which could act as a micro-switch. The first task could be achieved by any coordinate measurement machine, X - Y table or a precision robot. The probe however, is a sensing element which is distinctly different from other sensors included in the rest of the system. The probe normally provides an electrical signal when contact with the manufactured component is established, enabling the coordinates of the selected contact points to be accurately recorded [1]. The touch probe mostly used for this purpose relies on the breaking of electrical continuity on a kinematic mounting arrangement. Figure 1. shows the kinematic mount arrangement.

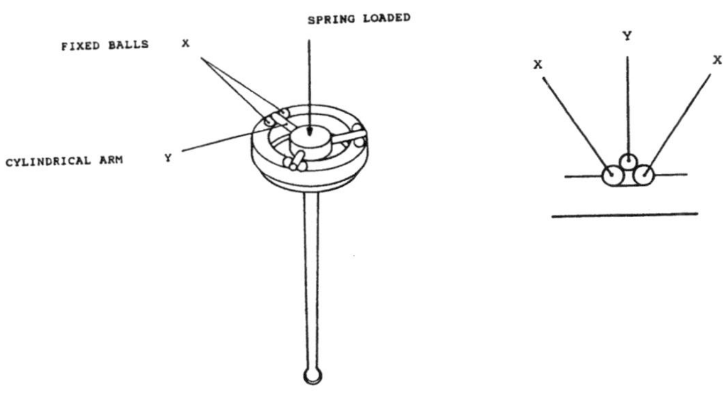

Fig.1. Kinematic mount arrangement.

The new probe developed for inspection purposes uses optical switching rather than mechanical switching. The basic components of the new probe are shown in Figure 2.

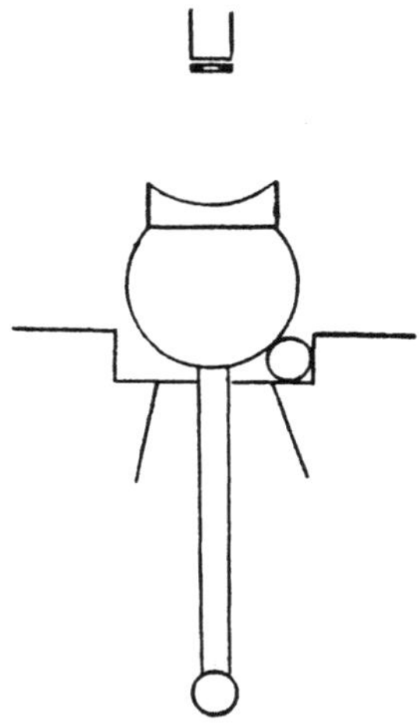

Fig.2. Basic components of the new probe.

The light is conveyed from an infra-red source through the optical fibre onto a concave mirror. The radius of curvature of the mirror is such that all of the light is reflected back into the fibre when the probe is in its undisturbed position. As the stylus is deflected, the light reflected back by the concave mirror starts to move away from the fibre core as shown in Figure 3.

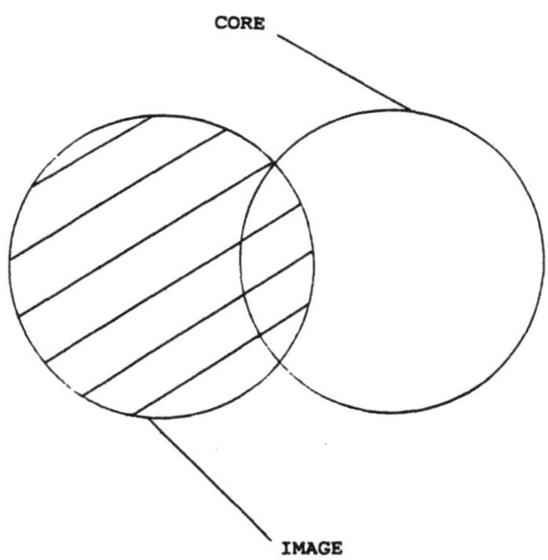

Fig.3. Fiber core and the reflected image.

This causes a modulation in the intensity of light which is deflected by the opto-electronic circuit employed. The change in the intensity of light against deflection from the rest position is shown in Figure 4.

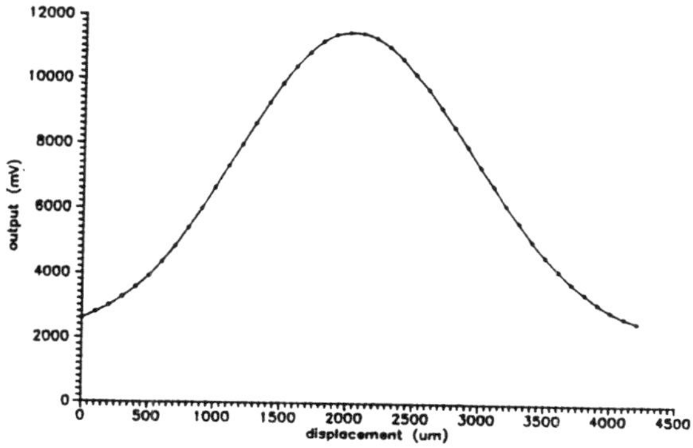

Fig.4. Intensity of light versus displacement.

There is a linear portion to the plot with a large gradient. This enables the user to employ the probe as both an anlogue or a digital contact sensor. The new inspection probe has been tested for repeatability on point measurements. The error in measurement was within one micron which is well within the accuracies of precision robots, most X - Y tables and CMM's.

By using fibres and light as the signal which is transmitted from the measurement point to the associated electronics, the effect of EMI have been eliminated [2,3]. Also such sensors, due to their passive nature could be used in hazardous environments[4]. The analogue measurement capability has been added to the digital capability of the conventional probe.

The work is supported by the British Technology Group.

4 References

Butler, C. and Shams, I. (1990) "Precision Coordinate Measurement Using a Fibre Touch Probe", 28th International MATADOR Conference.

Butler, C. and Jones, B.E. (1986) "Optical Techniques in

Metrology", Transducer/Temp Conference, London.
Butler, C., Gregoriou, G. and Shams, I. (1989) "Sensors in Automatic Inspection" Mechatronics Conference, Lancaster.
Jones, B.E. (1987) "Sensors Advance", J Phys E: Sci Instrum, 20(9), 1055.

A new three dimension piezoelectric transducer for diagnostic engineering

D.S. Shi and Y.T. Wang
Yan Shan University, HeBei, China
S.H. Miao
DaLian University of Science and Technology, DaLian, China

Abstract

A new type piezoelectric transducer with compression/shear sense is developed and was successfully tested for measuring three dimension accelerations in shock and vibration experiment.

Keyword : Acceleration Transducer, Diagnostic Engineering, Frequency Response, Piezoelectric Crystal.

1 INTRODUCTION

A current three dimension acceleration transducer normally consists of three compression piezoelectric elements and three sensing mass blocks assembled on a metal base. Each element respectively responses in one direction of Cartesian system. The consequent large size and mass of the transducer, however, inevitably and unexpectedly changes the mass distribution of the project vibrating system. In order to overcome the shortcoming, the features of the new type transducer are as follows :

(1) A tension/compression element and a shear element are adopted. The later has more advantages such as low noise and small thermoelectric effect caused by transient fluctuation of temperature.

(2) Only one sensing mass block is employed and located through a prefixed sleeve. It is accordingly smaller in size, more sensitive, and has wider frequency band.

2 STRUCTURE DESIGN AND MECHANISM OF THE PIEZOELECTRIC ELEMENT

The relationship between the piezoelectric charges and the induced stresses of a crystal may be written as

$$\begin{bmatrix} Q_{xx} \\ Q_{yy} \\ Q_{zz} \end{bmatrix} = \begin{bmatrix} d_{11} & d_{12} & d_{13} & d_{14} & d_{15} & d_{16} \\ d_{21} & d_{22} & d_{23} & d_{24} & d_{25} & d_{26} \\ d_{31} & d_{32} & d_{33} & d_{34} & d_{35} & d_{36} \end{bmatrix} \begin{bmatrix} \sigma_{xx} \\ \sigma_{yy} \\ \sigma_{zz} \\ \tau_{yz} \\ \tau_{zx} \\ \tau_{xy} \end{bmatrix} \quad (1)$$

If a piezoelectric crystal is cut at x=0 and polarized along x axis, for example, the elements d_{ij} can be determined, and the relationship follows

$$
\begin{bmatrix} Q_1 \\ Q_2 \\ Q_3 \end{bmatrix} = \begin{bmatrix} d_{11} & d_{12} & 0 & d_{14} & 0 & 0 \\ 0 & 0 & 0 & 0 & d_{25} & d_{26} \\ 0 & 0 & 0 & 0 & 0 & 0 \end{bmatrix} \begin{bmatrix} \sigma_{xx} \\ \sigma_{yy} \\ \sigma_{zz} \\ \tau_{yz} \\ \tau_{zx} \\ \tau_{xy} \end{bmatrix} . \qquad (2)
$$

Similarly , for a Barium Titanate ferroelectric ceramic polarized along z axis , Eqns. (1) may be simplified to

$$
\begin{bmatrix} Q_1 \\ Q_2 \\ Q_3 \end{bmatrix} = \begin{bmatrix} 0 & 0 & 0 & 0 & d_{15} & 0 \\ 0 & 0 & 0 & d_{24} & 0 & 0 \\ d_{31} & d_{32} & d_{33} & 0 & 0 & 0 \end{bmatrix} \begin{bmatrix} \sigma_{xx} \\ \sigma_{yy} \\ \sigma_{zz} \\ \tau_{yz} \\ \tau_{zx} \\ \tau_{xy} \end{bmatrix} \qquad (3)
$$

The mechanic structure of the transducer is illustrated in Fig. 1 . In addition to the single inertia mass block , it has three groups of piezoelectric elements . One of them consists of compression plates that response to the vertical vibration signal in z direction . The other two contain shear crystal plate to the signals in horizontal directions x and y . The crystal elements of each group are connected to the base by a conduct sheet . All the piezoelectric elements are fixed on the sensing mass block through a pre-loaded sleeve that is a thin wall cylinder and made of $C_{r_{18}}N_i\!-\!9T$ stainless steel or Beryllium Bronze . The adopted piezoelectric ceramic PZT - 5 has effect in both vertical and transverse directions . The piezoelectric effect function of the ceramic is expressed by Eqns. 3 .

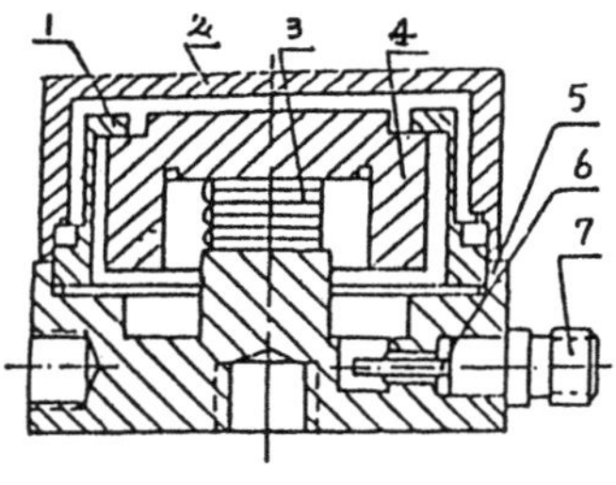

Fig.1. Mechanical structure of the transducer .

1_the pre-loaded sleeve, 2_the housing,
3_the piezoelectric plate, 4_the sensing mass block,
5_the base, 6_the lead, 7_the connector.

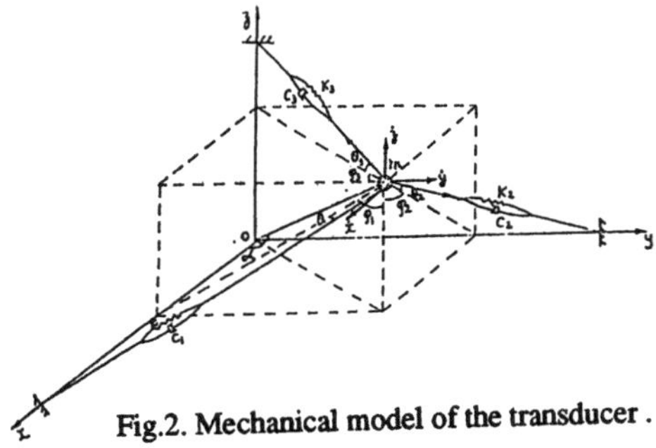

Fig.2. Mechanical model of the transducer .

The mechanical model of the three dimension transducer is illustrated in Fig . 2 . The dynamic equations of the sensing mass may therefor be set up as

$$m\ddot{x}_0 = -(K_1 \Delta x \sin\theta_1 - K_2 \Delta y \cos\theta_2 \sin\phi_2 - K_3 \Delta z \cos\theta_3 \sin\phi 3$$
$$+ C_1 V_{c1} \sin\theta_1 - C_2 V_{c2} \cos\theta_2 \sin\phi_2 - C_3 V_{c3} \cos\theta_3 \sin\phi 3),$$
$$m\ddot{y}_0 = -(K_2 \Delta y \sin\theta_2 - K_1 \Delta x \cos\theta_1 \sin\phi_1 - K_3 \Delta z \cos\theta_3 \cos\phi 3$$
$$+ C_2 V_{c2} \sin\theta_2 - C_1 V_{c1} \cos\theta_1 \sin\phi_1 - C_3 V_{c3} \cos\theta_3 \sin\phi 3),$$
$$m\ddot{z}_0 = -(K_3 \Delta z \sin\theta_3 - K_1 \Delta x \cos\theta_1 \cos\phi_1 - K_2 \Delta y \cos\theta_2 \cos\phi 2$$
$$+ C_3 V_{c3} \sin\theta_3 - C_1 V_{c1} \cos\theta_1 \cos\phi_1 - C_2 V_{c2} \cos\theta_2 \cos\phi 2).$$

It is clear that these equations are generally coupled each other , that is to say , the collected acceleration signal in one direction comes out not only from the contribution of dynamic motion in the same direction but also from the other two perpendicular directions . The arrangement of the design of the transducer , however , is that $\phi_1 = \phi_2 = 90^o$. Moreover assuming that $\Delta x \approx x$, $\Delta y \approx y$, $\Delta z \approx z$ are acceptable and $\theta_1 = \theta_2 = \theta_3 \approx 90^o$, the above equations are then reduced to

$$m_x \ddot{x}_0 + K_x x + C_x \dot{x} = 0 \tag{4}$$

$$m_y \ddot{y}_0 + K_y y + C_y \dot{y} = 0 \tag{5}$$

$$m_z \ddot{z}_0 + K_z z + C_z \dot{z} = 0 \tag{6}$$

Where K_x and K_y denote the shear modulus , K_z for the compression modulus ; C_x and C_y represent the shear damping coefficient , while C_z for the compression damping . Now it is interesting that these equations are independent from each other and thus the solution of the equations is very simple and familiar . For example , assuming the measured signal in z direction is sinusoidal , i.e. , $U_{in} = A_m \sin\omega t$, the steady solution of Eqn. (6) is

$$z = \frac{A_m \omega_z^2}{\sqrt{(\omega_{zn}^2 - \omega_z^2)^2 + 4\xi_z \omega_{zn}^2}} \sin(\omega_z t - \phi),$$

hence the frequency response of acceleration is

$$S_a = 1/\omega_{zn}^2 \sqrt{\left[1 - (\omega_z/\omega_{zn})^2\right]^2 + (2\xi_z \omega_z/\omega_{zn})^2}\ .$$

Since the relative displacement between the mass block and housing is negligible in the linear elastic region of the piezoelectric elements , it follows

$$F_z = K_z z.$$

It is known that under certain conditions the charges generated in the piezoelectric elements is proportional to the applied forces , that is

$$Q_z = n d_{ij} F_z,$$

here n is the number of the crystal plates adopted . The relationship between the charge sensitivity and frequency is consequently found

$$S_{Q_x} = Q_x/a_x = n_x K_x d_{24}/\omega_{xn}^2 \sqrt{\left[1 - (\omega_x/\omega_{xn})^2\right]^2 + (2\xi_x \omega_x/\omega_{xn})^2}$$

$$S_{Q_y} = Q_y/a_y = n_y K_y d_{24}/\omega_{yn}^2 \sqrt{\left[1 - (\omega_y/\omega_{yn})^2\right]^2 + (2\xi_y \omega_y/\omega_{yn})^2}$$

$$S_{Q_z} = Q_z/a_z = n_z K_z d_{33}/\omega_{zn}^2 \sqrt{\left[1 - (\omega_z/\omega_{zn})^2\right]^2 + (2\xi_z \omega_z/\omega_{zn})^2}\ .$$

The above analysis reveals that

(1) An appreciate structural design of transducer can greatly reduce the dynamic motion of the sensing mass block to very simple form .

(2) The charge sensitivities of the transducer are not equal to different directions . Those to x and y directions are much higher than that to z direction .

(3) $K_x = K_y < K_z$ because the shear modulus G in x and y directions is generally less the vertical compression modulus E . So that the operation frequency limitation in x,y in directions is different form that in z direction .

3 EXPERIMENT

The frequency responses of the transducer on every direction are tested by two measuring systems with different operation frequency regions . One with lower region from 0.5 HZ to 30 HZ is schematically shown in Fig . 3 . The other one with higher region from 30 HZ to 7000 HZ in Fig . 4 .

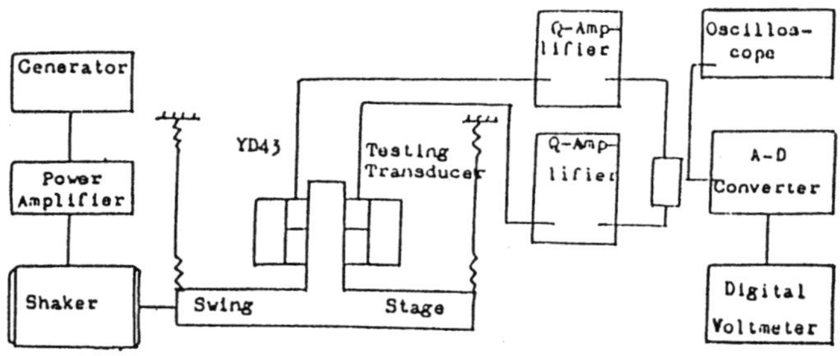

Fig.3. The system for lower F-response .

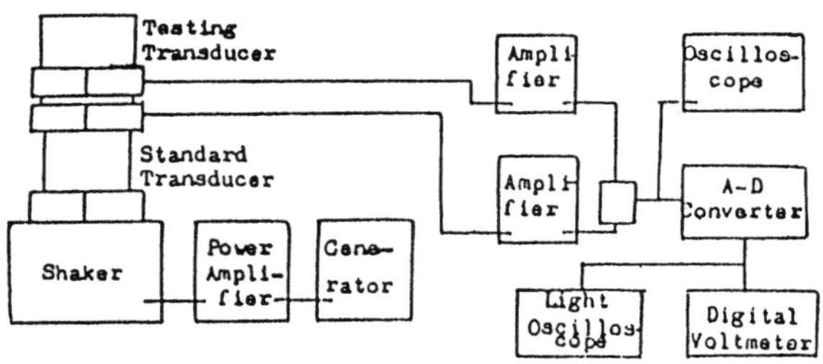

Fig.4. The system for higher F-response .

The result of sensitivity in lower frequency region is listed in Table 1 in comparison with the output of standard transducer .

Table 1.

Frequency (HZ)		0.6	0.8	1.0	1.5	2.0	5	10	15
x direction	output 1	50	50	50	50	50	50	50	50
	output 2	48	49	49	50	51	50	50	50
y direction	output 1	50	50	50	50	50	50	50	50
	output 2	50	49	50	50	51	50	50	50
z direction	output 1	50	50	50	50	50	50	50	50
	output 2	48	49	50	51	50	50	50	50

In the second column of the table output 1 represents the results from the standard transducer, while output 2 from the new transducer. The unit is mv. The frequency responses in x,y and z directions are sketched in Fig. 5. This transducer has much higher sensitivities, 600pc/g, 650pc/g and 400pc/g along axis x, y and z respectively. The accuracy of low frequency response to every direction is much better than that of the traditional transducer.

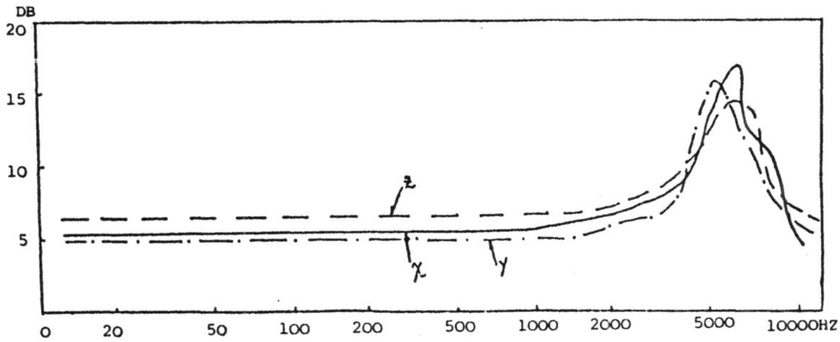

Fig.5. Characterised frequency response of the transducer.

4 CONCLUSION

Since only one sensing mass block is assembled and particularly the compression/shear piezoelectric technique is employed, the new transducer is structurally compact, small in size, lighter of mass and consequently has more accuracy of sensitivity. It is no doubt that the new transducer has a potential prospect in the future.

5 REFERENCES

Pointon, A. J. (1982) Piezoelectric Devices. **IEE PROC**, Vol. 129, No. 5.

Oaza Nagakute (1986) New technology of Sensors for Automative Application. **Sensor and Actuators**, pp. 181-193.

Neumeister, J. (1985) A Silicon Pressure Sensor using MOS Ring Oscillators. **Sensor and Actuators**, pp. 167-176.

Kowoalski, G. (1986) Miniature Pressure Sensor and Their Temperature. **Sensor and Actuators**, pp. 367-376.

Improving the prediction of condition using simple sensors

R.K. Stobart
Cambridge Consultants Ltd, Cambridge, England

Abstract
One approach to condition monitoring of industrial equipment is to install special purpose sensors to monitor the quantity of interest. The use of such sensors may complicate the installation of the equipment, add significantly to the cost or even depress reliability. Most modern industrial equipment is equipped with sensors for control purposes. Usually such sensors are specified for low cost and robustness and are read as part of the control process. A monitoring system which can make use of such sensors to extract detailed information adds very little to total system cost. Estimation methods offer the possibility of extracting system information from sensor signals. The monitored equipment is considered to be made up of a series of processes each with an input and output. Each input/output pair may be modelled. It is these models which permit comparison between "healthy" and current conditions. Reported results from the method are excellent and will be reviewed. In recent project work Cambridge Consultants Ltd has demonstrated that the techniques are applicable to diesel engines and provide a good initial diagnosis of a variety of faults.
Keywords : Model-based, Engine, Diesel, Estimation, Sensors

1 Introduction

Condition monitoring requires detailed information about the performance of equipment so that clues giving an early warning of a failure are noted. This objective tends to work against other pressures on equipment such as the need to keep costs down, or to keep reliability high. The very presence of sensors dedicated to monitoring can have the effect of diminishing overall system reliability.

On modern industrial equipment sensors are normally installed to support control, and so are available for more general use if required. What is needed is a simple computational scheme to support the more general use of control data.

In the 1970's the first significant steps were made in the application of *model based* techniques to the detection of "jump" phenomena in aerospace control systems (Willsky, 1976). Of particular interest was the detection of sudden sensor failure in control circuits. Later work (Pisano, 1980) was directed at running a real time model in parallel to firstly detect a sensor failure and upon detection, substitute the model value, so that contol could continue uninterrupted.

The success of these schemes relied on a phenomenon known as **analytical redundancy**. This simply means that the internal workings of the system are evidenced in the measurements made on the system and details can be extracted by appropriate signal processing.

The context of the application needs further consideration. Typical questions are

- Is a full diagnosis required?
- Will a "first level" diagnosis be acceptable?
- Is an indication that the monitored machine is just "off-design" satisfactory?

The answer will dictate the complexity of the scheme, but any of these considerations can lead to an estimation based method for condition assessment.

The subsequent sections of the paper develop the idea of estimation based condition monitoring where the emphasis is on getting more information from existing conventional sensors. The sections are as follows.

(1) An introduction to the estimation based methods.
(2) A review and comparison of estimation methods applied to some industrial systems.
(3) How the methods have been applied to diesel engines.
(4) How estimation methods fit into practical applications.

2 Parameter Estimation for Model Building in Industrial Systems

At the heart of estimation based methods is the need to build a simple model of the process to be monitored. This in turn relies on the availability of measurable inputs and outputs. The relationship between input and output is the desired model.

The majority of industrial systems have recognisable inputs and outputs. Consider the examples in Table 1.

Table 1. Input-Output pairs for some processes

Process	Input	Output
Motor-pump set	electrical power	pressure head
Diesel engine	fuel rack position	exhaust temperature
Heat exchanger	valve position	secondary fluid stream temperature
Gas turbine engine	fuel valve input	temperature of compressed air

The relationship between input and output is governed by the physical and chemical factors in the process. For most processes this relationship turns out to be quite simple. Consider as an example the diesel engine where fuel rack position is an input and exhaust temperature is an output. The behaviour is adequately described by a very simple model with the form,

present measurement = linear combination of old measurements
+ linear combination of old inputs

The estimation process needed to develop such a model is well documented (Norton, 1986). The parameters of the model need to be estimated, and this is done using a variation on Gaussian least squares. Notice that once the parameters have been estimated, past values of the data can be substituted in the model formula. The result is a *prediction* of the current measurement. This can be compared with the actual observation and the resulting *prediction error* is a measure of how well the current model fits the data.

The basis for applying estimated models is that the model is fitted at a time when the health of the system has been determined. The model is then compared with models or data from later in the life of the system. Changes in the parameters or an increase in the prediction error indicate that the system is no longer well represented by the model and is "off-design".

3 A Review of Estimation Approaches to Condition Assessment

Of the examples of the application of parameter estimation techniques two are particularly interesting (Geiger, 1984 & St Nold, 1987). The Motor Pump set (Geiger, 1984) is a very widely applied piece of process equipment. Geiger's method is based on a view of the motor pump set as an input-output process. At input is the electrical power required to drive the pump. At the output is the pump speed, head developed and the mass flow rate of fluid.

The successful application of the technique is based on the following steps.

(1) The problem is considered in terms of the governing equations.
(2) The governing equations are transformed into a form in which the important parameters of the motor pump set are isolated, including, for example hydraulic loss coefficients, friction factors, and motor flux linkage.
(3) These parameters are estimated on-line using the available input-output measurements.

Parameters estimates will vary because of measurement noise and factors not taken into account. The degree of variation needs to be learnt.

The decision step is based on differentiating between "normal" parameters and those derived from the system in an off-design condition. In this case the identification of which parameter is deviating gives the diagnosis directly. Geiger asserts that in this case the motor pump set should not need to be taught with fault examples. Instead it should learn during normal conditions what the typical variation in parameters is. It should then do a simple on-line significance test at regular intervals to determine whether the parameter estimate is now significantly worse than it was at the last test or at some previous "datum" test.

This approach is very attractive for the following reasons.

- it removes the need for a series of teaching tests.
- it overcomes the problem of variations between equipment. This can be a problem where a single fault threshold is being established for a large number of similar machines.
- the test can be reversed to allow the assessment of the effectiveness of a repair.

The second example still uses an estimation approach, but in a different role. The effective impedance of an AC induction motor is reduced to a form which is *linear in the parameters* or one in which the unknown parameters, appear singly rather than combined or raised to a power. In this form conventional parameter estimation can be applied and through some simple computation, values for main and leakage reactances, and rotor resistance can be obtained.

The measurements which need to be made are just voltage and current in each of two phases. The measurements may be made remotely from the motor, obviating the need for instrumentation close to or on the motor itself.

This method has proved very effective in detecting rotor temperature, and the paper's author anticipates application to the detection other faults which are temperature or air gap related. The success of the method relies on the motor seeing a varying load.

4 Estimation Methods Applied to Diesel Engines

The application of such methods to engines is a further variation on the basic technique. The first choice is of a series of input-output pairs to which models can be fitted. Such pairs will generally be easily measured, for example, exhaust temperature or manifold temperature paired with an input such as throttle position. The choice of low cost accessible sensors is an obvious advantage of the method.

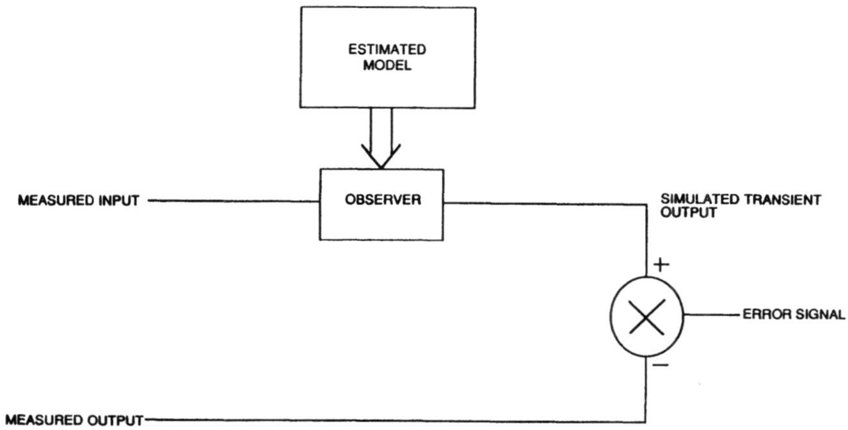

Figure 1. A scheme for comparing a model with reality

The input-output models are readily fitted and are then available to assess future engine performance. In the previous examples, the estimated parameters could be used directly since each carried a physical meaning. In the engine case, the underlying complexity of the system means that no simple mapping is possible. Instead an attempt must be made to assess the current system behaviour by using an original "healthy" model. A scheme for doing this is shown in Figure 1.

The model is estimated using the input-output data. It is then installed in an observer which in turn behaves like the system from which the data was originally derived. In this way it is possible to simulate the original healthy system and then compare the resulting signal with what is actually observed. The error or *residual*, gives a strong indication of a difference in behaviour.

For an example see Figure 2 in which the error is based on a model derived from exhaust temperature data in a turbocharged diesel engine. The peak resulted from a change in the flow of coolant to the intercooler and was accompanied by a change in manifold air temperature of less than 10°C. The peak is very clear and could be detected on the basis of thresholds. The change occurred over a period of about 20 seconds. In a realistic case of intercooler fouling such changes may occur over days or weeks. In this case treatment of the error signal over that time scale will give the required indication.

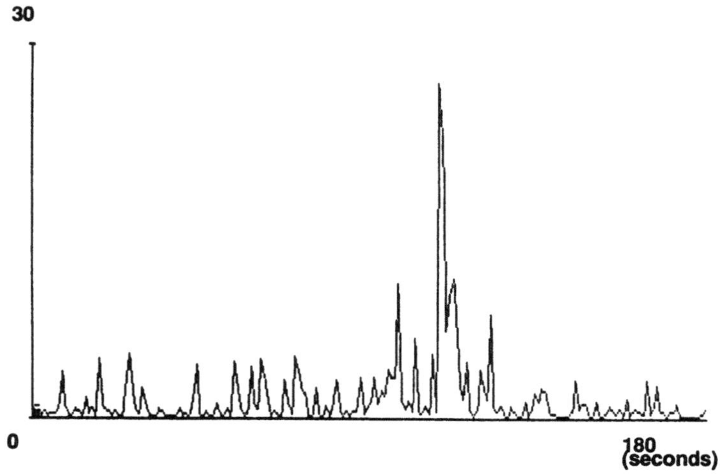

Figure 2. The error signal from an observer monitoring exhaust temperature

5 The Place of Estimation Methods in Condition Based Maintenance

An estimation approach to condition monitoring is able to borrow very substantially from work done in developing adaptive control. An adaptive *predictive* controller uses a system model to derive the control signal at each sample instant. At each sample instant an estimator running in parallel may update its own model. By comparing the *prediction error*, the quality of model currently being used for control can be assessed.

The understanding of on-line estimation has improved steadily in the last 10 years (for an example of improved methods see (Goodwin et al, 1985)). The most recent methods are based on matrix factorisation, and techniques to ensure responsiveness at times of sudden change while retaining stability in the parameter estimates at other times.

The user of such systems would not be aware that there was parameter estimation going on. To the designer of diagnostic systems such methods may offer the possibility for conducting a "first level" diagnosis prior to conducting a more detailed investigation. Consider two cases.

(1) In some systems tests on system components are costly either because the system must be stopped and production lost or because the effort to conduct the test is significant. A first level diagnosis would help justify the decision to make a more detailed investigation.

(2) In a computer based monitoring system with limited resources, the availability of a first level diagnosis would permit a better ranking of diagnostic investigation. The search through various possibilities could then be limited explicitly on the basis of the initial diagnoses. This is especially important in knowledge based diagnostic systems.

6 Conclusions

- Parameter estimation forms the basic technique for building simple but representative models of industrial systems.
- The data used to build models may be derived from existing sensors which in general will be of normal industrial standard.
- A variety of applications may be founded on the principle that models give clues about the inner workings of the system. Comparison between models forms the basis for off-design assessment.
- Estimation techniques may draw on the robust algorithms developed for industrial process control.

7 Acknowledgements

The research work into the application of estimation to diesel engines was supported by the MoD Procurement Executive, Royal Aerospace Establishment, Pyestock.

8 References

Geiger, G. (1984), Fault Identification of a Motor Pump System IFAC World Congress, Budapest, 1783-1788

St Nold (1987), Fault Detection in AC Drives by Process Parameter Estimation, IFAC World Congress, Munich, 399-404

Goodwin, G. C., Hill, D. J. & Palaniswami, M. (1985), Towards an Adaptive Robust Controller, Proceedings of IFAC Conference on Identification and System Parameter Estimation, York, England

Willsky, A. S. (1976), A Survey of Design Methods for Failure Detection in Dynamic Systems, Automatica, 12, 601-610

Pisano, A. D. (1980), Failure Indication and Corrective Action for Turboshaft Engines, Journal of the American Helicopter Society, 25, 36-42

Norton, J.P. (1986), An Introduction to Identification

A high accuracy test technology used for on-line measurement of abrasion wear in bearings

Z. Shengxuan and D. Wenrui
Yanshan University, Peoples Republic of China

Abstract
A novel technology used for on-line measurement of abrasion wear in bearing is presented. It is based on the principle of charge transfer and V/F conversion. The accuracy of the measurement method is higher than that of the conventional ones without any strict requirement to the measurement environment. It makes features of simple and reliable circuits, easy operation and digital display. It can be applied to other cases of engneering for micro-displacement measurement.
Keyword: Abrasion Wear, Charge Transfer, V/F Conversion, Bearing

1 Introduction

Abrasion wear is an appearance that makes materials wear off from the surface by rubbing, grinding, and other machenical actions. The shortcomings of the conventional methods are not directly reading value of abrasion wear, strict requirements for measurement environment(including temperature, humidity, surface treatment of material), poor repeatitive performance and impossible of on-line measurement and so on. For example, there are derrick cranes in a dock, their arms are about 50 meters long, the weights are several thousands tons, and the diameters of bearings are from 150 mm to 500 mm. Some axles and axle-sleeves are seriously worn after several years operation, there is no lubrication oil film between them, the machines are not able to operate normally. It is necessary to develop a new on-line measurement method to monitor the cranes condition at all times.

2 The charge transfer principle and its application in the measurement of abrasion wear

The physical process of charge injection, store, transfer and collection etc. completed by means of capacitance to realize signal processing is called the charge transfer principle. If this principle is used to complete wearing or micro-displacement measurement, a capacitor must be installed at the measurement position.

In general, the rotating angle of the derrick crane is less than 30°, as shown in Fig 1. The abrasion wear of the axle and axle–sleeve in this region is more serious. The electrode B of the capacitor is installed on the 30° radial line of the axle, another electrode A is installed on the axle–sleeve opposite the electrode B. About 5 mm thick silicone rubber is laid between electrode and axle (or axle–sleeve) to avoid a short circuit. More than 60° radial lines must be occupied by electrode A in order that the capacity does not change when axle rotates in 30°. The capacity C is

$$C = \xi_0 \frac{S}{d} \tag{1}$$

where ξ =8.854x10 F/m is the air dielectric constant, S is the effective area of electrode B, d is the distance between two electrodes(Sears et al 1980). If the variation of capacity C is measured, then the distance d or the wearing quantity is known.

Assume that the radius of axle is 75 mm, the distance d between the two electrodes at the beginning is 3 mm, the height of the electrodes A and B is selected as h=30mm. Therefore the area of electrode B is

$$S = hl = h.2\pi R\frac{30^o}{360^o} = \frac{\pi Rh}{6} \tag{2}$$

Substituting the values of h and R, we have

$$S = 1178.1mm^2$$

the capacity C is calculated by equation (1) and is given in table 1 when d is changed from 3 mm to 1 mm. It is shown that the capacity is rather small and less then 11 pf.

A practical capacitance sensor always has an earthed screen to protect the electrodes from interference of external electrical field and avoid ash, rain and snow etc. dropping into the sensor. This results in a three terminal sensor consisting of the sensing capacitor and the stray capacitance between the screen and the two electrodes. In many cases the value of the stray capacitance is much greater than that of the sensing capacitor and may fluctuate in value. This stray capacitance effect can be resolved by charge transfer transducer (Songming et al 1988).

The basic circuit of transducer is shown in Fig 2, where C is the sensing capacitor to be measured, C_{p1} and C_{p2} are stray capacitances, S_1 and S_2 are

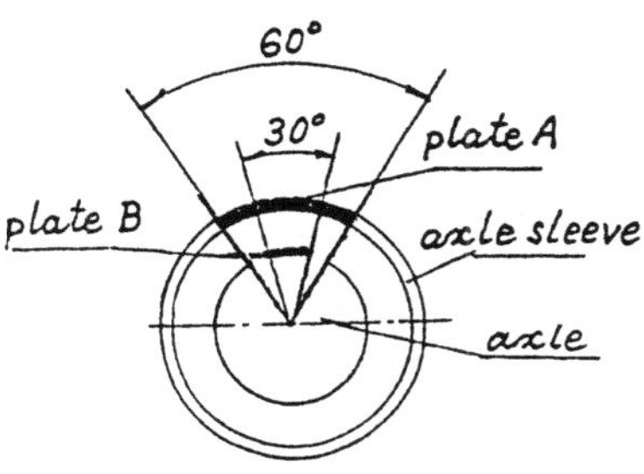

Fig. 1. The capacitance sensor

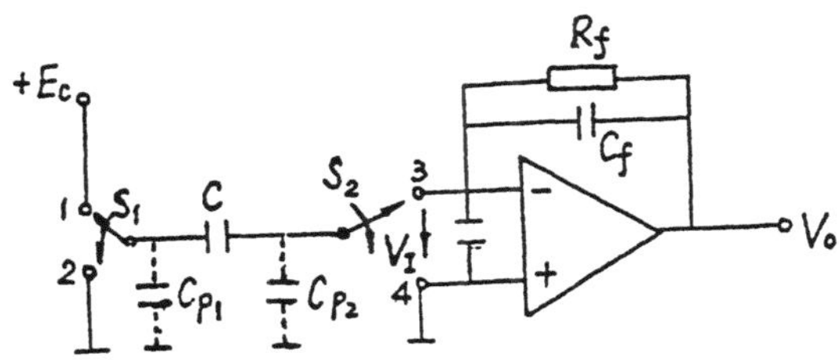

Fig. 2. The capacitance sensor circuit

electronic switches, C' is a decoupling capacitor(C'>>C), R_f and C_f consist of intergration circuit. At the beginning, S_1 is connected to voltage $+E_c$ and S_2 to ground. C and C_{p1} are charged to $+E_c$ after a short time. Then let S_1 switch to ground and S_2 to inverting input of operational amplifier, C and C_{p1} are discharged simultaneously. But the discharged current of C_{p1} does not flow into the amplifier, only the discharged current of C flows to the output of amplifier, i.e. there is only message to be measured in the output voltage V_o. The effect of the stray capacitance C_{p2} is considered as following: when S_2 is connected to terminal 3, the voltage of terminal 3 is almost equal to the voltage of terminal 4 provided an operational amplifier with good performance is selected. Therefore the voltage of C_{p2}(or charge in C_{p2}) almost does not change as the switch S_2 is changed from terminal 4 to 3, and it's effect can be ignored.

In order to quantitatively estimate the effect of C_{p2}, the total charge Q to be transfered during the discharge is calculated by

$$Q = CE_C + C_{p2}V_I$$

Supposing E =15V, V_I=1.5mV, we have

$$\frac{\delta Q/\delta C_{p2}}{\delta Q/\delta C} = \frac{V_I}{E_c} = 0.001$$

It can be seen that the effect of the stray capacitor C_{p2} in the output voltage V_o is nearly completely eliminated.

Because the inverting input(terminal 3)of the operational amplifier is at virtual ground, conducting resistance of S_1 and S_2 is very small, discharge time of C is very short (less than 10^{-9}s), so the discharged current can be treated as a successive impulse train and expressed as

$$I(t) = \sum_{n=0}^{\infty} \delta\left(t-nT_S\right) E_c C(t) \qquad (3)$$

where $T_s=1/f_s$ is the time interval between the current pulses(sampling interval) and C(t) is the sensing capacitor's signal. The Fourier transform of I(t) is

$$I(\omega) = f_S E_C \sum_{k=0}^{\infty} C\left(\omega-k\omega_S\right) \qquad (4)$$

where $\omega_s=2\pi f_s$ is the sampling angular frequency of the transducer and C(I) is the frequency spectrum of C(t) with a band-width of ω_m(see Fig 3). According to the sampling theorem, ω_s is chosen as

$$\omega_S \geq 2\omega_m \qquad (5)$$

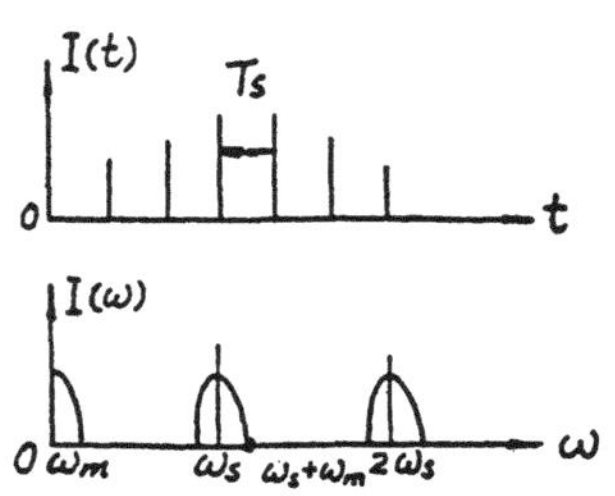

Fig. 3. Discharge current I(t) and frequency spectrum I(ω)

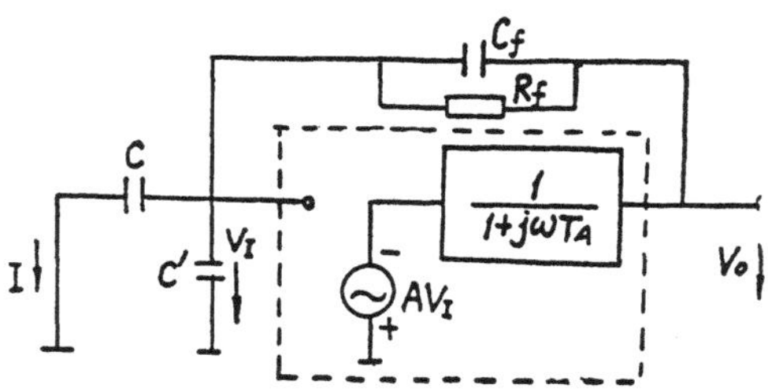

Fig. 4. Equivalent circuit of the capacitance sensor

and in order to recover the signal C(t), the circuit is designed with a low-pass band-width which is

$$B=\frac{1}{R_f C_f}<\omega_S-\omega_m \tag{6}$$

This means the frequency signal(in equation 4) of k>>0 are beyond the pass-band B. If sampling frequency f_s is rather high, the high frequency equivalent circuit of the operational amplifier(Fig 4)must be used. In Fig 4, T_A is an open loop time constant, A is open loop gain (usually A>>1), $AR_fC_f >> R_fC' + T_A$, and the transfer resistance function is

$$H_{(\omega)}=\frac{V_{O(\omega)}}{I_{(\omega)}}=\frac{A}{T_{A(C_f+C')}}\cdot\frac{1}{(j\omega)^2+\frac{AC_f}{T_{A(C_f+C')}}j\omega+\frac{A}{T_{A(C_f+C')}R_f}} \tag{7}$$

we define

$$\xi=\sqrt{\frac{AR_fC_f^2}{4T_{A(C_f+C')}}},\qquad \omega_O=\sqrt{\frac{A}{T_{A(C_f+C')}}}.$$

ξ is called the damping factor of the circuit, ω_o is called the natural frequency. Thus we obtain two poles of the second-order system to be

$$x_{1,2}=-\xi\omega_O\pm\omega_O\sqrt{\xi^2-1}$$

Equation (7) can be writen as

$$H_{(\omega)}=\frac{V_{O(\omega)}}{I_{(\omega)}}=\frac{R_f}{(j\omega/x_1-1)(j\omega/x_2-1)} \tag{8}$$

Combining equations (4) and (8) and considering the condition of equation (6), we have

$$V_{O(\omega)}=H_{(\omega)}I_{(\omega)}=\frac{F_SE_CR_fC_{(\omega)}}{(j\omega/x_1-1)(j\omega/x_2-1)} \tag{9}$$

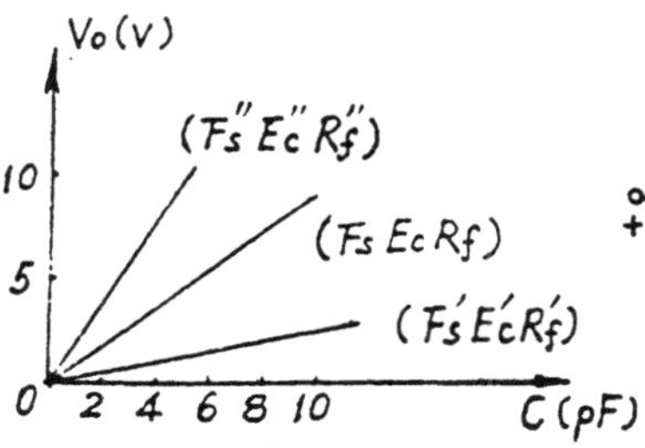

Fig. 5. Relationship between V_o and C

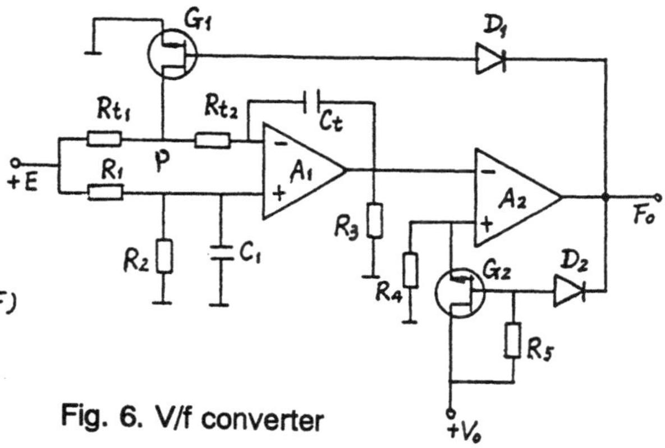

Fig. 6. V/f converter

Supposing the quantity of abrasion wear (or the distance between two electrodes) during each rotating cycle is a small step change, i.e. C(s)=c/s. According to the final value theorem of the Laplace transform, the steady state output voltage of the transducer is given by

$$V_O = f_S E_C R_f C \tag{10}$$

Choosing f_s =1.25 MHz, E_c=15V, R_f=51 kΩ, we have

$$V_O = 0.95625C \tag{11}$$

If the distance d changes from 3mm to 1mm(capacity c changes from 3.477pf to 10.431pf), function V=f(c) is drawn in Fig 5 by equation (11). It is a straight line through zero point and is verified by experiment. If the sampling frequency f_s, or/and the voltage E_c or/and the feedback resistance R_f is changed, the slope of the straight line is changed. Although V_o is proportional to C, because of equation(1), V_o is inversely proportional to the distance d. Thus we must use voltage to frequency (V/f) conversion principle to obtain digital display and a linear relationship between V_o and d.

3 V/f converter

V/f converter consists of intergrator, Schmitt trigger and electronic switch G_1 etc. as shown in Fig 6. A standard voltage source E connected to inverting input of operational amplifier A through resistance R_{t1} and R_{t2}, similarly connected to non-inverting input through R_1 and R_2. C_1 is a decoupling capacitor. If G_1 cuts off, C_t will be charged by E through R_{t1} and R_{t2}. If G_1 conducts, point P is connected to ground, the inverting input voltage is less than the non-inverting input voltage, and the capacitor C_t is discharged through R_{t2}.

The output voltage of the integrator controls the Schmitt trigger. Schmitt trigger consists of an operational amplifier A_2 and an electronic switch G_2. If the output of A_2 is at high level, the diode D_2 cuts off and G_2 conducts, the voltage V_0 come from Fig.2 is fed into non-inverting input of A_2 through G_2. If the output voltage of A_2 is below zero, D_2 conducts and G_2 cuts off, non-inverting input of A_2 is connected to ground. Clearly, the two limiting voltages of the Schmitt trigger are 0 and $+V_0$. The output frequency f_0 of the Schmitt trigger changes with V_0 as shown in Fig 7.

Choosing $R_{t1}=R_{t2}=Rt$, $R_1=2R_2$, the non-inverting input voltage of amplifier A_1 is E/3. The declining rate of output voltage (when C_t charges) and the ascending rate(when C_t discharges)are all equal to $E/3R_tC_t$. Because the conducting voltage of Schmitt trigger is V_0, we have

$$T_1 = T_2 = 3R_tC_t.\frac{V_O}{E}$$

or output frequency F_o is

$$F_O = \frac{1}{T_1 + T_2} = \frac{1}{6R_tC_t} \cdot \frac{E}{V_O} \tag{12}$$

Combining equations (1), (11) and (12), we have

$$F_O = \frac{Ed}{5.7375R_tC_t\xi_O S} \tag{13}$$

For example, if E=5.985V, $R_{t1}=R_{t2}=R_t$=10k ,C_t=1000pf, S=1178.1mm^2,

ε_0=8.854pf/m are given, the relation between output frequency F_0 and distance d is calculated by equation (13) as shown in table 1. If d changes by 1mm, the frequency F_0 will change by 10kHz. In other words, if frequency changes 1Hz, the distance must have been changed 0.1 micrometer. If the unit of frequency meter is micrometer, we can directly read out the change of the distance d or the quantity of abrasion wear in the bearing.

Table 1

d (mm)	C (pF)	V_0 (V)	F_0 (Hz)
3.0	3.477	3.325	30000
2.5	4.172	3.990	25000
2.0	5.215	4.987	20000
1.5	6.954	6.650	15000
1.0	10.431	9.975	10000

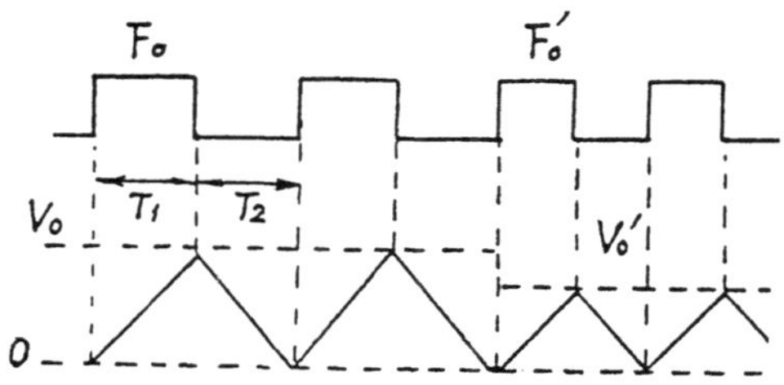

Fig. 7. Input and output waves of V/f converter

4 Conclusion

We have developed a very high accuracy test technology for on-line measurement of abrasion wear in bearings. It is based on the principle of charge transfer and voltage/frequency conversion. It is immune from stray capacitance effects and external electromagnetic interference. Provided the transducer is stationally installed, electronic switches and operational amplifiers with high quality are selected(Swerlien.R.L.1989, and Thomas. M et al, 1989), and suitable temperature compensation is used, the converting linearity of $+3\times10^{-4}$ and frequency error of 5×10^{-4} are obtained in the temperature range from −30 °C to +40 °C. The precision of micrometer order of magnitude is achieved. It is clear that this technology is able to apply to other micro-displacement measurement applications.

5 References

Natarajan.S.(1989). **Measurement of capacitances and their loss factor. Instrumentation and measurement**, IEEE. Dec. 1989.

Sears.F.W. Zemansky.M. and Young.H.D. College physics, 1980.

Songming.H., Robert.G.G., Andrzey.P. and Maurice.S.B. **A high frequency stray-immune capacitance transducer based on the charge transfer principle**. Instrumentation and measurement, IEEE. Sept.1988.

Swerlein.R.L. **Precision AC voltage measurements using digital sampling techniques.** Hewlett-Packard. April,1989.

Thomas.M and Higgius.Jr. **Analog output system design for a multifunction synthesizer.** Hewlett-Packard. Feb, 1989.

Machine diagnostics using the concepts of resolution ratio and automatic timing

D.M. Benenson, A. Soom and S.Y. Park
State University of New York at Buffalo, Buffalo, New York, USA
S.E. Wright
Electric Power Research Institute, Palo Alto, California, USA

Abstract
Noninvasive diagnostics have been developed and applied to determine the condition of circuit breakers. Vibration data are obtained from transducers placed at appropriate locations on a unit. Analysis of the vibration signatures leads to the characterization of a circuit breaker via a single index, the resolution ratio (RR). The analysis of the vibration signatures also includes the automatic timing (AT) of events occurring during operations of a unit. Both RR and AT are being integrated into a multifactor decision making process that will evaluate diagnostic inputs from several sources.
Keywords: Noninvasive Diagnostics, Circuit Breakers, Vibration Signatures, Maintenance, Reliability

1 Introduction

There are more than 40,000 power circuit breakers in the United States. These units play a key role in the protection and operation of electric utility transmission and distribution systems. While these devices operate infrequently, they must be reliable. The severe operating requirements, together with mechanical complexity, lead to the requirement of periodic mechanical and electrical inspections and overhauls. At such times, the operating mechanism is removed, the containment vessel (or tank) is opened, components are inspected, and the mechanisms are adjusted to predetermined standards. This task takes from two to ten days for a maintenance crew to complete. Thus, noninvasive diagnostic methods that are capable of detecting or predicting the onset and/or presence of malfunctions - in a reliable and cost effective manner - would be of substantial interest to the utilities and their ratepayers.

The objectives of this Electric Power Research Institute and New York State Energy Research and Development Authority project are severalfold: (1) to develop and test advanced noninvasive diagnostic procedures that can detect changes in the operating vibration (or acoustic) signatures of power circuit breakers, (2) to determine the mechanical condition of the units - based upon these changes in signatures, (3) to identify the defect causing the changes, if the unit is declared to be in abnormal condition, (4) to determine the times at which important events occur during operation, (5) to integrate the several diagnostics developed into a comprehensive, multifactor decision making process, and (6) to develop and test a portable diagnostic system - for use by utility personnel - that will carry out the signal processing and decision making procedures generated. This paper will address resolution ratio and automatic timing aspects.

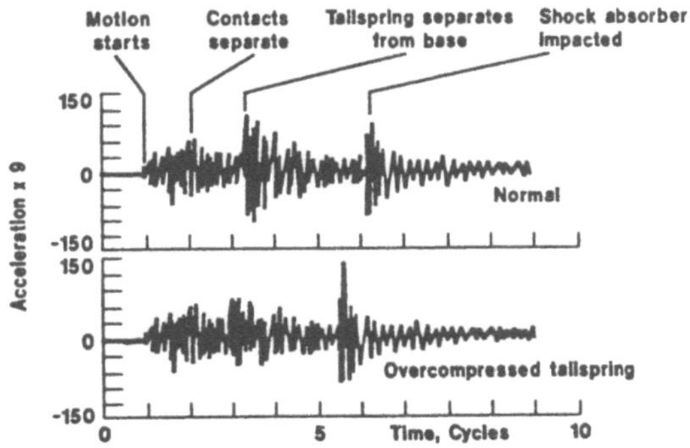

Fig. 1. Vibration signatures. Normal and overcompressed tailspring.

2 Resolution ratio

Information about the condition of a circuit breaker is embedded in its open (trip) and close vibration signatures. The trips and closes yield different information in respect to the mechanical operation of a unit. Both trip and close signatures, then, are employed in our analysis procedures. Vibration signatures are measured by transducers that are strategically placed on a unit - e.g., on the three-phase poles and at various locations on a given pole. Signatures for normal conditions and for an abnormal case (overcompressed tailspring) are shown in Fig. 1.

The signal bursts are typically windowed separately, in regions corresponding to significant mechanical events taking place during an operation (Fig. 2a). Analysis is then performed upon each of these individual time windows, and is compared with the corresponding analysis for normal (or healthy) signatures (Fig. 2b). This comparison enables the condition of a circuit breaker to be characterized by a single index, the resolution ratio - RR. Resolution ratio is particularly responsive to structural changes, e.g., defective shock absorber. In contrast, AT is particularly responsive to timing adjustments, e.g., overcompressed tailspring. Both techniques also overlap in their responses to defects and timing adjustments. Many analysis candidates are being evaluated at this time, including autocorrelation, short time energy, inverse filtering, short time spectra, and short time zero crossing rate.

Resolution ratio is defined as the ratio of (1) the numerical difference analysis of a circuit breaker in normal condition and that unit in test condition, to (2) the numerical difference analysis for that circuit breaker in normal condition. The numerator, (1) above, addresses the condition of the circuit breaker undergoing tests. The denominator, (2) above, addresses the repeatability and statistics of operations under normal conditions.

For a circuit breaker in normal condition, RR would be about unity. An RR in excess of about two would indicate that significant change has occurred in respect to the signatures that characterized the unit in normal condition, i.e., the characterizing normal signatures. The circuit breaker would then be declared to be in abnormal condition.

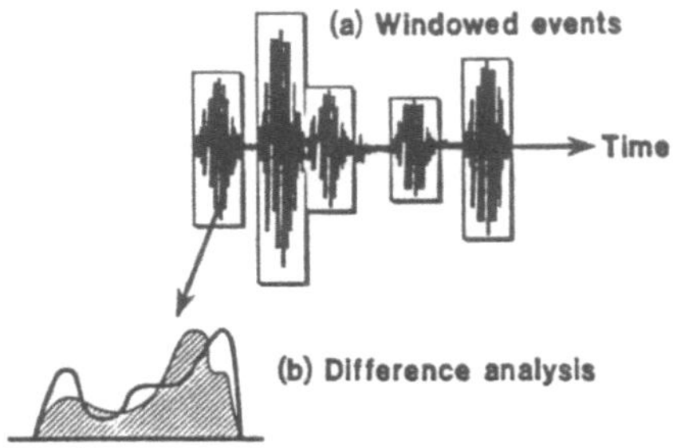

Fig. 2. Windowed events and difference analysis at a given event.

Table 1. Resolution ratios

Circuit Breaker*	Circuit Breaker Condition	RR
1 - GB	Normal	1.01
Closes	Eroded contacts	3.60
2 - GB	Normal	N/A
Trips	Damaged shock absorber	2.06
3 - 0	Normal	1.19
Closes	Shallow contacts	2.11
	Loose contacts	1.79
4 - 0	Normal	1.16
Closes	Misadjusted overtravel stop	6.07
5 - 0	Normal	0.93
Closes	Moving contacts removed	5.90
	Overcompressed tailspring	2.28
6 - 0	Normal	1.09
Closes	Minimum trip voltage	11.10
	Minimum oil pressure	4.47
	Minimum oil level	3.15

*: GB = gas blast; 0 = oil

3 Detection of abnormalities using RR

Circuit breaker tests have been carried out at many sites - utilities' switchyards, manufacturers' plants, and upon a unit in our laboratory at the University. Using the procedures cited in Section 2, RRs have been obtained for a number of oil and gas blast units - primarily involving trip and close operations. The tests have been conducted under unenergized and, more recently, energized-loaded (on-line) conditions.

Table 1 illustrates the RRs obtained for a variety of circuit breakers and for a range of defects and adjustments. Substantial values of RR are found - ranging from about 2 to around 11 - in the presence of defects and adjustments.

4 Automatic timing

Through our analysis procedures, the timing of events is obtained automatically from the vibration signatures (Fig. 3). The automatic timing method incorporates the effects of shot-to-shot repeatability. Further, closely spaced events - e.g., to around 0.1 cycle - are distinguished.

Timing of events offers another independent dimension for diagnosing the condition of a circuit breaker. As noted earlier, this approach is particularly sensitive to adjustments. The timing procedures automatically determine (1) the absolute timewise locations of events occurring during operation and (2) the time intervals between events and, as the condition of a circuit breaker changes, the corresponding changes of these intervals. The timing analysis is carried out using only the vibration signatures; no additional information is required.

5 Detection of abnormalities using AT

Table 2 presents the automatic timing of events occurring within an oil circuit breaker operated under normal conditions and with adjustments. The relatively low standard deviations for each time (generally less than 0.05 cycle) are strongly indicative of the repeatability of the (typically four) operations carried out for each condition. The accuracy of the timing analysis greatly exceeds the repeatability of circuit breaker

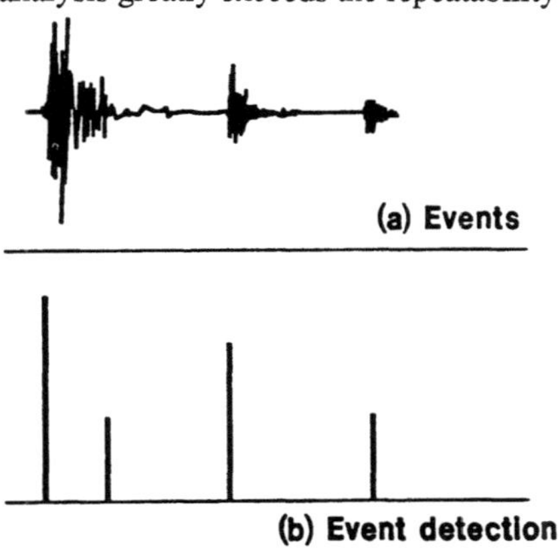

Fig. 3. Automatic timing of events.

Table 2. Automatic timing of events, cycles; circuit breaker 6

Condition	Closes T_s	T_{cm}	$T_{cm} - T_s$
Normal	3.37 (0.03)*	9.92 (0.03)	6.55 (0.01)
Trip voltage maximum	3.13 (0.06)	9.64 (0.05)	6.51 (0.01)
Trip voltage minimum	5.31 (0.13)	11.89 (0.07)	6.58 (0.07)
Operating pressure minimum	3.36 (0.02)	10.46 (0.03)	7.10 (0.01)

Condition	Trips T_s	T_{cp}	$T_{cp} - T_s$
Normal	0.71 (0.01)	5.47 (0.03)	4.75 (0.03)
Trip voltage maximum	0.66 (0.01)	5.38 (0.01)	4.72 (0.01)
Trip voltage minimum	1.04 (0.01)	5.76 (0.01)	4.72 (0.01)
Operating pressure minimum	0.71 (0.01)	5.43 (0.01)	4.72 (0.01)

*parentheses show standard deviations

T_s = time of start of motion; T_{cm} = time of contacts touch; T_{cp} = time of contacts part

operations. Detection of timing changes is clearly evident - in combinations of times for start of motion, contacts make or part, and intervals between the respective times.

6 Shot-to-shot operations

This aspect addresses two areas of interest: (1) the condition of the circuit breaker at its first operation and (2) the repeatability of several consecutive operations (e.g., four consecutive trips and closes). Characterization of a circuit breaker on its first operation is a matter of vital interest to the utilities. In essence, the first operation (e.g., switching) determines the reliability of the unit. The short-term repeatability (over several consecutive trips and closes) helps to establish the statistical envelope of this complex mechanical device.

Table 3 illustrates the RRs obtained for each of four consecutive trips and closes (the circuit breaker is in normal condition and is unenergized). The composite RRs are based upon the respective trip and close average signatures. Owing to the nonlinearities within the difference analysis, the average of the respective individual RRs is not equal to that of the respective average RRs. The RRs obtained for each of the individual shots are closely repeatable. Further, the initial shot (after the unit has been taken off line) is similar (in RR and in analysis) to the succeeding operations.

These analytical procedures, together with our automatic timing techniques (Sections 4 and 5), will enable the first, and most critical, operation of an on-line circuit breaker to be characterized accurately.

Table 3. RR - shot-to-shot operations; normal condition; unenergized; circuit breaker 7

Shot Number/RR	Closes	Trips
1	1.26	0.92
2	1.39	0.87
3	1.32	0.93
4	1.26	1.16
Composite (1-4)	1.14	0.93

7 Long-term repeatability

The University laboratory circuit breaker has served as a test bed - for such purposes as imposing defects and adjustments under controlled conditions, and for developing experimental techniques, instrumentation, and the portable diagnostic system. The unit was frequently returned to normal condition, following one or more series of tests under abnormal conditions.

Table 4 illustrates the RRs obtained over about a one-year period, when the unit was operated in normal condition (trips). Each RR is an average of four shots. The results show that reasonable repeatability was achieved during this time. The large excursions in RR are probably associated with the age of the unit, around fifty years, and the associated difficulties of maintaining settings over time. As a further consequence, the standard deviations of the long-term operations are greater than those of the shot-to-shot operations (Section 6).

8 Universality

Given the visually similar appearance of units of a given model and rating, it would seem reasonable to expect that the vibration signatures of these units - when in normal condition - would also be similar. If this concept were to be valid, analysis of the test condition signatures of such units would be simplified. For all circuit breakers of a given model and rating, then, one would be able to use the characterizing normal signatures, from any one unit, in the difference analysis procedure for all units of that model and rating

On this basis, and considering such a set of similar circuit breakers in normal condition and, further, employing, for all units, the characterizing normal signatures from one unit, the RRs of all the remaining units should be around unity. This concept has been tested over a number of cases. The RRs obtained in a typical study are shown in Table 5. Here, four similar units [circuit breakers 9-12] - the same model, and the same current and voltage ratings - were tested in normal condition. Circuit breaker 9 was used as the characterizing normal for units 10-12. Rather than obtaining RRs around unity for circuit breakers 10-12, large RRs were found, ranging from 4.0 to 6.3 (closes) and from 3.1 to 4.7 (trips).

The vibration signatures of such circuit breakers are roughly the same. . The large differences in RR found in Table 5 come about as a result of the sensitivity of our methods to changes. The results obtained are similar to those that would be expected from analysis of human electrocardiograms (ECGs). Generically, human ECGs exhibit rough similarity. Examination of their details, however, will generally demonstrate significant differences - changes - among individuals.

To date, then, and with both oil and gas blast (including puffers) circuit breakers, the concept of universality of signatures has not proved to be viable, i.e., units of the same model and rating are not, in detail, closely similar. The absence of universality comes

Table 4. Long-term repeatability; normal condition; trips; circuit breaker 8

Date	RR	Date	RR
February 1988	1.17	April 1988	1.53
February 1988	0.91	April 1988	1.44
March 1988	1.10	February 1989	0.94
April 1988	1.58	February 1989	1.18

Table 5. Universality tests (RR); normal condition; circuit breakers 9-12

Circuit breaker*/RR	Closes	Trips
10	6.35	4.27
11	6.32	4.75
12	4.04	3.15

*Circuit breaker 9 used as characterizing normal

about as a result of differences in the manufacturing process, in the respective installations on site, and in the various adjustments made to each unit.

The absence of universality is inconsequential, however, since we are interested in changes occurring in a given circuit breaker. Operationally, our analysis procedures, for example, allocate a separate, individual disk for each circuit breaker. Such procedures are employed in tests at the participating utilities and at manufacturers.

9 Multifactor decision making

Our diagnostic procedures currently incorporate resolution ratio and the automatic timing of events. These procedures are being integrated into a multifactor decision making process that will consider each of these inputs, and others, in assessing the condition of a circuit breaker and, as well, in identifying defects. The ability to resolve defects and adjustments for a given unit (high RRs for defects, as compared with values obtained under normal conditions, together with high shot-to-shot repeatability) and the accuracy of the automatic timing techniques make this approach viable. Multifactor decision making will be addressed in a subsequent paper.

Acknowledgements

The research is supported by the Electric Power Research Institute, Research Project 2747-1, and by the New York State Energy Research and Development Authority, Agreement No. 1859-EEED-TC-86.

The participation of Dr. Victor Demjanenko, and the assistance of Mr. Daniel P. Hess, Mr. Manoj K. Tangri, Mr. Stavros G. Vougioukas, and Mr. Jeffery C. Worst, State University of New York at Bufaflo, is acknowledged.

The authors acknowledge the support and encouragement of Mr. Stig Nilsson and Mr. Glenn Bates, Electric Power Research Institute; and Dr. Fred V. Strnisa, New York State Energy Research and Development Authority.

The efforts of Mr. Clayton Burns and Mr. Howard Lieberman, Niagara Mohawk Power Corporation; Mr. John W. Charlton and Mr. Dale George, New York State Electric and Gas Corporation; Mr. Shalom Zelingher, New York Power Authority; and Mr. Richard Rudman, Rochester Gas and Electric Corporation, are gratefully acknowledged.

The support and cooperation of Mr. David S. Johnson and ASEA Brown Boveri (ABB), Inc. (Greensburg, Pennsylvania); Mr. Gordon A. Wilson and McGraw-Edison Power Systems, Cooper Industries (Canonsburg, Pennsylvania); and Mr. William E. Harper and Siemens Energy and Automation, Inc. (Jackson, Mississippi) are acknowledged and are greatly appreciated.

Condition assessment and forecasting form plant diagnostic data with parèto model

Cempel Czeslaw
Poznan University of Technology, Piotrowo 3 Str. 60-965 Poznan, Poland

Abstract
The paper concerns with condition assessment and forecasting based on the plant results of vibration condition monitoring. It was possible due to elaboration and aplication of Pareto model of machine condition evolution, which may be observed in real plant condition.

Keywords: Condition Monitoring, Vibration, Condition Assessment, Forecasting, Pareto Model.

1 Introductions

There are three main task to be fulfilled by each diagnostic team in a plant; to find the best symptom for diagnostics, to assess the current machine condtition from the measured data, and to forecast this condition withim the time step of interest. For critical machinery like turbo-agregates there are good standards for allowable vibration amplitude and one can respect them and solve all the problem mentioned above. But for noncritical machinery like: fans, pumps, small electric motors, etc, due to random working conditions our confidence to known diagnostic standards has to be greatly diminished. But even than one can solve all the problems; i.e. to chose the symptom, to assess and forecast condition as below.

2 Method

Having M (5 for example) machines under the supervision and taking the measurements of the symptom S by every $\Delta\theta$ time step (~ 2 week for example) in the running time period θ_l (~ 1 year f.ex.), one has:

$$N = M \frac{\theta_l}{\Delta \theta} = \frac{5 \cdot 52}{2} = 130$$

observations of symptom S . Basing on this set of data one can calculate sympton reliability R(S) of our machinery set $R(S_i) = \frac{n_g(S_e > S_i)}{N}$ $i = 0,1...n$ = number of intervals of S.

Here S_e measured symptom value, S_i = interval values of symptom $i = 0,1,...n = 15 - 20$ intervals at the whole range of S. Presenting this empirical data on log - log plot i.e: $R_e(S) = f(S)$ (see Fig.1), one can find by linear regression technique the needed parameters of Pareto reliability model : $R(S) = (S_o/S)^{\gamma}$; i.e. minimal symptom value S_o(much better than from experiment) and also the value of the exponent γ.

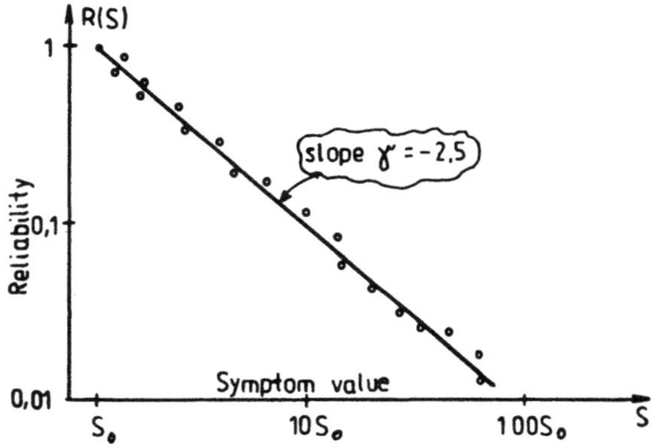

Fig.1. Empirical log-log symptom reliability plot for γ and S_o assessment

Knowing γ itself one can: 1°, to assess the fault orientation of given symptom, 2°, to assess plant maintenance policy in given time interval, 3° to assess maintenance quality of different maintenance teams. In all these cases: <u>the results are optimal when corresponding γ value is minimal.</u> In order to assess the condition of given machine we need to determine prebreakdown symptom value S_b adjusted to our plant. To do that one needs additional data on our machinery set i.e availability:

$$G = \frac{\sum \text{time up}}{\sum \text{time up} + \sum \text{time down}} \tag{1}$$

and prescribed maintenance policy at a plant -'A'.

It simply means what percent of machine set we allow to repair needlessly (false decission) in order to avoid unplaned breakdown. ('A' is usually taken as = 0,02 - 0,05 = 2 - 5%). Having these data one can calculate:

$$S_b/S_o = \sqrt[\gamma]{\frac{G}{A}}\ , \qquad (2)$$

and assess the condition of particular machine at which symptom value S_e was observed, as below:

$S_e < S_b$ - good condition ======> prognosis needed,

$S_e > S_b$ - faulty condition ======> shut down needed !!

If machine is in good condition one needs to know how long it will last. Hence the assessment of residual time to break-down (or shut-down) $\Delta\theta$ is needed. This may be calculated from the same model by the formula:

$$S_e/S_o = \sqrt[\gamma]{\frac{\theta_b}{\Delta\theta}}\ , \qquad (3)$$

as a fraction of time to break-down - θ_b, when S_e was measured (see Fig.2).

The both problems can be placed on one graph as below. Hence knowing the plant maintenance policy G/A one can assess symptom break down ratio S_b/S_o. Then having symptom value S_e measured on particular machine one can assess its condition and time to shut down $\Delta\theta$ in this case. It is worthwile to note that this procedure is plant oriented and allows to make improvements when more data are available.

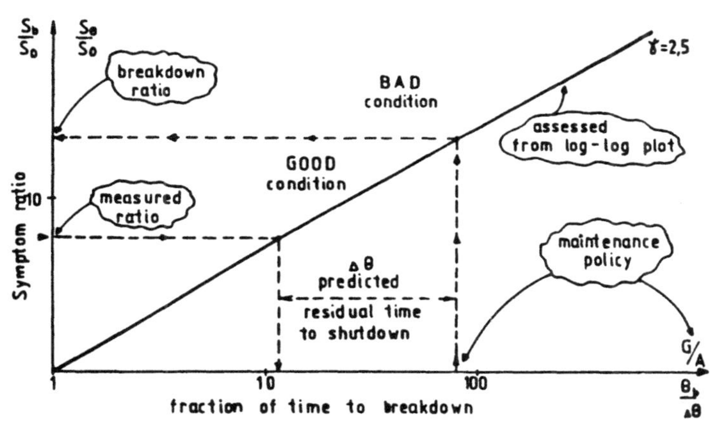

Fig. 2. Condition assessment and forecasting graph.

3 Conclusions

Summing up all what has been found receently [1] one can say that by the small processing of symptom values, gathered during routine monitoring of the machinery in given plant one can:

1) From the observation of machine set derive the conclusion on overage life time behaviour of a particular unit,
2) Assess the quality (fault orientation) of diagnostic symptom -S currently used for monitoring , and to choose the best one with minimal γ rule,
3) Determine plant oriented symptom shut-down value - S_b, and currently improve it when more data are available,
4) Determine the condition of given machine for current measured value of symptom S_e,
5) Assess the residual time to shut-down for the same symptom S_e,
6) Assess the maintenance policy in a plant and choose the best one, (with minimal γ rule),
7) Assess the maintenance execution by different maintenance teams and optimize it, (γ as above),
8) Differentiate and optimize process oriented diagnostics (departmental) and maintenance within the same plant, (γ as above).

4 References

1, Cempel C., Condition evolution of machines and its assessment from passive diagnostic experiment. Sent to **Mechanical System and Signal Processing.**

Drift-fault diagnostics using fault-curves fitted to changes in dynamic response

P.R. Drake
University of Wales College of Cardiff, Department of Electrical, Electronic and Systems Engineering, Cardiff, UK

Abstract
A drift-fault diagnosing method is presented based on fault-curves fitted to changes in a single characteristic of a system performance curve such as the peak of the step response. The method consists of fitting a fault-curve, for each potential fault in the system, to samples of the characteristic produced under different levels of the faults. Unknown faults are then diagnosed, irrespective of severity, by seeing which fault-curve is closest to the characteristic produced by the system under test. It is argued that this method is very appropriate for diagnosing faults detected by condition monitoring systems since these will be drift-faults. The method is demonstrated by simulation.
Keywords: Fault Diagnositics, Condition Monitoring, Drift Faults.

1 Introduction

Many fault diagnosing techniques are based on an analysis of a system's response to transient inputs in the time-domain e.g. step and impulse responses. A commonly used philosophy is one of trying to match the response produced under an unknown fault condition against a dictionary of responses produced under known fault conditions. Some of these techniques are described in detail by Williams (1985).

In general, these fault diagnosing techniques require the response to be sampled at certain fixed points in time, relative to the start of the input transient. The vector of values obtained at these points in time are then compared, in some way, with the vectors in the fault dictionary. This may cause problems when dealing with drift-faults which can occur with various levels of severity. This is because the times at which the characteristics of interest occur (such as deviation from tolerable response) for a particular fault, may be significantly different for different levels of that fault. In other words, the dynamic response of a faulty system will usually experience distortion in respect of time as well as amplitude.

The author proposes that one way of getting over this problem is to diagnose faults by using response characteristics which may be located without reference to time. Obviously, the maxima and minima in response curves are such characteristics. The method presented introduces fault-curves to define how a character drifts under different fault conditions.

This paper illustrates, by simulation, the application of fault-curves to the peak overshoot of the step response of an underdamped and stable system modelled by a third-order transfer function. The peak overshoot of the step response is defined by its amplitude and relative time of occurrence.

An advantage of this method over some other methods, particularly frequency response methods (Varghese et al. 1978), is that it can be applied to many types of normal operational signals without the need for any special test signals to be input to the system under test (SUT). This makes the method suitable for condition monitoring (CM) applications since ideally CM is performed in-line with the normal operation of the SUT (Institute of Production Engineers, 1987).

This drift-fault diagnosing method has another useful CM feature. As a performance characteristic is seen to drift along a fault-curve, early warning of the eventually intolerable fault level can be given. The fault-curves could be calibrated so that an estimate of the level of the drift-fault might be given with the warning.

The author has previously presented the application of fault-curves to system parameters, Drake and Williams (1989).

2 The Method

It is suggested here that if a curve is statistically fitted to a sample of a displaced characteristic, taken over a range of levels of a particular fault, then a system containing such a fault, irrespective of severity, will possess an instance of the charactristic on, or near to, that curve.

If curves are established for each of the possible system faults, then faults may be diagnosed by seeing which curve their instance of the characteristic is closest to - assuming that only single-faults can occur within the system. The method of calculating the shortest distance between a point and a curve is given in Appendix 1. These curves are referred to as fault-curves.

Intuitive reasoning suggests that the closer a faulty system's instance of the characteristic is to a particular fault-curve, then the more likely the actual fault is the fault corresponding to that fault-curve. Therefore, the order of closeness of the different curves, to the faulty system's instance of the performance characteristic, should indicate the relative likelihood of each of the corresponding faults. More precisely, the probability of each fault can be estimated as being inversely proportional to the distances, Drake and Williams (1989).

Establishing the fault-curves by the method of least-squares, rather than by an analysis of the system transfer function, means that a knowledge of the system transfer-function and its relationship with the system parameters is not required for the purpose of establishing the fault-curves.

Note, it would be wrong to assume that the characteristic will lie exactly on one of the curves, even if they were statistically perfect fits. This is because the non-faulty components will have small deviations in their parameter values within their tolerance ranges.

3 Testing The Method

3.1 The System Under Test

To test the effectiveness of the method it is applied to an underdamped and stable third order system with seven components (seven potential faults). The performance characteristic used is the peak overshoot of the system's step response. The system has the following transfer function:

$$H(s) = \frac{a_1 s + 1}{a_3 s^3 + a_2 s^2 \quad a_1 s + 1}$$

The components of the system are represented by system parameters, denoted by X1,X2,...,X7. Each of these parameters is assumed to have a simple multiplicative effect on the transfer function coefficients as follows:

$$a_3 = X1 \ . \ X2 \ . \ X3 \ . \ X5$$

$$a_2 = X1 \ . \ X2 \ . \ X4 \ . \ X6$$

$$a_1 = X1 \ . \ X3 \ . \ X4 \ . \ X7$$

Note, by using seven system parameters it is possible to cover all possible combinations of affecting and not affecting the transfer-function parameters. For the purpose of this exercise the a's are assigned the following values :
$a_3 = 200$; $a_2 = 50$; $a_1 = 10$.

3.2 The Fault-Curves

The first step in the test is to establish the peak overshoot of the step response of the SUT when each of the system parameters/components is subjected to the following faults: ±10%, ±20%, ±30%. Point to point line plots for the peaks are drawn in Figure 1. Fault-curves are fitted to the peaks for each faulty parameter, by using a proprietary statistics package, with the following results (P = peak of step response, T = time/seconds):

X1 faulty : $P = 4.87 - 0.20\ T$

X2 faulty : $P = 1.73\ T^{-0.03}$

X3 faulty : $P = 5.77 - 0.50\ T + 0.02\ T^2$

X4 faulty : $P = -2.59 + 0.26\ T$

X5 faulty : $P = 0.15\ T^{0.85}$

X6 faulty : $P = 8.99 - 0.83\ T + 0.02\ T^2$

X7 faulty : $P = 1/(0.90 - 0.02\ T)$

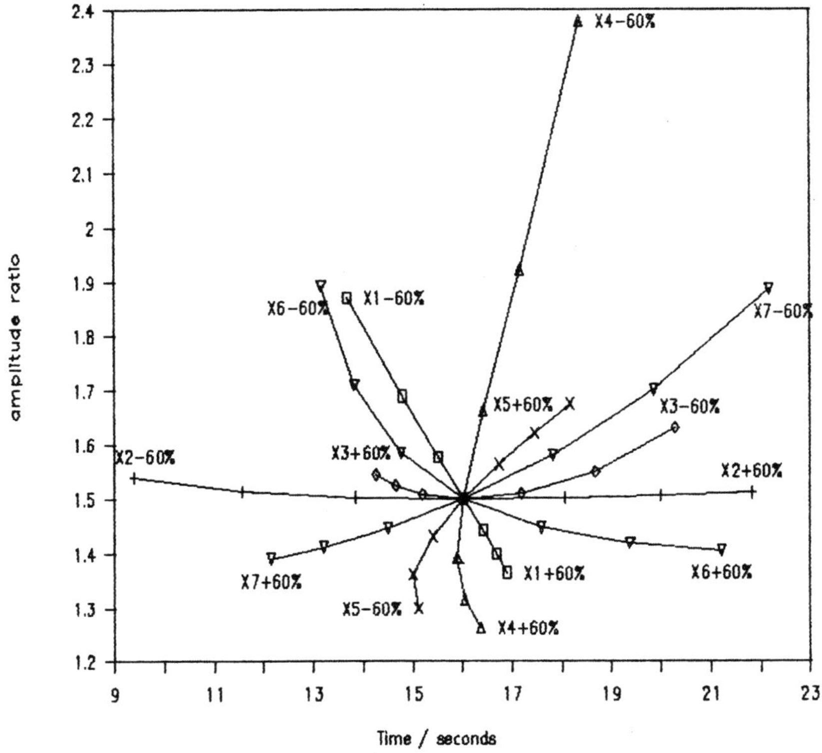

Figure 1 : Peak Overshoots of Step Responses

3.3 The Simulation Procedure

The fault diagnosing method is tested by computer simulation. In each of the trials one of the components in the system is subjected to a level of fault randomly sampled from a Normal distribution of fault levels with a mean value of zero and standard deviation of 10%. This means that 99.73% of the fault levels lie within the range ±30%. Only fault levels beyond ±10% are investigated i.e. only those above some notional level which causes the system performance to be unacceptable. At the same time, the other components are allowed to randomly vary their values within certain tolerance limits which are also specified by Normal distributions but with a standard deviation of 1/3%. This simulates manufacturers' tolerances of ±1% (on the assumption that such tolerances are specified as ± three standard deviations). 100 trials are performed per component i.e.a total of 700 faults are simulated.

The simulation procedure consists of calculating the time and amplitude ratio of the peak overshoot of the SUT's step response, by using the transfer function, and then seeing how close this point is to each of the seven fault-curves. The result is then to see the order in which the known fault is ranked. The results of these simulations are given below.

3.4 Results

faulty component	diagnosis rank 1	2	3	4	5	6	7
X1	94%	6%	0%	0%	0%	0%	0%
X2	74%	26%	0%	0%	0%	0%	0%
X3	72%	28%	0%	0%	0%	0%	0%
X4	100%	0%	0%	0%	0%	0%	0%
X5	98%	2%	0%	0%	0%	0%	0%
X6	100%	0%	0%	0%	0%	0%	0%
X7	100%	0%	0%	0%	0%	0%	0%
Average	91.14%	8.86%	0%	0%	0%	0%	0%

3.5 Discussion on Results

The results are very good except for faults in X2 and X3. The reasons for this are as follows. In the case of X2 the problem is that a large portion of its fault-curve is very closely followed by X3's - see Figures 1 at T = 15.8 to 17.4. As far as this particular application of fault-curves is concerned nothing can be done about this since what it means is that positive faults in X2 and negative faults in X3 have similar effects on the peak of the step response of the system. The consequence of this is that X2 faults are frequently diagnosed as X3 faults and X2 is ranked as the second most likely fault.

Of course, X3 experiences the same problem as X2 i.e. it is frequently mis-diagnosed as X2. X3's poor showing is further accounted for by the fact that the peak of the step response is less sensitive to changes in X3 than to changes in other parameters, this is particularly the case for positive X3 fault levels - see Figure 1. This means that many of the peak overshoots for X3 faults are going to lie relatively near to the nominal peak overshoot. In the region of this point the fault-curves are obviously very close together and fault diagnosing performance will therefore degrade.

Returning to the diagnosis of fault X2, since whenever it is misdiagnosed it is nearly always diagnosed as the second most likely fault, there is still useful information to carry forward into some larger fuzzy-logic/accumulated -evidence type fault-diagnosing scheme. Pieces of potentially useful information should not be wasted. If the next stage in the fault diagnosing process diagnosed the fault as, say, X7 or X2, then clearly there would be strong evidence that the fault was X2.

The results for X2 and X3 yield two conclusions about when the method may produce poor results. First, the method obviously begins to breakdown when different faults have similar effects upon the performance characteristic used for fault diagnosing purposes i.e. the characteristic lacks discrimination power. The second conclusion is that, usually, the higher the sensitivity of the observed performance characteristic to faults then the better the method will work. This is because, for most faults, the further the displacement from nominal the greater is the separation of the fault-curves. Unfortunately, this is not always true. The fault-curve for X6 approaches the fault-curvee for X1 as both parameters experience negative faults of increasing magnitude. However, this happens at high fault levels.

These two conclusions are to be expected since a general philosophy for selecting a system performance characteristic for fault-diagnosing purposes is that the characteristic should be sensitive to faults and it should discriminate between them. Therefore, the inability to produce very good diagnostics for X2 and X3 is not due to the method but due to the choice of performance characteristic chosen for detecting faults in X2 and X3.

Conclusion

It has been shown that the application of fault-curves to a single characteristic of a transient response can produce very good diagnostics. However, as with most fault-diagnosing methods, if the characteristic chosen is relatively insensitive or non-discriminatory for particular faults, then poor results will be produced for these faults.

References

Williams, J. Hywel (1985),
Transfer Function Techniques and Fault Location,
Research Studies Press Ltd., ISBN 0 86380 031 9.

Drake, P.R. and Williams, J.H. (1989),
System Diagnostics using Dynamic Fault Curves,
Proceedings of the Sixth Conference of the Irish Manufacturing Committee on AMT, Dublin City University, ISBN 1 872327 00 1.

Varghese, K.C., Williams, J.H., Towill, D.R. (1978),
Computer Aided Feature Selection for Enhanced Analogue System Fault Location, Pattern Recognition, Vol. 10, pp 265-280.

Institute of Production Engineers (1987),
Condition Monitoring, Production Engineer, June.

Appendix 1 : Calculation of Shortest Distance between the Peak Overshoot of a Step Response and a Fault-Curve.

Non-dimensionalised space is used in this measurement since a change of axis scale would cause a change in the distance calculated. Therefore, the following equation is used:

$$\text{normalised distance} = \sqrt{\left(\frac{t-T}{T}\right)^2 + \left(\frac{y-Y}{Y}\right)^2}$$

where: t and T = time/seconds, y and Y = amplitude ratio
(t,y) = peak overshoot of the step response,
(T,Y) = point on fault-curve.

The closest point on the fault-curve is then the one which minimises normalised distance (t,y,T,Y). Since Y = f(T) (i.e. the statistically fitted fault-curve), and t and y are known values, the distance can be described as a function of T. Therefore, the minimum distance occurs when the derivative of normalised distance (T) w.r.t. T = 0. This can be solved for T using Newton's Method (or any other appropriate method for deriving the roots of a polynomial) and Y is then f(T).

Fundamentals of linear time-varying plant for diagnostic engineering

J. McGhee and I.A. Henderson
Industrial Control Unit, University of Strathclyde, Glasgow, Scotland

Abstract
Using a classification of signals allows a division of time-varying systems by the form of parameter variation. Subsequently, the structure of time-varying impulse response functions provides the theoretical background upon which the diagnosis of first and second order time-varying systems depends. An illustration considers the improvement in step transient response of a second-order system through the addition of a single time-varying parameter.
Keywords: Time-varying Systems, Linear Drift Diagnosis, Transient Response Improvement.

1 Introduction

Plant behaviour is essentially time-varying. If means may be found to characterise the solution with time-variable attributes it becomes possible to predict plant behaviour. As prediction of performance deterioration or failure is a primary goal in condition monitoring, the importance of time-varying systems is evident. The solvability of time-varying systems is a representational problem and not an inherent feature of them (Wu, 1980). This oversimplifying conclusion is not really helpful in their solution. In the general case it is as difficult to find a transformation to reduce particular cases to a solvable class, as it is to obtain an explicit solution to the original form.

The basic dynamical structure of time-varying systems is given. It uses a classification of parameter or coefficient variation similar to that of signals (McGhee and others, 1986). The structure of impulse responses for time-varying dynamics is then presented as the basis for diagnosing the dynamic behaviour of first and second order time-varying systems. A second order system with extrinsic parameter variation is considered.

2 Parameters, structures and responses for time-varying systems

2.1 Coefficients and input/output structures

Considering the scalar form of single-input/single-output systems shows that the schematic structure of time-varying systems is the same

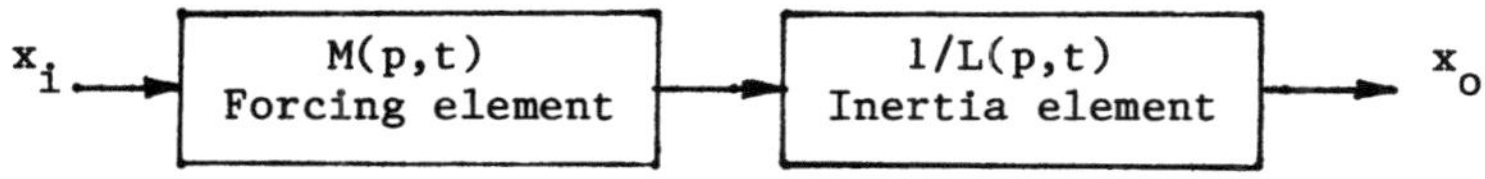

Fig. 1. A system structure with inertia and forcing elements

as in a stationary or constant coefficient system. This is clear when the inertia element/forcing element model of Fig. 1 (McGhee, 1985; Solodov, 1966) is used. The time-varying dynamic then has the form

$$L(p,t)x_o(t = M(p,t)x_i(t) \tag{1}$$

$$\text{where } L(p,t) = \sum_{r=0}^{n} a_r(t)p^r;\; M(p,t) = \sum_{q=0}^{m} b_q(t)p^q;\; p = d./dt;\; n>m \tag{2}$$

The coefficients, $a_r(t)$ and $b_q(t)$, which are functions of time, provide a means of grouping in time-varying systems. A main division for the elements concerns their physical or time-domain structure. Time-domain structure may be considered like the grouping of signals (McGhee and others, 1986). Primary classes may be deterministic, pseudo-random or random. As signals may be analogue or binary, then coefficients may also belong to these types. Hence, deterministic analogue or binary coefficients may be periodic, quasi-periodic or aperiodic. A linearly time-varying parameter is considered later.

2.2 Characteristic response for linear time-varying system models

The behaviour of systems is characterised through the effect of the system upon a known arbitrary signal. Time-domain behaviour is best described by the impulse response of the system. This description of linear system behaviour is far reaching, since it allows assessment of system response to arbitrary signals of any type.

Linear constant systems and time-varying systems have identical impulse response structures as the background given by Kaplan (1958; 1962) and Solodov (1966) verifies. Thus, it seems narrow to restrict any consideration to constant coefficient systems. Without any loss in generality due to its order, consider the linear, second order, time-varying system, whose equation is given by

$$a_2(t)d^2x_o/dt^2 + a_1(t)dx_o/dt + a_0(t)x_o = b_0(t)dx_i/dt + b_0(t)x_i \tag{3}$$

In its operational form, this is

$$L(p,t)x_o = M(p,t)x_i = f(t,\xi) \tag{4}$$

where the inertia element, $L(p,t)$, and the forcing element, $M(p,t)$, are given from eqns (1) and (2) with $n=2$ and $m=1$. Equation (3) is a proportional plus derivative controller equation with time-varying coefficients. A generalised forcing function, $f(t,\xi)$, is applied.

The first of three methods of obtaining impulse response functions is the "method of variation of constants". It assumes a knowledge of $x_1(t)$ and $x_2(t)$, the linearly independent solutions of the homogeneous version of eqn (3). The impulse response is then a Green's function $H(t,\xi) = h(t,\xi).1(t-\xi)$, where $1(t-\xi)$ is a unit step and the "weighting function", $h(t,\xi)$, is

$$h(t,\xi) = -\frac{[x_1(t)x_2(\xi) - x_2(t)x_1(\xi)]}{a_2(\xi)[x_1(\xi)dx_2(\xi)/dt - x_2(\xi)dx_1(\xi)/dt]} \quad (5)$$

A second method requires one solution, $x_1(t)$. A "quadrature" gives a second solution as

$$x_2(t) = x_1(t)\int \exp-\int[a_1(t)/a_2(t)]dt/x_1^2(t)dt \quad (6)$$

Using this in (5) gives another form for the impulse response as

$$h(t, \ )=\frac{x_1(t)x_1(\xi)}{a_2(\xi)} \ [\exp-\int a_1(t)/a_2(t)\ dt]\int_{\xi}^{t}\frac{\exp\int a_1(\xi)/a_2(\xi)d\xi}{x_1^2(\xi)}d\xi \quad (7)$$

A third method, which searches for an integrating factor, $x_1^*(t)$, for (3), terminates at the adjoint equation given as

$$d^2[a_2(t)x_1^*]/dt^2 - d[a_1(t)x_1^*]/dt + a_0(t)x_1^* = 0 \quad (8)$$

For arbitrary $f(t,\xi)$, equation (4) has a non-trivial solution. With $f(t,\xi)$ the impulse $\delta(t-\xi)$, $H(t-\xi)$ satisfies

$$a_2(t)d^2H/dt^2 + a_1(t)dH/dt + a_0(t)H = L(p,t)H(t,\xi) = \delta(t-\xi) \quad (9)$$

Multiply both sides by $f(\xi,\sigma)$, and integrate in $-\infty$ to $+\infty$. As $H(t,\xi)$ and $f(\xi,\sigma)$ are generalised functions, defined for all t and σ, take the inertia operator outside the integral to obtain

$$L(p,t)\int_{-\infty}^{+\infty} H(t,\xi)f(\xi,\sigma)d\xi = \int_{-\infty}^{t}\delta(t-\xi)f(\xi,\sigma)d\xi = f(t,\sigma) \quad (10)$$

Inserting $M(p,\xi)\delta(\xi-\sigma)$ for $f(\xi,\sigma)$ and using the sifting properties of the delta function eventually gives the impulse response of (4) as

$$W(t,\sigma) = -\frac{\partial}{\partial\sigma}[b_1(\sigma)H(t,\sigma)] + b_0(\sigma)H(t,\sigma) \quad (11)$$

Eqns. (3)-(11) constitute the cardinal elements of the theory of second-order time-varying systems for plant diagnostics.

For completeness consider the first order equation

$$a_1(t)dx_o/dt + a_0(t)x_o(t) = b_0(t)\ x_i(t) \quad (12)$$

If any solution, $x_1(t)$, of the homogeneous version of eqn (12) is known then a "quadrature" gives the solution for x_o as

$$x_o = x_1(t)\int_{-\infty}^{t}[b_0(\xi)x_i(\xi,\sigma)]/[a_1(\xi)x_1(\xi)]d\xi \quad (13)$$

This first order system has an impulse response which follows from eqn (13), with $x_i(\xi,\sigma) = \delta(\xi-\sigma)$, as

$$H(t,\sigma) = [b_0(\sigma)/a_1(\sigma)][\exp-\int_{\sigma}^{t}a_0(t)/a_1(t)dt]1(t-\sigma) \quad (14)$$

Initial conditions are accounted for using the following approach given by McGhee (1985). Substitute $y = x_o - x(0)$ in (3) and multiply by $x_1^*(t)$, an assumed known solution for (8). The equation in y obtained, whose left hand side is an exact differential, is then integrated taking account of initial conditions. These operations reduce the original eqn (3) in x(t) with initial conditions, to a first order equation in y(t) with initial conditions equal to zero. It is

$$[a_2x_1^*]dy/dt - [(da_2/dt - a_1)x_1^* + a_2dx_1^*/dt]y = \int_0^t [f(\sigma) - a_0(\sigma)x(0)]x_1^*(\sigma)d\sigma + C \quad (15)$$

with a constant of integration, $C=[a_2(t)x_1(t)(dx_o/dt)]_{t=0}$ (16)

Using the similarity between eqns (12) and (15) it can be shown that the impulse response of the homogeneous version of (15) is

$$y(t,\xi) = [\exp-\int_0^t a_0(t)/a_1(t)dt]1(t-\xi)/[a_2(\xi)x_1^*(\xi)] \quad (17)$$

In line with eqn (10) the final solution is given by

$$x_o = x(0) + \int_0^t y(t,\xi)\left\{\int_0^\xi [f(\sigma) - a_0(\sigma)x(0)]x_1(\sigma)d\sigma + C\right\}d\xi \quad (18)$$

Although the initial conditions are only proved for the second order case, a similar result may be obtained for the first order case.

3 An extrinsic variable parameter controller model

An important benefit of time-varying systems has been demonstrated. The work of McGhee (1985) extends an argument (Bittel and Siljak, 1970) that the transient response of systems may be improved by arranging that the damping is light for short times and heavy for large times. As the approach used by McGhee (1985) considers second order systems, which are reasonably representative of many systems, it fulfills the need for analytical veracity. In Fig. 2 a time-varying rate feedback converts a second order, constant coefficient system into a linear second order time-varying system whose closed loop dynamic is

$$d^2x_o/dt^2 + (1 + R_nt)dx_o/dt + K_nx_o = K_nx_i \quad (19)$$

McGhee (1985) has shown that the linearly independent solutions of

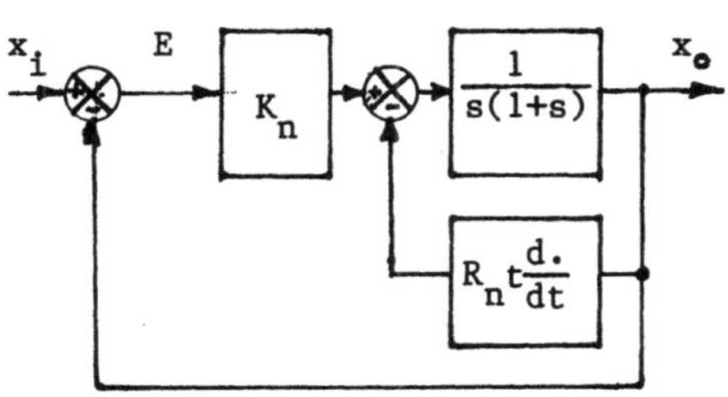

Fig. 2. A second order system with an extrinsic time-varying parameter

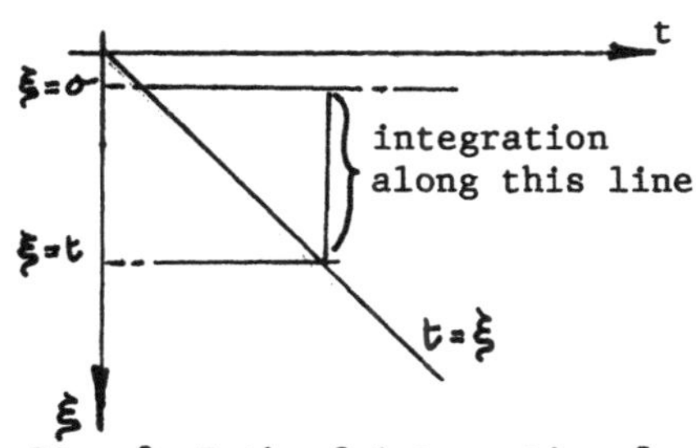

Fig. 3. Path of integration for response to arbitrary input

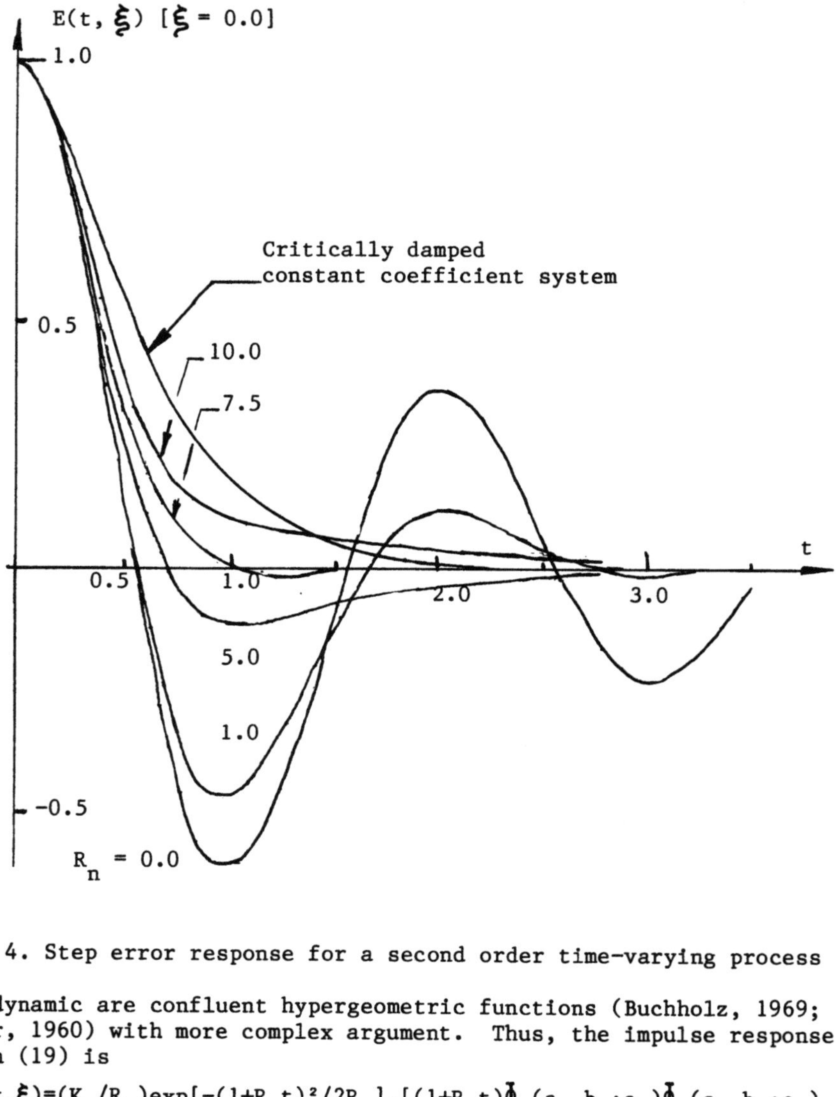

Fig. 4. Step error response for a second order time-varying process

this dynamic are confluent hypergeometric functions (Buchholz, 1969; Slater, 1960) with more complex argument. Thus, the impulse response of eqn (19) is

$$H(t,\xi)=(K_n/R_n)\exp[-(1+R_nt)^2/2R_n].[(1+R_nt)\Phi_1(a_1,b_1;c_1)\Phi_2(a_2,b_2;c_2)$$
$$-(1+R_n\xi)\Phi_3(a_3,b_3;c_3)\Phi_4(a_4,b_4;c_4)]1(t-\xi) \quad (20)$$

where $\Phi(a,b;c)$ is a confluent hypergeometric function with the parameters $a_1=a_3=1-(K_n/2R_n)$; $a_2=a_4=a_1-1/2$; $b_1=b_3=3/2$; $b_2=b_4=b_1-1/2$ $c_1=c_4=(1+R_nt)^2/2R_n$; $c_2=c_3=(1+R_n\xi)^2/2R_n$

Integrating eqn (19) as in Fig. 3 gives the step error responses, illustrated in Fig. 4, as

$$E(t,\xi)=(K_n/R_n)\exp[-(1+R_n t)^2/2R_n]x[(1+R_n t)\Phi_1(a_1,b_1;c_1).$$
$$\Phi_2(a_2,b_1;c_2)/R_n - (R_n/K_n)\Phi_3(a_3-1,b_2;c_3)\Phi_4(a_4,b_4;c_4)]1(t-\xi) \quad (20)$$

The second order system considered is important as it illustrates the behaviour of a system with linear drift. It can be seen from Fig. 4 that these systems exhibit a type of behaviour which is essentially different from, although similar to that of constant coefficient systems. The implications for plant monitoring are embodied in this behaviour.

4 Conclusions

A classification of linear time-varying systems by the time domain structure of their parameters introduced the range of possibilities, which they may give rise to. Adaptations of general time-varying response equations to first and second order cases revealed that integrating factors or linearly independent solutions to the dynamics are required for effective solution. An illustrative example of extrinsic parameter variation indicated the importance of time-varying systems. The variation considered could be equivalent to a representative linear parameter drift, which commonly occurs in many process plant. By indicating the possibility for an improvement of the transient response, the importance of time-varying compensation has been emphasised.

5 References

Bittel, R., and Siljak, D. (1970) An application of the Krylov-Bogoliubov method to linear time-varying systems. **Int.J.Cont.**, 11, 423-429.

Buchholz, H. (1969) **The Confluent Hypergeometric Function.** Springer-Verlag, Berlin.

Kaplan, W. (1958) **Ordinary Differential Equations.** Addison-Wesley, Reading, Mass.

Kaplan, W., (1962) **Operational Methods for Linear Systems.** Addison-Wesley, Reading Mass.

McGhee, J. (1985) Proportional Plus Time-Varying Rate Control of Second Order Processes. **CSCS6, 6th Int. Conf. Cont and Computers**, Polytechnical Institute of Bucharest, Bucharest, Romania.

McGhee, J. Henderson, I.A., Ibrahim, A.A. and Sankowski, D. (1986) Identification : Systems, Structures, Signals, Similarities. **Systems Science IX.**, Sept. 1986, Tech. University of Wroclaw, Poland.

Slater, L.J. (1960) **Confluent Hypergeometric Functions.** University Press, Cambridge.

Solodov, A.V. (1966) **Linear Automatic Control Systems with Varying Parameters.** Blackie, Glasgow.

Wu, M.Y. (1980) Solvability and Representation of Linear Time- Varying Systems, **Int. J. Cont.**, 31, 937-945.

An artificial intelligent vision system for inspection and monitoring

B.J. Griffiths
Manufacturing and Engineering Systems Department, Brunel University, England

Abstract
Vision Systems are being increasingly used in a variety of industrial situations to monitor and inspect manufacturing processes, components and equipment. This paper describes a hybrid vision system combining feature extraction with artificial intelligence for piece part recognition, orientation, tracking and inspection as well as the inspection of surface finish.
Keywords: Vision Systems, Artificial Intelligence, Inspection, Monitoring, Surface Finish.

1 Introduction

Vision Systems have been successfully employed on-line in manufacturing plants to solve difficult monitoring and control problems. Most commercial vision systems use feature extraction techniques to determine geometric based parameters of a part or scene. However, there are still situations where such approaches are limited. This paper describes a hybrid vision system combining feature extraction with artificial intelligence to solve some of these problems. This new system has been given the name WISE (Window Image Shape Extraction), chosen because it is 'wise' in that it acts in a way analogous to a human brain and that having located a part or feature it puts a window round it and confines the AI algorithm to that window.

2. Vision Systems

Vision Systems have been successfully employed for many years to recognise and identify artifacts. In engineering terms the most popular have been feature extraction based systems which are essentially feature calculation. Less well used or indeed understood are the AI or neural mimic approaches. The system used in this case is a hybrid one which combines in a modular manner subsets of both systems which can be matched together to provide the best solution to a vision problem. Conceptually the algorithms used in the AI technique mimic the operation of a group of neural brain cells. The mathematical technique used is called N-tupling which recognises shapes rather than geometric features. It is essentially a single layer neural net. The operating advantages can be seen in the table of Figure 1 where the AI vision systems was applied to the recognition of letters of the alphabet. The letters o,q,g, c,e,f, and b were trained using the system descriminators (the name for the training set). Letter o was trained such that it gave a maximum score of 127 for descriminator 1, letter q with descriminator 2 etc. All the letters were shown to the system again (the read cycle) and in each case the descriminator score was highest at the descriminator corresponding to its trained letter. The similarities between the letters o,g and c can be seen in the fact that the scores are higher than those for the dissimilar letters e and f.

2.3 The Hybrid WISE System

The operation of the system is as follows:
An image of the piece part is captured, edge traced and a window of pre-determined shape, size and configuration is placed around the image. This so far is standard feature extraction. The remaining analysis is then limited to the window and the N-tupling algorithm conducted by selecting an invariant sequence of random pixels and their bit pattern stored. This is repeated a number of times. The system is trained on a known image with the result that a bit pattern memory map is generated (a descriminator). During the recognition cycle the bit pattern memory map resulting from the unknown image is compared with the trained one and a comparison score recorded. As a new shape deviates from the trained one the score will decrease.

The mathematical technique employed is based on the method first proposed by Bledstow and Browning (1959) and developed in hardware by Alexander (1984). This hybrid software AI technique was developed as part of a low cost horizontal belt feeder for feeding piece parts to link with a robotic assembly cell (Elliott & Griffiths 1990). It has been successfully employed for the inspection and monitoring of a variety of piece parts and the remainder of this paper will describe its use in typical inspection and monitoring situations.

3. Piece Part Monitoring

The capability of the system is demonstrated by two parts, a simple two dimensional arrow shape and a three dimensional car rear light. Descriminators are trained corresponding to various orientations such that in the recognition phase the orientation is given by the greatest descriminator response.

The descriminator responses for the arrow shape are shown in figure 2. In this case three descriminators are shown and the responses are for 15° steps. The response is such that the full 360° orientating requirements can be met by a total of 8 descriminators with an interrogation time of 0.3s. The car rear light shown in figure 3 is a hollow cup shape of 160 x 80 x 46 mm made of plastic having a top with equal proportions of amber and red. One end is amber but at the other end it is totally clear. The only feature providing asymmetry from a top view is a small recess and a hole. This made orientation by shape alone very difficult. However, the differing colours give a grey scale pattern such that the single descriminator response is shown by the graph of figure 3. The steep response combined with a recognition speed of 0.1s shows the advantage of the technique over traditional feature extraction. The vision system can not only provide orientation ability but can also track parts passing through the field of view. The descriminator tracking response is shown in figure 4 for a controller case. In this case the window was held stationery and the part passed through the window view area. An alternative approach is to monitor the parts' progress by tracking its position as it passes through the camera total field of view.

4. Inspection

The system also allows inspection since an alteration to the image will be shown as a descriminator response attenuation. Although the individual discriminator responses will be reduced, the response pattern or histogram is unchanged and the orientation capability is unaltered. The level of the responses is an indication of quality as figure 5 shows where there is a relationship between response level and alteration to the arrow shape image. Each of the alterations to the arrow shape was different being due to such things as the central hole blanked off, the arrow head removed etc. This illustrates that the descriminators can also provide an inspection service in that they flag deteriorating quality. This is also illustrated by the descriminator response to a food label. When the label was correct the score was 128 but when there was no sell-by date the score reduced to 114. When one of the letters of the product title was removed the score dropped to 119 and with no price the score was 117. When the system was focused down to inspect only the sell by date, a score of 128 was recorded for a correct date. When the date was changed from the 19th to the 18th the score reduced to 120 and when the date was changed from the 18th to the 8th the score dropped to 107. With no the 18th to the 8th the score dropped to 107. With no number at all the score was 72. This example demonstrates the cascade inspection capabilities of a system where firstly the whole image is checked followed by the important subsets of the image in turn.

5. Surface Finish

A system has also been used for monitoring the light scattering patterns from coherent light directed onto engineering surfaces. A laser light was directed onto the surface and the light scattering pattern was monitored at different grey levels. These grey level slices are equivalent to a piece part in their own right. Thus a change to the grey level slice pattern will indicate a changing surface finish, texture or topography. This is illustrated in the graph of figure 6 which shows a single descriminator response for a variety of ground specimens. Specimen 1 represented the surface produced by a new sharp grinding wheel and specimen 10 the surface produced by a worn out grinding wheel. The system was trained on the surface produced by the worn out wheel at one particular grey level shape. Multi-level responses in combination with intensities will provide further views of the surface. The graph shows that the descriminator response can be used to monitor the state of the grinding wheel.

6. Concluding Remarks

A hybrid vision system has been described which combines traditional feature extraction with artificial intelligence. The system has been shown to be able to recognise, orientate, track and inspect a variety of piece parts and surfaces.

Note : Patents Are Pending on the WISE Vision System

7. References

Aleksander, I. **Artificial Vision For Robots**, Kogan Page

Beldsoe, W.W. & Browning, I. (1959), **Pattern Recognition and Reading by Machine**, proceedings of the Eastern Joint Computer Conference, page 255.

Elliott, D.N. & Griffiths, B.J. (1989) **A Low Cost Artificial Intelligence Vision System for Piece Part Recognition and Orientation**, International Journal of Production Research, Volume 28, No. 5 (to be published).

	Discriminator Number						
Letter	1	2	3	4	5	6	7
O	127	55	94	93	67	54	73
Q	50	127	45	46	40	28	43
G	92	47	127	99	82	53	75
C	92	46	103	127	88	63	74
E	58	36	82	74	127	85	88
F	47	29	59	58	86	127	68
B	51	35	67	60	87	66	127

Figure 1 Table of Descriminator responses for letters of the alphabet.

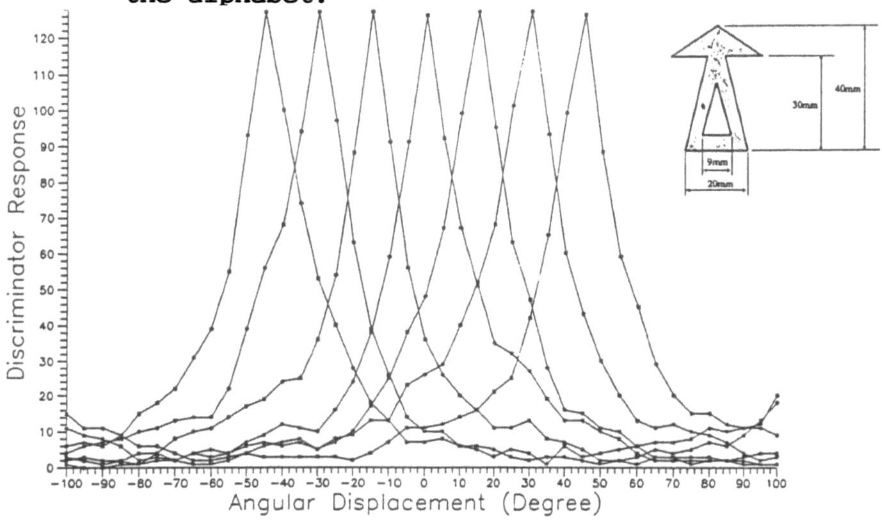

Figure 2 Descriminator responses for orientation of the arrow shape.

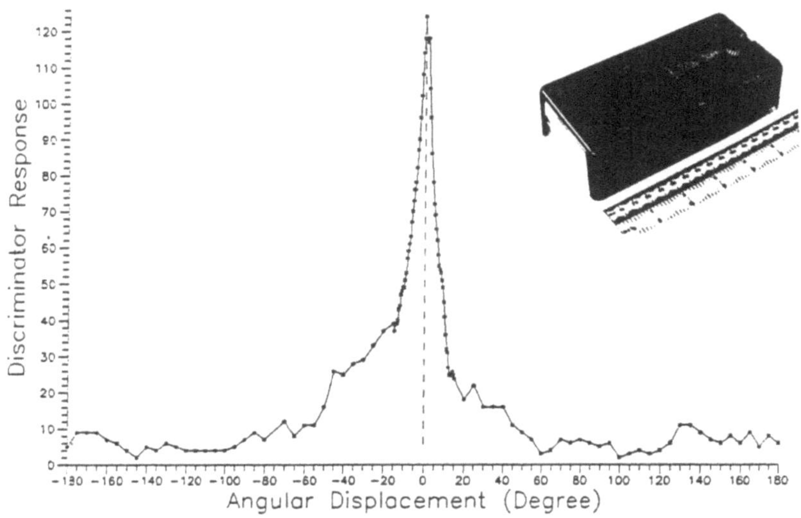

Figure 3 Descriminator responses for orientation of the car light.

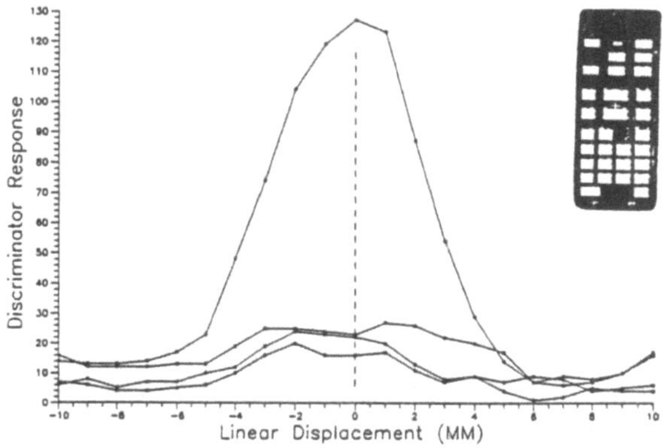

Figure 4 Tracking response.

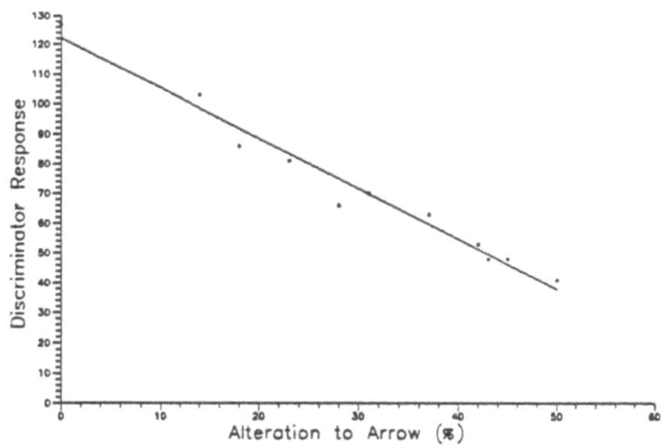

Figure 5 Inspection response for arrow shape.

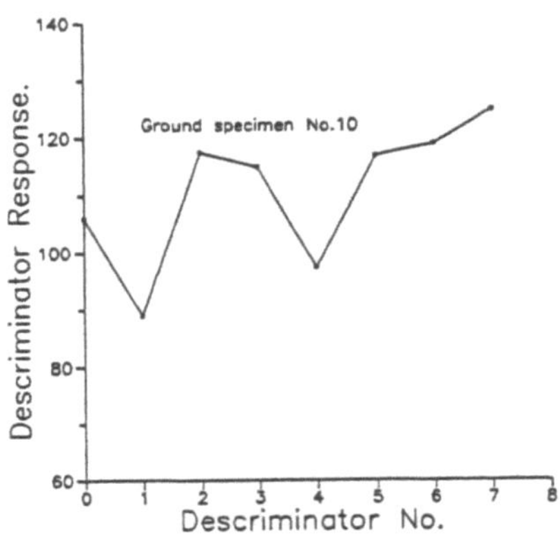

Figure 6 Descriminator responses to ground surfaces.

TEXYS: A real-time picture processing system for on-line quality control

M. Hajnal
FLEXYS Manufacturing Automation Co., Budapest
I. Lovanyi and A. Nagy
Technical University of Budapest, Hungary

Abstract

This paper gives a short description of an industrial quality control system based on real-time picture processing technique. It enlightens that the automatic inspection in the manufacturing processes is necessary and outlines a realized solution of this problem. Hardware and software of a dedicated system as well as the algorithms used are discussed. Finally a report on industrial application is given.

Keywords:Quality Control, Picture Processing, Line Scan Camera, Visual Information, Industrial Imaging, Inspection.

1 Introduction

Continuous real-time on-line inspection of running materials is an important factor in many industries. Since this operation is very labour intensive and the human inspection conventionally used can become extremely unreliable and rather expensive, automatic picture processing systems are needed. For this reason in the past few years real-time imaging using high resolution digital pictures has become a rapidly growing field in factory environment which involves fast decision making based on interpretation of these digital images. Programmable flexible automation including robotics is also a characteristic field of growing importance for real-time image processing, as reported Lovanyi (1986).

Prior to nowadays the evaluation of visual information for quality control was not feasible in many applications due to the inadequate performance of accessible systems in spatial resolution, number of grayscale levels, processing speed, storing capacity or last but not least, in cost. As until now image acquisition and processing at an acceptable rate for real-time applications wasn't available, an effort to solve this problem was made.

Note, that machine vision systems are already on the market, surveyed e.g. by Chin and Marlow (1982). Most of them operate on binary images only. The reason is the huge amount of data that every gray scale picture contains.

E.g.an ordinary image of 512x512 pixels of 8 bit gives over than 2 million bits of information. A method to get more computing power for a system is to use multiple or matrix processors. Many special purpose architectures have been proposed over the last years. There are several papers and books published on the subject e.g. by Fu (1984).

Realizing these facts the concept of our TEXtile inspection sYStem (TEXYS) is based on the following key items:

* The advent of VLSI technology made it possible to implement an IBM PC based modular image processing system with a hardwired vision processor executing many ofthe operations at video rate.
* Special sensors, mechanics and illumination systems have been developed providing extremely high resolution.
* The easy modification of the application level software makes it possible to solve various kind of industrial quality control problems.

2 System description

TEXYS is a high resolution, real-time industrial quality control system for running materials or products, which automatically inspect some visible quality features.It is capable to detect all defects of the material separable by a window scanning the digitized picture. Detecting means the just in time recognition of the type and location (position coordinates) of the defects. Moreover, a greater region of actual video data with 8 bit/pixel resolution can be stored in a temporary video buffer for more sophisticated off-line evaluation of interesting details.

A high resolution monitor can visualize the gray-scaled or binary video data and the results of feature finding tests on-line with inspection.

The sensor of TEXYS is a high resolution line-scan CCD camera, while the processing unit is an IBM PC based high speed video-rate hardwired processor. The system receives length measurement information,controls marking devices and the reflexiv or transparent lighting system.

The system consists of four principal components (Fig.1.):

High resolution CCD line-scan camera system,
IBM PC/AT based real-time hardwired image processor,
Testing mechanism & lighting system,
System and application software.

2.1 Line Scan Camera

Presently a Fairchild 3456-element line-scan camera Model CCD 1600R is used with bayonet-mount lenses with focal lengths of 25 and 50 mm. Interfacing for other kind of cameras is also possible, as well as one can use more than one camera.

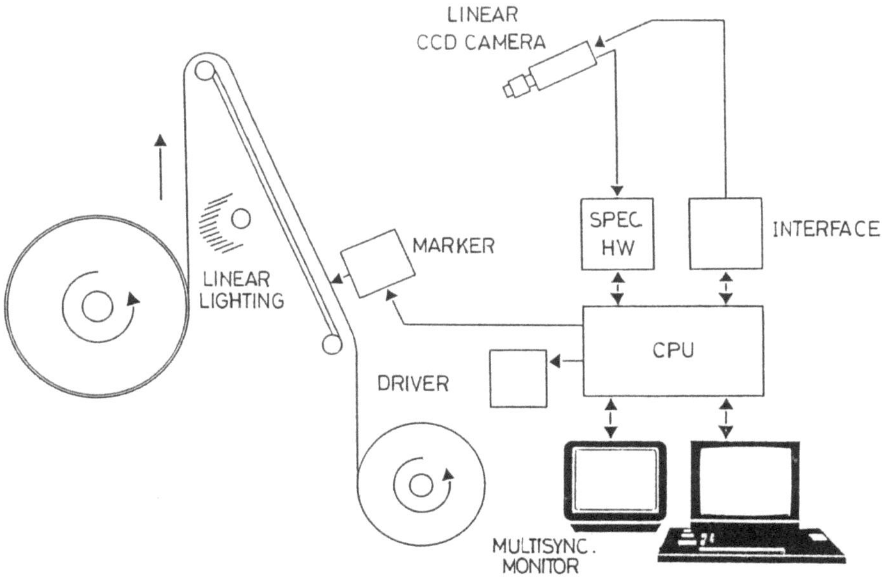

Fig.1. Structure of TEXYS

2.2 Image Processor

All system procedures are performed by an IBM PC/AT based microcomputer including the following units:

(i) Image Acquisition Unit:
- Analogue contrast enhance
- Exposure time: 0.7 ... 70 msec/line approx.
- Pixel processing rate: max. 10MHz
- Maximal speed of material (in m/s): cca. 1.5 x camera resolution (in mm)
- 8 bit A/D conversion
- Digital shading correction
- 10 bit D/A output (for illumination control)
- Operation is synchronised to the speed of material.

(ii) Real-Time Pattern Recognition Unit:
- Programmable 8 bit comparator for adaptive thresholding
- Texture analysis by a local window
- Feature detection with up to 8 parallel algorithms
- x,y coordinates of defects are stored
- Real-time histogram evaluation
- Temporary buffer for 8 bit/pixel or binary video and for the selected features.

(iii) IBM PC/AT (compatible) computer with: 1 Mbyte RAM, Enhanced Graphics Adaptor, High Resolution Colour Monitor, Printer, Hard Disk, 1.2 Mbyte Floppy Disk Drive, and at least two available slots.

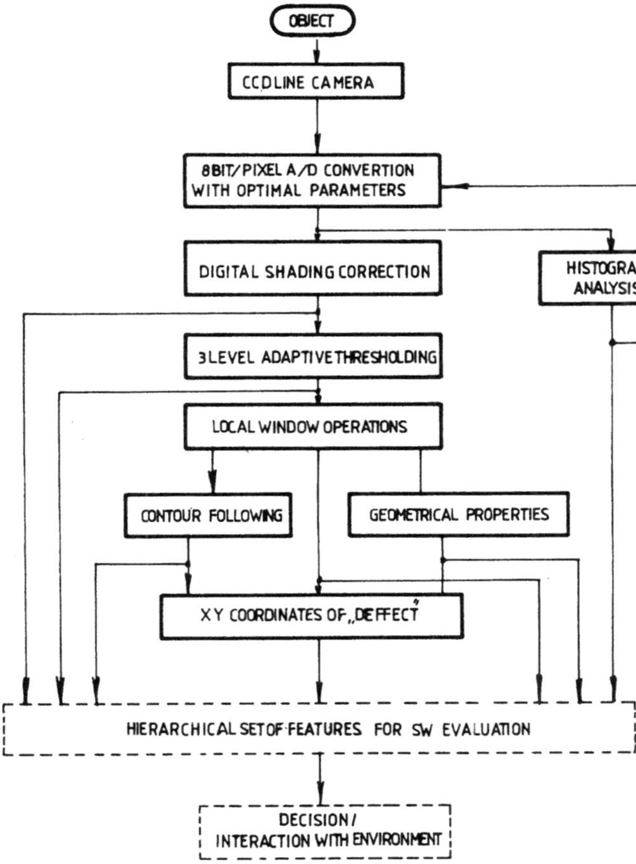

Fig.2. Software system of TEXYS

2.3 Testing Pad

It provides the mechanical movement of material. Features: Adjustable speed in the range of 0...1 m/s; Speed transducer with TTL pulse output; Adjustable homogeneous illumination (e.g. special fluorescent light source).

Note: Testing Pad is not included in TEXYS, as it is usually a customer-owned one, or delivered by other manufacturer. The above items constitute the basic requirements of this equipment.

3 Software and system functions

The modular, menu driven software consists of three layers (see Fig.2): I. Hardware handler level; II. Picture processing level; III. User level.

```
+---------+  +---------+  +---------+
| 1 1 1 1 |  |-1-1-1-1 |  |-1 0 1 0 |
| 1 1 1 1 |  | 0 0 0 0 |  |-1 0 1 0 |
| 1 1 1 1 |  | 1 1 1 1 |  |-1 0 1 0 |
| 1 1 1 1 |  | 0 0 0 0 |  |-1 0 1 0 |
+---------+  +---------+  +---------+
 averaging   horizontal     vertical
                  edge detection
+---------+  +---------+  +---------+
|-1-1-1 0 |  | 1 1 1 0 |  | 1 1 1 0 |
|-1-8-1 0 |  | 1 1 1 0 |  | 1 0 1 0 |
|-1-1-1 0 |  | 1 1 1 0 |  | 1 1 1 0 |
| 0 0 0 0 |  | 0 0 0 0 |  | 0 0 0 0 |
+---------+  +---------+  +---------+
 Laplace       erosion     dilatation
 gradient    PIXEL=0 IF   PIXEL=1 IF
               SUM<9        SUM>9
```

Fig.3. Convolution masks (examples)

The software performs the following main functions:

3.1 Calibration of optical system

By the visualization of line-scan camera information the user can bring into focus the whole optical system. Automatic setting of exposure time and illumination due to material transparency and material conditions is done, and the correction of the inhomogeneity of the whole optical system is also given.

3.2 System parameter set-up

The automatical system configuration takes into consideration the follwing parameters: camera resolution, set of features to be detected, type of material, etc.

3.3 Display options

Grayscale video, Binary video, User definable pseudo colour, Visualisation of features detected, Calibration curves, Alphanumerical information.

3.4 Real-time image processing

The heart of TEXYS is the wide range of fast hardwired video rate preprocessing algorithms based on previous works of Hajnal et al.(1980). Some of them are listed below: Shading correction, Adaptive thresholding, Filtering, Sharpening, Edge detection (by Laplace-, Sobel-, Hueckel operator), Erosion/dilatation, Mask matching, Evaluation of global geometric properties, User defined algorithms, etc.

Some examples are given on Fig.3. Algorithms can be chained and/or parallely executed. As the real-time feature inding unit can work up to 10 Mpixel/sec rate, parallel and/or serial connection of basic algorithms results in more complex features and greater window size.

Fig.4. Industrial application of TEXYS

3.5 System outputs

Outputs can be either messages or actions. Some of them: "Stop mechanism" message, "Defect detected" message, Type of defects, Coordinates of defects, Storage function, Display function.

4 Applications

A video inspection system of this kind was introduced at the BOBBIN SHOW (ATLANTA, USA September 1989), and was installed in a textile factory (SECOTEX Co.,Hungary) to inspect gray goods on an industrial inspection machine. Fig. 4 shows the quality control system in industrial environment. The system detects and classifies defects of different size, and gives protocol on defect type, size and location.

5 References

Chin, R.T. and Marlow, C.A. (1982) Automated visual inspection: a survey. IEEE Trans on Pattern Analysis and Machine Intelligence. PAMI-2, 557-573.

Fu, K.S. (1984) Special computer architectures for image analysis. Robotica, 2, 27-32.

Hajnal, M., Konyves-Toth, M. and Reti, T. (1980) Quality control in metallurgy by texture analysis. Proc. 6th IFAC/IFIP Intl. Conf. on Dig. Computer Application to Process Control. Pergamon Press, 291-300.

Lovanyi, I. (1986) Arhitectures of real-time robot vision systems. Proc. 10th Joint Symposium Tokai Univ., Techn. Univ. of Budapest, Tokyo, 553-562.

Moire photography in engineering measurement

C. Forno
NPL, Teddington, Middlesex

Abstract
By installing a slotted mask inside a camera lens its response can be tuned to resolve 300 lines/mm over the whole image field. The camera is used to record fine grid patterns applied to engineering structures and deformations are revealed as moiré fringes generated by changes in the pattern. The technique has been applied to large civil engineering structures as well as small components and automatic pattern analysis offers improved accuracy and speed.

1 Optical Techniques

There has never been a more opportune period for optics in the field of deformation measurement. Almost any problem from hot rotating components, to full-scale building structures can now be tackled. Some of the current techniques, of course, require the use of lasers, and NPL played a pioneering role in the development and associated data analysis of the two important fields of holographic and laser speckle interferometry. Despite such a rapid growth, there is a reluctance to exploit optical techniques by some engineers who instead favour a more conventional measurement approach, usually involving the ubiquitous resistance strain gauge. What then do optical methods for measuring deformation and strain have to offer?

- Measurements are made remotely, without contact with the surface.
- The whole of the field is examined, so that localised effects, such as cracking, can be identified.
- Quantitative measurements can be obtained. For instance, the wavelength of the laser light might provide the measurement scale.
- The sensitivity to deformation can be varied.

In general, there is an upper limit of the order of one or two square metres on the size of objects which can be studied by laser based methods. White-light techniques, however, are not so restricted. They are relatively straightforward to apply and often natural daylight can form an adequate substitute for artificial illumination. Moiré techniques fall into this white-light category. They have evolved from the familiar effect seen if two regular patterns are superimposed. When one pattern is located, either physically or optically on the surface of an engineering structure and a second master pattern is used to interrogate the first, the resulting moiré fringe map will depend on the shape and deformation of the surface.

There are two classes of moiré techniques used in deformation projected and applied pattern methods, the choice depends on whether the sensitivity vector is located in, or out of the plane defined by the surface of the structure.

Engineers commonly require deformation or strain to be expressed in terms of the in-plane components. Perhaps one of the most versatile in-plane methods to spring from this optical renaissance was started more than a decade ago and is called High Resolution Moiré Photography. At that time, a technique for monitoring very small displacements over large areas did not exist and reports of the unexpected failure of civil engineering structures, such as Ronan Point and a box girder bridge, gave the impetus to the successful development of this method.

High Resolution Moiré Photography[1,2] is a sensitive technique which uses a specially modified 35 mm camera tuned to record a fine dot pattern applied to the engineering structure. Usually the pattern consists of a regular array of dots running horizontally and vertically across the surface. When the camera is set at a pre-determined distance the demagnified image of the pattern will appear as 300 resolved elements per millimetre in the focal plane. Using fine grain film, a double exposure photograph is made before and after the object is deformed.

After the film has been processed the pattern detail in the image behaves as two orthogonal diffraction gratings. A beam of light directed on to the negative will be diffracted into four first-order directions and when the image is viewed in one of these diffracted beams, a moiré fringe pattern, or displacement map, is seen, with dark fringes corresponding to a movement of half the pitch of the object pattern. From the one negative two such displacement maps are generated, representing either the vertical or horizontal displacements, depending on which diffracted beam is being examined. In these maps the various features of material behaviour, such as cracking and slipping can be identified and a measurement of the fringe spacing and slope at different positions on the map provides the information that leads to local strain distributions.

This technique can measure strains smaller than 0.01% and displacements less than one twentieth of the pitch of the surface pattern. It has been used to monitor the deformation of a number of large engineering structures, including steel beams, pressure vessels, brick and concrete building structures as well as much smaller objects, such as carbon fibre composite specimens and nuclear engineering components maintained at high temperatures.

2 Examples

One large scale application, involving an ancestor of the box girder bridge, was recently carried out with this method. There are many mid-19th century stone bridges in Britain which although structurally sound, have not been subjected to any rigorous testing and so their behaviour and performance under heavy loads are largely speculative. As a result, the Transport and Road Research Laboratory instigated a study of Victorian bridges which included one redundant road bridge over a dried-up canal in Preston-on-the-Weald-Moor, Shropshire (Fig 1). This bridge was chosen after consideration of local environmental aspects, its redundancy and the high cost of maintaining it in a serviceable condition.

Nearly two hundred sheets of A3 size paper, bearing a 0.8 mm pitch grid pattern of white lines on a black ground, were bonded to the stone work using standard wallpaper paste. From a distance of almost 15 m, the whole 9 m wide surface was recorded by a single modified camera mounted on a tripod.

The test involved applying an ever increasing load to the roadway, by means of a massive hydraulic ram unit anchored to the ground. More than 200 tonnes force was applied before it eventually collapsed. The method was able to demonstrate that cracking, in fact, initiated close to the centre of the arch at almost one tenth of the failure load. Although it was considered before the test that the piers would remain stationary, it was revealed that under heavy loads there was a sideways displacement of 15 mm between the two. The recording process was conducted without interruption of the test programme, whereas a conventional theodolite survey carried out in parallel with the moiré photography, did inevitably introduce delays as surveying requires the operator to make repeated measurements. Additionally, heavy rain during the test produced instability at the survey stations resulted in unknown displacements.

Although this particular bridge experiment was completed in one day, occasionally testing may involve creep processes over long periods. In these circumstances it is still possible to record the first condition, remove the camera and then replace it at a later stage to record the second condition on a separate film frame. Repositioning the camera, for the second record, is not critical as small errors can be corrected when the negatives are superimposed during the analysis of the fringe maps. During one experiment when this working mode was adopted, successive comparisons were made of the complete facade of a pair of semi detached houses. Photographs were taken at approximately fortnightly intervals in order to monitor the effects of subsidence over a 3 month period. Little was identified in the way of displacements in this case, except a small horizontal slipping to the upper stories, which was not revealed by the usual traditional surveyors' levelling techniques (Fig 3).

A development of the technique, involving the application of very fine patterns, has extended its use to the examination of much smaller objects. In particular, a stencilling method for producing a 40 dots/mm pattern that can withstand repeated thermal cycling to at least 600°C in air, has found several applications in the measurement of thermal and mechanical strains in welded components for the nuclear industry. Apart from locating the camera at the closer 'tuned' distance of 0.5 m the procedure for generating moiré fringe maps from these patterns is unaltered.

The fine pattern is prepared by spraying a dispersion of titanium dioxide pigment in water through a 40 holes/mm grid fixed to the surface. Providing the surface is not touched, the pattern will remain intact for very long periods.

Although originally intended for studies at high temperature the stencilling facility has also found other applications. It has proved particularly suitable for monitoring the large deformations associated with fatigue. Where the structure is composed of a soft, or elastic material the use of this surface pattern, which is free from any supportive layer, will allow the uninhibited transfer of the material strain through to the surface. In this way the stencilled pattern has provided a means of measuring the true mechanical properties of plastics, where a conventional method, such as the use of resistance strain gauges is unsatisfactory.

In automating the analysis of the fringe patterns the method of phase-stepping has been adopted. Three or more patterns, each differing by equal phase changes, are recorded from pairs of processed negative with a digital frame store. By measuring the intensity at a point on each fringe map the displacement information can be retrieved at any position on the structure. Moreover, as the estimation of the local displacements can be as sensitive as 1/50 of a fringe, it reveals the presence of very small disturbances over large areas. Differentiation of the displacement information will yield local strain distributions.

REFERENCES

1 Burch, J M and Forno, C. A high sensitivity moiré grid technique for studying deformations in large objects. Opt. Engng. 14, No 2 (1975) Pp. 179-185.

2 Forno, C. Deformation measurement using high resolution moiré photography. Opt. Lasers in Eng. 8(1988) Pp 189-212.

Figure 1. Victorian bridge before testing, Shropshire.

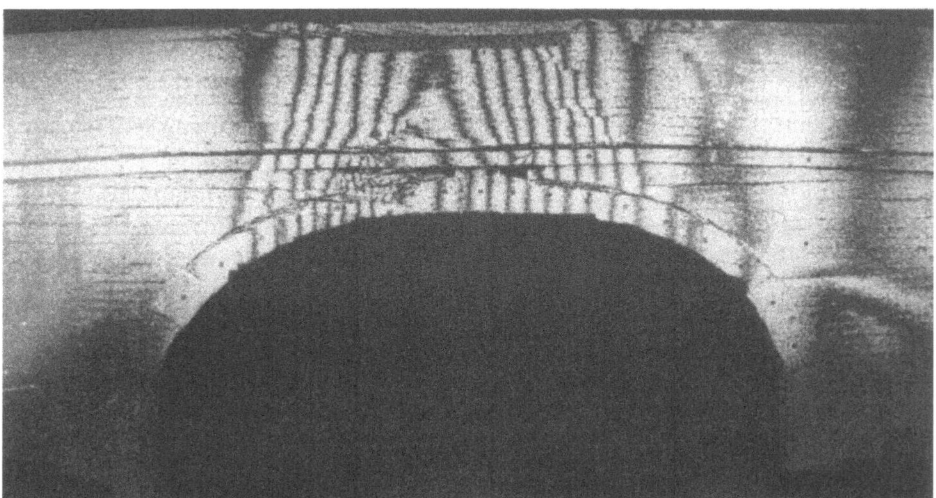

Figure 2. Y-displacement fringe map for bridge under load.

Photogrammetry in industry

R.A. Hunt
National Physical Laboratory, London, England

Abstract
The basic principals of photogrammetric survey will be presented with special reference to industrial measurement. Some illustrations will be given, based on the work of the NPL form and deformation section, of situations where this technique is particularly relevant.
Keywords: Photogrammetry, Engineering Survey, Form Measurement.

1 Introduction

Photogrammetry has been used for many years in the form of air survey for mapping relatively large objects like countries and continents. Since the war the advent of the digital computer has made possible the generalisation of the technique to the relatively precise measurement of engineering objects.

In essence photogrammetry consists of taking a number of photographs of the object of interest from different viewpoints and combining them to give the three dimensional form of the object.

Research at the National Physical Laboratory has been aimed at increasing the accuracy of the technique for industrial applications. Our target was to review each stage of the photogrammetric survey process and consider what may be done to improve the overall accuracy.

2 Conventional photogrammetric technique

The operations necessary in a conventional close range photogrammetric survey are given in Table 1. These operations will form the basis of the discussion of possible refinements of the technique.

Table 1. Conventional close range photogrammetry

1. Plan the experiment to obtain of the required accuracy.
2. Mark points on the object and position the cameras.
2. Take the photographs.
3. Measure the photographic images.
4. Compute the co-ordinates of points on the object.
5. Estimate the uncertainty of these co-ordinates.

In accurate photogrammetric work it is usual to use the so called "bundle adjustment" technique which is just the photogrammetrist's term for a global least squares adjustment of the data. An idea of what is involved in this is given by considering simple pinhole cameras.

If the camera has its pinhole at X0,Y0,Z0 and normal projection of this pinhole on the image plane is at x0,y0 then the image of a point at X,Y,Z is at x,y given by:

$$x = x0 - f.(X-X0)/(Z-Z0) \qquad ...(1)$$

and

$$y = y0 - f.(Y-Y0)/(Z-Z0) \qquad ...(2)$$

Where f is the distance from the pinhole to the image plane and the co-ordinate systems inside and outside the camera are assumed to be parallel to one another.

Thus for a typical case, where the shape of a radar antenna is being defined by the position of a thousand points on its surface as seen on four photographs there are at least three thousand unknowns X,Y,Z and up to four thousand measured values of x and y. So, even this minimal model of the system presents a very formidable least squares problem.

At NPL our software has been under continuous development during the life of the project, both to give quick and accurate estimates of the positions of the target points and to provide a method of estimating the expected uncertainty of the results corresponding to different arrangements of the camera positions and the "control" points used to establish the overall co-ordinate system.

3 Experimental refinements to increase accuracy

The ultimate aim of the project was to achieve the highest possible accuracy from the photogrammetric method and the following simple guidelines were found to make a considerable difference to the final uncertainty of the co-ordinate measurements.

3.1 Planning
The necessity for a good working plan of the experiment before undertaking any photography cannot be over emphasised. The chief contribution that the system designer can make to this process is to facilitate the use of the design software to such an extent that there is no temptation for the experimenter to go out on site and "play it by ear".

3.2 Targeting points of interest
In order to achieve any reasonable accuracy it is necessary to use highly convergent photography to get a good experimenal geometry (this would normally be revealed at the planning stage). This in turn precludes the use of intuitive sterioscopic matching. Thus it is necessary to decorate points of interest with high definition targets which have images that may be located on the photographs, either by eye or automatically, with the highest possible precision. The best targets for conventional photography appear to be concentric circles of retroreflecting material illuminated by a ring flash.

3.3 Photography
Using conventional cameras the best results are obtained with a good metric camera from a reputable manufacturer (we used Zeiss UMK10 cameras). Alternatively, if it is possible to use more unconventional imaging, the monocentric axicon camera developed by J.Burch and C.Forno at NPL gives an order of magnitude increase in accuracy but needs a very precise measuring machine to measure the plates. In all cases it is necessary to use very high definition emulsions e.g. Agfa 10E75 which is rather slow (about 0.5-1 ASA) on thick glass plates to avoid any changes in shape due to humidity variations.

3.4 Computation
The computation is carried out in two stages due to the nonlinear nature of the problem. First an initial estimation of the positions of the object points and the positions and orientations of the cameras, then a least squares refinement of these estimates. This least squares refinement is only possible because of explicit exploitation of systematic zeros in the design matrix.

3.5 Uncertainty estimates
An additional program is used to estimate the probable uncertainties of these estimates based on the last iteration of the least squares refinement. This type of

analysis relies on the random spread of the probability density function being sufficiently narrow that it lies within the region were the linear approximation to the photogrammetric equations is valid.

4 The NPL Monocentric Axicon Camera

An alternative approach to conventional photogrammetry is to redesign the camera lens and this was done by Jim Burch and Colin Forno at NPL. The lens used was essentially two concentric spheres of glass of different refractive index. This simple system is capable of such accurate manufacture that the image of a point source has no detectable distortion. The resulting image is a concentric ring pattern with a ring spacing of four micrometres. This image may be located on the photographic plate to better than 50nm on a 50mm plate which suggests an angular accuracy of one part in a million of the field of view. This has indeed been verified by estimating the positions of a set of test points using four photographs. This test survey agreed with a Moore 5Z three co-ordinate measuring machine to about two parts in a million which is at the limit of its performance. The device also has a very good effective depth of field giving sharp ring patterns for object points from 300mm to infinity.

5 Results

At NPL the conventional photogrammetric technique has been used for several industrial surveys and is capable of an accuracy of one part in one hundred thousand, see for example Oldfield (1984). The NPL camera is about ten times as accurate (see Burch and Forno 1984) but the necessity for a very accurate plate measuring machine makes it more restricted in its application.

As mentioned above the mathematical analysis is still being developed but the work on the production of initial estimates before the bundle adjustment has been published in Hunt (1984).

An earlier review of the work from a somewhat different point of view is to be found in Hunt and Oldfield (1987).

6 Acknowledgements

The work reported on here is the result of the collaborative effort of an interdisciplinary team of S.Brown, C.Forno, A.Kearney and S.Oldfield. I would also like to acknowledge M.Cox and G.Anthony who are working on the final form of the least squares program.

7 References

Burch, J.M. and Forno, C. (1984) Progress with the NPL centrax camera. **International Archives of Photogrammetry**, 25(5), 141-148.

Hunt, R.
A. (1984) Production of initial estimates of photogrammetric variables. **International Archives of Photogrammetry**, 25(5), 419-428.

Hunt, R.A. and Oldfield, S. (1987) Industrial photogrammetry at NPL. **Proceedings of the Second Industrial and Engineering Survey Conference**, Published by the Royal Institution of Chartered Surveyers, 241-249.

Oldfield, S. (1984) Photogrammetric determination of the form of a 10m diameter radio antenna. **International Archives of Photogrammetry**, 25(5), 590-596.

Energy management for conversion in oil rigs

D.J. Barua
Oil and Natural Gas Commission, Madras, India

Abstract
Oil rigs are powered with 3 to 10 mw of generation capacity. There are aprehensions in the mind of operators regarding the high risk of operational complications leading to high cost.If the risk factor is underplayed it is the total loss of rig as well as the reservoir.So far under such environment, rationalisation and conservations play a major role,as this upholds the basic philosophy of operational safety keeping a blind eye on conservation.But with some orientation of the operational and maintenance personal,consiedrable of conservation and rationalisation can be introduced.When an environment support a high consciousness,adding more consciousness will help the conservatio in the right spirit.

1. Introduction

The start up will be very convinent if we set the cost of oil and cost of finding the oil in the first frame of all reference. Next we look into the environment of energy before we deal with the oil exploration.There is no second openion that all our conventional energy sources are strained.Oil is gradually giving way to gas,coal to caol gassification (in-situ or before use) and all sort of non-conventional energies are attracting research.

All out effort on, to make nuclear cleanliest. But it is not definitely the most portable to be on our car,rails and peoples mode of transportation.

The conservation is also not the ultimate answer.But they have a very valid point.Conserve and give technology and the planet some more breathing time to get ready for the future.The question of saving the planet is more importent against the backdrop of pollution,greenhouse effect,ozone hole,eco-bio cycle disorder and finally the climatic chaos temperature rise.

This logic of the environmentologists is very deep rooted knock on the humen consciousness.If human civilisation did not try to improve its living by technology it would have lived 64000 AD with all the energy it has wasted sofar.Knoladge still would not have fallen short as all wisdom of thoughts are being reinvented and reused and reconfirned only, as the habit of listening and doing has no unidirectional motive among the races of this planet.

2 Cost energy use scenerio

a. Probable oil price in 90´s $ 18-22 per bbl
b. Probable gas price in 90´s $ 1.6-2 per Mcf.
c. Probable oil repacement by gas 5-15 %per year.
d. Probable cost of exploration in deepwater $ 22-26/bbl
e. Probable cost of exploration in arctic $ 26 per bbl
f. Finding and Developing cost.Minimum $ 5/ bbl Average $ 8-10/ bbl & Maximum $ 18-20 per bbl.
g.

Finding prob	Av.amt	Av.res. Irr.	c.dry hole	63%
25%	500,000	11.2%	non-cmmercial	14%
-	-	10.8	hydrocarbon 15%rr	12%
-	-	10.3	-do 15%rr less	11%
48	540,000	8.3		

h. Brief history of energy as the fundamental resourse.

Civilisation	Fuel used	Civilisation	Fuel used
Egyptian	Wood and animal oil	2000-3500	Nuclear,Non-
Hellenic	do	3500-	conventional
Pre industrial	Wood,coal,fossil		solar?to restart
Post industrial	Wood,coal,fossil Nuclear,Non-conventional		civilisation
Or Galactic Age	Supetech .		

*** In 1870 use of wood started declining and coal and oil started to pick up.*** In 1960-54 coal and oil started declining, Nuclear started to warm up.*** In 1900 Natural gas started to pick up and still in the trend of rise as replacement.*** Now non-conventionals and alternative energy sources are beig tried only.

Let us see this point from very close quarter.Replacing the six side of the cube as six elements influencing the outcome of the functional improvement of technology.This can also be considered as a browenian particle influenced by six different variables to reach the targeted effect.

Power up <-----General purpose * Specialisation----->Balancing
Scaleup<-----Humanisation *Automation-------->Minitaureing
Keep up <-----Efficiency * Redundency--------->Intermittent use

Also the problem of" extent of integration and dispertion" has its own set of culture to superimpose on the respective targets.

The final caution is that the problem of creation has many hardles then the problems of distribution without any loss of properties in transit and storage,and its" desired level of stisfaction".This is the bigining of all the failures,waste and accidents.

The systems of distriburtion that mostly culliminates in the political power has wasted so much that the technology is not in a position to recreate in the same pace.

Apart from this system we have many of our systems of

Design, operations, maintenance, management and communication those have similiar problems of distribution of" proper care and attention at controlled intervels."

The load rating normally depend on weight of the maximum length of pipe hanging from the Hook.This variation is because of the thickness of the pipe increases keeping in view the depth,expected type of pressure and the length to be set.

3000 Draw Works rating is a offshore feature as the change of water depth and the different casing policy is required in the sea environment.Weather and sea state is the influencing parameters.

3 Utility factors in drilling operation

Here is the display of the Utility Factors in % of different equipments calculated over a period of one month or 720-800 running hours.

Equipment	Engine	Drawworks	Pump	Rotary	Compressor	Auxillary
U.Factor	37-69	62-93	14-52	11-71	29-49	13-49

This case of utility factor is also very intersing as the power sizing goes through a trumatic cycle against such ratings shown earlier.But there is still some though should be left for the uncertainites in oil exploration and development.These data are from exploratory field.

For better power management this relations are useful.

a. DE = RT	1*	2**	3	4			
b. RT = 2SP	1	2*	3**	4			
c. DW = TC	1	2*	3**	4	* = good		
d. TC = 2DE	1	2*	3**	4	** = verygood		
e. DW = 6COMP	1	2	3*	4**	5	6	

The engine data related to heat rejection by cooling system and engine exhaust are tabulated and available. The rejected heat are used in water makers or heating/ cooling.

3.1 Engines

On the input side performance data and the fuel flow rate are used to complete the Energy calculations are available. Typical data for Cat D-399 (largest population) is presented here.

Percent Load	Engine power HP	MHP	KW	Gen power at .8PF KW	BSFC G/KW-H	FUELRATE GAL/H
100	988	1002	737	700	234	53.7
75	741	751	553	525	234	40.3
50	499	506	372	350	240	27.9
25	258	262	192	175	276	16.5
100	1195	1211	891	850	243	67.5
75	899	911	670	638	232	48.5
50	603	611	450	425	243	34.1
25	311	315	232	213	281	20.3

Among the mechanical, DC-DC and AC-SCR rigs SCR is the coveniet for all type of optimisation in power management. Hence it is replaceing faster the other two. AC-AC rigs and AC Hydraullic with higher mechronics is in the readyness to take off.

3.2 Wheather and operation factor

A) Rig Time -	Mild weather	Severe weather
a) Drilling + Tripping	57%	15.4%+ 11.8%
b) Logging + Cementing+Fishing	21%	5.5+11 %
c) Running Casing	7%	1.4 %
d) Running BOP & riser handling	15%	

B) General:
Waiting on weather 20.8% Mooring 8.3% Stack handling 8.3%
Others 4.9% Total 42.3%

We have presented the rig time of offshore rigs where activities of other kind are more yet there is more intensity of drilling compared to the land rigs where C replaced by transportation. The limitations imposed on drilling time(i.e. real productive time) are because of weather conditions. But a little understanding of the weather can help in improving the drilling time.

All these problems are mostly involved with the floater and semi submersibles. A jack up can be considered as a land rig as far as the power management is concerned. The requirement of power is much more in a floater and even more in a DP floater and semi submersibles where the power management becomes a real time job as in a DP ship and semi submersible power requirement for position maintenance itself requires another 8 to 20 mw.

The design of propellors influences greatly on the power requirement. Different designs and its parameters are profiled below for a better view of the whole complexity

Examples of Typical Open and Ducted Propellor

Types	No.of blades	Horsepower	Thrust lb	RPM	Diameter ft.
Open	3	1500	33000	150	10
Open	4	2100	48500	200	
Open	5	35000	-	240	15
Open	4	20000	-	235	13
Ducted	4	500	11300	400	4.8
Ducted	4	1000	24300	260	6.6
Ducted	4	1500	36400	240	7.9
Ducted	4	2200	55000	200	9.2
Ducted	4	3000	77200	161	11.2

The efficiency of propellers can be defined as ratio of thrust to torque.

$$\text{Efficiency} = (TV_A)/(2nQ)$$

A more meaningful measure of performence for dynamic positioning propeller is a static figure of merit.

$$C = \frac{(K_T)^{3/2}}{(pie^{3/2}) * K}$$

$$Eff = \frac{(K_T)}{pie*(K_Q)^{2/3} *2^{1/3}}$$

Figure of merit for several type of thrusters

Thruster type	Values of figure of merit
Ducted propeller	
Predominent direction	1.25
Reverse	0.4
Symmetris ducted propeller	1.01
Nonducted propeller in the predominent direction	0.88
Nonducted propeller with flat blades	0.71

General thruster properties those influences the power are

1. Thrust,T = dnDK

d = density of water. n = rotational speed of the propeller,RPM
D = diameter of the propeller. K = thrust coefficient.

Thrust is the product of torque applied to the propeller to make it rotate and force water to move through the blades.

2. Torque,Q = dnDK   K = Torque coefficient

The value of the thrust and torque coefficients are the functions of the physical chaaracterstics of of the propeller.

P = Pitch = 2*pie*r*tan.

r = radius at pitch   = angle of attack.

Efficiency is the ratio of the thrust to torque.
The performance of the propeller is given as a function of the P/D for constant rotational speed.

$$Effy = \frac{V\ K}{n\ D\ K} = J\frac{K}{K}$$

J = adventage ratio
The next importent parameter is the cavitation no.
This effect of cavitation are

1. reduction of thrust. 2. vibration. 3. erosion of the propeller and the duct. 4. generates noise interfearing with the acoustic referance and underwater probes.

The final element of the dynamic positioning system and probably the the one most likely to be overlooked is the power system.When thrusters are working in the ship,it requires more power than any other functions.

The of power on a DP-Drillship is between 10 to 20 MW.

Rig	power system	power units	Generating capacity	vac
Golmar	Diesei elec	10*800HP	6000KW	4160
445	do	5*2100KW 2*1050KW	12600kw	4160
Saipam	do	6*2000KVA	12000KVA	6300
Havdril	do		15000KVA	
Discover	do	6*2500KW 1*1050KW	16050KW	4160
470 471	do	5*2100KW 2*1400KW	13300KW	4160
709	do	7*2500KW 1*1400KW	23900KW	4160

Some of the difficulties of operating the power system can best be understood froma graphic representation of the generating capacity as a function on the on line units shows the generating capacity is the descrite multiple of the number of online units

If all the unit are of equal capacity then the stair-step function of the generating capacity has equal step size.

On the engine side a diesel engine has aminimum power level below which the engines does not operate efficiently.If prolonged on the minimum level the engine looses life faster.

Operator based power management watches the total power system loading.When loading reaches minimun he canstart one more unit or take a load off the line

To insreasc the load on the power system the operator must one of the user increases their power usage.

The user has the most control over its power usages in a dynamically positioning through its thrusters system.

But with such a large and complicated system can not be left to manual power managers as consiquences of slip and fatigue mistakes or ignorence.

The wise option left with us is to introduce autonatic power controls linked to a computer.PMS is the power managing system Supervising ,Monitoring ,Controlling,fault finding and announcing. There are some PMS having programmed online correction facilities catering services like switching off and on,commanding assignement correction,doing online data analysis and producing reports.

4 Conclution

PMS is expected to influence greately on

a. Reducing fuel consumption.
b. Efficient loading and unloading the generators,not to upset the operation due to overload trips or not run unloaded.
c. Reduces transition time in change overs.

d. Reduces operator fatigue keeping him fit for greater cotengies.
e. Keep monitoring the equipments and generation and shows trend, warns and some executes rudimentery corrections if programmed.

Some of the disignes now a days are take over as maintenance manager .It keeps on following the parameter, analyse them to produce a schedule and display for the maintenance team to take over.

Apert from these major power a floater of this dimension have

a.90-100 occupency living accomodation. b.HVAC.
c.Compressed air system. d.Bilge and ballast system.
e.Sae water cooling system. f.Fire services.
g.Material handling system.
h.Cementing and bulkhandling system.
i.Motion componsation system.
j.Safety and control instumentation.
k.Navigation and steering control system.

The requirement of power management what we have gone through on the drilling and production systems are laid down.The size, and the type of use and the coplexity of power management should have a good "design base " to react to all type of" add ons" arriving in the market.Most of the available system should respond to at least to the adoptation of conservation and control instrumentation.

Best available systems for 2000 ft of weter depth are

a. Subsea wells. b. Tension leg platforms.
c. Semifloaters. d. Turret vessels.

Best technology solution for cutting deilling and development cost.

a. Horizontal complition. b. Extended reach wells.
c MWM. d. Computer aided dicisions.
e. Top drive. f. Drilling automations.
g. Steerable botom hole systems.

Best technology solutions for cutting Production cost.

a. Remote control. b. Subsea comletion.
c. Multiphase flow. d. Enhance recovery.
e. Subsea saperation. f. Reservoir modelling.

** ** **

RELATIONSHIP BETWEEN FUNCTIONAL IMPROVEMENT OF TECHNOLOGY AND ITS TARGETED EFFECTS.

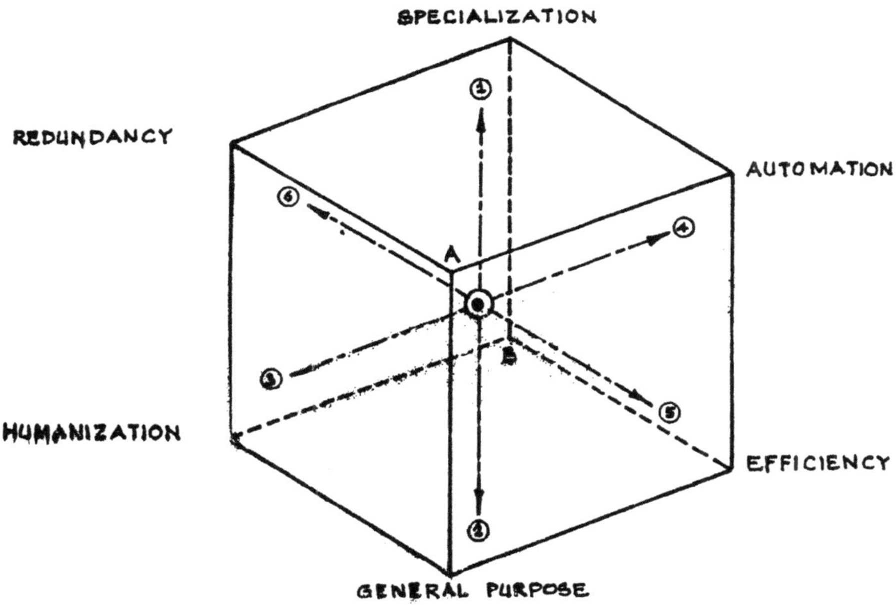

○ POSITION OF EXISTING (CONVENTIONAL) TECHNOLOGY.

DIRECTION OF FUNCTIONAL IMPROVEMENT OF TECHNOLOGY.

① POWER UP	VS	② BALANCING	
③ SCALE UP	VS	④ MINIATURIZATION	
⑤ KEEP UP	VS	⑥ INTERMITTENT USE	
A: INTEGRATION		B: DISPERSION	

SOURCE: "A VIEW ON THE ENERGY TECHNOLOGIES FOR THE 21ST CENTURY".

Continuing education/training on computerized automation

A. Yamamoto
General Manager, Training Center, Yokogawa Electric Corporation, Tokyo, Japan

Abstract
Nowadays, most "high-tech" industries are operated and managed by computer-integrated automation systems. Industrialized countries must provide training facilities and continuing education, refresher courses and training programs -- to familialize older engineers and technicians who are already working in such industries with the latest computer and control system technology, and also train new employees in job-specific skills.

Training choices range from generalized academic training or supplementary courses to courses provided by manufacturers of computers and control systems. The latter are frequently more desirable, because trainees can learn to use job-related hardware and software. This paper introduces some typical course curricula suitable for continuing education and training in computerized automation.

1 The Era of Automation

Industrialized countries have experienced rapid technological development over the last forty years. Automation has been the most important factor in changing the social climate. Modern, automated industries can produce high quality products in quite large volumes at far lower cost than with the manual methods of the past. Moreover, new computer-integrated flexible manufacturing systems support just-in-time small-volume multi-product manufacturing with fast turnaround between products.

Older workers in such industries were initially apprehensive about automation invading their jobs. But their situation has improved. Most workers could adapt to the changing work environment by continuing education, training or skill-development programs. Those who acquired new knowledge and skills, to tame the invader (automation), succeeded in changing their jobs from hard work to smart work. Increased productivity blessed them with higher wages and shorter working hours. Japanese average working hours have been reduced by one third, compared with forty years ago. Most modern jobs rely on brain rather than muscle, so modern society must support innovation in industry by continuing education/training for engineers and technicians.

2 Knowlege/Skill Development

Automation systems, whether large or small, contain many complex instruments. To obtain the optimum system performance, users must have a certain level of theoretical training, backed up by practical instrumentation and control-related knowledge and skills -- which can be acquired through intensive training courses. Training programs must therefore be organized as a suitable combination of lectures and hands-on laboratory work.

As modern automation systems are mostly configured from microcomputer-based instruments with inter-system data communication, the training course should support the specific computer hardware and software used in the system. Computer hardware and software technology progresses very rapidly, so the latest model (or version) must be used for the hands-on training. In this respect, system manufacturers' training centers have an advantage in that they can easily update their training facilities. Large manufacturers can also provide a wide range of video-based material that will facilitate training and enhance understanding.

Better training can contribute to innovation and growth in industry. Some suggested course curricula are shown below.

(These courses are taught in one of the largest process control and factory automation companies in Japan, Yokogawa Electric Corporation, which now offers more than 60 different training courses and trains more than 6000 people per year.)

3 Typical Course Curricula

Training courses should be organised to cover the total control system hierarchy in present-day industries:

(1) Process measurement fundamentals.
Sensor, measurement and signal conditioning techniques for process variables -- i.e. flow, pressure, level, temperature and analytical quantities.

(2) Automatic control techniques.
Appreciation of fundamental process characteristics and response. Single loop feedback control with proportional, integral and derivative actions.

(3) Distributed microcomputer control systems.
Configurations with several tens to hundreds of control loops, CRT operator console, data communication lines, and microcomputer-integrated field control stations.

(4) Factory automation computers.
Computerized automation systems suitable for manufacturing discrete products. BASIC programming allows flexible batch/daily/hourly operation changes.

(5) Supervisory control computer systems.
A hierarchical control system configured with a supervisory computer. Hardware setup, software generation, operation and maintenance techniques.

3.1 Process measurement fundamentals
This course gives the participants fundamental knowledge of process measurement technology. Trainees learn how to measure process variables as key control data. Yokogawa Training Center conducts five-day courses with the following syllabuses:
- Principles of process measurements
- Flow measurement theory and laboratory work
- Pressure measurement theory and laboratory work
- Level measurement theory and laboratory work
- Temperature measurement theory and laboratory work

3.2 Automatic control techniques
To help the trainees acquire fundamental knowledge and skills in process control, Yokogawa Training Centre arranges five-day courses with lectures and laboratory work using single-loop controllers and simulators.
- Process characteristics and simulation of response
- Single loop controller and its control actions
- Feedback control with P, I, and D actions
- Optimum PID parameter tuning techniques
- Cascade, ratio and feedforward control setup

3.3 Distributed microcomputer control system
This course is conducted with our distributed microcomputer control system which is now very popular for modern process automation. It is ideal for trainees to learn on systems similar to those installed in their factories. The Yokogawa Training Center courses consist of an operators' course, system engineering course, and programming course.

(1) Operators' course
- System configuration overview
- Operator station functions and laboratory work
- Feedback control functions and laboratory work
- Sequence control functions and laboratory work
- Utility functions and laboratory work

(2) System engineering course
- Hardware configuration
- System setup procedures
- Cascade control builder
- Complex control builder
- Sequence control builder
- Graphic panel display builder

(3) Programming course
- BASIC commands and statements
- Program initiation
- Process data read/write
- I/O card programming
- File-oriented programming
- Program linking

3.4 Factory automation computer
This course is intended for development of computer integrated flexible manufacturing system operators, programmers and system engineers. The contents are:
- How to start up the factory automation computer
- BASIC commands and statements
- Manufacturing line controller I/O statements and real-time statements
- Graphics builder
- Sequence control builder
- Database management and report generator

3.5 Supervisory computer control system
The supervisory computer system is at the top of the control hierarchy, so this course is more for specialists than the other courses described above. The course curriculum consists of four parts: fundamental, advanced, programmers' and maintenance courses.

(1) Fundamental course
- Hardware overview
- Operating system overview
- File system and its related commands
- C language programming
- OS shell commands and shell script creation

(2) Advanced course
- Data cell overview and builder
- Data cell access programming

- Cell maintenance and analysis utility
- Graphics creation
- Data display/entry programming

(3) Programmers course
- File manipulation -- input/output
- Process creation, execution, wait & termination
- Interprocess communication
  - Interrupt signals
  - Synchronisation by semaphore
  - Shared memory

(4) Maintenance course
- System and CPU outline and configuration
- System operation fundamentals
- Command explanation and operation
- Disk/tape storage outline, configuration, operation and self-testing
- Terminal outline, on-line operation, configuration and self-testing
- Utility functions and operation

4 Conclusion

People are the most important asset of any industry, because their contribution determines its success. As the development of their knowledge and skills is one of the key investment items for management, tuition fees (around $200 per person per day) are borne by the trainees' company, since they want enjoy the good fruits of automation. Some governments provide "skill development funds" that can also be applied as financial supports. Such investments are surely repaid to the company/industry in increased productivity. Continuing education/training really facilitates a nation's modernization and industrialization.

References:
- P. F. Drucker, "The New Realities", Harper & Row Publishers, Inc., New York 1989
- J. Naisbit & P. Aburdene, "Megatrends 2000", William Morrow & Co, Inc., New York 1990

Forecasting the change of operating condition for machinery via fuzzy classification

Chen Yuedong and Qu Liangsheng
Research Laboratory of Machinery Surveillance and Diagnostics, Xian Jiaotong University, Xian, China

Abstract
In this paper, with the aid of fuzzy classification to vibration data of machine, it is easy to evaluate the degree of variation of operating condition in a certain time interval in the present or in the past. By the integration of fuzzy classification method and optimization method, in this paper, a new method forecasting the operating trend has been put forward. As for faults, especially those change in relation to some regularity, by means of this method, field engineers can obtain a better knowledge about the operating condition, correspondingly, it gives a good guidance to set up the timetable of overhaul and shutdown the machine set. In the end, through the field application, we achieve the conclusion that this method is satisfactory.
Keywords: Fuzzy classification, Forecast, Rotor, Optimization.

1 Introduction

In the process of surveillance machine sets, one important task is to determine the operating condition . Some machines,especially some large rotating machine sets, there exist so many vibration phenomena and lots of causes resulting in changes of vibration condition, so far, it is difficult to decide what result in the change of operating condition in practice. In order to find the abnormal situation in time, some features that reflect the machine sets vibration condition are often in use. In this paper, by the integration of fuzzy classification method and optimization method, a new method forecasting the operating trend has been put forward and corresponding examples are shown.

2 Fuzzy dynamic classification

Let A be the subset of the universe of discourse U, $A \in F(U)$, F(U) represents all fuzzy subsets that belong to the universe of discourse U. Let $X=\{x_1, x_2, \ldots, x_n\}$ be the object sets for clustering, each object x_i has m order fuzzy features. $A_{ik1} \in F(U_{k1})$, k1=1,2,...,m. According to the property of rotating machinery,the similarity coefficient between object x and y chosed is

$$R = \frac{\sum_{k=1}^{m}(x_{ik1} \wedge x_{jk1})}{\sum_{k=1}^{m}(x_{ik1} \vee x_{jk1})} \qquad (1)$$

where $\vee$, $\wedge$ are maximum and minimum operator. $a \vee b=\max(a,b)$, $a \wedge b=\min(a,b)$.

3 Trend analysis

3.1 The set up of forecasting equation

Let $\{x_1\}$, $\{x_2\}$, ..., $\{x_n\}$, $\{x_{n+1}\}$ be the known series and $\{x_{n+2}\}$ be the forecasted one. In view of the fact that the similarity coefficient reflects the degree of closeness between two series, we have:

$$R_{k1,i}=R_{k1-1,i}\phi_1+R_{k1-2,i}\phi_2+\cdots+R_{k1-n,i}\phi_n+\varepsilon_i \qquad (2)$$

where $k1>n$, $i=1,2,\ldots$, $k1-1$. Corresponding to one step forecast, $k1=n+1$, the equation (2) can be written as:

$$\begin{aligned}
R_{n+1,1} &= R_{n,1}\phi_1+R_{n-1,1}\phi_2+\cdots+R_{1,1}\phi_n+\varepsilon_1 \\
R_{n+1,2} &= R_{n,2}\phi_1+R_{n-1,2}\phi_2+\cdots+R_{1,2}\phi_n+\varepsilon_2 \qquad (3) \\
&\cdots\cdots \\
R_{n+1,n} &= R_{n,n}\phi_1+R_{n-1,n}\phi_2+\cdots+R_{1,n}\phi_n+\varepsilon_n
\end{aligned}$$

For arbitrary i, sloving the above equation by means of least square method, Corresponding to the (2), let k=n+2, then the estimate value is

$$\hat{R}_{n+2,i}=R_{n+1,i}\phi_1+R_{n,i}\phi_2+\cdots+R_{2,i}\phi_n$$

$$i=1,2,\ \ldots\ ,n+1$$

3.2 The optimization of forecasted values

About similarity coefficient, it should meet the condition: $R_{ii}=1$, $R_{ij}<1$, $i\neq j$. Let forecasted similarity coefficient $\hat{S}$ be

$$\hat{S}=\{\hat{R}_{n+2,1},\ \hat{R}_{n+2,2},\ \ldots\ ,\hat{R}_{n+2,n+1}\}$$

In general, the interval between sampling data in one series is the same, so does the interval between different series. According to the property of rotating machine sets that the operation condition changes in continuation, as for a mechanical system that operates in steady, with the known series and the forecasted series $\{x_{n+2}\}$, we have

$$\left|\hat{R}_{n+2,n+2-l} - R_{n+1,n+1-l}\right| \leqslant \max\left|R_{j,j+l} - R_{j+1,j+l+1}\right| \qquad (4)$$

$j+l \leqslant n \qquad j=1,2,\ldots,n; \; l=1,2,\ldots,n$

Since the forecasted values above have a great relationship with the historical data, which reflects the property of operating condition, it is proper to press the unknown values close to the forecasted values, so the objective function can be set up as follows: $f = \|\hat{S}^* - \hat{S}\|$
where $\hat{S}^*$ is unknown similarity coefficient matrix,

$$\hat{S}^* = (R^*_{n+2,1},\ R^*_{n+2,2},\ \ldots,\ R^*_{n+2,n+1})$$

using vector two norm, the equation is written as :

$$f = \left(\sum_{i=1}^{n+1}(R^*_{n+2,n+2-i} - \hat{R}_{n+2,n+2-i})^2\right)^{1/2}$$

as a result, the optimization model is
min

$$f = \|\hat{S}^* - \hat{S}\|$$

constrained condition:

$$R^*_{n+2,n+2-i} < 1 \qquad i=1,2,\ \ldots\ ,n+1$$

$$\left|R^*_{n+2,n+2-l} - R_{n+1,n+1-l}\right| \leqslant \max\left|R_{j,j+l} - R_{j+1,j+l+1}\right|$$

$j+l \leqslant n; \; j=1,2,\ldots,n; \; l=1,2,\ldots,n$

In this paper, the synthetic constrained dual-descent method (SCDD) is applied to solve the above equation.

3.3 The setup of similarity coefficient matrix and its solution
Because the similarity coefficient between itself is equal to 1, so we set up new similarity matrix $(R^*_{ij})_{n+2,n+2}$
where
$R^*_{ij} = R_{ij} \qquad i \leqslant n+1,\ j \leqslant n+1$
$R^*_{ij} = R^*_{ij} \qquad i=n+2,\ j=1,2,\ldots,n+1$
By applying the transitive closure method, it is easy for us to obtain corresponded equivalent matrix.
Out of question, applying different partition threshold value, different clustering result is obtained, if the partition threshold values are determined relying on real condition, it is easy for us to forecast the operating trend and very useful in the aspect of guideline of overhaul.

4 The choose of proper values

In general, in the process of vibration surveillance, if only amplitude spectrum in frequency domain is used, lots of information is losted. In practice, some faults have the similar property in the amplitude spectrum diagram. Sometimes, the vibration caused by some faults is highly directional, if only one transducer in one cross sector is mounted, it is difficult to obtain the right information. On the premise that two transducer have been mounted perpendicularity in one cross sector, in this paper, two directional signal is compounded in frequency domain. We denote as:

$$X_i = PX_i \sin(i\Delta\omega t + \theta_i)$$
$$Y_i = PY_i \sin(i\Delta\omega t + \varphi_i)$$

Where $\Delta\omega = 2\pi/N'\Delta T$, ΔT is the sampling interval, N' is the sampling number. Through the composition of two signals above according to the frequency, we get the equation of the ellipse. Using major axis, minor axis and inclination angle of major axis represent the equation of the ellipse, symbolically,
$\{LP_{ik}, SP_{ik}, \theta_{ik}\}$ i=1,2,... ,n+1; k=1,2,..., m, where Lp is the length of major axis; SP is the length of minor axis; θ is the inclination angle of major axis; n+1 is the number of the objects, in order to reflect the real vibration , we chose the maximum values when frequency is less then 0.8x together with values for 1x,2x,3x,4x. So m is equal to 5.

Similarity coefficient is

$$R_{ij} = \frac{\sum_{k=1}^{m}(llp_{ik} \wedge llp_{jk}) + \sum_{k=1}^{m}(ssp_{ik} \wedge ssp_{jk}) + \sum_{k=2}^{m} cc_{ijk} * (\theta\theta_{ik} \wedge \theta\theta_{jk})}{\sum_{k=1}^{m}(llp_{ik} \vee llp_{jk}) + \sum_{k=1}^{m}(ssp_{ik} \vee ssp_{jk}) + \sum_{k=2}^{m} cc_{ijk} * (\theta\theta_{ik} \vee \theta\theta_{jk})}$$

where

$$llp_{ik} = LP_{ik} / \sum_{k=1}^{m} LP_{ik}; \quad ssp_{ik} = SP_{ik} / \sum_{k=1}^{m} LP_{ik} \qquad k=1,2,\ldots,m \quad i=1,2,\ldots,n+1$$

$$\theta\theta_{ik} = (\theta_{ik}/2\pi)^2 \qquad k=2,3,\ldots,m$$

$$cc_{ijk} = (LP_{ik} + LP_{jk}) / (\sum_{k=1}^{m} LP_{ik} + \sum_{k=1}^{m} LP_{jk}) \qquad i=1,2,\ldots,n+1$$

5 Example

Table 1 is a fuzzy equivalent matrix of a compressor. The data were measured by eddy current transducer with an interval of one month. The first four groups have been obtained before overhaul, but the last four, taken after overhaul. From table 1, when partition level we taken is greater than 0.46 and less than 0.791, the first four can be grouped in one cluster, and the last four in another. Especially, the minimum value among the first four is equal to 0.886, but the minimum value among the last four is equal to 0.791. What this table reflects is just coincident with the real situation. Before overhaul, the vibration was greater but deterministic, i.e. the random factor that influenced the vibration amplitude was relatively small. After overhaul, the situation was just opposite. It is easy for us to concluded that the similarity degree among the first four groups of data be greater than that of the last four groups.

Table 1. A fuzzy equivalent matrix of one compressor

1	2	3	4	5	6	7	8
1.0	0.893	0.893	0.886	0.453	0.453	0.453	0.453
	1.0	0.893	0.893	0.453	0.453	0.453	0.453
		1.0	0.893	0.453	0.453	0.453	0.453
			1.0	0.453	0.453	0.453	0.453
				1.0	0.826	0.791	0.817
					1.0	0.791	0.817
						1.0	0.791
							1.0

In order to evaluate the quality of overhaul, taking data before and after overhaul to set up fuzzy equivalent matrix. From the change degree of data in the matrix, it has a great help to decide the effect of overhaul and the degree of stability of operating condition before and after overhaul. Then, by means of vibration energy analysis, the decrease degree of vibration level after overhaul is determined.

Table 2. The real values

1	2	3	4	5	6	7	8	9
0.848	0.904	0.904	0.848	0.926	0.904	0.881	0.906	1.0
0.823	0.824	0.824	0.805	0.898	0.889	0.905	0.941	1.0
0.737	0.728	0.728	0.728	0.728	0.728	0.728	0.728	1.0

Table 3. The forecasted values

1	2	3	4	5	6	7	8	9
0.848	0.904	0.926	0.848	0.904	0.904	0.882	0.883	1.0
0.823	0.824	0.824	0.798	0.862	0.874	0.874	0.874	1.0
0.728	0.769	0.778	0.778	0.778	0.729	0.778	0.778	1.0

In order to know the trend of operating condition, chosing the data came in practice with one month interval and from different unit. Table 2 lists the real values, table 3 lists the forecasted values. From the table 2,3, the relative error is less then 7.5% . this is enough to meet the practical needs. Comparing the forecasted values with the historical values, the change degree of operating condition in next future is decided. If we have found some faults about machine sets, it will have a great help to decide the developing velocity of faults. If a great different existed between the forecasted values and historical data, we should notice whether some regularity existed in historical data and find the faults as soon as possible.

6 Conclusion

By fuzzy dynamic classification, the change degree of operating condition is determined, especially, it is easy to find thee abnormal condition of machine in time. Combining other analysis methods, we can evaluate the quality of overhaul. The forecasted method recommended here has a great benefit to perform predictive maintenance and prevent great faults.

7 References

Qu Liangsheng, He Zhengjia. (1986) Machinery Fault Diagnostics. Shanghai Science and Technology Press.

Chen Yuedong, Qu Liangsheng, Liu xong. (1989) A New Method Evaluating Operating Condition for Rotating Machinery. 1st International Machinery Monitoring and Diagnostic Conference. U. S. A.

Chen Yuedong, Qu Liangsheng, He Zhengjia. (1989) Fuzzy classification of vibration signals and its application to rotating machinery. Journal of Xi'an Jiaotong University, 6, 49-56.

Homosource ANC and its application to the machinery diagnosis

Xiao Li and Liangsheng Qu
Research Laboratory of Machinery Surveillance and Diagnostics, Xian Jiaotong University, Xian, Peoples Republic of China

Abstract
A new form of Adaptive Noise Cancelling (ANC) technique named homosource ANC technique is proposed in this paper. By extending the concept of "signal source" and "noise source", homosource ANC technique is constructed based on a traditional ANC system. It not only can cancell the background noise as usual but also can extract the faulty feature of the machine. For machinery diagnosis, the dynamic signal generated from a working machine can be divided into two parts: one is corresponding to the failure which is considered as "signal source" here and the other is concerned with the normal condition and background noise. Taking the signal of the machine in normal working condition as the "noise source" and putting into the reference input of an ANC system, the output of this system is mainly the signal relating to failure of the machine. In this way, the background noise is canceled. furthermore, the faulty feature of the machine is extracted.

This kind of ANC technique is applied together with the spectral analysis to detect the local damage of rolling element bearing and to monitor the oil whip of rotating rotor system as well as the working condition of gear box. Satisfied results proved that the homosource ANC technique is effective in machinery diagnosis.
Keywords: Diagnosis, Monitoring, Adaptive Noise Cancelling.

1 Introduction

The machinery diagnosis based on vibration signal analysis is mainly concerned with the extraction of those features which are related to the faulty state of machine. In many real cases, it is often the problem that the presence of excessive background noise. In common sense, it is an obstacle to signal analysis in every research field. Facing this problem, various kinds of signal processing techniques have been used successfully to improve the signal to noise ratio (SNR). One of them as widely known, is Adaptive Noise Cancelling technique. However, in machinery diagnosis, signals related to the faulty state are often buried not only in background noise but also in signals related to the right state. For solving this problem , a new form of ANC technique---homosource ANC is proposed in this paper. It has no difference in construction between conventional ANC except for selection of reference input signals. The concept of homosource ANC is discussed in section 2.

The application of homosource ANC to extract defective signal in machinery diagnosis is expressed in section 3. Rolling element bearing, gear box and rotor system are dealt with. The power spectrum analysis was used subsequently to detect machinery failure and meanwhile to verify the effectiveness of homosource ANC. Results before and after using homosource ANC were given by means of graphics. According to the analysis and experiment some conclusions were given out in section 4.

2 Concept of Homosource ANC

The basic construction of ANC is shown in fig. 2-1. In general, the application of ANC is for eliminating background noise and improving the signal to noise ration (SNR). Sensor 1 is located near machinery component to be measured. The signal S and noise N1 which is transmitted from outside are picked up simultaneously and then put into the primary input of ANC system. Sensor 2 is mounted far from the measured point to pickup background noise N2.

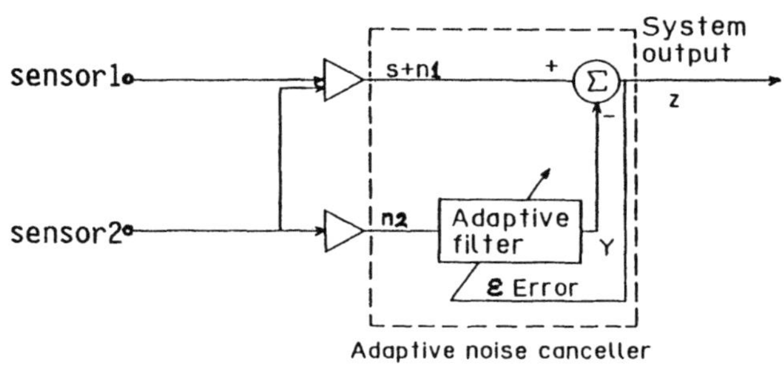

Fig. 2-1. Basic construction of ANC

N2 is put into reference input of ANC system in the same time. Since N1, N2 are usually correlated with each other and uncorrelated with S, then the output of ANC system is,

$$Z = S + N1 - Y \tag{2-1}$$

then

$$Z^2 = S^2 + (N1-Y)^2 + 2S(N1-Y) \tag{2-2}$$

taking expectation of both sides

$$\begin{aligned} E[Z^2] &= E[S^2] + E[(N1-Y)^2] + 2E[S(N1-Y)] \\ &= E[S^2] + E[(N1-Y)^2] \end{aligned} \tag{2-3}$$

adjusting the parameters of adaptive filter (weights) to minimize the output power E[Z],

$$\begin{aligned} \text{Min}E[Z^2] &= \text{Min } E[S^2] + \text{Min } E[(N1-Y)^2] \\ &= E[S^2] + \text{Min}E[(N1-Y)^2] \end{aligned} \quad (2\text{-}4)$$

then the output Z of ANC is the best estimation of signal S in the sense of least mean square.
In the field of machinery diagnosis, signal s can often be written as below:

$$S = S1 + S2 \quad (2\text{-}5)$$

and S1 is the signal component caused by normal condition of machine, S2 is the signal component caused by failure condition of machine. For the purpose of machinery diagnosis, S2 is very expected to be extracted from S and N1 especially in the early stage of machine fault. In this case, only one sensor (sensor1) can be used instead of two as mentioned above. The signal is picked up by sensor 1 before hand when the machine working in normal condition and then recorded in tape recorder. It is also consists of two parts,

$$S' = S1' + N' \quad (2\text{-}6)$$

here $S1'$ is signal , N' is background noise.

In procedure of diagnosis, the signal picked up by sensor1 is put into the primary input of ANC system. Then following result is obtained by the same inference process as mentioned above,

$$\text{Min } E[Z^2] = E[S2^2] + \text{Min } E[(S1+N1-Y)^2] \quad (2\text{-}7)$$

It means that the output of ANC is the best estimation of the failure signal S in the least mean square sense.

Since the primary input and the reference input are coming from same measuring point but not in same time, this kind of ANC is named homosource ANC here. After the processing of homosource ANC, the output will represent the faulty condition of machinery quite well even for slight fault. Bellow are its applications on diagnosis of rolling element bearing, gear box and rotor bearing system.

3 The application of homosource ANC to the machinery diagnosis

In the field of condition monitoring and diagnosis of machinery, there have been some successful examples of ANC application, but they all focused on eliminating the disturbance of background noise. In this paper, homosource ANC is used for not only eliminating the influence of background noise but extracting weak faulty information as well.

3.1 To detect local damage of rolling element bearing

The local damage on rolling element bearing (e.g. damage on surface of ball, outer race and inner race) will excite periodical impulsive vibration when it is on working. The frequency and spectral components of this impulsive vibration signal are mainly depend on parameters of bearing and rotational speed. Table 3-1 gives out the calculating equations of them.

Table 3-1. Frequency features of the bearing with local damage

Position of damage	Impulsive frequency	Spectral components
Inner race	zf_i	mzf_i, $mzf_i \pm nf_r$
Outer race	zf_c	mzf_c
Ball	$2f_b$	$2mf_b$, $2mf_b \pm nf_c$

f_c: Rotating speed of bearing cage,
f_b: Self-rotating speed of ball,
f_i: Speed of the ball relative to outer ring,

If the damage is quite weak, the impulsive signals are relative weak and are often buried in the normal vibration signals of bearing. This will make the feature extraction of fault difficult. In this case, homosource ANC technique is used. The type of experimental bearing is 7011C (SEIKO JAPAN), rotating speed of shaft is 3000 rpm, load is 500N. The slight curve is made on surface of the ball, inner race and outer race respectively. As an example, the power spectrum before (above) and after (bellow) homosource ANC processing for the damage on ball surface is shown in fig. 3-1. The spectrum components related to local damage are denoted on graphics. It can be seen from graph that after the processing of homosource ANC, many spectral harmonics of impulsive signal appeared and can be easily distinguished. The position and number of local damage was then determined via theoretical equations listed in table 3-1.

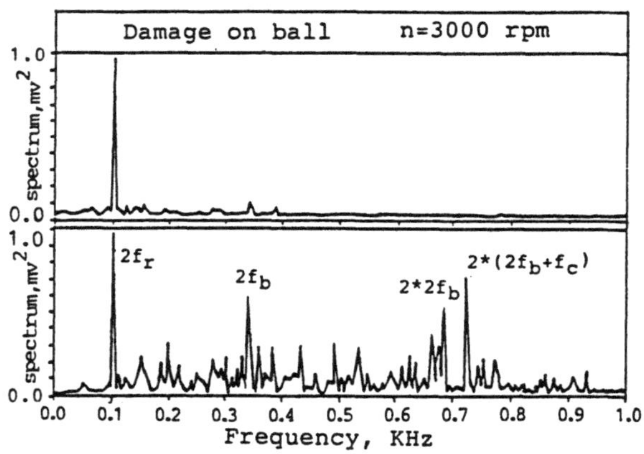

Fig.3-1. Power spectrum comparison

3.2 To monitor the oil whip of rotor system

The homosource ANC was used to monitor the working condition of a gear box in an oil refinery plant. The power transmitted by this pair of gears is 1980 KW and the meshing frequency is 6.85 KHz. The power spectrum of vibration signal is shown in fig.3-2(a) when it was working in a normal condition. The spectral component of vibration signal was recorded to a tape recorder before monitoring and then played back to the reference input of homosource ANC when working process started. The failure condition of the gear box arisen because a few of condensed vapour mixed into the oil lubrication system. The lubrication state and the vibration of the gear box changed a little. A low frequency component (1601 Hz) appeared, as shown in fig.3-2(b), on power spectrum not apparently. It is quite difficult to make diagnosis according to this directly. By the help of homosource ANC technique, the spectrum was given in fig.3-2(c). It can be seen apparently that the low frequency component reflecting the faulty characters appeared more clearly than before. This make the monitoring and diagnosing of gear box more early.

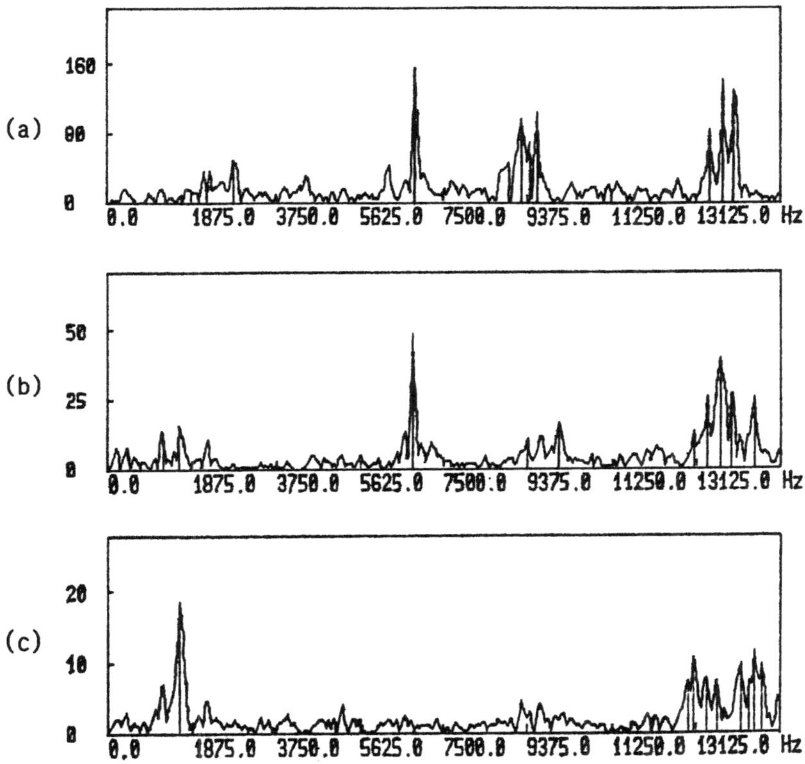

Fig. 3-2. Power spectrum of gear box

3.3 Working condition monitoring of gear box

In a rotor-journal bearing system, the self excited vibration of bearing is very unexpected. The oil whip is one of this self excited vibration and the monitoring of it in large scale rotating machinery is quite necessary. The most simple and common used method is to measure the relative displacement between rotor and bearing by mounting two sensors perpendicular to each other as shown in fig.3-3a, and then to synthesize the rotor orbit for monitoring.

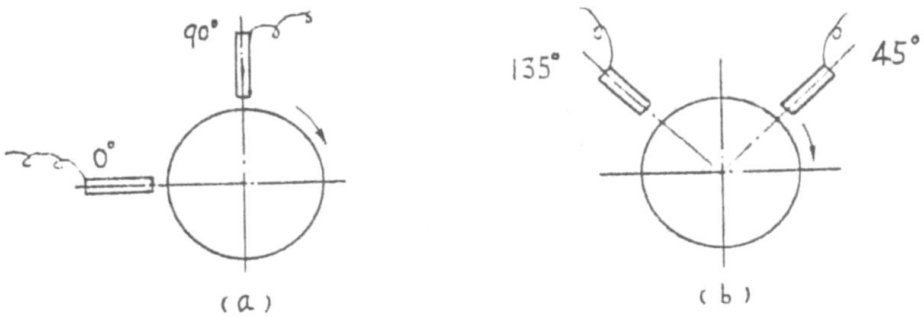

Fig. 3-3a Mounting position of sensors

Bellow is an example for monitoring oil whip of a large scale compressor with power of 11100 KW and rotating speed of 10314 rpm. The normal vibration signals picked up by two sensors in the meantime were recorded and prepared to be as the reference input of homosource ANC system. At the beginning, the extent of oil whip was quite small and the recognition of it is therefore difficult. If synthesizing the rotor orbit directly by two channel vibration signals, the rotor orbit would be confusion as shown in fig.3-3b. Filtering the vibration signals by homosource ANC system at first and then to synthesize the rotor orbit, the track of oil whip were seen clearly as shown in fig.3-3c.

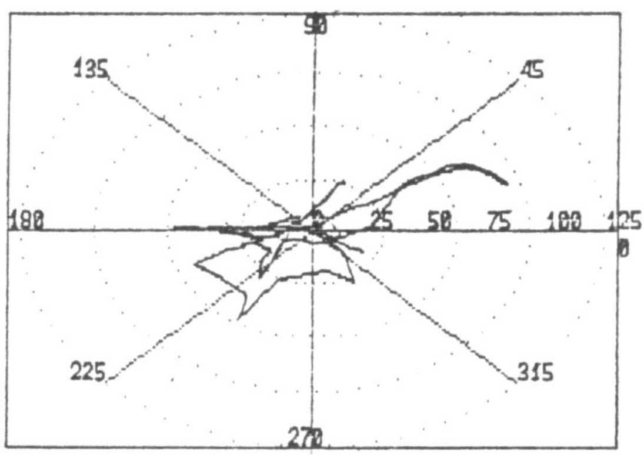

(a)

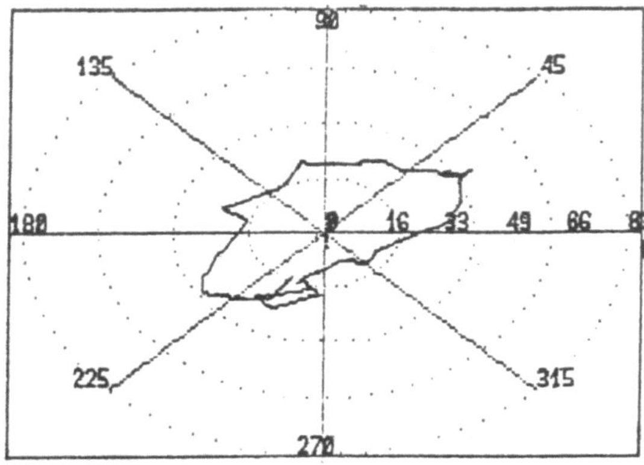

(b)

Fig. 3-3. Rotating orbit of the shaft

4 Conclusions

According to analysis and application of homosource ANC it is known that the homosource ANC has the same construction but different way of application. In the field of machinery diagnosis, it not only can depress the background noise but also can extract faulty features of the machine provided that the failure information and the normal information mixed in the form of addition. It is suitable for diagnosis of local damage of rolling element bearing and also for the working condition monitoring and diagnosis of gear box and rotor bearing system.

5. References

B. Widrow, J. R. Glover, J. M. Mccool, J. Kaunitz, C. S. Williams, R. H. Hearn, J. R. Zeidler, E. Dong and R. C. Goodlin (1975) Adaptive Noise Cancelling: Principles and Application. Proc. of IEEE, 63, 1692-1716.

G. K. Chaturvedi, D. W. Thomas (1982) Bearing Fault Detection Using Adaptive Noise Cancelling. Trans. of ASME Journal of Mechanical Design, 104, 391-405.

C. C. Tan, B. Dawson(1987) An Adaptive Noise Cancellation Technique For The Enhancement of Vibration Signatures From Rolling Element Bearing: A Feasibility Study. Proc. of IMEKO, 2nd Symposium of Technical Diagnostics, 19-35.

T. Igarashi, H. Hamada (1981) Reseach on vibration and sound of defective rolling bearing (1st report, vibration of ball with a defect). Trans. of JSME, 47, 1327-1336.

The phase information in the rotor vibration behaviour analysis

Xiong Liu and Liangsheng Qu
Research Laboratory of Machinery Surveillance and Diagnostics, Xian Jiaotong University, Xian, Peoples Republic of China

Abstract
Nowadays the industry is accustomed to analyze the rotor vibration behaviour via the magnitudes of the vibration components. The phase information, especially the relationships among phases of the vibration components are often overlooked. In this paper, the improved FFT method is used, so that the precise values of the phase angles of the signal components can be obtained. A new rotor surveillance method in which, the initial phase differences among harmonics and the phase differences among different measuring points have been considered is provided. The calculating formulas are derived. The vibration conditions of the misaligned rotors are analyzed. Their changes due to the misalignment in the vertical and the horizontal directions in comparison with the ideal alignment condition can be revealed distinctly by means of the above method. So far it is difficult to distinguish these conditions from the ordinary power spectrum analysis. The above method has been used in chemical plants in China in order to monitor the change of the compressor operating condition. The surveillance examples show that this method is more sensitive to monitor the operating condition of the large rotating machinery in practice.
Keywords: Rotor, Phase, Harmonic, Alignment Condition.

1 Introduction

Nowadays, the transverse vibration of the rotor is monitored through two orthogonally mounted eddy current transducers. The results of signal processing about rotor vibration make it clear that the transverse vibration of the rotor consists of sub-harmonics, first harmonic and higher harmonics. Figure 1 shows the vibration magnitude spectrum of an air-compressor rotor. The change of the vibration behaviour of the rotor in general is monitored via the change of the vibration amplitudes of the harmonic components in these magnitude spectra. Occasionally the phase difference between the first harmonic and an assigned notch on the rotor is taken into

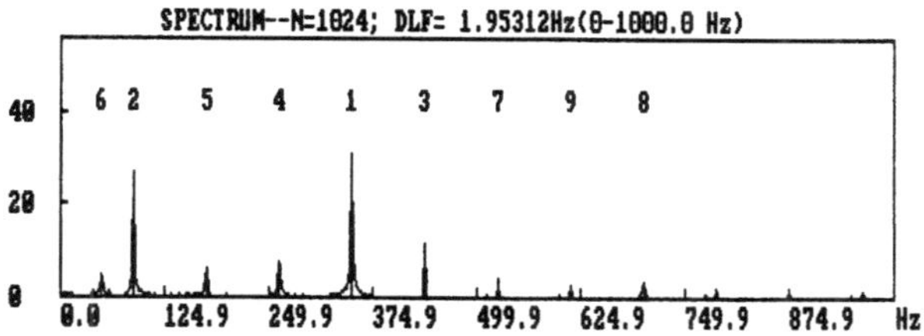

No.	rpm	amp.	No.	rpm	amp.	No.	rpm	amp.
1	20976	30.781	2	5273	26.782	3	26249	11.831
4	15703	7.399	5	10546	6.474	6	2929	5.131
7	31523	4.410	8	42070	3.710	9	36796	2.797

Fig.1. The vibration spectrum of an air-compressor.

account. However, the phase difference between higher harmonics and the assigned notch, the phase relationships among harmonics and the phase relationships among the different measuring points are often overlooked.

2 The improved FFT method and the phase determination

The accurate determination of the rotor vibration parameters is the prerequisite in rotor surveillance and diagnosis. Because the conventional spectrum can only provide the values in the discrete points, the peak amplitudes, frequencies and phases of the vibration components can not be accurately obtained from the conventional spectrum with the measured frequencies between the discrete points of the FFT spectrum, especially the error of the phase is too great to be used. The errors depend upon the relative location of the measured frequency between the discrete points. In order to obtain the accurate values of the rotor vibration parameters, the improved FFT method is adopted and the calculating formulae are:

$$f = \frac{1}{N\Delta}\left(L + \frac{\alpha}{1+\alpha}\right); \qquad \text{where } \alpha = \frac{|S(L+1)|}{|S(L)|}; \tag{1}$$

$$A = \frac{2\pi\,|S(L)|}{N|\mathrm{Sin}\,\pi\delta|}; \qquad \text{where } \delta = \frac{\alpha}{1+\alpha}; \tag{2a}$$

$$A = \frac{2\pi(1-\delta)\,|S(L+1)|}{N|\mathrm{Sin}\,\pi(1-\delta)|} \tag{2b}$$

$$\phi = \mathrm{Phase}[S(L)] - a\delta + \frac{\pi}{2} \tag{3a}$$

$$\phi = \mathrm{Phase}[S(L+1)] - a(\delta-1) + \frac{\pi}{2} \tag{3b}$$

where Δ is the sampling interval, N is the sampling number, S(L) and S(L+1) are the amplitudes of the two neighbouring spectrum lines, $a = \pi \frac{(N-1)}{N}$, $\frac{L}{N\Delta} < f < \frac{L+1}{N\Delta}$ The (2b) and (3b) are recommended using only when |S(L+1)| is larger than |S(L)|.

Table 1 shows a comparison among the ordinary FFT method, ZOOM FFT method and the improved FFT method. The accuracy of frequency is enhanced by means of ZOOM FFT method, but the accuracy of amplitude and phase is not improved. The tested signal is a sinusoid with a maximum amplitude of 10 and a frequency of 80.4Hz. From the table, it is obvious that the satisfactory value of the phase angle can not be obtained via both the conventional FFT method and the ZOOM FFT algorithm. By means of the improved FFT algorithm(AFFT), the results can satisfy the analysis requirement.

Table 1. A comparison of different processing methods

Instrument or software	Frequency interval	Frequency (Hz)	Amplitude	Error of phase
Signal processor	0.25	80.50	9.69	129.9°
	0.50	80.50	9.86	108.3°
(ZOOM FFT)	0.0156	80.44	8.79	9.1°
Signal processor	0.25	80.50	9.62	126.7°
	0.50	80.50	9.93	107.9°
Software	0.25	80.50	9.62	124.8°
	0.50	80.50	9.93	113.6°
Improved software	0.25	80.44	10.03	1.1°
	0.50	80.44	10.04	0.3°

3 Phase relationships

As stated above, a lot of higher harmonic components exist in the vibration signal of the rotor. Without doubt, the phase relationship between two arbitrary harmonics contains the information of vibration behaviour of the rotor. Figure 2 shows the phase difference between the first harmonic and the second harmonic. The phase angles can be obtained via the improved FFT algorithm. Let φ_1 be the phase angle of the first harmonic, let φ_i be the phase angle of the ith-harmonic, let φ_i' be the phase difference between the first harmonic and ith-harmonic.
then

$$\varphi_i' = \varphi_i - i\varphi_1 \tag{4}$$

where i=1,2,...,n

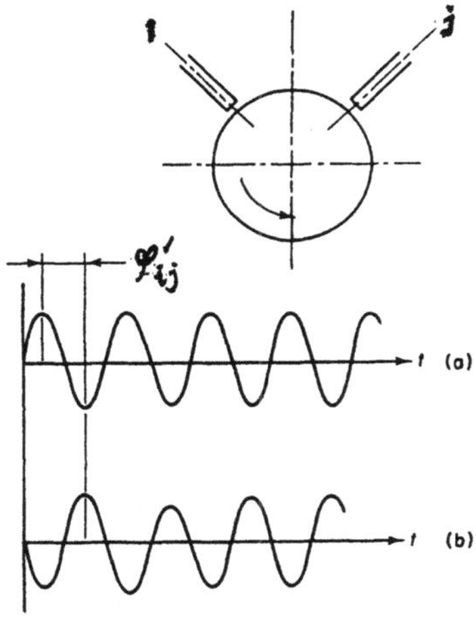

Fig.2. The phase difference between first and second harmonics.

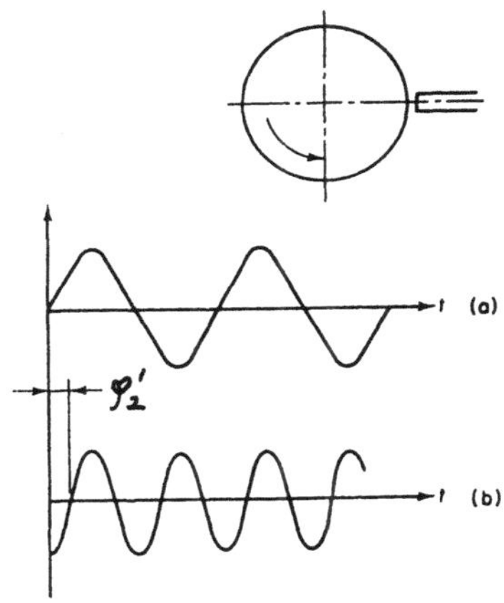

Fig.3. The relative phase of the same order between two signals.

Two probes are recommended at each radial bearing for the vibration monitor of the rotor in one bearing section. In order to monitor the overall vibration behaviour of a rotor, at least four probes are installed. The relative phases among the same order of these picked-up signals of the probes can provide much information about the rotor vibration behaviour. Figure 3 shows the relative phase of the same order between two picked-up signals. The formula to calculate the phase differences among different measuring points is as follow:

$$\varphi'_{ij} = \varphi_{ij} - \varphi_{i1} \tag{5}$$

where φ_{ij} is the phase angle of the i-th order from the picked-up signal of j-th probe, φ'_{ij} is the phase difference of the i-th order between the j-th probe and the first probe.

It can be accurately obtained by means of ordinary FFT algorithm due to the same amount of error included in the phase angles among all probes for monitor one rotor.

4 Vibration conditions of misalignment rotor analysis

The phase analysis methods are used to study the vibration behaviour about the different alignment conditions of the rotor on the rotor test bench. Figure 4 shows the distribution of the probes. Three alignment conditions of the rotor are prepared. First, the rotor alignment is ideal, which is named condition "N". Second, the rotor mislignment is in the horizontal direction, which is named condition "ΔH". Third, the rotor misalignment is in vertical direction which, is named condition "ΔV".

Figure 5 shows the phase differences of the second order among

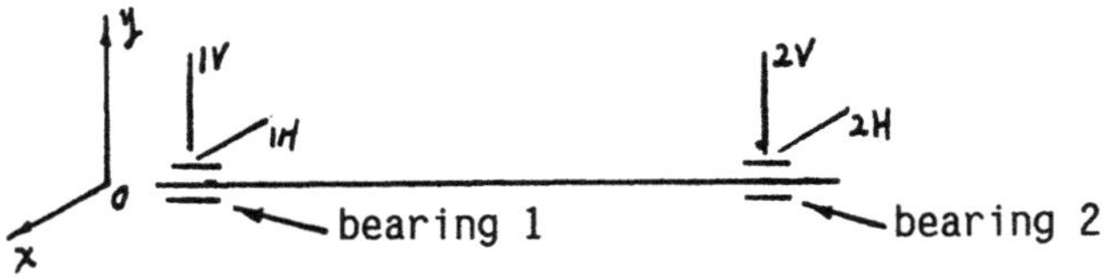

Fig.4. The distribution of probes in the rotor test bench.

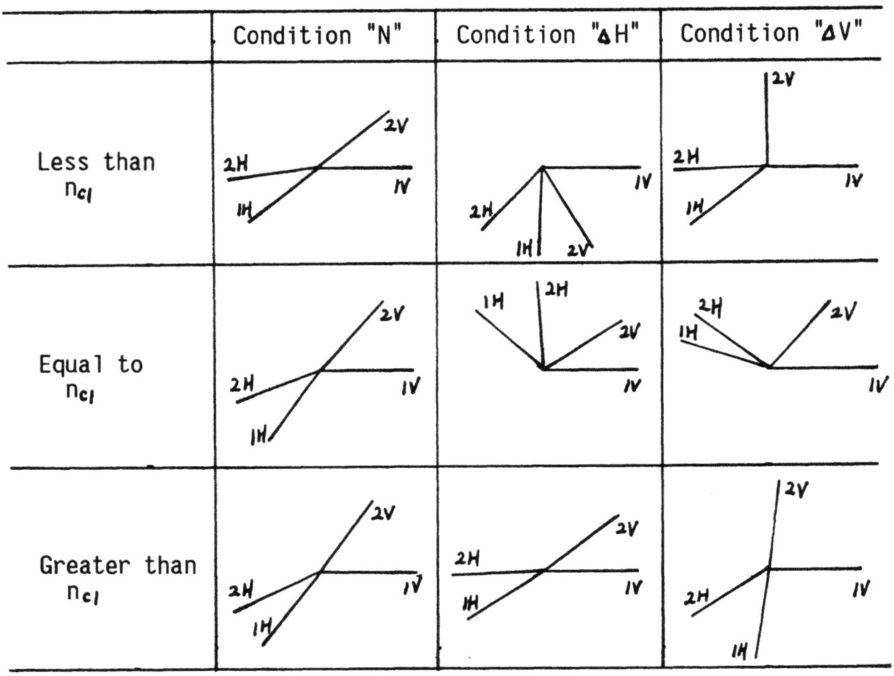

Fig.5. The phase differences of the second order among different measuring points.

different measuring points of three conditions at less than n_{c1}, n_{c1} and greater than n_{c1}, where n_{c1} is the first critical speed of the rotor. There is no clear difference among three speeds with the ideal alignment. But in both the condition "ΔH" and the the condition "ΔV", the clear distinction exists between less than n_{c1} and greater than n_{c1}. This conclusion provides us an approach to determine the rotor alignment condition.

Figure 6 shows the phase differences among harmonics under both condition "ΔH" and the condition "ΔV" at less than the first critical speed. It is easy to distinguish between the condition "ΔH" and "ΔV" through the figure 6. But it is difficult to distinguish these conditions via the ordinary power spectrum.

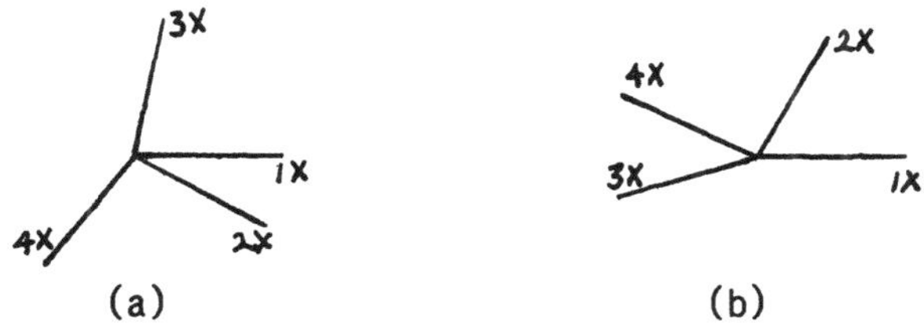

Fig.6. The phase differences among harmonics.
(a) Condition "ΔH", (b) Condition "ΔV".

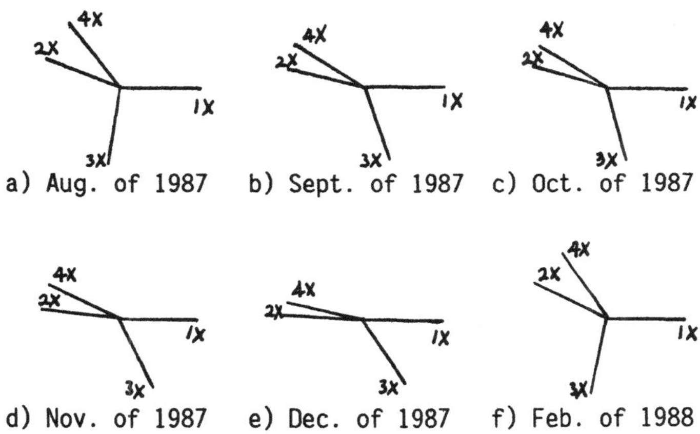

Fig.7. The change of the phase differences among harmonics of an air-compressor in 7 months.

5 The compressor vibration analysis

The air-compressor is one of the critical equipment in the chemical plant. Figure 7 shows the change of the phase differences among harmonics of the air-compressor vibration through 7 months of operation. The change of operating speed in these months is very small. The compressor set was shut down in January 1988 and again in February. From the diagram it is clear that the phase differences between first harmonic and both second and forth harmonics increase with the passage of time, but the phase difference between second and forth harmonics decreases. After the repair in January 1988, the phase differences among harmonics in February 1988 is very similar to that in August 1987. This case indicates that the vibration behaviour of the air-compressor through the repair is recovered. From the diagram of the amplitude trend in the same time it can not be obtained.

Figure 8a shows the phase differences of the first order among different measuring points of a co_2-compressor. Since the rotor is operating at very high speed(13400r/min), it is sensitive to any small imbalance and in figure 8a, the phase differences between 1H and 1V, 2H and 2V are equal to 90° approximately, the phase differences between 1H and 2H, 1V and 2V are equal to 180° approximately, where 1H and 1V represent the probes in the horizontal and vertical directions of one end, 2H and 2V represent the probes in the horizontal and vertical directions of the other end. Figure 8b shows the phase differences of the first order among different measuring points of the air-compressor. In comparison with the that shown in figure 8a there exists the clear difference. This is due to the change of the properties of exciting sources. The air-compressor operates at speed of 5300r/min. The vibration excited by imbalance does not occupy the protruding position and the first order component are excited not only the imbalance force but also other exciting sources.

Fig.8. The phase differences among different measuring points. (a) A co_2-compressor, (b) An air-compressor.

6 Conclusion

An improved FFT algorithm can provide us the precise values of the phase angles of the signal components. It is the base to utilize the phase information. The phase differences among harmonics make the relationships among the signal components clear. It is more sensitive to monitor the operating condition. By means of the phase differences among the different measuring points, the picked-up signals of the different measuring points are combined closely. It can provide much more information about rotor vibration behaviuor. The test results about the alignment conditions show that the phase analysis is a powerful method to distinguish the alignment conditions. Some new phenomena can be obtained through the phase analysis. Of course, other utilization of the phase information need to be further developed.

7 References

Liangsheng Qu, Xiong Liu, Gerard Peyronne and Yaodong Chen, (1989), The holospectrum: a new method for rotor surveillance and diagnosis, Mechanical Systems and Signal Processing, 3, 255-268.

J. Grandke (1983), Interpolation algorithms for discrete fourier transform of weighted signals, IEEE Transaction IM-32, 350-355.

L. Fox Randall (1983), Comparative phase measurements aid vibration analysis, Proceeding of Machinery Vibration Monitoring and Analysis Seminar and Meeting, 117-122..p164

In-situ measurement of cylindricity

Xu Qiuming, Li Jun and Yuan Hongxiang
Changsha Institute of Technology, Changsha, Peoples Republic of China

Abstract
A new method is developed for in-situ measurement of cylindricity of workpiece by using an extended principle of 3-point roundness measurement. Different from the measurement performed on a dedicated measuring instrument, the workpiece form error is measured right on the working position by utilizing the spindle rotation and feed motion of the machine tool. In the present paper, described is a method by which the measurements are processed so as to obtain the substantial rotating error of the spindle, the straightness error of the feed motion and the workpiece form error simultaneously. Simulation and evaluations of the cylindricity error are carried out and some valuable conclusions are obtained.
Keywords: Cylindricity, Form Error, Measurement.

1 Introduction

The need to improve the machining accuracy of machine tools has become increasingly significant; better accuracy of both size and shape is required. Cylindricity is one of the most complex parameters of all engineering measurements and it is expected to find a new way for high-precision measurement. Although some beneficial research works on the measurement and evaluation of cylindricity have been done by many scholars, the practical implementation of an in-situ high-precision measurement system has not yet been accomplished because of the strict mathematical model of cylindricity, the accuracy of measuring probes, the influence of various disturbances, and other factors.

In this paper, an attempt is made to develop a new method of measuring the cylindricity error of workpiece on a machine tool based on the application of a roundness measurement principle called the 3-point method. Instead of measuring the workpiece off-line, the measurement is

*Finance supported by The Chinese National Nature Science Foundation.

carried out in situ with the use of the spindle rotation and the feed motion of the machine tool. By applying the error seperation theory, the workpiece cylindricity error, the substantial rotating error of spindle and the straightness error of the feed motion can be acquired simultaneously. The algorithm is programmed in a microcomputer and some valuable conclusions are obtained.

2 The mathematical model of the cylindricity of workpiece

A cylinder encountered in practice is imperfect; this is due to three deviation factors: the irregularity of the radius of each successive cross-section (referred to as "the radius deviation"), the out-of-roundness error and the curvature of axis. In this paper, they are disignated as $r_0(z)$ $r_1(\theta,z)$ and $r_2(\theta,z)$ respectively. Fig. 1 shows the mathematical model of the form error of a cylinder. The cylindricity error of workpiece measured is described as following:

$$r(\theta,z)=r_0(z)+r_1(\theta,z)+r_2(\theta,z) \tag{1}$$

where θ: the angle on the cross-section of workpiece
z: the point on the axis of workpiece

And $r_2(\theta,z)$ can be approximated by the orthogonal multinomials $P_j(z)$ $(j=0,\ 1,\ 2,\ \cdots)$:

$$r_2(\theta,z)=\sum_{j=2}^{\infty}(e_{xj}\cos\theta+e_{yj}\sin\theta)P_j(z) \tag{2}$$

In the algorithm, the Legendre multinomials are chosen, which can be written as following:

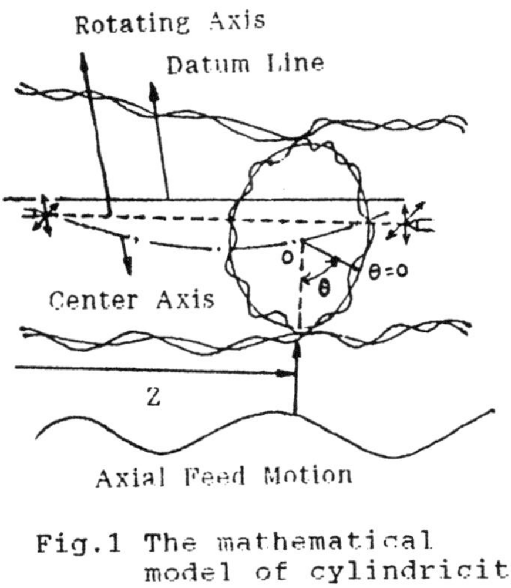

Fig.1 The mathematical model of cylindricity

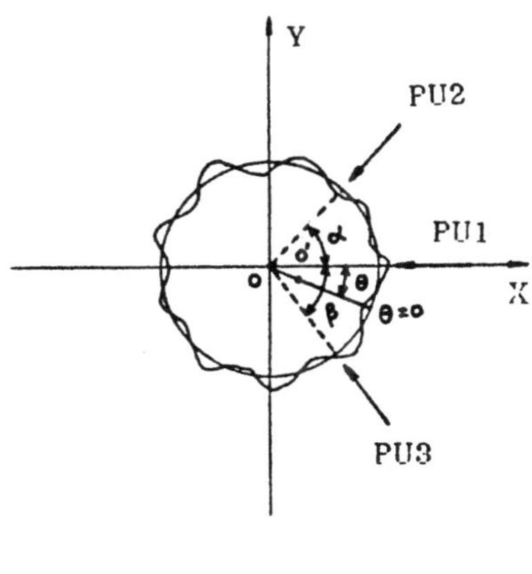

Fig. 2 Principle of the 3-point method

$$P_j(z)=\frac{1}{2^j j!}\frac{d^j}{dz^j}[(z^2-1)^j] \quad (j=0,\ 1,\ 2,\ \cdots) \tag{3}$$

In the measuring system, the errors are detected by the probes which are clamped on the measuring equipment. They include not only the cylindricity error but also the radial error motion of the rotating axis and the straightness error of the axial feed motion of the probes.

3 The principle of the in-situ measurement of cylindricity

3.1 Analysis of roundness measurement by the 3-point method

The principle of the 3-point method is described in Fig. 2. Three probes are placed around the workpiece. Point o is the crosspoint of the sensitive directions of the three probes, and point o' is the instantaneous center of rotation. The angles between the three probes are defined as α and β.

The outputs of the three probes 1, 2, 3, disignated as $y_1(\theta,z)$, $y_2(\theta,z)$ and $y_3(\theta,z)$ respectively, form the following measuring equations:

$$y_1(\theta,z)=r(\theta,z)+x(\theta,z)-S_x(z) \tag{4a}$$

$$y_2(\theta,z)=r(\theta+\alpha,z)+x(\theta,z)\cos\alpha+y(\theta,z)\sin\alpha-S_x(z)\cos\alpha -S_y(z)\sin\alpha \tag{4b}$$

$$y_3(\theta,z)=r(\theta-\beta,z)+x(\theta,z)\cos\beta-y(\theta,z)\sin\beta-S_x(z)\cos\beta +S_y(z)\sin\beta \tag{4c}$$

where $x(\theta,z)$, $y(\theta,z)$: components of the rotating error of the spindle on coordinate axes X, Y

$S_x(z)$, $S_y(z)$: components of the straightness error of the axial feed motion on coordinate axes X, Y

By elimination, we can get the result as following:

$$\begin{aligned} y(\theta,z)&=y_1(\theta,z)+a\cdot y_2(\theta,z)+b\cdot y_3(\theta,z)\\ &=r(\theta,z)+a\cdot r(\theta+\alpha,z)+b\cdot r(\theta-\beta,z)\\ &=(1+a+b)r_0(z)+r_1(\theta,z)+a\cdot r_1(\theta+\alpha,z)+b\cdot r_1(\theta-\beta,z) \end{aligned} \tag{5}$$

where $a=-\sin\beta/\sin(\alpha+\beta)$
$b=-\sin\alpha/\sin(\alpha+\beta)$

In the discrete system, equation (5) can be writeen as following:

$$y(n,z_i)=(1+a+b)r_0(z_i)+r_1(n,z_i)+a\cdot r_1(n+m_1,z_i) +b\cdot r_1(n-m_2,z_i) \tag{6}$$

where $n=0,\ 1,\ 2,\ \cdots,\ N-1$
$m_1=\alpha\cdot N/2\pi$, $m_2=\beta\cdot N/2\pi$, integers

i: the number of each cross-section measured
N: the sampling number of one revolution of spindle axis

Take the N-point DFT of equation (6):

$$Y(k,z_i)=F(k)\cdot R(k,z_i) \tag{7}$$

where $k=0,\ 1,\ 2,\cdots,\ N-1$
$F(k)=1+a\cdot e^{jk\alpha}+b\cdot e^{-jk\beta}$: weight function
$Y(k,z_i)$: DFT of $y(n,z_i)$
$R(k,z_i)$: DFT of $r_1(n,z_i)$ and $r_0(z_i)$

From equation (7), we get:

$$R(k,z_i)=Y(k,z_i)/F(k) \tag{8}$$

Let $R(1,z_i)=0$ and take IDFT of $R(k,z_i)$, $r_0(z_i)$ and $r_1(\theta,z_i)$ can be obtained.

3.2 Analysis of the method for measuring the straightness error of axial feed motion

The measurement method of straightness which is called STRP method is described in Fig. 3. Three probes are placed on the tool post which moves along the axial direction.

The outputs of the three probes 1, 4, 5, disignated as $y_1(z)$, $y_4(z)$ and $y_5(z)$ respectively, form the following equations:

$$\begin{aligned} y_1(z)&=r(z)-S_x(z)\\ y_4(z)&=r(z-l_1)-S_x(z)-l_1\cdot q(z)\\ y_5(z)&=r(z+l_2)-S_x(z)+l_2\cdot q(z) \end{aligned} \tag{9}$$

where $r(z)$: the form error of the workpiece
$S_x(z)$: the translational component on the X-axis of the straightness error of the axial feed motion
$q(z)$: the inclined component of the straightness

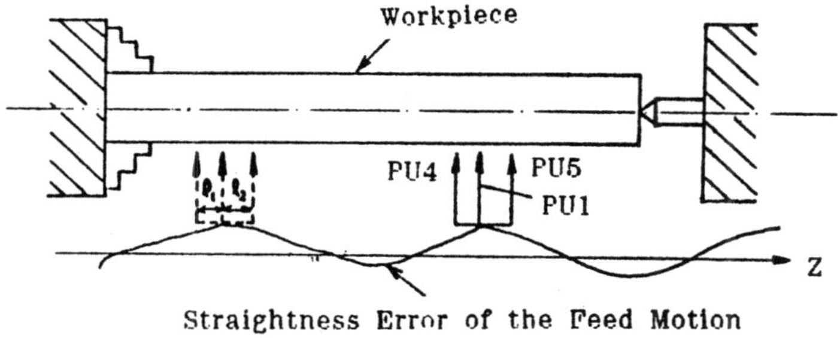

Fig. 3 Principle of the STRP method

error of the axial feed motion
l_1, l_2 : the distances among the three probes

In the discrete system, equation (9) can be written as follows:

$$
\begin{aligned}
y_1(n) &= r(n) - S_x(n) \\
y_4(n) &= r(n-n_1) - S_x(n) - n_1 \cdot \Delta l \cdot q(n) \\
y_5(n) &= r(n+n_2) - S_x(n) + n_2 \cdot \Delta l \cdot q(n)
\end{aligned} \tag{10}
$$

where $n=0, 1, 2, \cdots, N_0-1$
$n_1 = l_1/\Delta l$, $n_2 = l_2/\Delta l$, integers
Δl: the distance between two sampling positions
N_0: the number of cross-sections measured

Using the similar method of data processing as above, $s_x(z)$ can be obtained.

3.3 Analysis of the method for measuring the curvation of axis

Now, we discuss the sampling data of probe 1. Let θ be 0° and 90° respectively, equation (4a) will become as follows:

$$
\begin{aligned}
y_1(0°,z) &= r_0(z) + r_1(0°,z) + \sum_{j=2} e_{xj} P_j(z) + x(0°,z) - S_x(z) \\
y_1(90°,z) &= r_0(z) + r_1(90°,z) + \sum_{j=2} e_{yj} P_j(z) - S_x(z)
\end{aligned} \tag{11}
$$

Form equation (11), we get:

$$
\begin{aligned}
y_1(0°,z) - r_0(z) - r_1(0°,z) + S_x(z) &= \sum_{j=2} e_{xj} P_j(z) + x(0°,z) \\
y_1(90°,z) - r_0(z) - r_1(90°,z) + S_x(z) &= \sum_{j=2} e_{yj} P_j(z)
\end{aligned} \tag{12}
$$

Considering a generatrix of a cylindrical workpiece, the error which is caused by the slight shift of a rigid axis is linear, so the order of $x(0°,z)$ is not more than one. After carrying out data processing discussed earlier, we can know $r_0(z)$, $r_1(\theta,z)$ and $s_x(z)$. Apply the best square approximation method to equation (12), e_{xj} and e_{yj} can be acquired. So the curvation of axis is obtained.

4 Parameters selection and experimental measurement system

The parameters include the angles α and β, the distances l_1 and l_2 between the individual probes, the sampling number around each workpiece cross-section N and the number of cross-sections measured N_0. In the measurement system, we select them as follows: $\alpha=53°6'15''$, $\beta=59°3'45''$, $l_1=25mm$, $l_2=30mm$, N=128, $N_0=32$. Parameters selection is based on the following factors:

(a) Accord with the Nyquist Law.

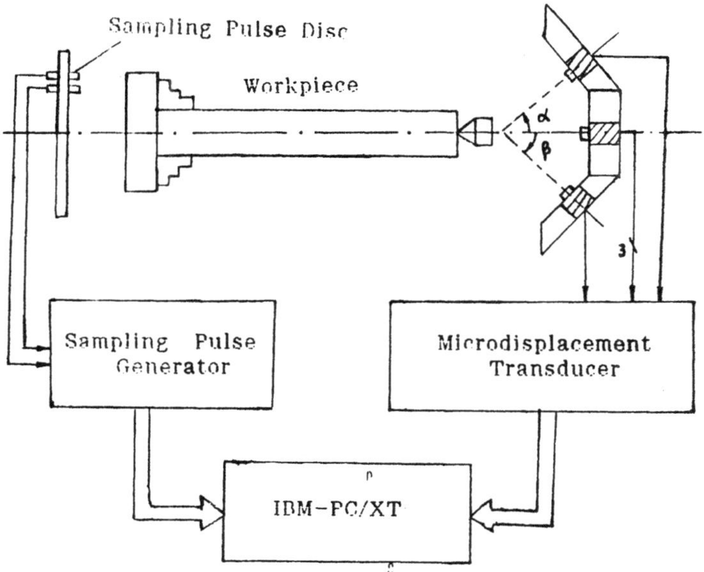

Fig. 4 The diagram of the measuring system

(b) $N=2\pi/(2\pi, \alpha, \beta)$. $(2\pi, \alpha, \beta)$ means the maximum common divider of 2π, α and β.
(c) $(N, m_1, m_2)=1$; $(n_1, n_2)=1$.
(d) Balance the weights of output data of the probes.
(e) Make it convenient to demarcate the probes.
(f) Try to decrease the calculation error of computer.
(g) Try to decrease the influence caused by error transmission of output data of the probes.

Fig. 4 shows the principle diagram of the measurement system of cylindricity. S1-255 lathe and five digital capacitance-based probes are used in the system. The workpiece is set and held by the chuck so it is rotated, while the probes are fed axially.

5 Results

The measurement precision of the workpiece roundness and the spindle rotating error is 0.02μm respectively; the measurement precision of the straightness error of workpiece and the axial feed motion is 0.05μm respectively.

Optimization techniques are applied to evaluate the cylindricity. Compared with the Least Square Method, the Mininum Zone Method is superior in the accuracy of the cylindricity error and inferior in the evaluation time.

The precision of the in-situ measurement of the workpiece cylindricity is expected to be 0.1μm.

6 Acknowledgement

The authors wish to acknowledge the helpful discussions and contributions made by Zhuang Ling, Changsha Institute of Technology; and Han Zhenyu, Xi'an Jiaotong University.

7 References

D. J. Whitehouse (1976) Some aspects of error seperation tcchniques in surface metrology, J. Phys. E., Sci. Instrum., 9, 531.

Han Zhenyu, Li Jun, Yuan Hongxiang (1989) In-situ measurement of cylindricity and data processing, Thesis for M. Sci. of Changsha Inst. of Tech..

Li Jun, Yuan Hongxiang, Dia Yingzhou (1989) An on-line roundness measuring technique, Chinese Journal of Sci. Instrum., Vol. 10, No. 1.

Yuan Hongxiang, Li Jun, Zhang Fong and Pan Peiyuan (1989) Roundness on-line measurement and error compensation technique in ultra precision machining, Chinese Journal of Sci. Instrum., Vol. 2, No. 1.

Three-dimensional component inspection using a profile projector equipped with a fibre optic sensor

Qingping Yang and C. Butler
BCMM, Brunel University, London, England

Abstract
In many cases, a contribution to machine condition monitoring can effectively be carried out by dimensional monitoring of manufactured components. This paper describes a novel method, based on the use of an optical fibre sensor, to achieve the three dimensional inspection of parts on a traditional profile projector. The new system can be operated either manually using a joystick, or automatically via a computer aided controller, offering a measurement accuracy of 10 microns with a resolution of 1 micron and repeatability of 5 microns. Application of the technique improves the capability of profile projectors at relatively low cost, and can also be applied to any similar measuring machines.
Keywords: Optical Fibre Sensor, Profile Projector, Automated Inspection, 3-D Inspection, Triangulation, Edge Detection

1 Introduction

Automated inspection plays an important role in advanced manufacturing and Computer Integrated Manufacturing (CIM). Inspection constitutes an operational part of quality control and represents one of five basic functions in manufacturing[1987, Groover]. Current automated inspection equipment, however, is very expensive compared with the traditional, manual alternatives[1989, Butler]. Automated manufacturing is an evolving technology which began many decades ago, automated inspection is evolving in its wake. We can now achieve a great deal to meet the demands for better inspection and quality by reorganizing the systems in order to make better use of the technology already available[1988, Brook]. It can be expected that with the latest technology, the traditional measuring machines will be able to offer particular benefits at relatively low cost and remain competitive for some considerable time due to the current high cost of special automated inspection equipment.

A profile projector is a traditional measuring machine which magnifies the image profile of components onto a screen. This allows easy and accurate inspection and has long been used in industry for quality control. Although X,Y axes of the table of some latest profile projectors have been motorized, almost invariably they are two dimensional profile measuring machines. Together with the restriction on types of components, this has greatly limited their use in many practical inspection circumstances.

Three dimensional automated inspection has been achieved on a profile projector using a novel method based on an optical fibre sensor. The new system, with three axes of the table being motorized, can be operated either manually using a joystick, or automatically via a computer aided controller, offering a measurement accuracy of 10 microns and high performances at a low cost. The modifications are such as to retain the conventional mode of operation permitting direct observation of a magnified view of the component. The structure of edge profiles (viewed directly) and three dimensional measurements (taken automatically) provide valuable information for machine monitoring and quality control.

2 Operating principle

2.1 The optical fibre sensor

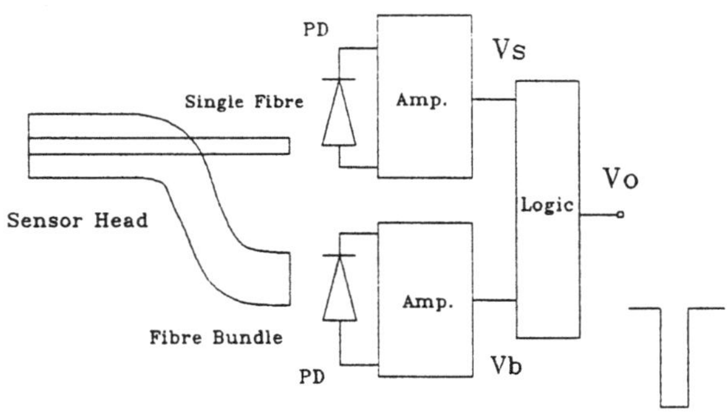

Fig.1. The optical fibre sensor

As shown in Fig 1, the optical fibre sensor consists of one single fibre located centrally in a fibre bundle. This is followed by two photodiodes and signal processing circuitry. The optical system includes a 'knife edge' which leads to a semicircular image being produced on the component, which is enlarged and appears on the profile projector screen. The sensor head is positioned very close to the straight edge portion of the semicircle on the screen.

The signals, Vs and Vb, detected by the single fibre and the fibre bundle respectively, will vary as shown in Fig. 2 as the stage is translated. They are compared using a comparator and a negative pulse is generated when they are equal. A pulse will be generated, therefore, if any edge crosses the diameter in any direction. The different movement directions give the same results because of the symmetrical arrangement of the sensor head. The negative pulse output by the optical fibre sensor indicates the detection of an edge, either the part edge, which occurs when the table moves in X-Y plane, or the semicircle image edge, which occurs when the table moves in Z axis(see Fig.2).

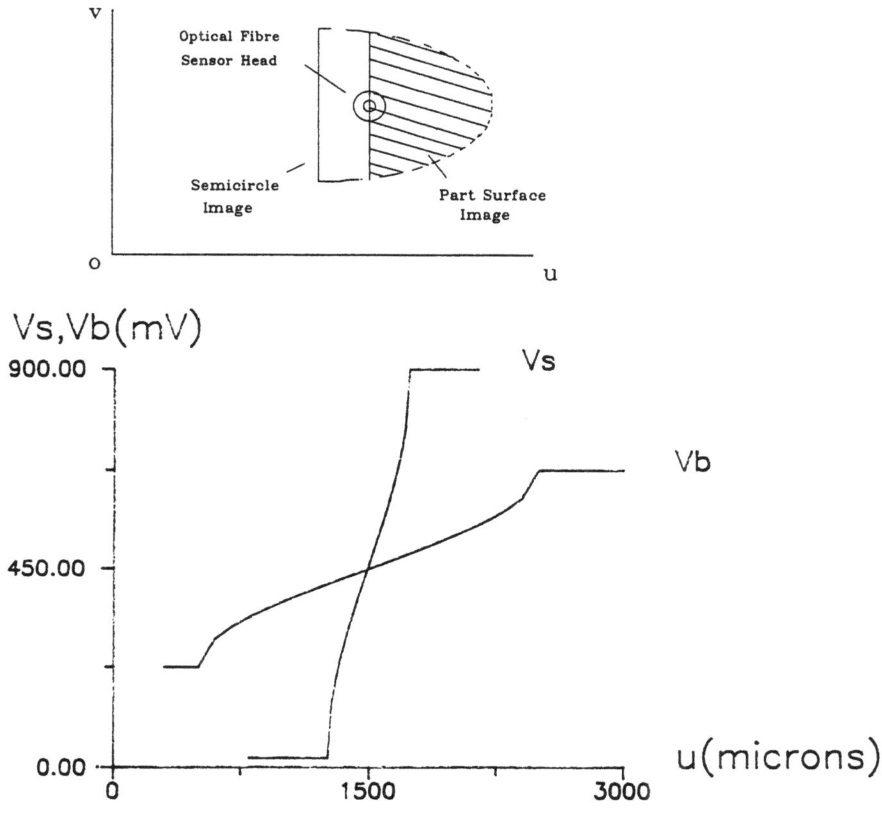

Fig.2. Vs and Vb, the signals detected by the single fibre and the fibre bundle

2.2 The optical fibre sensor aided 3-D profile projector
The system is shown in Fig. 3. The light emitted from the source, passes through the lenses and a 'knife-edge' image

is produced on the surface of the part. Provided the part has a scattering surface, this image produces a further image through the lenses and mirrors of the profile projector, and is detected on the screen by the sensor head. A digital readout device (DRO) (Heidenhain type), displays the X,Y,Z axes displacements measured by three optical grating type linear encoders and provides an RS232C interface to a Personal Computer. The PC is also used to control three stepper motors installed on the stage of the profile projector.

When the table moves in the X,Y plane, the negative pulses corresponding to the detection of the edge of the component will 'freeze' the readings on the DRO and send

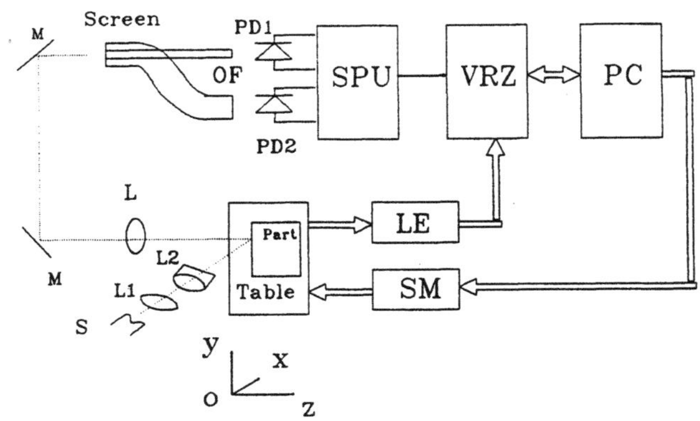

Fig.3. Schematic diagram of the optical fibre sensor aided 3-D profile projector. S-Light source; L1, L2, L-Lens; M-Mirror; OF-Optical fibres; SPU-Signal processing unit; VRZ-Digital readout device; PC-Personal computer; LE-Linear encoders; SM-Stepping Motors.

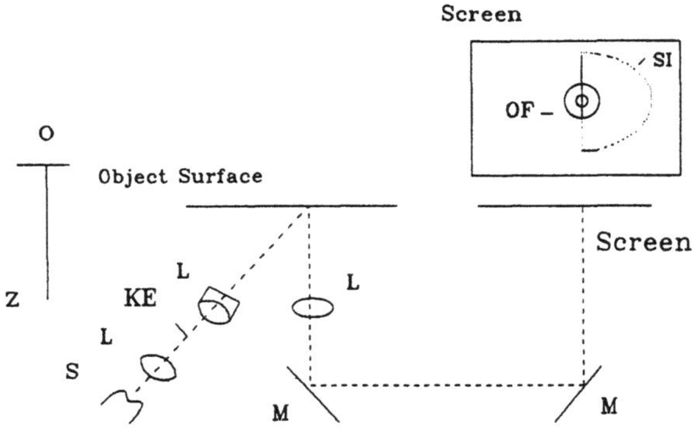

Fig.4. The depth information is derived from the well known triangulation principle. KE-Knife edge; SI-Semicircle image.

the X,Y,Z coordinates of the measured point on the edge to PC. When the table moves in in the Z axis, however, the semicirclar image itself will move, and the edge of the semicircle rather than the part edge crosses the diameter of the sensor termination. It again freezes the readings of the DRO and sends them to PC. The measured point, instead of being on the part edge, is on the component surface. The imaging properties have been designed such that the sensor head is very close to the chord of the semicircle when the part surface is in focus. The depth information is derived, therefore, from the well known triangulation principle(see Fig.4). The sensitivity depends on the angle of incidence of the illuminating beam, the scattering properties of the surface, and the efficiency of the optical system and sensor.

A simple test program has been developed on the PC to store the measurement points required, the features to be measured and also the sequence. The system is capable of measuring features in 3-D space and computes the actual dimensions. Existing software used on conventional CNC coordinate measuring machines, with some modifications, could also be employed.

3 Results

A demonstration system of the optical fibre sensor aided profile projector has been built at the Centre for Manufacturing Metrology, Brunel University. An example workpiece, which has been measured, is shown in Fig. 5.

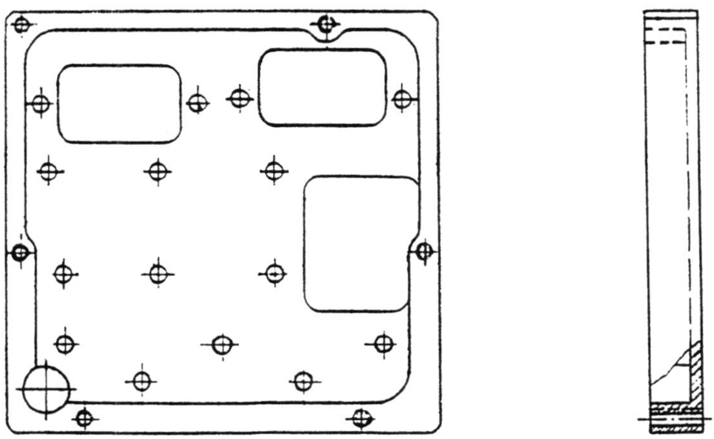

Fig.5. An example workpiece measured on the demonstation system.

The measurements indicate an accuracy of 10 microns, with a resolution of 1 micron and a repeatability of 5 microns.

4 Conclusion

The new, optical fibre sensor assisted, 3-D profile projector has been realized and provides high performances at relatively low cost. The work has succeeded in greatly improving the capability of profile projectors, and the method can be applied to many similar machines. A miniaturised head, based on the same principles, could be used on a conventional coordinate measuring machine. Furthermore, it has been shown that the application of recently developed technology can enhance the performance and versatility of traditional measuring machines at modest cost. The alternative approach of designing completely new non-contact systems, based on machine vision systems for example, has resulted in technically successful products. Unfortunately, many such products have not been commercially successful due to the perceived risk associated with a large capital investment in new technology.

5 Acknowledgement

We would like to thank Dr. C. Mardapittas and Mr. G. Gregoriou, for their technical assistance, and Mr F. Gooch for his help on mechanical modification of the original system.

6 References

Brook, Richard A. (1988) Automatic Inspection in Industry Today, in **SPIE In-process Optical Measurement**, Vol. 1012, pp. 1-6.

Butler, C. (1989) **Future Prospects for Automatic Inspection**, Brunel University, London.

Groover, Mikell P. (1987) **Automation, Production Systems, and Computer Integrated Manufacturing.** Prentice-Hall, Englewood Cliffs, NJ.

Author Index

This index details the authors of each individual paper. The numbers refer to the first page of the relevant paper.

Ahmed, M.M. 247
Allenby, G. 155
Allwood, R.J. 253
d'Amato, E. 148
Andrews, J.D. 225
Ansari, Q. 207
Ardjmand, N. 260
Ashman, D. 14
Au, J. 62

Badi, M.N.M. 20
Barua, D.J. 212,442
Benenson, D.M. 396
Bennouna, M. 74
Berther, T. 26
Blount, G.N. 367
Boness, R.J. 38
Bouhelier, C. 62
Burrows, J.H. 326
Butler, C. 373,483

Chand, J. 218
Charles, A.J. 305
Chen, Y.D. 455
Chen, Z. 234
Cooper, C.N. 253
Cowan, M.P. 86
Cowley, A. 14
Cross, P. 162
Cumming, A.C.D. 1
Czeslaw, C. 403

Dai, W.R. 390
Davies, A. 44
Davies, P. 26
Dong, W.P. 56
Drake, P.R. 407

d'Emilia, G. 148

Forno, C. 432

Gianotti, M. 266
Griffiths, B.J. 420
Grosset, E. 62

Hajnal, M. 426
Hale, K.F. 343
Henderson, I.A. 349,414
Hensey, J. 169
Ho, H.K. 275
Hu, C. 234
Hunt, R.A. 437
Hunt, T.M. 92

Jawad, A.K. 311
Jiang, X. 281
Johnson, D.E. 20
Jones, B.E. 8,343
Jones, M.H. 98
Jones, N.B. 317

Khalil, A.I. 114,311
Khan, A.R. 225
Khatwa,R. 108
Kristnamurthy, R. 49

Lee, S.S.G. 275
Lenaghan, J.T. 114
Li, C.Y. 355
Li, H. 361
Li, J. 361,476
Li, L. 62
Li, X. 461
Lim, G.H. 240
Lim, M.H. 240
Lim, M.K. 275
Liu, X. 468
Loudon, G.H. 317
Lovanyi, I. 426
Luelf, G. 126

Maiden, J.D. 74
Maillard, A. 62
Mardapittas, A. 62
McBride, S.L. 38
McGhee, J. 349,414
McMahon, R.J. 176
Miao, S.H. 379

Midha, P.S. 80
Milne, R. 266
Molinari, P. 266
Moore, T.N. 68
Morris, E.J. 182

Nagy, A. 426
Nair, M.N.K. 169
Nicholls, C. 132

Oak, B.G. 293
Ohga, K. 299
Ooi, T.H. 240

Pandian, R. 197,212
Park, S.Y. 396
Parker, E.W. 14
Penny, J.E.T. 74
Pozzetti, A. 266
Pringle, R.D. 247

Qu, L. 455,461,468
Qureshi, A.F. 207
Qureshi, M.A. 207

Raghunandan, M. 49
Rahbery, M.A. 367
Ranic, Z.M. 260
Rao, B.V.A. 197
Rao, D.A. 32
Rao, M.A. 32
Rao, R.B.K.N. xi
Reif, Z.F. 68
Roy, K.R.D. 32

Sehmi, A.S. 305,317
Shams, I. 373
Sharp, J.M. 142
Shi, D.S. 379
Sobczyk, M. 38
Soom, A. 396
Souquet, P. 38
Stobart, R.K. 384

Taibi, S. 74
Tominaga, Y. 299
Trmal, G.J. 20,80

Upshall, P. 207

Vogel, R. 126

Wale, G.D. 253
Walton, M.H. 253
Wang, S. 281
Wang, W.Y. 56
Wang, Y.T. 379
Wenman, R.A. 86
Williams, J.H. 44
Wong,Y.W. 275
Wright, S.E. 396

Xu, Q. 476

Yamamoto, A. 450
Yang, L. 361
Yang, Q. 483
Yang, Z. 299
Yuan, H. 476

Zhang, C.R. 355
Zheng, S.X. 390
Zhu, C.B. 80

MIX
Papier aus verantwortungsvollen Quellen
Paper from responsible sources
FSC® C105338

If you have any concerns about our products,
you can contact us on
ProductSafety@springernature.com

In case Publisher is established outside the EU,
the EU authorized representative is:
Springer Nature Customer Service Center GmbH
Europaplatz 3, 69115 Heidelberg, Germany

Printed by Libri Plureos GmbH
in Hamburg, Germany